D1299785

Conservation Biology

George W. Cox

Professor Emeritus - San Diego State University

Conservation Biology
second edition

concepts and applications

Boston   Burr Ridge, IL   Dubuque, IA   Madison, WI   New York   San Francisco   St. Louis
Bangkok   Bogotá   Caracas   Lisbon   London   Madrid
Mexico City   Milan   New Delhi   Seoul   Singapore   Sydney   Taipei   Toronto

McGraw-Hill Higher Education

A Division of The McGraw-Hill Companies

Project Team

Editor *Marge Kemp*
Developmental Editor *Kathy Loewenberg*
Production Editor *Kay J. Brimeyer*
Marketing Manager *Tom Lyon*
Designer *Kaye Farmer*
Art Editor *Jennifer L. Osmanski*
Photo Editor *Nicole Widmyer*
Advertising Coordinator *Heather Wagner*
Permissions Coordinator *Mavis M. Oeth*

President and Chief Executive Officer *Beverly Kolz*
Vice President, Director of Editorial *Kevin Kane*
Vice President, Sales and Market Expansion *Virginia S. Moffat*
Vice President, Director of Production *Colleen A. Yonda*
Director of Marketing *Craig S. Marty*
National Sales Manager *Douglas J. DiNardo*
Executive Editor *Michael D. Lange*
Advertising Manager *Janelle Keeffer*
Production Editorial Manager *Renée Menne*
Publishing Services Manager *Karen J. Slaght*
Royalty/Permissions Manager *Connie Allendorf*

Cover credit © Siegfried Eigstler/Tony Stone Images

Chapter opening photos for chapters 1–5, 9–12, 15, 26, 28–30 by G.W. Cox;
6, 7: U.S. Foresty Service; 8: © Doug Sherman; 13: © Richard O. Bierregaard, Jr.;
14, 16–19, 22, 24, 25: Digital Stock and Corel CD's.

The credits section for this book begins on page 351 and
is considered an extension of the copyright page.

Copyright © 1997 The McGraw Hill Companies, Inc.
All rights reserved

Library of Congress Catalog Card Number: 96–60380

ISBN 0–697–21814–7

No part of this publication may be reproduced, stored in a retrieval
system, or transmitted, in any form or by any means, electronic,
mechanical, photocopying, recording, or otherwise, without the
prior written permission of the publisher.

Printed in the United States of America.
2460 Kerper Boulevard, Dubuque, IA 52001

10 9 8 7 6 5 4 3

To Darla

---

for helping in field studies throughout the world,
and for patience, encouragement and love during
the preparations of this textbook

brief contents

contents

---

part two
Terrestrial Ecosystems   51

chapter 17
Oceanic Ecosystems 167

chapter 18
Coastal Marine Ecosystems 178

part five
Special Problems of Aquatic Ecosystems 189

chapter 19
Marine Mammals and Birds 190

part six
Special Problems at the Biosphere Level 201

chapter 20
Management of Exotic Species 202

chapter 21
Disruption of Migrations 211

part seven
Conservation Theory and Practice 245

preface

The Earth is now in a critical period for the survival of its natural ecosystems and their plant and animal members. The continuing growth of human populations, through consumption of resources and discharge of the resulting wastes, is modifying ecosystem processes on a global scale. Global warming, stratospheric ozone depletion, acid deposition, tropical deforestation, desertification, accelerating soil erosion, chemical pollution—these and other large-scale threats are combining to threaten a massive loss of biodiversity. Yet our ultimate well-being depends on health of the global processes that are experiencing these massive changes—the myriad interactions among plants, animals, microorganisms, and their abiotic surroundings that provide us with food, fiber, medicine, pure water, and clean air. The crisis of survival of natural ecosystems and their members is a crisis of survival of the human species.

This global challenge requires our personal involvement. As responsible citizens, we must educate ourselves about the nature of this crisis, and about the choices that must be made in meeting it. An understanding of basic ecology has become a requirement of every educated person.

Each of us is now faced with decisions, on an almost daily basis, about environmental issues relating to the global ecosystem. Should we buy only tuna that were caught by dolphin-safe techniques, eat hamburgers made from beef that was not produced on tropical pastures created by deforestation, or use paper products instead of plastic and styrofoam items that might pollute the oceans? Should we support preservation of old-growth forests in the Pacific Northwest, taxation of sugar to aid the restoration of the Florida Everglades, and removal of exotic species from national and state parks—even the removal of mountain goats from Olympic National Park? Should we oppose efforts to weaken the Endangered Species Act and the Clean Water Act, to reduce populations of wolves and grizzly bears in Alaska in efforts to increase moose and caribou populations for hunters, or to divert more water from the Platte River so that greatly reduced flow reaches the portion of the river in Nebraska where migratory cranes stop in the spring? Should we support measures to reduce power plant emissions that contribute to acid rain, to ban the use of chlorofluorocarbons and other chemicals that damage the stratospheric ozone layer, and to reduce the use of fossil fuels that contribute to the greenhouse

effect and global warming? Should we continue to require shrimp fishermen to use devices that prevent sea turtles from entering their nets, to protect the mountain lion from hunting in California, or to support an international ban on trade in elephant ivory? These, and countless issues like them, demand the collective decision of society.

The goal of this text is to introduce the reader to the nature of biodiversity in its broadest sense, to the threats to its survival that are intensifying daily, and to ecologically sound approaches to conserving biodiversity. The earth's biodiversity is not only the remarkable product of billions of years of evolution. It is our survival system.

Conservation Biology: Concepts and Applications is based on the course "Conservation of Wildlife" that I have taught at San Diego State University since the early 1960s. Its content has evolved considerably over the years, although remaining concentrated on the global ecosystem and its health. This same course was the stimulus for *Readings in Conservation Ecology*, first published in 1969 by Appleton-Century-Crofts. The students in this course (in some cases now the offspring of earlier students!) thus deserve credit for helping to maintain their instructor's interest in conservation ecology, especially over a long period when conservation was placed on the back burner by society at large. My sincere appreciation goes to all those students.

New to This Edition

The first edition of this text, titled *CONSERVATION ECOLOGY: BIOSPHERE AND BIOSURVIVAL,* emphasized the ecosystem concept and its application to conservation issues. This second edition has broadened in coverage, and includes several new chapters. Two topics of growing scientific concern have received full chapter treatment: habitat fragmentation and global climate change. The examination of societal issues has also been broadened and strengthened by inclusion of a chapter on environmental ethics and by the expansion of other chapters to include discussions of public policy relating to conservation of biodiversity and of efforts to achieve sustainability in the relation of human life within the biosphere. A chapter dealing cohesively with ecological principles relevant to conservation of biodiversity has been placed early in the text, and the nature and origin of biodiversity have received expanded coverage. Because of these changes, I have retitled the text *CONSERVATION BIOLOGY: CONCEPTS AND APPLICATIONS*. Its coverage retains an essential emphasis on ecological aspects of conservation science, but now gives fuller coverage to fields with which ecology interfaces, including biological fields such as systematics and genetics, as well as rapidly growing interdisciplines such as ecological economics and environmental ethics. Thus, this text represents a comprehensive examination of the discipline now generally known as *Conservation Biology*.

Acknowledgments

I thank the many colleagues, including several former students, who have reviewed chapters or groups of chapters: Edith B. Allen, Ellen T. Bauder, A. T. Bergerud, Lawrence J. Blus, Mark S. Boyce, Beth Braker, Richard W. Braithwaite, Bayard Brattstrom, John Celecia, Charles F. Cooper, Deborah M. Dexter, Thomas A. Ebert, David A. Farris, Christopher G. Gakahu, Renatte K. Hageman, Rick A. Hopkins, Don Hunsaker II, Jodee Hunt, Stuart H. Hurlbert, Joseph R. Jehl Jr., David W. Johnston, Steve Kingswood, Barbara E. Kus, Michael Kutilek, Stephen Leatherwood, Leroy R. McClenaghan, Dale R. McCullough, Pierre Mineau, Dieter Muller-Dombois, David A. Munro, Harry M. Ohlendorf, Dwain Parrack, William F. Perrin, Louis F. Pitelka, Donald B. Porcella, Philip R. Pryde, Keith Ronald, Victor B. Scheffer, J. Michael Scott, Robert C. Stebbins, Douglas H. Strong, William Toone, Ross A. Virginia, David Western, Kathy S. Williams, Susan L. Williams, Joy B. Zedler, and Paul H. Zedler.

In addition, I thank the individuals who helped me obtain illustrations for the text. These include Anthony F. Amos, Spencer Beebe, Charles Birkeland, Richard O. Birregaard Jr., Richard W. Braithwaite, James J. Brett, David S. Brookshire, Lincoln Brower, Wendy M. Brown, Graeme Chapman, Walter Courtenay, A. T. Cringan, Don Despain, Deborah Dexter, John Francis, Donald W. Kaufman, Joseph R. Jehl Jr., Peter Johnson, Janet Jorgenson, Charles D. Keeling, Stephen Leatherwood, Thomas E. Lovejoy, A. W. Maki, Peter Meserve, Robert Mesta, J. P. Myers, David A. Norton, Bruce Nyden, Storrs Olson, David R. Parsons, Donald B. Porcella, Rolf O. Peterson, Dan Reed, Scott Robinson, John D. Schroer, W. Roy Siegfried, Robert L. Smith, Tom Stehn, Brent Stewart, Harrison B. Tordoff, H. W. Vogelmann, LaRue Wells, Dorn Whitmore, Susan Williams, Thomas D. Williams, and David Zoutendyk.

I would especially like to thank the hard-working editorial and production staff at Wm. C. Brown publishers—particularly Marge Kemp, Kathy Loewenberg, Kaye Farmer, Kay J. Brimeyer, Mavis Oeth, Nicole Widmyer, and Jennifer Osmanski.

George W. Cox

one

Basic Concepts

C onservation biology is a new science that has drawn together scientists and environmentalists in basic and applied studies of biodiversity. This introductory section examines the nature of this emerging field, and traces the history of conservation activism in North America. Following this, we survey basic principles of ecology, with emphasis on the concept of the ecosystem and its central role in conservation management. Finally, we examine biodiversity in detail and consider the processes of extinction that are leading to a biodiversity crisis.

chapter

1

Conservation Biology: Emergence of a Discipline

Outline

T hroughout history, times of crisis often have been times of advance in human capabilities. In Western history, crises have fostered advances in science and technology. Crises also have stimulated the formation of new institutions, both governmental and private. In the 1990s, conservation biology has emerged as a vital new branch of science directed at an environmental crisis: the human population stands on the verge of causing the massive extinction of species and loss of ecosystems throughout the biosphere. What is this newly organized branch of science? Can it contribute to the formulation of policies for protection of the earth's biotic heritage and the management of ecological resources such as forests, rangelands, fisheries, and game animals?

Conservation biology is the scientific study of biodiversity and its management for sustainable human welfare. It seeks to understand how the rich variety of plant and animal life around us arose, how it has been maintained by natural processes, and how we can utilize this resource sustainably. It is not content with understanding the patterns of biodiversity and their origin; systematics and evolutionary biology effectively fill this niche. Nor is it concerned only with encouraging the protection of biodiversity: this niche is filled by a variety of activist conservation organizations. Conservation biology is unique in focusing on the interaction of humans with biodiversity, seeking to reveal the aspects of this interaction that are significant to human interests, and striving to answer questions about how these aspects can be managed for sustainable human benefit.

But a crisis does exist, and some authors have contended that conservation biology has the primary role of preventing the loss of biodiversity. Soulé (1985) states that the mission of conservation biology is . . . to provide principles and tools for preserving biological diversity, and Primack (1993) describes it as . . . the new, multidisciplinary science that has developed to deal with the crisis confronting biological diversity. Brussard (1995) recently has affirmed that conservation biology has the specific goal of preserving biodiversity.

In this chapter we shall first describe the relation of conservation biology to the fields of science, humanities, and human affairs with which it interfaces. Next, we shall examine the biodiversity crisis and consider how this crisis has given rise to the conservation biology movement. Finally, we shall define the goals of conservation biology in the study and management of biodiversity.

Conservation Biology is the science of survival for thousands of species, including the American bald eagle.

FIELDS WITH WHICH CONSERVATION BIOLOGY INTERFACES

Conservation biology is engaged in activities that involve the diverse, often conflicting, interests of complex human society. Thus, it must draw knowledge and use methods not only from biology, but also from the physical sciences, engineering, economics, sociology, law, history, and political science. Conservation biology has both basic and applied aspects as well. In the area of basic science, our knowledge of how both natural and human-dominated ecosystems function is still primitive. Thus, a major goal of conservation biology must be to examine these systems in detail and reveal the ways in which biodiversity is important to their function. Ultimately, this knowledge will be of great value to humanity. As an applied science, however, conservation biology functions to obtain data and devise methods to manage ecological systems to achieve goals that society has defined. Through the political process, for example, society may decide that national parks should be managed to minimize the risk of disastrous fires, that pollution of streams and lakes be reduced, that productivity of game or fish populations increase, or that extinction of species be prevented. Conservation biologists then must use their knowledge of ecological processes to develop management procedures to obtain the desired goal.

From the start, conservation biology has emphasized an interdisciplinary approach that draws on many basic and applied disciplines within biology (Brussard 1991) (Fig. 1.1). Perhaps the closest relationships are with ecology, the science concerned with relationships of organisms with each other and with their non-living environment. Some branches of ecology deal with particular groups of organisms; these include plant ecology, animal ecology, and microbial ecology. Others, such as terrestrial ecology, freshwater ecology, and marine ecology, focus on particular environments. Still others, such as organismal, population, community, and ecosystem ecology, are concerned with different levels of organization. Other fields of biology, however, are essential to conservation biology. Captive breeding and reintroduction of animals, for example, require the expertise of ecologists, endocrine physiologists, geneticists, animal behaviorists, wildlife managers, veterinarians, and biopark managers. Conservation biology thus is drawing together scientists from many subfields of biology that have grown apart over the past century.

Moreover, to achieve a defined goal such as the protection and recovery of an endangered species, conservation biologists must work in an even broader, real-world context of societal institutions and laws. They must understand the biology of the species, the ways in which human activities are likely to affect its environment, and how efforts to preserve the species can be carried out under the constraints of laws, available funds, and public opinion. Like a doctor of medicine, a successful conservation biologist must be able to integrate the areas of research, diagnosis, and treatment. The global crises that humanity now confronts, in fact, reflect the lack of effective integration of scientific research, monitoring of environmental change, and corrective management.

Conservation biology thus interacts with many nonbiological fields of the physical sciences, social sciences, and humanities

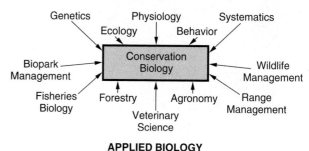

FIGURE 1.1

Conservation biology has close ties to many basic and applied disciplines within biology.

(Fig. 1.2). These fields are as important to sound environmental management as is the core science of conservation biology. An individual conservation project may require collaboration of individuals with expertise in many of these areas. Ecosystem restoration efforts, for example, may require active participation of individuals skilled in biopreserve design, ecological systems modeling, ecological economics, and environmental law, as well as advice from scientists and engineers in plant ecology, agronomy, hydrology, geology, and civil engineering.

An interface field of particular interest is that with the humanities. This is the arena in which society debates the merits of alternative values: environmental ethics, or in its more popular form, environmentalism. **Environmental ethics** examines moral values relating to the natural environment. Leopold (1949), in his essay on the land ethic, stimulated modern interest in environmental ethics when he recommended,

> A thing is right when it tends to preserve the integrity, stability, and beauty of the biotic community. It is wrong when it tends otherwise.

Leopold's writings have become a powerful catalyst for the field of environmental ethics, now one of the most active interfaces of science and the humanities (Schrader-Frechette 1981; Rolston 1988).

Environmentalism is the popular expression of environmental ethics. It is a societal movement, in the words of Scheffer (1991),

> . . . toward understanding humankind's natural bases of support while continuously applying what is learned toward perpetuating those bases.

In other words, environmentalism is the effort to live in harmony within the global ecosystem. Environmentalism is thus a philosophical conviction that the future of humankind depends on the establishment of a sound relationship between humans and the natural world. This conviction provides the political support for many efforts of conservation biology. Many, if not most, conservation biologists are environmentalists at heart, although to do their job right, they must also be objective scientists. Since environmental policy is the responsibility of all of society, environmental ethics and environmentalism are important

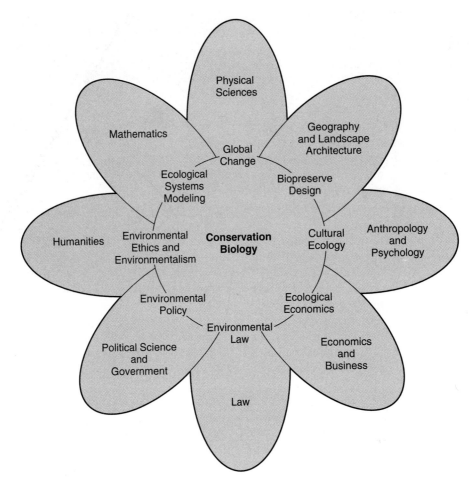

FIGURE 1.2

Conservation biology interfaces with many nonbiological fields of the physical sciences, social sciences, and humanities.

arenas of discussion, but arenas distinct from those of science. We shall examine the field of environmental ethics in more detail in Chapter 28.

Thus, the key goals of conservation biology are to gain scientific knowledge and to apply it to the management of biotic diversity. Management implies three things. First, how can we identify and protect the values that biodiversity possesses? Second, how can we modify biodiversity to increase its sustainable benefits to humanity? Third, how can we restore it where it has been degraded to the detriment of human interests? Knowledge thus must be put to work in programs of protection, sustained use, and restoration of ecological systems. These efforts must go hand-in-hand with continuing studies. The unfortunate truth is that conservation actions must often be undertaken when knowledge is still rudimentary. The critical state of the global environment demands that the best information available be put to work at once.

CONSERVATION BIOLOGY AND BIODIVERSITY

If conservation biology is concerned with biodiversity, what is biodiversity? Edward O. Wilson (1992) states, Biodiversity is our most valuable but least appreciated resource. In its broadest sense,

biodiversity is the richness of the biosphere in genetically distinct organisms and the systems they compose. Biodiversity thus spans a hierarchy from the subspecific level to the level of the biosphere. In this hierarchy, two levels are of primary focal importance: diversity of species and diversity of ecosystem types. We shall examine biodiversity in detail in Chapter 3.

Biodiversity is being lost rapidly in the modern world. Every year a few of the large and conspicuous species, together with thousands of the less conspicuous and often undescribed and unnamed species, are being lost. Ecosystems are being lost as well. In the United States alone, Noss et al. (1995) identify more than 30 types of ecosystems that are critically endangered, having suffered greater than 98% loss (Fig. 1.3). In addition, 58 additional ecosystem types were considered endangered and 38 threatened, having been reduced in area by 85%–98% and 70%–84%, respectively.

THE BIODIVERSITY CRISIS AND THE ORIGIN OF CONSERVATION BIOLOGY

Why has conservation biology become a center of interest and concern so recently? Basically, it is because many scientists and environmentalists perceive that the loss of biodiversity has

FIGURE 1.3

Midwestern oak savanna like that at Agassiz Dunes, Minnesota is one of the critically endangered ecosystems of the United States, having been reduced to less than 2% of its original area.

(Photo by G. W. Cox)

FIGURE 1.4

Clear-cutting in western North America has severe impacts on stream habitats that are home to endangered migratory fish.

(Photo by G. W. Cox)

reached crisis proportions, and even characterize conservation biology as a "crisis discipline" (Soulé 1985; Meffe & Carroll 1994). They believe that existing institutions have not been able to deal adequately with several problem areas involving biodiversity. Three major problem areas exist.

First, the system of legal protection for endangered species has not proved adequate to handle the need on a species-by-species basis (Hutto, et al. 1987; Scott et al. 1987, 1993). Furthermore, the development of programs for the recovery of endangered species has not only been slow, but many are weak in their ecological foundation and vague in their goals (Tear et al. 1995). The regulatory structure of federal and state laws has become complex and has often proved exceedingly burdensome both to private landowners and scientists (Mann & Plummer 1995).

Second, management policies for public forests, range-lands, and parks have emphasized the exploitation of resources and the promotion of commercial activities. These have often lacked a sound ecosystem basis. As President Clinton's Secretary of Interior, Bruce Babbitt, has stated,

> Traditionally, the American West has been something of a third-world economy based on resource extraction (Baden & O'Brien 1994).

Examples of recent political issues in this area have involved sustainable harvests of timber in national forests and the interaction of timber harvesting practices with stream fisheries (Fig. 1.4). After establishing national parks in the Pacific Northwest, in Babbitt's (1995) words:

> Congress then opened the door to relentless exploitation of the rest of the land. In the ensuing orgy of clear-cutting, the original forest expanse was reduced by 80 percent, leaving a few protected islands in a sea of devastation.

Achieving sustainable management of these public lands and their resources is clearly vital to our national economic health.

Third, at the international level, the strategy of protecting biodiversity by creating strictly protected national parks does not work well in many developing countries. David Western, Director of the Kenya Wildlife Service, and many other park administrators recognize that new approaches are needed—approaches that extensively involve local peoples in biodiversity management (Western & White 1994).

What kinds of people have the biodiversity crisis drawn together in the new conservation biology movement? A diverse lot, they come mainly from three areas: traditional ecology and systematics, wildlife management, and neoenvironmentalism. In each of these three areas, major trends have occurred over the past 25 years, focusing attention on applied aspects of biodiversity. Ecology and systematics, as basic branches of biology, have long been concerned with theoretical aspects of biodiversity. For most of this century they have pursued basic descriptive and theoretical studies. Over the past 25 years, however, both fields have become more concerned with applied issues such as causes and rates of extinction, protection of endangered species, and the values of biodiversity resources to society.

Wildlife management, which emerged as a scientific field in the 1930s under the influence of Aldo Leopold (See Chapter 2), initially concentrated its attention almost entirely on game animals. In recent decades, it also has redirected its interest, giving greater attention to management of nongame species and the total wildland ecosystem. Many wildlife scientists, in fact, consider that they are the original conservation biologists. Teer (1988), for example, noted:

> The new [Conservation Biology] society is either naive about what has been done by conservation biologists and managers in the past or chooses to ignore it.

Neoenvironmentalism, the environmentalist movement of the 1960s, brought many individuals with a passion for preservation of biodiversity into the conservation biology arena. In recent years, this movement has seen a shift from emphasis on local and national issues to greater emphasis on global issues. Environmentalists, who advocate specific ways of living in harmony with nature are, in a sense, the activist arm of conservation biology. Their philosophies range from a sort of scientific environmentalism, like that defined by Scheffer (1991), to a strong "right to life" for all organisms and species (Regan 1983).

The need for an organized Society for Conservation Biology began to be recognized in the late 1970s and 1980s. Interest in such a society was strengthened by a 1978 conference on conservation biology in San Diego, California (Soulé 1987). The society itself was incorporated in 1985. It has experienced the most rapid growth in membership of any recently formed scientific organization, and now has over 4,000 members. The journal *Conservation Biology* was initiated in 1987. The journal is now issued bimonthly, and carries a diverse content of scientific articles, reviews, editorials, book reviews, and conservation news notes. Several other new journals also address issues in conservation biology and in interface fields such as global change, ecological economics, and environmental ethics.

Major issues in the developmental phase of this young science are the role of values in guiding research, and the proper role of conservation biology in societal affairs. Should conservation biology strive to be value-free in the conduct of research, and provide only value-free scientific data and expert prediction to society at large? Or should conservation biology act as a strong advocate for the preservation of biodiversity, both through research and political activism? This issue is examined in Reading 1.1.

RESPONSIBILITIES OF CONSERVATION BIOLOGY

Conservation biology is thus faced with a major question: "What is the proper role of the core science of conservation biology in the formulation of public policy relating to biodiversity?" Public biodiversity policy has been driven by varied and changing forces. Some species have acquired a strong "charisma," leading the public to support efforts by environmentalists to protect them. Such species include the striped and spinner dolphins that swim with schools of tuna in the eastern tropical Pacific, the harp seals that bear pups with white pelts prized by the fur industry, and "noble predators" such as the gray wolf and mountain lion. In California, the mountain lion was protected against hunting by a ballot initiative proposition passed by the state's voters.

Other species have been managed largely for harvest, and this demand sometimes has led to policies that affect other components of biodiversity through activities such as clear-cutting, grazing, or predator control. The wolf population reduction program undertaken by the State of Alaska in 1994 and 1995, for example, was intended to increase moose and caribou populations for the state's hunters.

Still other policies have been based on scientific study of the role of species in their ecosystems. Reintroduction of a native species, the gray wolf, to the Greater Yellowstone Ecosystem was initiated after extensive studies concluded that some ecological benefits (to vegetation) and few economic detriments (to ranching and hunting) would result (Cook 1993).

Many conservation scientists feel that conservation biology has the kinds of central functions that science played in the Yellowstone wolf project. These functions fall into the following three major areas:

1. Clarify the ecological and economic values of biodiversity.

2. Develop technologies to protect, restore, and manage biodiversity.

3. Provide responsible opinion about future trends in biodiversity and their significance to humanity.

The first of these tasks is to clarify the ecological and economic values of biodiversity; that is, find ways to evaluate the importance of biodiversity in ecological terms and, ultimately, in economic terms. To resolve issues involving preservation of species and their ecosystems, we must be able to define both the benefits and costs of maintaining particular levels of biodiversity. The benefits associated with biodiversity are potentially quite varied. First, they include the importance of biodiversity to basic ecosystem functions. Second, they include utilitarian benefits to productive human activities. Third, they include nonmaterial benefits to human well-being.

We know very little about the ecological significance of biodiversity, or even about the ecological importance of some of the most abundant members of major ecosystems. The passenger pigeon, for example, was once the most abundant bird of the eastern forests of North America (*See* Reading 5.1). What was the ecological impact of the extinction of this species? How did this loss affect the eastern forests in North America? We have little idea, yet we know that the forest ecosystem did not collapse. Nevertheless, many people argue that loss of much scarcer species might lead to serious ecological disturbance.

Determining the ecological importance of biodiversity requires new techniques and a much more experimental approach. Many basic questions remain unanswered. How important is species diversity to ecosystem function? How much redundancy exists in functional roles of species? How important and prevalent are keystone species? How important are diversity gradients across habitats to landscape function? Will biodiversity provide a buffer for ecosystem function in the face of global change? We shall examine these questions in more detail in our examination of global ecosystem types and their conservation issues.

The economic values of biodiversity need to be determined, as well. Some conservation biologists resist this objective, arguing that biodiversity has intrinsic "right-to-exist" value that cannot, and perhaps should not, be valued in dollar terms (e.g., Sagoff 1988; Ehrenfeld 1993). The strength of this view rests on the number of people who affirm it, however, and, as argued, this should allow quantitative assessment. The democratic political system is designed to develop policy by considering the costs and benefits of various actions for society at large (although not

IS CONSERVATION BIOLOGY
A VALUE–LADEN SCIENCE?

The recent, rapid emergence of conservation biology means that it is still in the throes of defining itself. Should it be a field that promotes conservation of biodiversity based on a set of assumed values of biodiversity? Or should it be a rigorous science that seeks to provide objective answers to questions about how to achieve the goals of biodiversity management set by society? This issue presents a major challenge to the field today, as noted by Brussard (1995), in an essay as the society's outgoing president:

> . . . we will face challenges on two sides—one urging the [Society for Conservation Biology] to take a stronger advocacy role, the other claiming that conservation biology, because of its stated agenda of conserving biological diversity, is not really a science.

Many prominent conservation biologists characterize the field as a "value-laden science" (Meffe & Carroll 1994). By this they mean that conservation biology accepts certain things as intrinsically "good," among them biotic diversity, ecological complexity, and continuity of evolutionary lineages (Soulé 1985). This emphasis reflects the philosophies of the scientists, environmentalists, and environmentalist-scientists that have led the development of the field. Based on these and other values, Brussard (1995) states,

> . . . we know more than enough to assert that humankind's excesses of reproduction and consumption and our habit of maximizing short-term economic gains at the expense of

natural capital, if not curtailed promptly, will result in worldwide ecological armageddon.

Others (Schrader-Frechette & McCoy 1993) contend that ecology, including conservation ecology, is inescapably value-laden. They point out that values of some sort (perhaps even the most crass values, such as money) determine what questions ecologists undertake to study, and that the methods they use often contain major value judgments about what does or does not provide a definitive answer to the research question.

Although scientists cannot entirely escape the influence of values like those mentioned above, they are obliged to be as unbiased as possible in the design and interpretation of their studies. Most conservation biologists (e.g., Hagan 1995) believe that a clear distinction should be maintained between conservation biology, as a science that seeks unbiased answers to environmental questions, and areas of society in which explicit values and goals are assumed. In business, for example, the goal of maximizing profit is assumed. Similarly, environmentalists espouse specific goals, such as protecting all species and ecosystems. Conservation biologists, in fact, find themselves dealing with questions on which different value-holding segments of society have strongly different views. Hagan (1995) recommends that as individuals, conservation biologists maintain a clear distinction between what they say as practitioners of science and what they say as holders of particular sets of moral values, economic philosophies, and political opinions. In other words, as a science, conservation biology should be as rigorous as basic ecology or the research arms of other applied sciences, such as medical or agronomic research. In none of these fields are assumed intrinsic values encouraged as determinants of the outcome of experiments or tests.

Bunnell and Dupuis (1995), looking at conservation biology from their professional experience in the field of wildlife management, believe that

> The value-driven and crisis-oriented nature of conservation biology removes it from the desired, value-neutral character of science writ large, and exposes it to both scientific criticism and the volatility of society.

They suggest that conservation biology has only escaped serious conflict so far because the values held by many in the field correspond to those institutionalized in the U. S. Endangered Species Act and other pieces of environmental legislation.

The controversy rests largely on the question of how conservation biology should address the scientific and policy aspects of environmental affairs. In principle, most would agree that the core science of conservation biology should be practiced free of pressure from societal values. Values cannot be ignored in the areas where conservation biology interfaces with society—the arenas in which public policy is determined and put into practice. In these arenas, the values that are important to society must be assessed, and the assessment must be objective and complete, not one biased by the values that one group within society considers as "intrinsic." Conservation biologists have an obligation to bring the best scientific information into the process of determining environmental policy. Organizations such as the Society for Conservation Biology have a responsibility to provide summary knowledge and expert opinion to the public and to policymakers. And, as individuals, conservation biologists have the right and responsibility to express and encourage their own personal values.

necessarily by a rigid cost/benefit analysis computed in dollars). The costs of preserving biodiversity can clearly be great. Therefore, we need to know the value of all of the benefits that preserved biodiversity provides.

We are just beginning, however, to find ways to measure the economic values of many biodiversity resources (Fig. 1.5). The nature of the resources themselves ranges from specific harvestable materials, to utilitarian ecological services such as biological control (Cairns & Pratt 1995), to information resources such as genetic codes, and to conditions that stimulate aesthetic

appreciation (Ehrenfeld 1976; Rolston 1988). In Chapter 30 we shall explore techniques that conservation economics is developing to measure these values.

Refining technologies to protect, restore, and manage biodiversity is the second area of responsibility of conservation biology. These are the real "problem-solving" functions of conservation biology. Examples of these functions include designing effective systems of preserves using computers to store and analyze biodiversity data, devising ways to restore ecosystems, and developing strategies for using biodiversity resources in a sustainable manner.

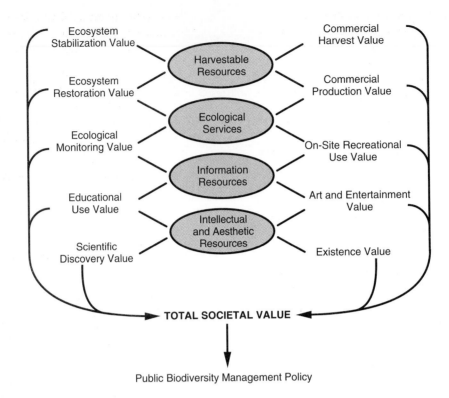

Biodiversity Values

FIGURE 1.5.

Many different categories of economic value exist for biodiversity, all of which should be taken into account in the development of public biodiversity management policy.

Restoring and rehabilitating ecosystems is perhaps the most rapidly evolving activity within conservation biology (Cairns 1991). Restoration draws on the remaining biodiversity of a region to recreate systems that have utilitarian or nonutilitarian values. Rehabilitating less severely disturbed ecosystems is easier. We shall consider these efforts in several of the chapters that examine major ecosystem types.

Devising systems for sustainable use of biodiversity is another critical effort of conservation biology. In many places in the developing world, near-pristine systems survive. But here, human poverty and population growth are often so great that human utilization of these systems is unavoidable. The challenge becomes one of developing ways for humans to use biodiversity in a sustainable manner.

Finally, the third major responsibility of conservation biology is to help society look to the future. We need to identify the probable trends in biodiversity that will occur due to processes such as continued deforestation, global climate change, and invasions of exotic species. And, of course, we must try to predict what these trends in biodiversity will mean for human welfare. Global climate change will present one of the greatest challenges to conservation biologists over the next century. We shall explore the implications of global change for conservation management throughout this text, but especially in Chapter 23.

Thus, conservation biology has many challenges, enough to guarantee the field a strong place in science long into the future. But all of these challenges were really issued years ago. Although one of Aldo Leopold's major accomplishments was to put wildlife management on a more scientific, ecological basis, his philosophy was grounded in the concept of the ecosystem and extended to all species. In "The Conservation Ethic," an essay published in 1933, and included in slightly revised form in the 1949 book, *Sand County Almanac,* Aldo Leopold wrote:

> Given then, the knowledge and the desire, the idea of controlled wild culture or 'management' can be applied not only to quail and trout, but to *any living thing* from bloodroots to Bell's vireos . . . A rare bird or flower need remain no rarer than the people willing to venture their skill in *building it a habitat.*

After 62 years, conservation biology is seriously taking up his challenge to manage biodiversity.

Summary

1. Conservation biology is the scientific study of biodiversity and its management for sustainable human welfare. Conservation biology is intrinsically interdisciplinary, interfacing with many basic and applied areas within biology, as well as with many other fields of science, social science, and humanities.

2. The focus of conservation biology is biodiversity, the richness of the biosphere in life, viewed in a hierarchy from the infraspecific level to the level of the biosphere. The biodiversity crisis has stimulated the emergence of this new science, bringing scientists from ecology, systematic biology, and wildlife management together with environmentalists to seek ways to preserve, restore, and manage biodiversity.

3. The extent to which intrinsic values, such as the view that diversity, complexity, and evolutionary continuity are "good," is an issue on which conservation biologists are divided.

4. The major responsibilities of this new science include clarifying the ecological and economic values of biodiversity and developing techniques to protect, restore, and manage biodiversity.

5. In public policy development, conservation biologists have the major responsibility of bringing current knowledge and expert opinion about biodiversity issues and their significance to humanity into the public arena.

Questions for Discussion

1. Should our present inability to identify positive economic values for a species be considered adequate reason to allow it to decline to extinction?

2. A recent popular book states that relying solely on experts to determine biodiversity policy is as absurd as using public referenda to decide which species are endangered, and recommends that science should resume its proper place: providing support for policymakers. Do you agree or disagree with this view?

3. Should conservation biologists, speaking as scientists or as representatives of a scientific society, endorse specific political groups or proposed actions? If so, under what conditions? If not, why not?

Suggested Reading

Hagan, J. M. 1995. Environmentalism and the science of conservation biology. *Conservation Biology* 9:975–76. A concise discussion of the question of value-free versus value-driven research in conservation biology.

Mann, C. L. and M. L. Plummer. 1995. *Noah's choice. The future of endangered species.* Alfred A. Knopf, New York. An examination of the scientific and societal issues relating to operation of the U.S. Federal Endangered Species Program.

Noss, R. F. and A. Y. Cooperrider (Eds). 1994. *Saving nature's legacy: Protecting and restoring biodiversity.* Island Press, Washington, DC. A broad survey of biodiversity and strategies for its preservation.

chapter

2

History of Conservation

onservation of natural ecosystems and their plant and animal inhabitants has a long and complicated history. Modern conservation attitudes and practices have evolved largely within the context of western society, and have been molded definitively by the major political, economic, and intellectual revolutions that western society has experienced. These forces continue to shape the practice of conservation worldwide.

In North America, societal concern for conservation of wildlife and natural environments has been cyclical (Strong 1988; Nash 1989a). Periods of conservation activism have occurred at times when crises due to intensified human impact on nature have coincided with breakthroughs in our understanding of the natural world and its significance to human well-being. In this chapter, we shall examine some of the major influences that have shaped conservation activity in western society, and consider the major phases of conservation activism that we have experienced. Finally, we shall look ahead and speculate how major trends in environmental conditions may shape the next conservation movement.

Aldo Leopold provided both the scientific and the ethical grounding for conservation of ecosystems and their member species.

Background of Western Conservation Attitudes

Western viewpoints about conservation are conditioned by basic western philosophy, rooted in the Judeo-Christian view of man and nature, and conditioned by political, economic, and intellectual attitudes that grew out of the democratic revolutions of the 1700s and 1800s (White 1967).

Judeo-Christian religious philosophy combines two ideas: the right of exploitation, and the responsibility of stewardship. A fundamental Judeo-Christian belief, outlined in the biblical book of Genesis, is that nature was created to serve the human race, so that the exploitation of nature is thus a legitimate and natural pursuit. Judeo-Christianity does not endow the environment and its inhabitants with protective spirits that prohibit exploitation or that must be appeased before the environment is exploited. In the 1600s, this view found strong support in the philosophy of René Descartes, who argued that humans were "masters and possessors of nature" (Nash 1989b). On the other hand, Judeo-Christian dogma also asserts that human beings are only temporary occupants of the earth, and must be responsible stewards of the natural heritage on which future generations depend. Although many members of western societies may not espouse the "master of nature" view in a biblically fundamentalist way, Judeo-Christian philosophy creates a permissive attitude toward exploitation of nature that is different from that in many eastern societies.

Until the 1700s, however, the rights to, and rewards of, exploitation of the natural world lay largely in the hands of an elite aristocracy. The democratic revolutions of the late 1700s, including the American Revolution of 1775–1783 and the French Revolution of 1789–1799, triggered a restructuring of the framework of society throughout most western societies. With this change came increased access of individuals to productive resources, and increased ability to use them for improving economic and social status. The legitimate right of exploiting nature was now extended to individuals at large in society.

At nearly the same time, the mid-1700s, the Industrial and Scientific Revolution began. On the industrial side, machines and sources of energy were harnessed to enable resources to be processed on new scales and with increased efficiency. With the steam engine came railroads and steamships that opened new regions to exploitation. On the scientific side, discoveries led to a revolutionary change in the basic concept of nature: a world in which all was created and overseen by God was replaced by a world that functioned according to the operation of basic laws of physics and chemistry. The Industrial and Scientific Revolutions gave individuals an enormously expanded ability to exploit resources and create material wealth.

Finally, the 1800s were the culmination of a period of worldwide spread of western culture through colonialism and establishment of world trade. The western system of environmental exploitation was thus spread widely, so that it became the operational system even in areas where the basic philosophical view of humans and nature was quite different (White 1967).

Conservation: Cycles of Crisis and Activity

In North America, five major eras of conservation activism have occurred in the past century and a half. These were periods of social and political activism, each about 10 to 15 years in duration, when concern about the natural environment spread through much of society, enabling new laws and institutions to be created. These revolutionary periods share four important features. Each, first of all, was triggered by patterns of national growth and development that reached some limit of structure and function of regional ecosystems. Second, each represented a time of emergence of basic new ideas in ecology and evolution. Third, in each period, one or more key naturalist-writers emerged as influential spokespersons that dramatized the issues and approaches to their resolution (Strong 1988). Finally, each period resulted in the establishment of new conservation institutions.

Disappearance of the Eastern Wilderness, 1850–1865

By the middle decades of the 19th century, European settlement was well into its third century. The landscape of eastern North America had experienced its greatest degree of agricultural clearing and ecological transformation. Coupled with clearing of the land was the disappearance of the original forest, referred to as the "wilderness," together with many of the larger forms of wildlife. What remained was a cleared, tamed, urban and agricultural landscape. Small farms covered the landscape in New England and the middle Atlantic states, and cotton and tobacco plantations the lands of the southern states. Recognition of this transformation brought with it a nostalgia for the wilderness as it had originally existed. In 1850, for example, James Fenimore Cooper had just finished the last of the *Leatherstocking Tales,* a popular series of novels set in eastern frontier days, then over a century earlier.

At the same time, biology was undergoing a revolutionary change in its view of the natural world—the replacement of a static, creationist view of life by an evolving, mechanistic view. This change is best exemplified by the emergence of the **theory of evolution by natural selection,** presented jointly by Charles Darwin and Alfred Wallace to the Linnean Society of London in 1858. Darwin's *Origin of Species,* which documented the process exhaustively, appeared in 1859. The concept of natural selection replaced the creationist view of the origin of living species with a mechanistic process of interactions within nature. Moreover, it placed humans within the world of nature rather than separate from nature. The evolutionary view also opened the eyes of many to the fact that change in the environment, including change caused by humans, could bring about the extinction of many kinds of organisms, as the fossil record demonstrated. The intellectual dynamism of this era infused all of the natural sciences. Other noted scientists of the period—not all evolutionists—included the paleontologist Louis Agassiz and the botanist Asa Gray.

The taming of the eastern landscape stimulated an aesthetic appreciation of the natural areas that remained, as well as a nostalgia for the wilderness that had disappeared. The aesthetic view of nature was evident in the work of Ralph Waldo Emerson (1803–1882), one of the period's most influential speakers and essayists. Emerson espoused the philosophy of **transcendentalism,** the rejection of material goals and a seeking of harmony and beauty through the contemplation of nature. Emerson's protégé, Henry David Thoreau (1817–1862), adopted and lived this philosophy (Fig. 2.1). He lived simply, observed nature and society in detail, and faithfully recorded his observations and interpretations. For 24 years, Thoreau maintained a journal, begun at Emerson's suggestion, in which he recorded notes from material he had read, personal observations, philosophical musings, and poetry. Drawing on this journal, Thoreau published many newspaper and magazine articles, but, during his lifetime, only two books, both with transcendentalist themes. *A Week on the Concord and Merrimack Rivers,* published in 1849, used a boat trip by Thoreau and his brother as a unifying vehicle for transcendentalist musings. The book sold very poorly, and Thoreau ended up with over 700 of the 1,000 copies printed. *Walden,* published in 1854, was the account of 26 months of living at Walden Pond, near Concord, Massachusetts. This book, a modest but short-lived publishing success at the time, described the virtue of a simple, contemplative life, close to nature. Two other books, *Excursions* and *The Maine Woods,* were published shortly after his death, with the editorial aid of Ralph Waldo Emerson, Emerson's sister Sophia, and William E. Channing, a close friend. His journals, in edited form, were eventually published as a series of 14 volumes (Torrey & Allen 1906). Thoreau also compiled an extensive set of notebooks, most of which remain unpublished, on the lives of Indians and other aboriginal peoples.

Thoreau sought meaning both by philosophical contemplation and scientific study of nature. His early journal entries are strongly philosophical. "The fact will one day flower out into a truth," he wrote in his journal in 1837. In 1856, as his appreciation of nature matured, he lamented the disappearance of the New England wilderness and its wildlife, leaving "a tamer, and as it were, emasculated country," and wished "to know an entire heaven and an entire earth." Later, in 1859, he advocated the establishment of parks, suggesting that every township

> should have a park, or rather a primitive forest, of five
> hundred or a thousand acres, . . . for instruction and
> recreation.

By his later years, Thoreau had become, through his own study, a perceptive ecologist (although the term *ecology* had not yet been coined), and his journal records many scientific observations. His observations of the kinds of trees that predominated in various woodlots near Concord, for example, led him in 1860 to recognize forest succession, anticipating a major concept that was not formally stated until more than 35 years later.

Thoreau's idea of park development was, in fact, carried forward by others. Frederick Law Olmsted (1822–1903) grew up in a rural, farming area near New York City, acquiring a

FIGURE 2.1

Henry D. Thoreau, who died two months short of his 45th birthday, found meaning in the natural world of New England, which was rapidly disappearing through human settlement and clearing for farms.

knowledge of farming, horticulture, and landscaping. A visit to Europe stirred an interest in parks, and in the relation of people to the landscapes they inhabited. Ultimately, by winning a competition among landscape architects for the best design, he gained the opportunity to create Central Park in New York City. Carried out between 1857 and 1861, the Central Park project was a deliberate effort to restore an example of eastern wilderness within the city: "a specimen of God's handiwork" to serve the equivalent of "a month or two in the White Mountains, or the Adirondacks. . . ." Olmsted participated in the development of other urban and nonurban parks in the East, and in 1864 helped the effort to gain protection by the State of California for Yosemite Valley, an action that proved to be the first step in western park development.

This period saw the awakening of an appreciation of the value of nature, its dynamic character, and the need for stewardship of the natural world by humanity. In a sense, the period culminated with the appearance in 1864 of the book *Man and Nature* by George Perkins Marsh. This book described the worldwide ecological impact of humans on the natural environment. Although it did not influence events of the time, it profoundly influenced those of later conservation eras.

Closing of the Western Frontier, 1890–1905

By the end of the 19th century, western settlement in the United States and Canada had reached the Pacific Ocean. With the recognition that further territorial expansion in mainland North America was impossible came the realization that the forest, water, and mineral resources of North America were finite. Dramatic effects of careless exploitation of some of these vital resources also drew society's attention to the need for conservation. Disastrous fires raged through wastefully logged forests of the Lake States, destroying towns and killing thousands of persons (Biswell 1989). The effects of severe blizzards and droughts also emphasized the degree to which a healthy national economy depended on a benign environment.

In science, the 1890s saw the emergence of ecology as a recognized branch of biology. Ecology gained recognition, as distinct from natural history, with the growth of a body of theory about interactions in nature. The process of **biotic succession** was the most dramatic example of such a theory. Henry C. Cowles (1869–1939), a botanist at the University of Chicago, studied the sequence of plant communities that occupied the successively higher and older beach terraces at the southern end of Lake Michigan. These beachlines represent lake levels that existed at various times during the recession of Pleistocene glaciers over the past 10,000 or so years. From this time sequence, Cowles inferred the process of plant succession and its causes (Cowles 1899). Others soon recognized that succession was also shown by animal communities. Succession revealed that natural communities were dynamic and changing on a time scale that allowed them to be influenced for the better or worse by human activities. It also provided a theoretical framework that proved to be applicable in management of resources such as forests, rangeland, and wildlife.

The importance of living resources, and the need for their sound management, was stated most eloquently by the naturalist-writer John Muir (1838–1914). Muir (Fig. 2.2), born in Scotland, came to the midwestern United States as a boy. He was strongly influenced by the writings of Emerson and Thoreau (Fleck 1985). The forests and mountains of the West, particularly of California and Alaska, ultimately captured his interest. Muir was appalled at the damage being done to the California Sierras by grazing, logging, and burning, and turned to popular writing as a means of raising public concern about these issues. Through books such as *The Mountains of California* (1894) and *Our National Parks* (1901), he stimulated an interest in the natural history of the western mountains. He actively promoted federal programs for protecting lands and forests. As the final statement in his book on national parks, he argued,

> . . . God has cared for these trees, . . . but He cannot save them from fools—only Uncle Sam can do that.

Muir's activism and love for the California Sierras led to the establishment of the Sierra Club, of which he was the first president.

FIGURE 2.2

John Muir, known as "John o' the Mountains," was an articulate proponent of the preservation of wilderness, and influenced the conservation efforts of President Theodore Roosevelt. Here, both men are seen at Glacier Point, overlooking Yosemite Valley.

(Courtesy of the Bancroft Library, University of California, Berkeley)

This period saw the establishment of several federal agencies devoted to resource management. In 1891, a system of national forests was established under the administration of the United States Department of Interior, and, in 1898, a Division of Forestry was created within the United States Department of Agriculture. The first director of the Division of Forestry was Gifford Pinchot (1865–1946), a forester trained in Europe, where forest management had a long history. In 1905, the national forests themselves were transferred to the United States Department of Agriculture, and the present administrative structure of the **United States Forest Service** was established. Between 1889 and 1897, Presidents Benjamin Harrison and Grover Cleveland designated 36 million acres of national forest. Between 1901 and 1909, Theodore Roosevelt added 148 million acres, creating a system of 150 national forests throughout the United States.

Another new resource agency was the **United States Bureau of Reclamation,** created in 1902, in response to the need for improved management of land and water resources. The **United States National Wildlife Refuge** system was also initiated with the establishment of Pelican Island Refuge in Florida in 1903. This refuge provided for the protection of breeding colonies of egrets and other water birds. These birds had been exploited heavily for their plumes, which were used in the production of ladies' hats.

This period saw enormous expansion of the system of **United States National Parks and Monuments.** Prior to 1890, the only national park was Yellowstone, created in 1872. Sequoia, Yosemite, and Mt. Rainier National Parks were created in the 1890s, and some 20 national monuments, including Mt. Lassen and the Grand Canyon, were designated, initiating a phase of national park development that continued for three decades.

Closing of the western frontier thus forced recognition of the dynamic nature of living resources, the emergence of a new science—ecology—that could provide the theory needed to manage them, and the creation of major federal institutions to oversee resource management at a national scale.

The Dust Bowl Era, 1930–1940

Following the turn of the century, crop farming spread onto marginal lands in areas such as the Tennessee Valley and the Great Plains, spurred by population growth, poorly regulated land settlement schemes, and, in 1929, the collapse of the urban-industrial economy that signaled the start of the Great Depression. In the 1930s, water erosion and flooding plagued the midwestern and southern states, while droughts and wind erosion afflicted the central plains. The Dust Bowl of the Texas and Oklahoma panhandles, southeastern Colorado, and southwestern Kansas came to symbolize these conditions, which reached their extreme on Black Sunday, April 14, 1935, in one of the largest and most severe dust storms on record (Fig. 2.3). Agricultural development had reached the limit of tolerance of the land to farming practices that did not recognize the intimate interactions between the living and nonliving components of the environment.

The 1930s were also a time of widespread recognition of a new concept in ecology, that of the **ecosystem.** The concept itself had begun to crystalize in the 1920s, with growing recognition by ecologists that the living and nonliving components of natural systems influenced each other in a reciprocal fashion. The activities of organisms modified their nonliving environment, as well as being influenced by its physical and chemical conditions. By the 1930s, recognition of this concept was widespread. Aldo Leopold (1933a) wrote, for example, that civilization depended on

> . . . *mutual and interdependent cooperation* between human animals, other animals, plants, and soil,

a clear, humanistic statement of the ecosystem concept. In 1935, a British ecologist, A. G. Tansley, coined the term *ecosystem,* noting that,

> Though the organisms may claim our primary interest, . . . we cannot separate them from their special environment, with which they form one physical system.

Scientists in both basic and applied areas of ecology thus recognized a deeper level of dynamics in the natural environment, specifically the powerful influences that living organisms, including humans, could have on the physical landscape.

FIGURE 2.3

The dust storms of Black Sunday, April 14, 1935, symbolized the misuse of land and water resources that precipitated the conservation crisis of the 1930s.

(USDA Soil Conservation Service photo by Thomas G. Meier)

This period saw the emergence of scientific wildlife management, largely through the effort and inspiration of Aldo Leopold (1887–1962) (Hedgepeth 1989). Trained as a forester, Leopold (Fig. 2.4) spent his early career with the United States Forest Service in Arizona and New Mexico. There, he gained field experience in wilderness areas with large game populations. Gradually, his interests shifted to the management of wildlife in forests and other ecosystems. At first, his approach was traditional, espousing predator control as a primary technique to encourage game species (Flader 1974). Gradually, however, Leopold and many others began to appreciate the intricacy of relationships among the living and nonliving components of nature. This appreciation brought with it a new view of the role of predators and predation in natural communities (Dunlap 1985). After leaving the Forest Service, Leopold devoted most of his attention to game ecology, bringing to this field an ecosystems philosophy. This approach is perhaps best expressed in Leopold's (1949) essay, *Thinking Like a Mountain,* in which he describes how deer overpopulation could decimate the natural vegetation and writes,

> I now suspect that just as a deer lives in mortal fear of its wolves, so does a mountain live in mortal fear of its deer.

Leopold became professor of game management at the University of Wisconsin. The textbook, *Game Management,* published in 1933 (Leopold 1933b), is still a basic reference for students of wildlife ecology. Leopold also supported the establishment of wilderness areas, and in 1935 was one of the founders of the

FIGURE 2.4

Aldo Leopold, regarded as the father of scientific wildlife management, saw the concept of the ecosystem as central to management of nature.

(Photo courtesy of the University of Wisconsin–Madison Archives)

Wilderness Society. In the eyes of environmentalists, one of his most important contributions was to begin the effort to define an ethical relationship between humans and the rest of nature (Flader 1987), an effort embodied in his early essay on "The Conservation Ethic" (Leopold 1933a).

Paul B. Sears, however, was perhaps the most influential writer-naturalist of this era. Sears (1891–1989) was born in Ohio and attended Ohio Wesleyan University, where one spring day in 1913 he found his route to classes blocked by the flooding Olentangy River. His awareness of how humans were mistreating soil and water resources, in fact, might have been awakened by this event. Sears is best remembered for the book *Deserts on the March,* published in 1935 and in print continuously since then. In this book he dramatized the crises of land and water mismanagement that much of North America faced in the 1930s, including the "sea of swirling, swishing liquid mud" of his college experience and the "slow, chilling, and pervasive horror" of the dust storms of the Great Plains. Sears later chaired the Yale Conservation Program, the first graduate program in conservation science in the United States.

Under the presidency of Franklin Roosevelt, with Harold Ickes as Secretary of Interior, the mid-1930s saw the establishment of federal institutions aimed at the crises of water and land management (Strong 1988). The **Tennessee Valley Authority,** organized in 1933, sought to revitalize one of the most impoverished and degraded regions of the United States by impoundments to control flooding and produce hydroelectric power, coupled with soil conservation and reforestation programs to rehabilitate the land. The **Soil Conservation Service,** formed in 1935, was also created to devise and promote techniques of evaluating land capability and controlling erosion.

The dust bowl era revealed patterns of ecological disruptions on a scale greater than ever before, and showed that the physical environment was not an unmanageable control over the living world, but one both affecting and affected by the plants and animals occupying it. It firmly established the need for an ecosystem approach to the resolution of conservation problems.

The Explosion of Population and Environmental Pollution, 1960–1975

World War II suddenly diverted attention from conservation issues. It also initiated an era of unparalleled economic expansion, and spawned explosive growth of technology and human population. Technology provided "Better Living through Chemistry" in the form of synthetic fibers, plastics, inorganic pesticides, leaded fuels, and many other products that sooner or later caused the introduction of toxic and biologically active materials into the environment. The availability of cheap petroleum encouraged the growth of a society centered on the internal combustion engine. To this explosion of industrial technology was added that of the human population, in the postwar "baby boom." The result was exponential growth in the pollution of air, land, and water by chemicals and chemical wastes. Soon, natural mechanisms of detoxification of pollutants and of homeostatic adjustment to their effects were overwhelmed by these inputs. The limits of ecosystem function were being challenged by the chemical products of human activity.

Postwar technologies had also given ecology new ways of working with the dynamics of ecosystems. Geiger counters for tracing the movement of radioisotopes, spectrometers for analyzing the composition of organic and inorganic matter, radiometers for measuring the wavelength and intensity of solar radiation, calorimeters for determining energy content of organic materials, and computers that permitted rapid analysis of masses of data enabled ecologists to follow the flows of energy and cycles of chemical substances through complex pathways in ecosystems. **Ecosystem analysis,** using these tools, led the way in studies of environmental pollutants and their ecological effects. Stimulated largely by the pioneering efforts of Eugene and Howard Odum (1955) in the 1950s, many ecologists concentrated on measuring and modeling the flow of energy through whole ecosystems during the 1960s. At the Hubbard Brook Experimental Forest in New Hampshire, other pioneering studies by F. H. Bormann and Gene Likens (1967) stimulated interest in studies of nutrient cycling in whole ecosystems. These efforts culminated in the decade-long International Biological Program, begun in 1966, one goal of which was to develop predictive mathematical models of ecosystem energy and nutrient dynamics. Although this ambitious goal was not fully attained, ecological modeling emerged as an important tool in the study of environmental pollution.

Rachel Carson (1907–1964), a writer by training and a naturalist by avocation, was editor-in-chief of publications of the United States Fish and Wildlife Service during this era (Fig. 2.5). Originally, her interests had centered on the sea, and she had written several books about marine science, the most popular being *The Sea Around Us,* published in 1951. In the 1950s, however, her work with the Fish and Wildlife Service made her

FIGURE 2.5

Rachel Carson drew public attention to the problem of environmental contamination by pesticides through publication of *Silent Spring* in 1963.

(Erich Hartmann/Magnum)

aware of the growing impacts of environmental pollution, particularly by pesticides, on wildlife. She brought her concerns about pesticides to public attention in 1962 in the book *Silent Spring*. In the preface to this book she warned of the possible fate of wildlife:

> On the mornings that had once throbbed with the dawn chorus of robins, catbirds, doves, jays, wrens and scores of other bird voices there was now no sound; only silence lay over the fields and woods and marsh.

This book, which proved to be a best-seller, was at first regarded as heretical by many biologists. The concerns that it raised, however, proved to be real, and an unparalleled effort to investigate the effects of environmental pollution began. Newly crystallized concepts of energy flow and nutrient cycling played a key role in revealing how pesticides and many other pollutants exerted their impacts. The near extinction of the peregrine falcon in North America due to pesticide pollution, and declines of populations of many other species, also raised concern for endangered species in general.

A second influential scientist and writer of the period was Paul Ehrlich, professor of biology at Stanford University. Ehrlich's concerns centered on the growth of human popula-

tions as the ultimate driving force behind environmental pollution. A charismatic speaker and prolific writer, Ehrlich expressed his gravest prognosis in the book, *The Population Bomb*, published in 1968. Ehrlich warned:

> Cancer is an uncontrolled multiplication of cells; the population explosion is an uncontrolled multiplication of people.

Television, radio, and newspapers brought environmental issues to public attention with greater impact than in previous environmental eras. Ultimately, the efforts of environmentalists such as Barry Commoner, David Brower, and many others extended concern to almost every facet of human relation to the natural environment. Even in competition with the enormous issues of the Vietnam War, environmental quality became a major societal priority. **Environmentalism,** both as a philosophy of living and as a strong political force, emerged from its roots in philosophy, ecology, and the appreciation of nature (Scheffer 1989). Environmentalists led the drive that culminated in the **Wilderness Act** of 1964, which established an effective system of wilderness areas in national forests, the national park system, and national wildlife refuges. As a result of concern over pollution, the **U.S. Environmental Protection Agency** was established in 1970, and charged with maintaining an environment safe for humans and wildlife, specifically by regulating the discharge of toxic materials into air, water, and soil. Other major items of federal legislation included the **National Environmental Policy Act** of 1970, which established the requirement for environmental impact assessment for projects involving federal support or approval, and the **Endangered Species Conservation Act** of 1973, which created the current system of designation, protection, and recovery of endangered species.

The era of pollution and population brought the realization that human intrusion into ecosystem processes could lead to catastrophic indirect effects, many of which were not intuitively predictable. It also raised the issue of endangerment of species, setting the stage for the concerns about biodiversity that have come to the fore in the last decade of the 20th century.

The Biodiversity Crisis, 1990–????

Human activities are now exceeding the capacity of most ecosystems to sustain their member species and maintain their integrity of function. Human population growth, urban sprawl, and the spread of industrial agriculture are transforming vast areas of natural ecosystems. Overharvest of forests and fisheries is degrading still larger areas of land and water ecosystems. Massive urban-industrial discharges of sulfur and nitrogen oxides, greenhouse gases, and chlorofluorocarbons are modifying the atmosphere, leading to regional or global problems such as acid deposition, stratospheric ozone depletion, and the beginnings of global climatic warming. Tropical deforestation and overgrazing of arid lands are changing water and energy relationships of vast areas of the land surface. Directly, and by promoting climatic change, these processes are promoting forest declines in temperate and subarctic regions, intensifying desertification in tropical

EDWARD O. WILSON: BIODIVERSITY AND BIOPHILIA

Born in Alabama in 1929, to parents that divorced when he was quite young, Edward O. Wilson (Reading Fig. 2.1) spent an almost rootless childhood in a succession of locations scattered from Florida, west to Mississippi, and north to Washington, DC. The commonality that he found early was a love of life in the woods, swamps, and waters of the Southeast, insects and reptiles in particular. His early encounters with biodiversity were sometimes harsh: he lost the sight of his right eye when it was pierced by the dorsal spine of a pinfish he jerked from the water when fishing, and he was bitten by a pygmy rattlesnake he had handled carelessly.

Notwithstanding these episodes, his passion for nature grew, and became focused on ants, a group of animals for which his slightly limited vision was no handicap. He attended the University of Alabama, where he began to study these insects in earnest, carrying his studies there through his master's degree. After a year at the University of Tennessee he moved to Harvard University for his doctoral studies. At Harvard, he entered an academic world that not only expanded his intellectual horizons, but gave him opportunity to travel to the tropics, first in Mexico and the West Indies, and later throughout the world. Finishing his doctorate, he stayed at Harvard in a position that did not offer the possibility of tenure. His promise as a scholar and offers from other universities, however, led Harvard to grant him a full faculty position.

Wilson's studies of an ancient, diverse, and widespread insect group gave him insight into a wide range of topics beyond those of ant taxonomy and systematics. He saw geographical patterns in the evolution of the family, and these led him to collaborate with Robert MacArthur on the development of rigorous biogeographic theory. With his student, Daniel Simberloff, he pioneered the experimental study of biogeographic

READING FIGURE 2.1

Through detailed observations in nature, combined with a search for patterns and underlying causes, Edward O. Wilson has pioneered the fields of sociobiology and island biogeography, and has awakened both scientists and the public to the values of biodiversity. With Wilson is "Dacie" (*Daceton armigerum*), a predaceous ant of the Venezuelan tropical rain forest canopy.

(*Harvard Gazette.* Photo by Jon Chase)

processes on islands. With other colleagues he investigated the social biology of ants, an interest that led to the emergence of sociobiology as an important field of behavioral and evolutionary ecology.

His observations of change in the environment of the southeastern United States, over a period of half a century, and of the rapid changes that have occurred throughout the tropics in the last few decades, have raised his concern about the rapid loss of biodiversity. He sees a deep human affinity for life, *biophilia,* that is evident in the many ways that human cultures relate intimately to living organisms. He sees value in biodiversity, both intrinsic value that has resulted from the long span of evolutionary history, and utilitarian

value in the products and services that biodiversity provides. He has become the foremost advocate for preservation of biodiversity, particularly through books and essays. He admonishes

> We should judge every scrap of biodiversity as priceless while we learn to use it and come to understand what it means to humanity.

Wilson is author, coauthor, or editor of 18 books, two of which have received Pulitzer Prizes. He has been elected to the National Academy of Sciences, and is a recipient of the National Medal of Science and many national and international prizes for his accomplishments in science.

---

and subtropical areas, and increasing erosion of soil in mountain areas. Several of these processes have positive feedbacks that are likely to increase the severity of disturbance. Global changes such as these compound the problems of conserving biotic diversity in the rapidly declining areas of natural ecosystems that now remain. The combination of these changes threatens a catastrophic loss of biotic diversity throughout the biosphere.

Edward O. Wilson (1929–) has emerged as a powerful spokesperson for the value of biodiversity, and has led efforts to bring the biodiversity crisis to public awareness. His 1992 book, *The Diversity of Life,* is an eloquent account of the evolutionary origins of biodiversity, the threats that humans have created to its survival, and of the importance of biodiversity to human welfare (*See* Reading 2.1).

The 1990s have seen the emergence of powerful new computer-based technologies that show promise of helping conservation biologists deal with this crisis. Computerized databases, known as **geographic information systems (GIS),** can store detailed data on environmental and biodiversity attributes on a state-to-national scale. By analyzing these attributes, maps showing the need and suitability of sites for biodiversity preserves can be prepared. Improved mathematical models of population dynamics have also appeared. These so-called **spatially explicit models** are capable of describing the dynamics of individual subpopulations of endangered species, taking into account factors such as the age structure and birth-death relations of individuals in each subpopulation, as well as the frequency of movement of individuals among the various subpopulations. These models are giving much improved predictions about the viability of endangered species, and about the design of habitat features needed to assure their survival.

The realization that a rich national heritage is in danger has also led to efforts to improve institutions concerned with biodiversity. It is perhaps too early to tell which institutions will prove to be the most effective and enduring. By the mid-1990s, almost all states had developed **Natural Heritage Programs** designed to inventory biodiversity and store the data in geographic information systems. In most states, these programs were coupled with **Gap Analysis Programs,** in which data in these geographic information systems are used to identify the "gaps" in biodiversity protection—the components of state biodiversity that remain inadequately protected.

At the federal level, an important development was the creation in 1993 of the **National Biological Service (NBS)** by consolidation of research personnel of seven bureaus within the Department of Interior. The mission of the NBS is to work with both private and governmental organizations to achieve sound management and conservation of the nation's biodiversity resources.

Finally, the early years of the 1990s have seen the rapid growth of conservation biology as a branch of science, together with that of the **Society for Conservation Biology** (*See* Chapter 1).

Looking into the Future

History should prepare us to look to the future, but we can do this only speculatively in the arena of global ecology. Given the weakness of efforts implemented in the 1990s to restrain causes of global change that have been clearly identified, it seems likely that the impacts of many of these—fossil fuel use, release of industrial chemicals, tropical deforestation, and mismanagement of arid lands—will continue to grow. Global change may contain positive feedbacks, such as arctic warming that increases decomposition and the release of CO_2 from tundra soils. Humanity thus will soon be faced with realities of global change like those now only suggested by computer models of the global ecosystem (*See* Chapter 23).

Summary

1. Environmental concern has been concentrated in periods of environmental crisis and scientific revolution. Since the mid-1800s, North America has seen five such periods, each with unique issues, influential writers and spokespersons, major advances in science, and the emergence of new societal institutions. In the mid-1800s, concern arose over loss of the eastern North American wilderness. Henry Thoreau and others drew attention to this loss. This era saw the emergence of the theory of evolution by natural selection and the first efforts to establish natural parks in North America.

2. From 1890 through 1905, closing of the western frontier drew attention to issues of renewable resources, particularly forests and wildlife. John Muir was a proponent of federal conservation efforts. Growth of the field of ecology led to new insight into renewable resource management. The U.S. Forest Service and National Wildlife Refuges date from this era.

3. In the 1930s, mismanagement of soil and water resources led to the Dust Bowl crisis. Paul Sears and Aldo Leopold dramatized this crisis, and ecologists employed the new concept of the ecosystem in its resolution. The Soil Conservation Service and Tennessee Valley Authority were established, and scientific wildlife management became a recognized branch of applied biology.

4. From 1960 through 1975, the population and pollution explosions spawned concern about the health of the biosphere. Rachel Carson and Paul Ehrlich raised concern about the environment to a new high, and ecosystem ecologists found new tools to study energy flow and chemical cycling in ecosystems. The U.S. Environmental Protection Agency was established, and laws were enacted to protect endangered species and require environmental impact analysis.

5. In the 1990s we face the catastrophic loss of biodiversity. Warnings of this loss have been sounded by Edward O. Wilson and others. The new science of conservation biology has applied computer methodologies to assessment and protection of biodiversity. State Natural Heritage Programs and the new U.S. National Biological Service are making efforts to safeguard biodiversity. Global change is likely to create new environmental challenges in the early 2000s.

QUESTIONS FOR DISCUSSION

1. Do you agree with the hypothesis that a cyclic concern brings conservation issues to the fore every 25–30 years or so? If so, what is the basis for this periodicity?

2. Do you believe that the institutions created during the environmental era of the 1960s and early 1970s have made long-lasting improvements in the North American environment?

3. Looking to the future, what do you think might be critical environmental issues in A.D. 2025 or so?

SUGGESTED READING

Leopold, A. 1949. *Sand county almanac.* Oxford Univ. Press, New York. Classic essays by the most influential figure in the North American conservation movement.

Scheffer, V. B. 1991. *The shaping of environmentalism in America.* Univ. Washington Press, Seattle. A thoughtful tracing of the origin and maturation of the neoenvironmentalist movement.

Strong, D. H. 1988. *Dreamers & defenders: American conservationists.* University of Nebraska Press, Lincoln, NE. Biographies of ten of the most influential individuals in conservation in the United States from the mid-1800s through the 1960s.

chapter

3

Populations, Communities, and Ecosystems

C onservation efforts require working with populations of species, and with the communities and ecosystems of which they are members. In the past, managers and conservationists have often centered their efforts on individual species and their populations. Now, however, conservation biologists realize that individual species cannot be considered in a vacuum, and that their relations with other species, in communities, and with the total biotic and physical environment, in ecosystems, must be considered. In this chapter, we shall review basic concepts of population, community, and ecosystem ecology that are essential to conservation practice.

A biotic community, comprising populations of many species, interacts with its abiotic habitat to form an ecosystem, as exemplified by this mountain lake.

Outline

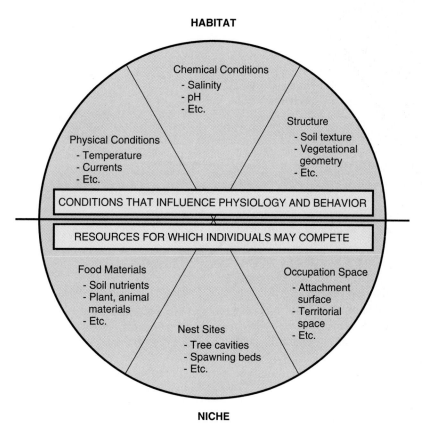

HABITAT

Chemical Conditions
- Salinity
- pH
- Etc.

Structure
- Soil texture
- Vegetational geometry
- Etc.

Physical Conditions
- Temperature
- Currents
- Etc.

CONDITIONS THAT INFLUENCE PHYSIOLOGY AND BEHAVIOR

RESOURCES FOR WHICH INDIVIDUALS MAY COMPETE

Food Materials
- Soil nutrients
- Plant, animal materials
- Etc.

Occupation Space
- Attachment surface
- Territorial space
- Etc.

Nest Sites
- Tree cavities
- Spawning beds
- Etc.

NICHE

FIGURE 3.1

The habitat of a species is the total set of conditions under which it can conduct its life activities, and its niche is the resources it must exploit for growth and reproduction.

POPULATION CONCEPTS

The coexisting and interacting individuals of a species constitute a **population.** In higher plants and animals, these individuals vary in size, age, sex, genotype, and other ecologically important features. They require favorable conditions of the environment, such as temperature and humidity, in order to carry out their essential life activities. They also require resources such as food and nest sites to enable them to grow and reproduce. Thus, a species can be characterized according to its required **habitat,** the total set of conditions under which it can conduct its life activities, and its **niche,** the resources that it must exploit for growth and reproduction (Fig. 3.1).

Populations of species are not distributed uniformly over their geographical ranges. Often they consist of semi-isolated units among which occasional dispersal of individuals occurs. Some of these subpopulations may even disappear and reappear through time due to local extinctions and later recolonizations. A geographical pattern of this type is termed a **metapopulation.** Ecologists now recognize that regional populations of many species are structured in this fashion.

A discrete population tends to grow in numbers as a result of **reproduction** and **immigration,** and decrease in numbers through **mortality** and **emigration.** Many factors may influence these processes. Ecologists recognize that some of these factors tend to increase their intensity of negative influence on population growth as the population density increases. These are termed **direct density dependent factors,** and are usually biotic relationships such as competition or predation. Other factors affect populations with an intensity unrelated to the population density. These are termed **density independent factors,** and are usually factors of the abiotic environment, such as favorable or unfavorable weather. All populations are affected by both kinds of factors.

Populations present all degrees of variation in stability of numbers and patterns of growth and decline. Many species of large body size and long life tend to show some degree of **population regulation,** in which the relaxation and intensification of direct density dependent factors tend to maintain numbers within a narrow range of densities. The level around which such populations fluctuate is often termed the **carrying capacity.** Other species, particularly small, short-lived forms, tend to fluctuate widely in numbers, their populations growing rapidly when conditions are favorable and resources abundant, and declining precipitously when conditions deteriorate or resources become exhausted. In the past, ecologists tended to view regulation of populations closely about carrying capacity as being the typical case. Now, they recognize that a high degree of regulation is probably the exception, and that disequilibrium relationships between populations and carrying capacity are the norm.

COMMUNITY CONCEPTS

The populations of species that occur together and interact with each other form the **biotic community.** Any such assemblage is a community, but ecologists often use this term to refer to a **community type,** an assemblage that has a characteristic composition and structure, and that can be recognized in different locations. The beech-maple forest of eastern North America, for example, is an example of a community type. Whether community types are anything more than arbitrary divisions of a potentially continuous variation of composition and structure of communities along habitat gradients is argued, sometimes heatedly, by ecologists (McIntosh 1995). Most, however, now believe that communities probably do vary continuously and gradually along habitat gradients, but that the designation of community types is a useful practice, mainly to aid in communication among ecologists themselves. The sharp breaks that one sees in community structure and composition over the landscape thus reflect sharp change in habitat conditions, as along the ridge of a mountain, or the effects of disturbance, such as forest fire.

Communities are dynamic entities, changing through time in response to changing climate, changing structure of the landscape, and changing regional biota. Communities also change as a result of their own internal dynamics, and their interaction with habitat conditions, a process known as **biotic succession.** Biotic succession can be illustrated by the sequence of communities that occurs on a newly exposed rock outcrop (Fig. 3.2). Initially, organisms such as lichens and mosses colonize the bare rock surface, forming a pioneer community. These simple organisms trap soil particles carried by wind and water, and contribute to the breakdown of the rock surface by chemical action. Eventually, they accumulate enough soil to permit the growth of annual grasses and broad-leaved herbs. These species, in turn, accelerate breakdown of the rock by penetrating their roots into crevices, and forming a layer of dead plant litter that is an even more effective trap for soil. In time, perennial herbaceous plants invade. As the soil system develops, still larger plants such as shrubs and trees appear, and many of the pioneer and early successional species drop out. In some cases, this process leads toward a theoretical community of species that can reproduce and replace themselves indefinitely, a stable **climax community.** This process is termed **primary succession,** since it takes place on a site that was not previously occupied by a community. Succession that follows a disturbance, such as a forest fire or the clearing of land for farming, is termed **secondary succession.** Secondary succession usually proceeds faster than primary succession in terrestrial ecosystems, since well-developed soil is often already present.

Succession rarely, if ever, reaches a stable climax, however. In the later stages of succession, change may simply become very slow, and before a true climax is achieved, some disturbance may strike the community. A fire may occur and initiate secondary succession. Disease, insect outbreaks, or an invasion of a new species may perturb the community, triggering a successional response. In some instances, what seem to be climax species can bring about the degradation of their habitat, causing the system to undergo succession anew. In addition, a

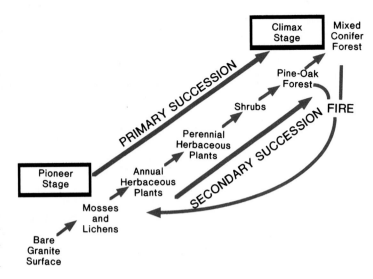

FIGURE 3.2

A schematic diagram of biotic succession beginning on bare rock and leading toward a mature forest that eventually acquires old-growth characteristics.

long-term shift in climate may favor a slightly different group of climax species. Ecologists now recognize that primary succession does not often lead to a highly predictable, stable climax, and that secondary succession rarely restores the exact system that experienced disturbance.

Biotic succession varies in pattern and duration in different ecosystems (MacMahon 1981). In chaparral shrublands and subarctic conifer forests, for example, the successional sequence may be short and the dominant plants of late succession may be the same species that resprout or germinate immediately after disturbance. Species of the same growth form, such as woody shrubs, may dominate the entire sequence, as well. In deciduous forests, on the other hand, succession may be an extended sequence with several fairly distinct successional stages. In this case, early successional communities may be dominated by species of life forms quite different from those late in succession.

If biotic succession is a process that can occur under constant climatic and geologic conditions, what are its causes? The most important processes are themselves biotic. Ecologists now recognize four biotic mechanisms of succession (Fig. 3.3), which can act alone or in combinations (Connell & Slatyer 1977; Pickett et al. 1987). The first of these is **differential dispersal.** Plant and animal species differ enormously in their dispersal capabilities. Some, such as winged insects and plants with windblown seeds, are able to colonize new or vacant sites very quickly. Others with slower dispersal mechanisms arrive later. As species with different dispersal capabilities establish themselves, the composition of the community changes. A second mechanism is **differential survival.** Many kinds of species may initially reach and colonize a new area of habitat. In time, however, only some of these are able to tolerate the range of habitat conditions that occurs. As species drop out, the composition of the community changes. Third, some species modify the conditions of the habitat in a way that favors the establishment of other species, a process termed **facilitation.** Pioneer species may accelerate the

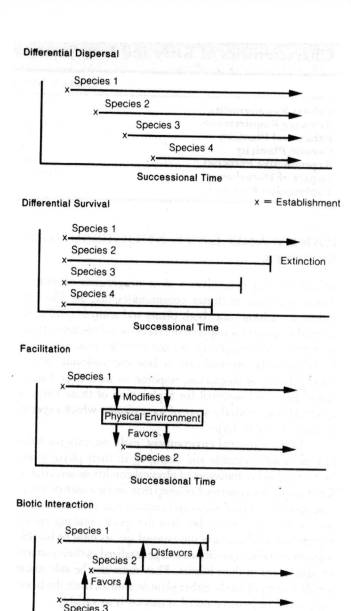

Differential Dispersal

Species 1
Species 2
Species 3
Species 4

Successional Time

Differential Survival

x = Establishment

Species 1
Species 2
Species 3
Species 4

Extinction

Successional Time

Facilitation

Species 1

Modifies

Physical Environment

Favors

Species 2

Successional Time

Biotic Interaction

Species 1

Disfavors

Species 2

Favors

Species 3

Successional Time

FIGURE 3.3

Diagrammatic representation of the four major mechanisms of biotic succession.

physical and chemical weathering of terrestrial substrates and aquatic sediments, aiding purely abiotic processes of weathering in the buildup of soil. Facilitation is likely a major mechanism in the example of succession on bare rock. Fourth, direct positive or negative **biotic interactions** may favor the establishment or hasten the disappearance of species. Establishment of a certain plant, for example, will favor herbivores that feed on it. The appearance of a predator may cause the decline of prey species unable to tolerate heavy predation. The key feature of biotic succession is that its mechanisms are primarily internal ecosystem processes, rather than changes in external factors such as climate, physiography, or regional biota.

FIGURE 3.4

The concept of the ecosystem applies to small islands of soil, together with plants, animals, and microorganisms, that develop on granitic rock outcrops in many parts of the world.

(Photo by G. W. Cox)

In real situations, not surprisingly, several or all of these mechanisms operate, and external controlling factors also show change. In a lake, for example, successional processes are active, but geological processes also act to transport sediment into the lake and reduce its depth, eventually filling it. The changes that occur in lake ecosystems through their geological history are not purely the result of biotic succession.

THE ECOSYSTEM CONCEPT

An ecosystem is a unit of the environment made up of living and nonliving components that interact with interchanges of nutrients and energy. The living portion of the ecosystem is the community of organisms. The nonliving components of the ecosystem make up the physical **habitat** in which these organisms exist, and with which they interact. As Evans (1956) noted long ago, the ecosystem is the basic unit of ecology; appreciation of this fact is central to conservation ecology.

A lake provides an easily appreciated example of an ecosystem. The living components of the lake ecosystem can be classified in various ways, depending on the objective of a particular study. In studies of energy and nutrient relations, five major components are often distinguished: **producers,** the organisms (usually green plants) that store energy in new organic matter by photosynthesis or chemosynthesis; **herbivores,** the organisms that feed on producers; **carnivores,** the animals that feed on herbivores; **top carnivores,** the animals that feed on carnivores; and **decomposers,** the organisms that consume dead organic matter produced by all organisms (including themselves). In a lake, producers include forms such as **phytoplankton,** the tiny single-celled or colonial plants that live floating free in the water, filamentous algae that live attached to submerged surfaces, duckweeds that live floating on the surface, and larger flowering plants that are rooted in the bottom. Herbivores

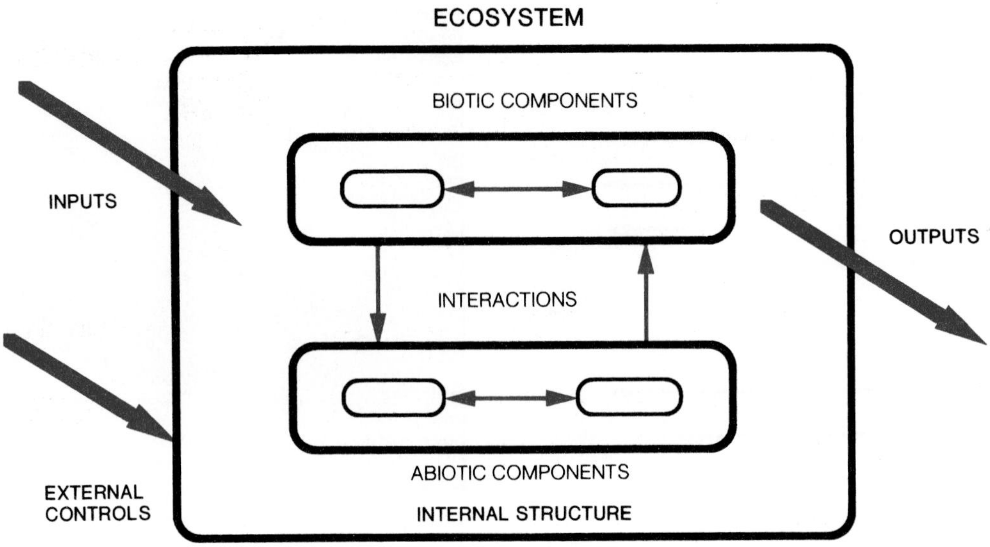

ECOSYSTEM

FIGURE 3.5

Ecosystems possess major interactions within and between their living and nonliving components. External forces exert controlling influences on the ecosystem, as well. Ecosystems are not closed systems; both materials and energy may enter and leave.

include **zooplankton,** the small free-swimming organisms that feed on phytoplankton, along with snails, turtles, and other plant-feeders. Carnivores consist of predatory invertebrates and fish, and top carnivores of forms such as fish-eating birds. The major nonliving components of the lake ecosystem are the water mass, with its particular physical and chemical conditions, the sediments forming the lake bottom, and the air space above the surface within which organisms functionally connected with the lake system are active.

The concept of the ecosystem can be applied to environmental units of greatly differing size and permanence. A tiny temporary pond is an ecosystem, as is a large body of water such as Lake Michigan or the cold upwelling waters that surround the continent of Antarctica. Terrestrial ecosystems can range in scale from small islands of soil and life on bare granitic outcrops (Fig. 3.4) to vast areas such as Yellowstone National Park and its surrounding areas of national forest, which together form the Greater Yellowstone Ecosystem. The **biosphere,** the zone of the earth's surface occupied or influenced by living organisms, is an ecosystem as well.

The key feature of an ecosystem is the interaction among these living and nonliving components. These interactions, in fact, provide the functional integration that enables specific ecosystems to be distinguished. Within an ecosystem the interactions among various living and nonliving components are strong; across the boundaries of different ecosystems they are weaker. Interactions are of four basic types (Fig. 3.5). Some may be of a purely physical or chemical nature, such as change in water density with change in temperature, or the chemical reactions by which acids are neutralized by buffering minerals in the water. Others, such as predation, competition, and mutualism among organisms, are purely biotic. Still others occur between living and nonliving components. The nonliving components influence organisms; increased acidity, salinity, or turbidity of

lake waters, for example, can kill certain organisms and benefit others. Finally, the organisms themselves strongly influence their physical habitat. In a lake, for example, the amount of photosynthetic production in the surface waters determines how much dead matter falls into the deep water zone. Decomposition of this dead matter, in turn, determines how much oxygen is used up, and how much remains for the use of other organisms of deep waters. In fact, much of the chemistry of the lake waters and sediments is influenced by the activities of the lake organisms. Appreciation of the magnitude of the influences of organisms on their physical environment, during the 1920s, was really the key factor in the emergence of the ecosystem concept.

These interactions exist not only at the scale of a lake ecosystem, but at the scale of the biosphere. That living organisms affect the physical environment at the global scale is evidenced by the fact that the composition of the atmosphere, the chemistry of the oceans, the structure of soils, and the composition of marine sediments are all determined primarily by life processes. Increasingly, the activities of humans are affecting these components of the biosphere, which, in their turn, influence the biosphere's biotic processes.

The relationships that exist within an ecosystem are the result of interaction through evolutionary time as well. Evolution has adapted groups of organisms to utilize different resources available in the physical and chemical environment. Some producer organisms use light energy for **photosynthesis,** others use energy from chemical reactions to carry on **chemosynthesis,** for example. The remarkable ecosystems of the deep ocean thermal vents depend on chemosynthetic production. The living members of ecosystems have also become adapted to each other through patterns of mutual evolutionary adjustment, or **coevolution.** The degree of this adjustment depends on the length and intensity of their interaction. In some cases the interaction is species specific, as for certain orchids and

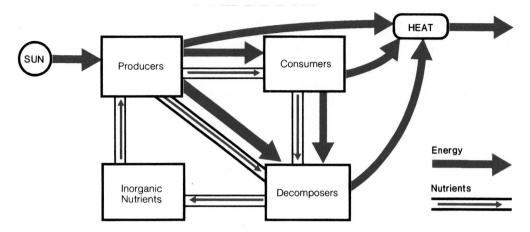

FIGURE 3.6

Energy moves through an ecosystem as a one-way flow, entering as solar radiation and leaving as heat radiated back into space. Nutrients, on the other hand, are capable of moving in true cycles.

the specialized insect that pollinates them. In other cases it is quite generalized, as for hummingbirds and the various flowers that they visit and pollinate. Through such evolution, the biotic structure of the ecosystem tends to adjust so that energy is more effectively captured and transformed, nutrients are retained and recycled more efficiently, and biotic interactions become more beneficial to the participants.

Many kinds of factors also act as external controls on ecosystems, influencing conditions within an ecosystem without being influenced significantly by the ecosystem itself. The regional climate, or **macroclimate,** is one of the most obvious external controls. Regional weather systems impose seasonal patterns of temperature, precipitation, solar radiation, wind, and other factors that strongly influence the conditions that prevail within an ecosystem such as a lake. Regional geology and geography also set limits on the development of ecosystems. The type of bedrock in the watershed of a lake, for example, influences the chemistry of the lake waters. The available regional biota—the kinds of plants and animals that exist in a particular geographical area—also places limits on the nature of the living community that develops. A newly formed lake, for example, will become inhabited by species that are present in nearby aquatic ecosystems, not those of lakes on distant continents, even though many of those species might be able to flourish. In the case of ecosystems of small and medium scale, humans are also an external influence. Humans add materials such as sediment and sewage to lakes, and remove other materials, such as water and fish. At the scale of regional ecosystems and the biosphere, however, humans are inextricable components of the system. Indeed, this is the fact that makes global environmental change a threat to human welfare.

Despite the importance of their internal interactions, ecosystems are never completely closed systems. In all instances, ecosystems experience inflows and outflows of energy and matter. In the case of a lake, light energy enters the surface waters and an equal amount of energy eventually leaves the ecosystem in some fashion, much of it as heat radiation from the lake sur-

face. Dissolved and particulate materials enter the lake from the watershed, and quantities of these materials often leave through streams that flow from the lake. In some ecosystems, such as streams, the inflows and outflows are the dominant factors of ecosystem function. Not even the biosphere is a closed system; solar energy enters and heat energy is radiated back to space.

The external controls and internal interactions of an ecosystem combine to produce a certain **ecosystem structure.** The macroclimate interacts with internal components of the ecosystem to create an internal **microclimate.** In a lake, the microclimate constitutes the regime of temperature, water density, light, and current pattern. Regional geology and geography interact with internal physical and biotic factors to create specific conditions of water and sediment chemistry. From the regional biota, certain species are sorted out by physical and biotic challenges to become the members of the lake community. These species develop certain patterns of abundance, biomass, stratification, and seasonality. Where human influences impact a lake, many structural conditions may be modified as well.

Internal relationships are not only structural, but involve many forms of **ecosystem dynamics.** Most important among these are the processes of **energy flow** and **nutrient cycling.** Important also, especially for conservation ecology, are the regulatory processes that determine inertia, stability, and resilience.

Energy and nutrients move through an ecosystem in basically different fashions (Fig. 3.6). Energy flows through the system in a one-way fashion, entering almost entirely as solar radiation that is captured and stored by photosynthesis in the chemical energy of organic molecules, and leaving as heat energy released by the metabolic breakdown of organic molecules. Although the pathway of flow may branch in a complex fashion as chemical energy is passed from organism to organism along food chains, once energy has been converted to heat it cannot be recycled into the chemical energy of organic molecules. Nutrients, on the other hand, can move in a truly cyclical manner, the same atoms of nitrogen or phosphorus passing back and forth from the organisms of a lake to the water mass many times.

FIGURE 3.7

The California chaparral shows high resilience, and recovers quickly after fire. After only 6 months, this stand is recovering by resprouting of many perennials and by the germination of both annuals and perennials.

(Photo by G. W. Cox)

Nutrient cycles in ecosystems are not perfect, however. Quantities of nutrients enter the active cycle from outside the system, and other quantities are lost from the system, so that an ecosystem may be envisioned as having a certain **nutrient capital.** This nutrient capital may increase or decrease, depending largely on the actions of external controls, increasingly those exerted by human activity.

The dynamics of ecosystems confer particular patterns of inertia, stability, and resilience (Westman 1986). **Inertia** is the ability of an ecosystem to *resist* change in the face of a disrupting external influence. In this sense, ecosystems are somewhat like organisms; they have a certain ability to regulate their internal conditions in the face of external challenges of the environment. Many desert ecosystems, for example, are quite stable in the face of highly variable macroclimatic conditions; a few years of unusual heat, cold, drought, or dampness cause little change in structure. Many stream ecosystems, on the other hand, are very unstable; flood and drought conditions may cause enormous change in conditions from year to year. **Stability,** on the other hand, is *constancy* of ecosystem conditions as influenced by internal interactions. The arctic tundra, where population cycles of lemmings and their predators create great year-to-year variation in biotic conditions, exemplifies ecosystems of low stability.

Resilience, in contrast, is the ability of an ecosystem to *recover* from a change in structure caused by some disturbance (Westman 1986). The chaparral shrublands of California are highly resilient in the face of fire. Following the destruction of aboveground material by a wildfire, many woody plants sprout from specialized root crowns and regrow rapidly (Fig. 3.7). Other plants are stimulated to germinate profusely in the early postfire environment. Within just a few years, the chaparral community has returned to a state much like that before the fire. Similarly, many stream ecosystems are highly resilient, and are able to recover from the effects of drought or flooding in a short

time. Most desert ecosystems, however, are not very resilient. If the aboveground parts of the dominant plants are killed over a large area, centuries may be required for reestablishment of the original vegetation. Similarly, semiarid grasslands, stressed by overgrazing or drought, may be converted to desert scrub in a matter of a few years. An understanding of the degree of stability and resilience of various ecosystems is essential to their wise management by man.

EQUILIBRIUM AND DISTURBANCE IN ECOSYSTEMS

Ecologists have long viewed ecosystems as being equilibrium systems, first and foremost. According to this view, an ecosystem possesses a certain optimal structure at which relatively stable flows of nutrients and energy occur among its members. In a popular sense, the ecosystem possesses a "balance of nature." Disturbance that upsets this equilibrium is followed by a type of healing process, through which the system returns to its optimal state, its balance of nature restored. This viewpoint suggests that the most important relationships are those that involve interactions between ecosystem components at equilibrium; everything else is relatively unimportant noise.

In recent decades, ecologists have found that equilibrium conditions in most ecosystems are rare or short-lived. Most systems are in a state of change in structure and dynamics, in some cases gradual, in other cases rapid. Disturbance is often a natural and dominant influence in ecosystem dynamics, as it is in terrestrial ecosystems prone to fire or aquatic ecosystems prone to flooding. For such ecosystems, in fact, disturbance may be the factor that maintains such features as high biotic diversity and a high yield of materials useful to humans (Botkin 1990).

Furthermore, the idea of an optimal equilibrium structure is itself a myth. It is unlikely that any system returns to its original state following some disturbance; the renewed ecosystem may be similar, but it may often be quite different in structure and function. Thus, no specific balance of nature exists. Many different balances might occur, assuming that the system reaches an equilibrium at all. This is not to say that all is chaos, and that stability and resilience are meaningless concepts, but that the external controls on ecosystems are themselves highly variable, changing, and complex.

In spite of the fact that disturbance is natural, and that many different equilibria are possible, not all equilibria are equally beneficial from a human perspective. Some forms of disturbance, such as the introduction of an exotic species of plant or animal, may disrupt coevolved relationships among native species, leading to their extinction. Other disturbances, such as pollution, may kill the carnivores and top carnivores of the ecosystem, which may be species that are valuable to humans for food or recreation. Still other disturbances, such as acid rain, may flush away nutrients, leaving a less fertile and less productive ecosystem. All of these ecosystems may approach a new equilibrium, but the conditions of that equilibrium may be less favorable to human welfare.

ECOSYSTEMS AND LANDSCAPES

Different types of ecosystems form a mosaic across the landscape. Since these ecosystem units are open systems, they interact with each other through the flows of materials and the movements of organisms across their borders. The importance of these processes has only recently been given its deserved recognition, with the emergence of the field of landscape ecology (Forman & Godron 1981; Naveh & Lieberman 1984). Landscape ecology examines how the juxtaposition of ecosystems, both natural ecosystems and those dominated by humans, influences their individual function and that of the landscape as a whole.

In continental areas, for example, terrestrial ecosystems often exist in a mosaic with ponds, lakes, and streams. The terrestrial systems form the watersheds for the aquatic systems, so that how these watersheds are managed influences the quantities of water, nutrients, and pollutants that flow into the aquatic systems. Likewise, how the aquatic systems are managed influences the kinds of species that are able to live in the adjacent land systems. The kinds and abundance of fish in a river, for example, influence the presence and numbers of animals such as ospreys and otters. Even in the ocean, different types of ecosystems are influenced by currents and animal migrations. The influx of warm tropical water into areas of normally cold, upwelling water in the eastern Pacific—an **El Niño**—can catastrophically reduce populations of kelps, fish, and seabirds. The seasonal migrations of larger marine mammals and fish likewise affect the activity and abundance of their prey.

Human activities are increasingly important in structuring landscapes, both terrestrial and aquatic. New types of ecosystems created by human activity, such as cultivated cropland, landscaped residential land, and dredged waterways, border or surround natural ecosystems. In eastern North America, for example, fragments of native forest exist as islands in agricultural and urban landscapes. Thus, to employ an ecosystem perspective in conservation ecology, we must consider not only the interactions within ecosystems, but how the different ecosystems, both natural and human-made, interact in a landscape.

THE NECESSITY OF THE ECOSYSTEM APPROACH

Well-intentioned actions that do not recognize the dynamic interactions that occur among the components of an ecosystem, and the influence of adjacent ecosystems on each other, can destroy the object of the conservation effort (Walker 1989). Protection of an area against disturbances, such as fire, that are integral to the dynamics of an ecosystem, or centering management policy on a particular species with inadequate consideration of its effects on other species, are examples of well-meaning mismanagement (White & Bratton 1980).

This principle is illustrated in many of the wildlife parks of East Africa, where a rich assortment of ecosystems and species may exist in a small area (See Reading 3.1). The principle also applies to terrestrial and aquatic environments elsewhere, including North America (Wagner 1977). Overprotection of deer and elk often leads to their overpopulation, resulting in depletion of food plants, starvation of animals, and long-term degradation of habitat. In other instances, simple protection, such as that afforded the California condor, has proved inadequate to offset the problems created by deterioration of the ecosystem. This species, in fact, provides a good example of lack of an ecosystem perspective in conservation efforts (See Chapter 25). In the chapters that follow, we shall encounter many other examples of management problems created by the lack of an ecosystem perspective.

THE BIOSPHERE AND GLOBAL CHANGE

The highest ecological level of organization is that of the biosphere. At this level of organization, the only major inflow is that of solar radiation and the only major outflow is that of heat radiation to space. This flux of energy, together with a small flow of heat energy from the earth's core to its surface, is, in a sense, the earth's macroclimate. Over several billion years, these energy sources, interacting with the primeval geochemistry of the land, surface waters, and atmosphere, have produced the conditions of climate, soil structure, water chemistry, and biodiversity.

Human activities now impact the biosphere at large, influencing basic interactions at this global scale (Schlesinger 1991). Burning of fossil fuels releases increasing quantities of CO_2, sulfur and nitrogen oxides, and heavy metals into the atmosphere. The release of industrially produced halogen compounds, such as the chlorofluorocarbons, and the widespread use of nitrogen fertilizers that are partially returned to the atmosphere as nitrogen oxides, are influencing the chemical composition of the atmosphere (Lester & Myers 1989). Deforestation and overgrazing of arid lands are directly changing the biota of the continents and modifying the energy exchange characteristics of the land surface (Myers 1988). Wind and water erosion of crop monocultures and overgrazed rangelands are increasing sediment inflow to aquatic ecosystems and dust input to the atmosphere. Reduced river discharges to the oceans are promoting coastal erosion and depriving coastal marine waters of nutrients. Nutrients from disturbed landscapes and from urban sewage discharges are destructively over-enriching other aquatic ecosystems. Natural ecosystems are being replaced by farmland and urban landscapes, and major trophic components of aquatic ecosystems are being exploited with growing intensity.

These changes, all increasing in magnitude, pose a growing threat to the diversity of life on earth. Changes in climate, oceanic circulation patterns, and the water budgets of rivers and lakes will likely occur more rapidly than at any time in earth's history (Lester & Myers 1989). At the same time, the ability of terrestrial plants and animals to shift their geographic ranges will be increasingly constrained by developed landscapes (Mintzer 1988). The limits of stability and resilience of all of the earth's ecosystems will certainly be tested throughout the biosphere within the next century.

A clear example of the need for an ecosystem view in conservation ecology is provided by many of the wildlife parks of East Africa (Boughey 1963). In Hwange National Park, Zimbabwe, for example, a mosaic of woodlands, savannas, grasslands, and other communities exists in an area of relatively homogeneous soils and climate (Reading Fig. 3.1A). Each of these plant communities is the favored habitat of

certain large animal species, so that the park supports a rich fauna. What is responsible for this diversity of plant and animal life? Much of the diversity of large animals obviously results from the vegetational diversity within the park. Analysis of biotic interactions within the park ecosystem, however, shows that the vegetational diversity is largely created by the activities of large animals and by the influence of these activities on the frequency of fire.

The scenario for this relationship is complex. Elephants, which favor the denser woodlands, kill many trees by stripping their bark for food, or simply by pushing them over to gain access to the foliage. When this activity goes on for a long time, or is carried out by many animals, the density of large trees is reduced and the woodland becomes more open. Initially, this reduces the tendency for one tree species to dominate the woodland canopy, increasing

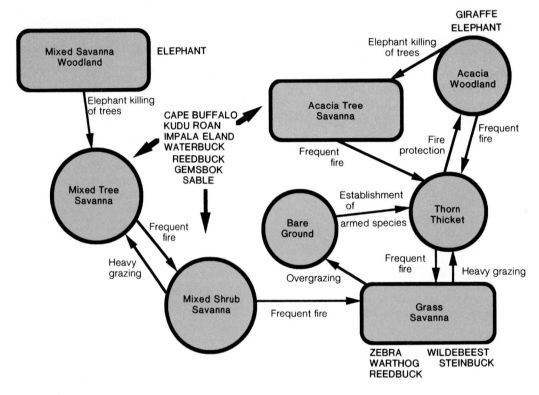

READING FIGURE 3.1A

Many East African wildlife areas show a complex interaction between animals, vegetation, and fire that favors the maintenance of a mosaic of plant communities, each with its associated species of large animals. Communities identified by rectangles are most closely adjusted to soils and climate (mixed savanna woodland), frequent fire (grass savanna), and intense animal effects (acacia tree savanna).

READING FIGURE 3.1B
Impala and many other ungulates occupy East African savanna woodlands that are maintained by the interaction of the vegetation with browsing and grazing mammals and fire.

(Photo by G. W. Cox)

READING FIGURE 3.1C
Heavy elephant browsing on trees such as the mopane, *Colospermum mopane*, can maintain the vegetation as shrub savanna, as in this location in Chobe National Park, Botswana.

(Photo by G. W. Cox)

tree diversity. Eventually, however, it converts the community into a much more open savanna woodland. Here, increased light reaches ground level and permits the growth of a dense grass and herb layer. Many species of grazing and browsing animals, such as impala, greater kudu, and sable antelope, frequent the savanna woodland (Reading Fig. 3.1B). The intensity of feeding by these species determines how much grass and herb material remains into the dry season to become fuel for wildfires. If the populations of these animals are high, fires will be infrequent and low in intensity. If their populations are low, fires are likely to be frequent and intense, killing many of the remaining trees and transforming the vegetation into a shrub savanna. Shrub savannas, too, are the preferred habitat of some species, such as Cape buffalo and gemsbok. Elephants also frequent certain areas of shrub savanna, browsing the woody plants so frequently and heavily that they cannot grow into trees, even in the absence of fire (Reading Fig. 3.1C). If fire continues to be frequent, however, it may eliminate virtually all woody plants, creating an open grassland that will become the home of still other grazing animals such as wildebeest, zebra, and gazelles. When these animals are numerous, their feeding may favor the invasion of spiny acacias that are resistant to the feeding activities of grazers. In time, this may result in a thicket or woodland dominated by acacias, the preferred habitat of species such as the giraffe.

Thus, the diversity of large animals depends on the existence of a mosaic of distinct plant communities, but this mosaic, in turn, depends on the activities of the various large animals themselves. Strictly protective management is inadequate for such a situation. A policy of complete fire control, for example, might quickly lead to the disappearance of the savanna woodlands, shrub savannas, and grasslands, and of the animals that occupy these communities. Overprotection of elephants might lead to the loss of the denser woodlands, and excessive protection of the animals of open grasslands might lead to the transformation of grasslands into thorn woodlands, ultimately causing the grassland animals to disappear. To be successful, management of areas such as this must recognize and maintain the basic interactions of the ecosystem (McNaughton 1989).

Humans as Members of the Global Ecosystem

As we noted earlier, humans act as external controls on small-scale ecosystems, but are an integral component of regional ecosystems and of the biosphere as a whole. As the human population grows, and as its use of natural resources also expands, our interactions with other living and nonliving components of the global ecosystem intensify. The impressive accomplishments of our technology sometimes lead us to imagine that we are becoming more independent of the rest of the global ecosystem. In reality, however, our dependence on the rest of the global ecosystem is increasing. Abnormal seasonal patterns of regional rainfall, temperature, ocean upwelling, and other conditions now cause major disturbances in the world economy. Slow, progressive changes in global climate, sea level, area of forests and deserts, and other features threaten to cause major impacts on human food and fiber production within one or two decades. Conservation must thus be a global effort. In the following chapters, our examination of varied topics gives particular attention to international efforts to implement conservation practices.

The reality of human membership in the global ecosystem means that, to be successful, conservation strategy must address the human as well as the nonhuman components of the global ecosystem. The grandeur of the tropical forests, the teeming wildlife of the African plains, and the wealth of fish and invertebrate life of tropical reefs cannot be preserved without solving the problems of the human populations that seek to exploit these ecosystems for the bare necessities of life. Consequently, we shall emphasize the need to couple conservation with improvement of economic conditions for impoverished human populations.

Countering the threats of global environmental change and continued growth of human populations is the challenge of the next two decades. If this challenge is not met, the year 2010 will find the biosphere a biologically, culturally, and economically impoverished ecosystem, perhaps analogous to a heavily polluted lake.

Summary

1. Populations consist of the individuals of a species, varied in their characteristics, that occur together. Populations grow as a result of reproduction and immigration, and decline through mortality and emigration. Populations show varying degrees of regulation, but rarely are they in close equilibrium with the carrying capacity of the environment for very long. In a certain region, a species may exist as a metapopulation consisting of many semi-isolated subpopulations among which individuals occasionally disperse.

2. Communities are made up of the species that occur together and interact. As a result of interactions among these species, and of interactions between them and their physical habitat, the process of biotic succession occurs.

3. An ecosystem, the community together with its nonliving habitat, is influenced both by external controls and internal interactions. Ecosystems are affected to differing degrees by disturbance, and show varying degrees of inertia, stability, and resilience. The ecosystems that occupy a landscape are linked by the flow of materials and movement of organisms across their boundaries.

4. Landscapes such as those of East African game parks illustrate the importance of interactions of climate, soils, fire, plants, and animals, and thus the need to employ an ecosystem perspective in conservation.

5. The global nature of environmental change, together with its roots in human activity, indicate that conservation must be a global effort that simultaneously addresses human population growth and economic development.

Questions for Discussion

1. Do you think the views of ecologists about stability and equilibrium in populations and ecosystems have been colored by the results of experimental studies carried out under confined, constant conditions?

2. Review the recovery or protection efforts being carried out for some species (e.g., the California condor) that you have read about recently. How do these efforts exemplify, or fail to exemplify, an ecosystem approach to conservation of the species?

3. Consider the characteristics of a state, provincial, or national park near where you live. How have conditions in this park been affected by climatic change and by natural and human disturbance over the past 10,000 years?

Suggested Reading

Botkin, D. B. 1990. *Discordant harmonies: A new ecology for the twenty-first century*. Oxford Univ. Press, New York, NY. An analysis of the changing views of ecologists about regulation, stability, and equilibrium relationships in nature, and the significance of these changed views for conservation.

Mills, L. S., M. E. Soulé, and D. F. Doak. 1993. The keystone-species concept in ecology and conservation. *BioScience* **43**:219–24. A summary of the ways that the influences of key species may strongly influence the composition and dynamics of the entire ecosystem.

Sprugel, D. G. 1991. Distance, equilibrium, and environmental variability: What is "natural" vegetation in a changing environment? *Biological Conservation* **58**:1–18. A discussion of the forces that operate on different time scales to create disequilibrium in ecosystems.

Global Biodiversity

As we noted in Chapter 1, the science of conservation biology is concerned with **biodiversity,** the richness of the biosphere in genetically distinct organisms and the ecological systems they compose. The central goals of conservation biology are to understand the nature, origin, and values of biodiversity, to evaluate the threats to its survival, and to devise effective ways to manage biodiversity for sustained human welfare.

In this chapter we shall examine first the various aspects of biodiversity, and then survey patterns of diversity throughout the biosphere. Finally, we shall consider the major theories of how biodiversity originates and is maintained.

Diversity among zebras: one example of biotic richness that is being threatened by human impacts within the biosphere.

Outline

COMPONENTS OF BIODIVERSITY

Biodiversity includes not only species and their attributes, but biological richness at several hierarchical levels of complexity from local populations to the biosphere as a whole (Fig. 4.1). These levels include infraspecific, species-level, interspecific, and higher-level relationships. Biodiversity has both evolutionary aspects (Wheeler 1995), identified along the left side of Figure 4.1, and ecological aspects, identified along the right side of this figure.

The Biodiversity Hierarchy

At the base of the hierarchy, diversity is expressed as patterns of variability within populations. In evolutionary terms, local populations differ greatly in the amount of variability of their gene pools. In some cases, all or most of the genes of the individuals can be represented by only one allele, so that individuals are genetically identical or nearly so. In other cases, many of the genes are represented by several or many alleles, and almost every individual has a genotype that is unique. We know that genetic variability within populations is constantly being acted on by natural selection. Genetic variability constitutes, in effect, the insurance policy of the population against extinction due to changing environmental conditions that may favor new patterns of genetic adaptation. We shall examine this aspect of biodiversity in detail in Chapter 25.

In ecological terms, local populations also exhibit great variability in demographic structure, their makeup in individuals of different age, body size, and sex, together with the pattern of dispersion of these individuals in space. These features influence the potential survivorship and reproduction of the population, and thus its vulnerability to extinction by chance and by fluctuations in environmental conditions. Whether or not an endangered species survives some period of stress is influenced by whether the individuals are young or old, if the sex ratio is balanced or unbalanced, and if the individuals are scattered widely or concentrated in areas where they can easily encounter each other. We shall consider this aspect of biodiversity in Chapter 26.

At the species level, we find evolutionary patterns of diversity on a broader scale. The overall population of a species often consists of taxonomically distinguishable **subspecies** that occupy nonoverlapping geographical ranges, but share the essential characteristics of the species, such as the ability to interbreed. The criteria for recognizing subspecies are usually consistent differences in morphology or behavior. The extent to which species show subspecies varies considerably, depending in part on the extent to which taxonomists have examined them. For widespread species of birds and butterflies, many subspecies are often recognized. The familiar song sparrow, which occurs over much of North America, for example, has 31 subspecies.

A new concept, that of the **evolutionarily significant unit (ESU)**, also applies at this level. An ESU is a geographically isolated portion of the species population that also has a high level of genetic difference from other subpopulations of the species, whether or not this difference is recognized taxonomically. Populations of a species of fish in different lakes would be

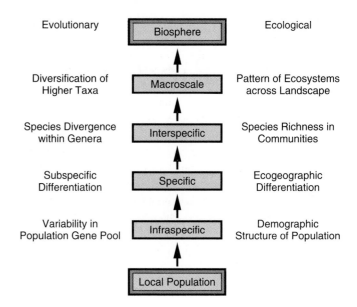

FIGURE 4.1

Evolutionary and ecological components of biodiversity.

an example of geographically distinct populations. Such populations might show no significant differences in genetic characteristics, however, and thus would not represent different ESUs. If the lake populations fell into two groups, with strong genetic differences between but not within groups, two ESUs would be recognized.

Ecologically, diversity at the species level is also evident as **ecogeographic variation,** the differentiation of local populations in genetic adaptations or responses to habitat conditions. Sometimes these are evident as relatively distinct **ecotypes** that occur in different habitats, but more often they appear as gradual gradients of genetic characteristics, or **ecoclines,** along gradients of habitat conditions. Many species of perennial grasses that occur from southern Canada to northern Mexico, for example, show ecoclines of morphological and physiological characteristics related to day length and climate change along this latitudinal gradient. In regions with areas of soils derived from igneous rocks rich in magnesium, known as **serpentine soils,** many plants show ecotypes physiologically adapted to serpentine and nonserpentine soils. Most ecotypes and ecoclines go unnamed by taxonomists, but some may be recognized as subspecies, if consistent taxonomic criteria can be defined to distinguish them.

Higher in the biodiversity hierarchy, at the interspecific level, evolutionary processes lead to diverse patterns among closely related species. Some evolutionary lineages show a tendency to form numerous species, others only a few species. Genera thus vary greatly in the numbers of species they contain, and in the patterns of coexistence and adaptive radiation among these congeners. Some genera contain many species that show slight differentiation and tend to occupy different habitats or nonoverlapping ranges, whereas other genera contain species with striking differences in features that permit them to coexist.

In western North America, several species of small *Eutamias* chipmunks, which vary only slightly in appearance, occupy nonoverlapping habitat belts from sagebrush desert to high mountain forest. In contrast, in the Galápagos Islands, several finches of the genus *Geospiza*, which differ strikingly in beak size and shape, coexist in the same habitats.

Also at the interspecific level, interaction of species integrates them into biotic communities, the living components of ecosystems. Ecosystems vary greatly in number of species, and in the abundances, biomasses, and spatial dispersion patterns of these species. Biodiversity includes ecosystems at all levels of scale. In aquatic environments these range from tiny temporary ponds to large freshwater lakes and oceanic basins such as the Gulf of Alaska. On land, they range from oases of forest at desert springs to the forests of the Amazon Basin.

At the highest level in the biodiversity hierarchy, we can recognize patterns of macroevolution: the diversity that exists in genera, families, orders, and phyla. These taxa reflect processes of speciation, adaptive radiation, and extinction over long periods of geological time. As one examines progressively higher taxonomic levels, one sees more and more basic differences in the ways that organisms have found to survive and diversify through evolutionary time. The differences that appear at these higher levels are thus of correspondingly greater importance for conservation biology.

Similarly, at the highest ecological level, we recognize regional landscapes that show a mosaic of different ecosystems. These ecosystems are interspersed in different ways. Landscapes shaped by continental glaciation, for example, exhibit a mosaic of terrestrial ecosystems, marshlands, and lakes. Riparian woodlands follow rivers as they penetrate toward their headwaters in grasslands and deserts. Forests, brushlands, and meadows form patchwork patterns on mountain slopes of varying steepness and directional exposure. Reefs and lagoons create a complex offshore seascape along tropical coastlines. The number and kinds of ecosystems within a biogeographic region, along with their patterns of juxtaposition, are probably as important to ecological function on a regional level as is the species make-up of each ecosystem type.

Endemism

Endemic taxa are forms that have evolved their uniqueness in, and are restricted to, a particular region. Endemism can exist at any taxonomic level from subspecies to phylum. An additional aspect of biodiversity is therefore the degree of endemicity in the biota of a particular region. Geographical areas that have been isolated for a long period of evolutionary time often show high levels of plant and animal endemism. Examples include oceanic islands, freshwater lakes in continental rift basins, and land areas bounded by high mountain ranges. Endemism is high in long-isolated areas such as Madagascar, in remote oceanic archipelagos such as the Galápagos, in ancient lakes such as Baikal (*See* Reading 4.1), and in mountain-bounded continental areas such as Chile.

Conservation concern is usually greatest for the endemic taxa of a region, since their loss in that region is equivalent to their loss globally. Taxa endemic at the level of the genus or higher are generally regarded as being of greatest conservation priority. The lowest level of endemicity and conservation priority correspond to subspecies and evolutionarily significant units.

Estimating the Total Number of Species

Much of the quantitative evaluation of biodiversity has centered on species. What constitutes a species, however, influences any estimate of their number. Over most of the past half century, the dominant concept was that of the **biological species,** a group of actually or potentially interbreeding individuals that is reproductively isolated from other such groups. As we have noted, a biological species may have several or many subspecies. Although these subspecies can be distinguished by various characteristics, their actual or presumed capacity to interbreed means that all belong to one biological species.

In recent decades, however, biologists have found that interbreeding between what seem to be clear biological species is sometimes frequent. Many now regard the criterion for a biological species to be a reproductive pattern that maintains cohesion of a species' characteristics, rather than one of strict reproductive isolation.

An emerging species concept is that of the **phylogenetic species,** a group of individuals that is distinct in its characteristics and has a common ancestry. Although most subspecies of the biological species have a common ancestor, all may not be equally distinct. When many subspecies exist, some may be much more distinct than others as a result of longer, separate ancestry. Thus, greater acceptance of the phylogenetic species concept is likely to elevate many subspecies to the level of species, increasing the total number of recognized species.

Estimating the total number of species on earth is extremely difficult, because in most taxa only a fraction of the species that exist have been described (Pimm et al. 1995). Many estimates of the total number of species in various groups of organisms are based on the opinions of taxonomic specialists in those groups and their knowledge of the rate at which new species are being discovered.

Other estimates are based on extrapolations of various sorts (May 1988, 1990; Williams & Gaston 1994). A very controversial extrapolation by Erwin (1982), for example, was based on the results of insecticidal fogging of the canopies of trees in tropical forests. From 19 trees of one species, he obtained 1,200 species of beetles, of which he estimated that 162 were probably restricted to that tree species. By extrapolation, the 50,000 or so tree species in tropical forests might therefore hold 8 million species of beetles. Assuming that beetles are about 40% of all canopy arthropods, and that canopy arthropods are about two-thirds of all tropical forest arthropods, leads to an estimate of 30 million species. This extrapolation has been roundly criticized by various people (e.g., Gaston 1991; Stork 1993; Gaston & Hudson 1994), but serves to show the high uncertainty that exists about the number of species on earth.

L ake Baikal (Reading Fig. 4.1) is the most distinctive lake ecosystem on earth (Nijhoff 1979). The lake, which has existed for about 25 million years, occupies a basin formed by a rift in the Asian continental plate. Its depth of 1,620 m, combined with a surface area of 31,500 square kilometers, gives it the greatest volume of any lake on earth. Its volume equals, in fact, one-fifth of the earth's fresh water. The lake is low in primary productivity and has extraordinarily clear water. A black and white metal disk (known as a **Secchi Disk**) can be seen to a depth of 36 m.

Even more unusual is the biota of the lake. About 2,700 species of aquatic plants and animals are known from the lake, 84% of these being endemic. The lake contains hundreds of species of planktonic diatoms, and over 100 species of ostracods, 250 species of amphipods, and 70 species of mollusks. An endemic family of sponges, a group of organisms rare in fresh water, occurs in the lake. Among vertebrates, the lake possesses an endemic seal, the nerpa, and an important whitefish, the Baikal omul. Both the seal and the omul have been exploited heavily.

The greatest threat to Lake Baikal's health is pollution from industries and deforestation. Several wood processing plants are located at the south end of the lake, and rafting of logs to these plants has polluted the lake with sunken timber. The plants, which produce a wide range of lumber, cellulose, and paper products, discharge 250,000 m³ of treated wastes into the lake daily, although alternative means of disposal exist. On the east side of the lake, a large industrial complex on the Selenga River produces petroleum, sulfates, chlorides, and thermal effluent. Wastes from these plants are discharged into the river and carried into Lake Baikal. Completion of the Baikal-Amur railroad at the north end of the lake has led to tanker transport of petroleum and other chemicals from production areas to railway loading facilities, creating additional pollution. Due to these forms of pollution, the Baikal omul reportedly has ceased to spawn (Stewart 1990).

A governmental commission developed plans to eliminate industrial pollution, log rafting, and watershed deforestation by 1995, and an International Centre for the Ecological Study of Lake Baikal was established (Galazii 1991). The breakup of the Soviet Union, leading to the economic crisis that subsequently gripped Russia, has prevented much of the pollution control from being implemented.

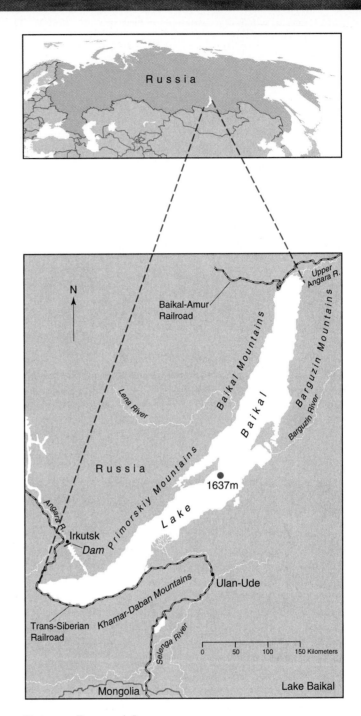

READING FIGURE 4.1

Lake Baikal, Russia, is one of the oldest and largest freshwater lakes on earth, and has an endemic flora and fauna richer than any other lake.

(Reprinted from P. Nijhoff, 1979, "Lake Baikal Endangered by Pollution" in *Environmental Conservation*, 6:111–115 with kind permission from Elsevier Science—NL, Sara Burgerhartstraat 25, 1055 KV Amsterdam, the Netherlands.)

TABLE 4.1

Global Species Diversity

KINGDOM	SPECIES DESCRIBED	MARINE	FRESHWATER	TERRESTRIAL	SYMBIOTIC	TOTAL SPECIES PROBABLE
(Viruses)*	(5,000)				(5,000)	(500,000)
Monera						
	4,000	1,000	1,000	1,000	1,000	400,000
Protista	80,000	40,000	30,000	5,000	5,000	400,000
Fungi	70,000	1,000	1,000	60,000	8,000	1,000,000
Plantae						
Algae	22,000	10,000	10,000	2,000		50,000
Lichens, mosses	31,000		50	30,950		90,000
Ferns and relatives	12,000		100	11,900		24,000
Gymnosperms	500			500		520
Angiosperms	250,000	100	2,500	247,000	400	300,000
Animalia						
Invertebrates						
Various phyla (29)	123,866	81,086	13,725	10,280	18,775	861,296
Arthropoda	870,500	22,300	24,700	821,300	400	8,900,000
Chordata						
Urochordata	1,250	1,250				1,500
Cephalochordata	25	25				35
Vertebrata						
Fish	21,732	13,321	8,411			30,000
Amphibians	4,000		3,900	100		4,200
Reptiles	6,550	50	300	6,200		6,700
Birds	9,672			9,672		9,800
Mammals	4,327	200	10	4,117		4,500
TOTAL	**1,510,910**	**167,232**	**90,146**	**961,569**	**31,975**	**12,062,451**

*Not included in totals

RICHNESS OF LIFE ON EARTH

What is the magnitude of biodiversity of these types? At the species level, recent estimates suggest that the number of described species is about 1.5 million (Table 4.1), and that a "working figure" of 12.5 million total species may exist in the biosphere (World Conservation Monitoring Centre 1992). Others offer estimates that range from 5 to 50 million.

At the species level, most biodiversity is terrestrial. Nearly a million species have been described from the terrestrial environment. The vast majority of terrestrial species are arthropods, and most of them are believed to live in tropical forests. The figure for the terrestrial environment is several times that for the marine environment. Described species diversity in the marine environment exceeds that in fresh waters, mostly because of the large numbers of phyla that have many marine species but fail to penetrate fresh waters to any major degree.

Biodiversity is unevenly distributed in both terrestrial and aquatic environments. We can recognize a number of **biodiversity hotspots,** where unusual concentrations of species, and usually of endemic taxa, occur (Fig. 4.2). These hotspots tend to be located where biogeographic isolation has favored speciation and adaptive radiation over a long period of evolutionary time, where biogeographic connections have mixed the biotas of different continents, or perhaps where long continuity of favorable environmental conditions has allowed evolution to proceed with few setbacks for a long period of geological time.

In the terrestrial environment, about 14 major hotspots can be identified in tropical forest regions (Myers 1988) and 4 in regions of Mediterranean climate (Myers 1990). These hotspots contain about 20% of the earth's species, but cover only about half of 1% of the earth's land area. Madagascar, for example, is estimated to have about 4,900 endemic higher plants and 234 endemic species of reptiles. The very small Cape Region of South Africa, one of the several Mediterranean climate regions of the world, has perhaps 6,000 species of endemic plants.

In fresh waters, ancient lakes, many of them located in continental rift valleys, have unusually rich biotas. Such lakes include Lake Baikal in Russia (*See* Reading 4.1), several African rift valley lakes, and Lake Titicaca in South America. Three major tropical rivers, the Amazon, Zaire, and Mekong qualify as hotspots of flowing fresh waters (*See* Chapter 16).

In the marine environment, diversity is greatest in coral reef communities, and hotspots can be recognized in areas such as the Great Barrier Reef. The shallow waters of the Caribbean region, another marine hotspot, are home to about 10% of the world's marine fish (Robins 1991). Oceanic upwelling areas, although not necessarily rich in numbers of species, support

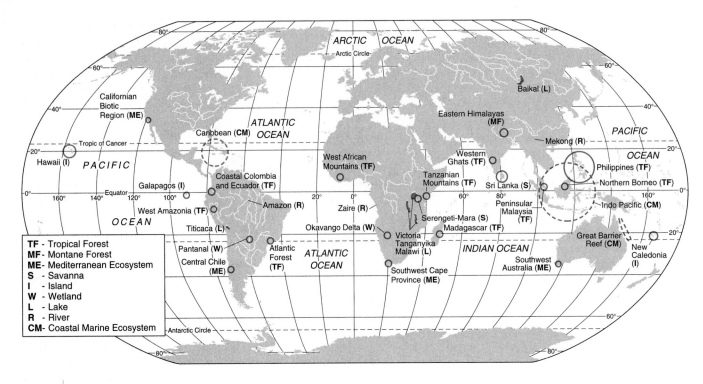

FIGURE 4.2

Global biodiversity hotspots, where unusual concentrations of species, and usually of endemic taxa, occur.

complex food webs of invertebrates, fish, marine mammals, and seabirds. Diversity may also be much higher on the deep ocean floor than previously suspected (Grassle 1991).

In addition to hotspots, major gradients of biodiversity can be identified. These gradients vary greatly in detail from one taxonomic group to another, and some taxa show gradients quite contrary to the general pattern. Two main patterns can be recognized. First, a strong tendency exists for species diversity to be high in the tropics and to decline toward the poles. This pattern is followed by many terrestrial, freshwater, and marine taxa. Even in the ocean depths (Rex et al. 1993) diversity is highest in the tropics. Second, in terrestrial environments, diversity tends to be highest in regions where mountain systems, such as the North American Rocky Mountains, create a landscape differentiated in elevation, rainfall pattern, slope aspect and steepness, soil conditions, and lake and stream conditions. From such areas, diversity declines toward less differentiated continental regions.

The enormous diversity at the species level is multiplied many-fold when infraspecific variations are included. Ecogeographic variation is shown by almost every widespread plant and animal species. In plants, ecoclines often occur along strong local gradients, such as gradients of soil moisture at the edges of lakes or streams. Plant ecologists have also documented the rapid origin of ecotypes in areas of strong natural selection, such as populations growing on mine tailings with high concentrations of heavy metals, or populations in areas subject to heavy grazing. In fact, detailed study of the genetic and demographic makeup of populations reveals that essentially every population is distinct.

Likewise, above the species level, diversity is multiplied by the many interactions of species with each other through competition, mutualism, and feeding relationships. These biotic interactions, together with interactions of species with their physical environment, create enormous diversity at the level of ecosystems. Quantifying such diversity is difficult, however, because no really objective criteria exist for recognizing distinct ecosystem "types." Most ecologists, in fact, believe that variation in ecosystem structure is potentially continuous, and that no two ecosystems are identical, so that diversity at this level is essentially infinite.

Systems of classification are needed, however, to enable the designation and mapping of ecosystems for conservation purposes. Several systems for classifying major ecosystem types have been developed. In the United States, Bailey (1978) developed a hierarchical classification of **ecoregions,** or major terrestrial ecosystem types, defined on the basis of climate, dominant plant life form, region of the country, and vegetational composition. Recently, this system has been extended to the earth as a whole, with both continental and oceanic ecoregions defined (Bailey 1995). Cowardin et al. (1979) created a hierarchical classification of aquatic habitats based largely on physical conditions, general vegetational features, and salinity.

On the global scale, Udvardy (1975) developed a classification of **biogeographic provinces** based on 14 major land and freshwater biomes and 8 biogeographic realms. **Biomes** are ecosystem types defined by dominant plant life forms such as "temperate grasslands," or by landscape features such as "mountains and highlands." **Biogeographic realms** correspond to major continental units that have been centers of evolutionary independence, such as the "neotropical realm." Within each realm, a number of provinces are also recognized, so that altogether 193 biogeographic provinces are defined. This system has become the basis for global conservation planning by international organizations such as the United Nations Educational, Scientific, and Cultural Organization (UNESCO). Hayden et al. (1984) proposed a similar classification of marine biogeographic provinces. This system includes a series of deep oceanic and coastal oceanic realms, again divided into provinces.

MEASURES OF DIVERSITY

Species diversity in a particular region consists of three aspects. First, individual communities possess a certain richness in species, an aspect termed **alpha diversity.** Alpha diversity is often considered to have two components: number of species and evenness of abundances of those species, with diversity increasing as either number or evenness increases. The common-sense logic of these two aspects can be illustrated by a simple example. To a human walking through a woodland, the richness of trees would seem low if only a few species were present, and especially if all but one of these species were very scarce. The woodland would seem to be a stand of the one common species. Richness would seem to increase if many more species were added, even if each was scarce, because the observer would see more different species. If no new species were added, but the ones present were made equal in numbers, the woodland would also appear richer, looking quite unlike a single-species stand. Ecologists have developed mathematical indices that measure the combined aspects of number and evenness of species, as well as evenness itself (Cox 1996). For more general discussion, however, we can consider alpha diversity to be more or less equivalent to number of species.

A second aspect of species diversity, **beta diversity,** is the degree of change in species from one community type to another. In a Mediterranean landscape in California, for example, one may have grassland, chaparral, and woodland communities intermixed, with a high degree of change in species from one community to another. In a forested landscape in Ohio, on the other hand, the change in species from floodplain to hillside to ridge-top may be much less.

Finally, for the region as a whole, overall richness in species, **gamma diversity,** reflects the number of community types present, the alpha diversity of each, and the pattern of beta diversity from type to type.

DETERMINANTS OF BIODIVERSITY

What factors determine species diversity? The answer depends on whether one is seeking a proximate or ultimate explanation. A proximate explanation concerns the factors that operate in ecological time to influence the number of species that co-exist in a community and whether or not these species can also exist in neighboring community types. An ultimate explanation involves the evolutionary and biogeographic processes that have determined patterns of speciation, extinction, and geographic dispersal of species, and thus, the richness of species in a given region.

Proximately, several features appear to contribute to diversity in a community. Communities of moist, tropical habitats appear to have a greater range of types of resources than communities of the most nearly comparable habitats of higher latitudes, thus permitting coexistence of more species that utilize these resources. In tropical forests, for example, the nocturnal resources available to bats include insects, fruit, and nectar, whereas only insects are available in abundance in most temperate and boreal forests. Thus, a much richer bat fauna can occur in tropical forests.

Another factor that often promotes diversity within communities is a moderate level of disturbance. This idea, the **intermediate disturbance hypothesis,** is simply that too little disturbance allows certain highly competitive species to take over the entire system, excluding many other species (Petraitis et al. 1989). Too much disturbance permits only a few specialized, disturbance-tolerant species to exist. Diversity is maximized where disturbance is strong enough to create openings for establishment of successional species, but not so strong as to exclude all late successional species. For example, high species diversity is often shown in intertidal situations with intermediate wave shock. Where there is severe shock, no predators can survive and mussels take over all substrates. Where there is little wave shock, many predators can be active, eliminating herbivorous invertebrates and allowing algae to dominate all substrates. But where wave shock is moderate, a mixed community of predators, mussels, and algae can exist.

Several ultimate determinants of diversity also have been hypothesized. These fall into two classes: equilibrium and nonequilibrium hypotheses. One, a nonequilibrium idea, is the **catastrophism hypothesis.** This idea holds that evolution tends to increase diversity by speciation and adaptive radiation more or less continuously through evolutionary time. Periodically, however, major regional or global catastrophes cause massive extinctions. Examples of such catastrophes include asteroid impacts, massive volcanism, episodes of continental glaciation, or major changes in sea level that affect vast areas of low-lying lands or shallow seas. Tropical regions, including the oceans, may have been less subject to some types of catastrophes, such as those related to glacial cycles. This may account for the higher diversity seen in the tropics than at high latitudes.

Equilibrium theories suggest that under particular environmental conditions a balance develops between the appearance of new species and the extinction of existing species (Fig. 4.3). For continental areas this balance is largely between the origin of species by speciation and their disappearance by extinction. The conditions that determine at what level these processes balance involve factors such as the number and effectiveness of geographical barriers that can isolate portions of ancestral species. In oceanic archipelagos, the primary processes may be colonization of species from a continental source area and extinction of species already resident on the islands. Even here, though, speciation also may contribute, as it evidently has in archipelagos such as the Galápagos and Hawaiian Islands.

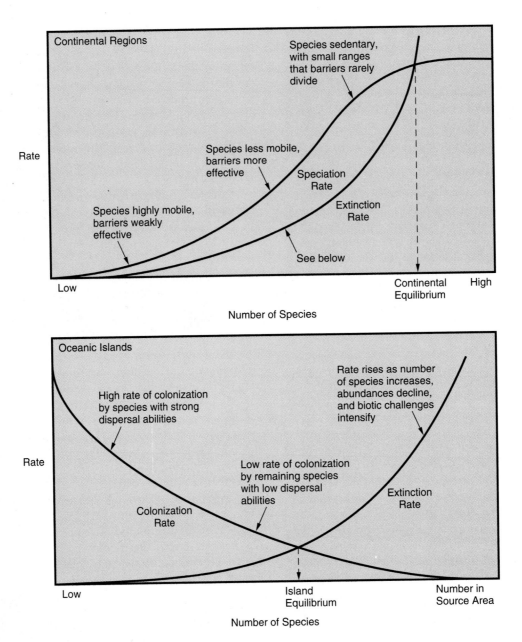

FIGURE 4.3
Equilibrium theories of biodiversity origin suggest that under particular environmental conditions a balance develops between the appearance of new species by speciation or colonization and the extinction of existing species.

PUBLIC PERCEPTION OF THE BIODIVERSITY CRISIS

As we shall see in the next chapter, many scientists believe that the earth is approaching a biodiversity crisis comparable to those that decimated life several times during evolutionary history. The human population continues to grow explosively, and a greater and greater fraction of the earth is being converted to intensive human use. The accelerating impact of human activity on global climate means that even areas set aside for protection of biodiversity will experience severe stress. Two of the responsibilities of conservation biology are to devise ways to protect biodiversity from direct destruction on the short run, and to meet the challenges of global change in the first decades of the 2000s.

To do this, the public must be educated to the significance of biodiversity for human welfare. In spite of this impending crisis, popular concern with biodiversity remains limited. Overt public concern tends to center on a few species and ecosystems. At the species level, "charismatic" plants and animals, such as the sequoia, the grizzly bear, and the killer whale, dominate public attention. Indeed, some congressmen who have supported U. S. federal endangered species legislation suggest that their concern was only with such species.

At the ecosystem level, public concern tends to center on "TV special" ecosystems and their plant and animal members, such as the savannas of the Masai Mara in Kenya (Fig. 4.4), the krill-based food web of marine mammals, birds, fish, and squid in Antarctic waters (Fig. 4.5), and the old-growth rain forests of

FIGURE 4.4

Biodiversity: The savannas of Masai Mara National Park in Kenya support the richest fauna of large mammals on earth.

(Photo by G. W. Cox)

FIGURE 4.5

Biodiversity: Enormous populations of marine mammals, birds, fish, and squid depend on the high primary production of the Antarctic upwelling zone.

(Photo by Stephen Leatherwood)

FIGURE 4.6

Biodiversity: Old-growth rain forests of Pacific Rim National Park, British Columbia, have been protected in only a few areas.

(Photo by G. W. Cox)

FIGURE 4.7

Biodiversity: California vernal pools are a microecosystem rich in endemic plants and temporary pond animals.

(Photo by G. W. Cox)

the Pacific Northwest (Fig. 4.6). The public readily identifies these aspects of biodiversity as "world class" features that humanity should protect.

Biodiversity also includes thousands of less conspicuous species and the less spectacular ecosystems they form; every region has some examples of such ecosystems. California vernal pools are a good example (Fig. 4.7). Vernal pools are temporary ponds that form during wet periods of winter and spring, holding water for periods of a few days to several weeks. Many people, including many developers and even some prominent politicians, consider vernal pools to be little more than "mud puddles." In southern California, however, nearly 50 species of vascular plants

are restricted to this microhabitat (Zedler 1987). Many of these plants begin to grow when the pools are full of water, and then flower after the basins dry. Many unique invertebrates and amphibians also inhabit these pools, including the remarkable colonial ciliate protozoan, *Systylis hoffi* (Cox 1982). The treelike colonies of this protozoan reach a centimeter in height. *Systylis* is known from only four locations on earth, and in the United States only from vernal pools near San Diego.

Nevertheless, many people feel that biodiversity at all levels is a heritage that should be preserved as fully as possible, both for its utilitarian and nonutilitarian values. Slightly over half of people surveyed in both developed and developing countries in

1992 considered the loss of animal and plant species to be a "very serious" problem (Bloom 1995). Many very valuable aspects of biodiversity have been degraded because their value was overlooked in the rush to exploit other resources. Unsustainable timber harvest and hydropower dams that block salmon migrations in the Pacific Northwest are good examples of such short-sightedness. Other aspects of biodiversity have been lost because their utilitarian or nonutilitarian values, although real, are difficult to measure, permitting the political process to ignore them. Thus, another area of responsibility in conservation biology is to reveal the full importance of biodiversity and develop ways to manage it sustainably. We shall examine these questions in detail in Chapters 29 and 30.

SUMMARY

1. Biodiversity involves both evolutionary and ecological aspects, and spans a hierarchy from infraspecific features through species-level and interspecific relations to macroevolutionary patterns and landscape-level dispersion of ecosystems. Endemic taxa—those that are restricted to the area in which they evolved their uniqueness—are of particular concern to conservation biologists.

2. About 1.5 million species have been described and named. Estimates of the total number of species on earth are informed guesses, but 12.5 million appears to be a reasonable estimate. Most species are terrestrial, and most of these occur in tropical forests. Various biodiversity hotspots—areas rich in endemic forms—occur in terrestrial, freshwater, and marine environments.

3. Diversity can be assessed as the richness of specific communities in species (alpha diversity), the extent of change in species from one community to another (beta diversity), and the overall diversity of a region, based on number of community types and the extent of change from one to another (gamma diversity).

4. Ultimate factors determining biodiversity are those favoring speciation and colonization of isolated regions, on the one hand, and causing extinctions, perhaps catastrophically, on the other. Biodiversity tends to decline from the tropics poleward in most groups of organisms, and to reach high levels in regions of differentiated topography. Intermediate levels of disturbance tend to favor high alpha diversity.

5. Biodiversity has major utilitarian and nonutilitarian values, many of which have been ignored or unassessed in political and economic processes of resource exploitation.

QUESTIONS FOR DISCUSSION

1. Considering the hierarchical aspects of biodiversity and the aspect of endemism, what do you think are the biodiversity components of greatest and least importance for conservation effort?

2. Consider one of the world's biodiversity hotspots. What are the evolutionary and ecological factors that might be responsible for the high level of biodiversity in that location?

3. If the remarkable biodiversity of life in certain locations, such as Madagascar or Lake Baikal in Russia, is of global significance, what responsibilities do you think people in other countries have for helping to protect it?

SUGGESTED READING

Stork, N. E. 1993. How many species are there? *Biodiversity and Conservation* **2**:215–32. A concise, nonmathematical discussion of estimation techniques and estimates of global species diversity.

Wilson, E. O. 1992. *The diversity of life*. Harvard University Press, Cambridge, MA. A highly readable account of the origin, magnitude, and importance of biodiversity by an influential systematic biologist.

World Conservation Monitoring Centre. 1992. *Global biodiversity*. Status of the Earth's living resources. Chapman and Hall, London. An illustrated compendium of data on global biodiversity.

Processes of Extinction

E	**xtinction** is the termination of an evolutionary lineage by demographic failure or by the loss of its genetic distinctiveness. Extinction is one of the several fates (Fig. 5.1) that an evolutionary lineage can experience (Temple 1986). **Speciation,** first of all, may split one evolutionary lineage into two or more daughter lineages. An evolutionary lineage may also undergo gradual change through natural selection until it has become so different that taxonomists no longer regard it as the same taxonomic unit, a process termed **anagenesis.**

Extinction may result either from demographic failure or genetic swamping. **Demographic failure** is the disappearance of a population by the death of all its members. A catastrophic event, such as a volcanic eruption or a hurricane, could kill all of the individuals of a particular group, or some condition could prevent reproduction of members of the group. Demographic failure could also be the final event in a long sequence of population decline due to an excess of mortality over reproduction. Extensive interbreeding can lead to the absorption of one form in the gene pool of a more abundant related form. This process, known as **genetic swamping,** can thus cause the disappearance of a distinctive evolutionary line.

Inasmuch as extinction is a normal evolutionary event, the goal of conservation biology is not to abolish it but simply to prevent it from running far ahead of the rate of speciation. As we shall see, however, this is a considerable challenge. The threat of massive extinction throughout the biosphere has stimulated great concern, as evidenced by several recent books (Wilson 1992; Kaufman & Mallory 1993; Lawton & May 1995).

The passenger pigeon, once North America's most abundant bird, passed into extinction because of human actions.

Outline

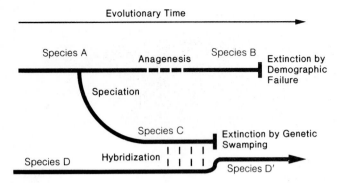

FIGURE 5.1

Speciation, extinction by reproductive failure, extinction by genetic swamping, and transformation of species in evolutionary (anagenesis) time.

EXTINCTION AS A NORMAL PROCESS

Over long periods of evolutionary time, it is likely that the rates of speciation and extinction have balanced each other. The evolutionary lifetime of marine invertebrates has been estimated at 1–10 million years (May et al. 1995). For a marine invertebrate fauna of half a million species or so (*See* Chapter 4) this suggests normal extinction and speciation rates of 0.05–0.5 species annually. For birds and mammals, individual species are thought to endure for perhaps 0.5–2 million years (Fisher et al. 1969; King 1980). For birds, King (1980) estimated that the normal extinction rate (and a similar speciation rate) is about one species every 83.3 years (King 1980).

Evolutionary biologists have long recognized that extinction rates have not been constant through evolutionary time, although how strongly pulsed they are remains a point of great controversy. Some scientists maintain that widespread extinction resulted from catastrophic events such as collisions of asteroids with the earth (Alvarez et al. 1980). The most recent of such events may have occurred at the end of the Cretaceous period, about 60 to 75 million years ago, causing the demise of the last of the dinosaurs, the large flying and swimming reptiles, certain groups of mollusks, and many forms of marine plankton. If such a pattern has been general, it also suggests that extinctions were quite rare during intervening periods. Others argue that the apparent sudden disappearance of large numbers of species is an artifact of the fossil record, and that these forms actually became extinct gradually over periods of several million years—periods that appear instantaneous due to the incompleteness of the fossil record (Officer & Drake 1983; Archibald & Bryant 1990).

EARLY HUMANS AND EXTINCTIONS

With development of the ability to use tools and fire, humans began to exert a major impact on other occupants of the biosphere. This influence increased further with the development of agriculture, and with the resulting transformation of the landscape.

We are just beginning to understand the impact of prehistoric human populations on plant and animal life. Consider, for example, the rich fauna of large mammals, birds, and reptiles that survived in North America until about 11,000 years ago. This fauna, now represented by skeletons recovered from sites such as the La Brea tar pits and displayed in the Los Angeles County Museum of Natural History, included 31 genera of large mammals, such as mastodons, mammoths, saber-toothed tigers, camels, ground sloths, and many other forms, all now extinct. Some 22 genera of birds and several groups of reptiles also disappeared. Were these species still in existence, North America would have a large animal fauna similar to that of Africa at present, or perhaps even richer. As it is, only 14 genera of large mammals now survive in North America.

Why these animals disappeared has long been a topic of interest to paleoecologists. Some have attributed the extinctions to changing climates at the end of the last glacial period. The forms in question experienced as many as ten glacial advances and retreats, however, and survived the climatic swings that accompanied them.

Martin (1973, 1984), however, has championed the hypothesis that these extinctions resulted from the sudden arrival of humans in the New World. Over the earth, periods of Pleistocene and Recent extinction correspond closely to the arrival of human populations (Fig. 5.2). In South America, massive extinctions occurred almost simultaneously with those in North America. In Australia, extinctions were heavy and began somewhat earlier. In island regions such as Madagascar, New Zealand, and the West Indies, extinctions were much more recent, occurring only 4,000 to 400 years ago. In Africa and Asia, extinctions were light and very ancient. Martin (1973) concluded that in all areas except Africa and Asia, man appeared suddenly as a skilled hunter, encountered a naive fauna of animals, and caused their extinction by overkill. In Africa and Asia, where gradual cultural and biological evolution slowly transformed humans into skilled hunters, animals had time to adapt by evolution to the new predator, and thus survived.

A model of this process in the New World was suggested by Mosimann and Martin (1975). This scenario assumes that a group of hunters arrived at the vicinity of Edmonton, Alberta, about 11,200 years ago, having made their way through a land corridor from Alaska created by retreat of the continental glaciers (Fig. 5.3). Encountering a rich fauna of native game animals, this population grew and spread rapidly, killing off many species behind the advancing population front. The spread was so rapid that humans reached the Isthmus of Panama about 10,930 years ago and Tierra del Fuego about 10,500 years ago, accomplishing the colonization of North and South America in only 700 years. This short period allowed no evolutionary adjustment by many species, and resulted in their extinction.

Extinction of many of the large mammals doubtless influenced the natural vegetation and caused the disappearance of many plant species. Central American forests contain a number of trees with large-seeded fruits like those now eaten and dispersed by elephants and other large animals in Africa and Asia (Janzen & Martin 1982). Today, these trees are not abundant, but they may have been much more common when mastodons, ground sloths, and other large Pleistocene species were present to disperse their seeds.

Throughout the islands of the Pacific Ocean, including New Zealand, the arrival of humans apparently led to the rapid extinction of many animals, mostly birds (Steadman 1995). Between about 1500 B.C. and A.D. 400 the islands of

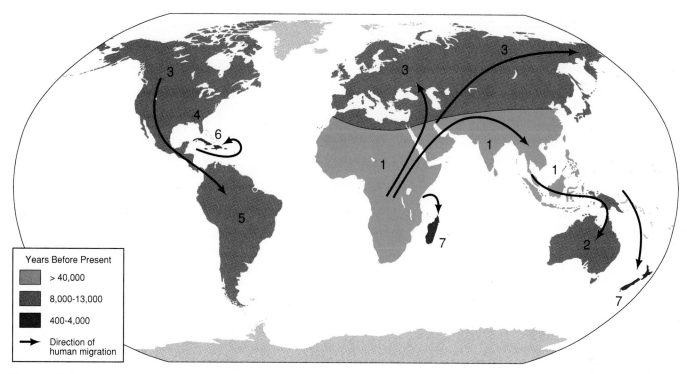

FIGURE 5.2

World pattern of Pleistocene extinction of large animals, and the time of appearance of hunting man.

(Modified from Martin 1970)

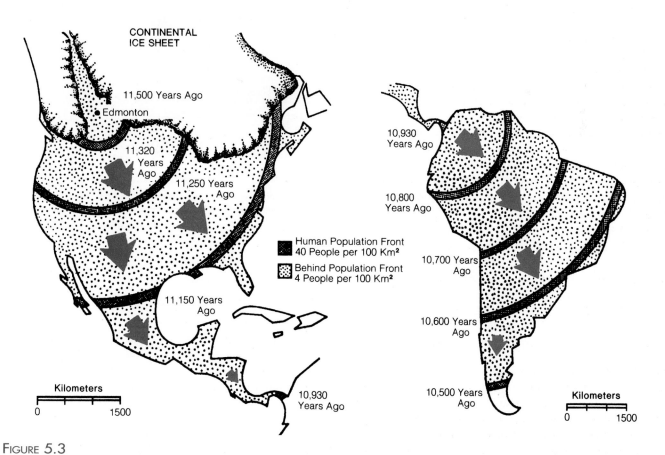

FIGURE 5.3

A model of colonization of the New World by Paleoindian hunters, coupled with extinction of much of the megafauna by overkill.

(Reprinted with permission from P. S. Martin, 1973, "The Discovery of America" in *Science*, 179:969–74. Copyright © 1973 American Association for the Advancement of Science, Washington, DC.)

TABLE 5.1

BIRD GROUP	SURVIVING SPECIES TOTAL	SURVIVING SPECIES ENDANGERED	POLYNESIAN EXTINCTIONS	POST-EUROPEAN EXTINCTIONS
Ducks and Geese	2	2	8	
Ibis			3	
Hawks and Eagles	1	1	2	
Owls	1		3	
Rails, Coots	2	2	9	1
Crows	1	1	2	
Thrushes	2	2	2	
Flycatchers	1			
Honeyeaters	1	1	2	4
Finches	14	10	29	8
Total	**24**	**19**	**60**	**13**

The Present Hawaiian Native Land Bird Fauna, Together with Extinctions During the Polynesian Period and During the Period Since Western Contact

Source: Data from D. W. Steadman, "Prehistoric Extinctions of Pacific Island Birds: Biodiversity Meets Zooarcheology." *Science* **267:**1123–31. Washington, DC: American Association for the Advancement of Science; and various other sources.

Polynesia and Micronesia were colonized by humans, who eventually reached New Zealand in about A.D. 800. The bone remains of birds in cave deposits and archeological sites document both the presence and extinction of numerous species of seabirds and land birds. Hunting by people, predation by the dogs, pigs, and rats that accompanied them, and habitat destruction by the clearing of areas for farming led to the extinction. Flightless birds, especially rails, were susceptible to extinctions. Steadman (1995) estimates that perhaps 8,000 island populations of birds, representing more than 2,000 species, were eliminated.

In the Hawaiian Islands, which were colonized by Polynesians about A.D. 400, many native birds were driven to extinction (Olson & James 1984; Steadman 1995). Some 60 species that became extinct during the Polynesian period have now been identified (Table 5.1). Included are flightless species of geese, ibis, and rails, together with flying species of several other families (Fig. 5.4). Again, many of these species were probably easy prey to humans and other mammals because of the previous absence of any such predators.

RECENT EXTINCTIONS

Since about A.D. 1600, many more extinctions have been recorded (Table 5.2). As a percentage of the number of existing species, losses have been greatest for mammals and birds. In all likelihood, many other undescribed and unnamed species have also disappeared, mostly through the destruction of tropical forests (*See* Chapter 9).

Humans have continued to be the major agent of species extinction. Several factors can cause populations of plants and animals to decline to extinction or extreme endangerment. Even if these do not directly cause extinction, however, a number of other factors come into play when populations are reduced to less than 50–100 individuals, and these may be the final causes of extinction.

FIGURE 5.4

An artist's reconstruction of a flightless ibis and a flightless rail that became extinct during the Polynesian period in the Hawaiian Islands.

(Painting by Douglas Pratt; photo courtesy of the Bishop Museum, Honolulu, Hawaii)

Deterministic Processes and Decline to Extinction

Deterministic processes are stresses that affect populations in direct, detrimental ways by increasing mortality and reducing reproduction, thus causing the population to decline toward zero. Direct killing, habitat destruction, increased predation or competition, disease, and genetic swamping are examples of deterministic factors that may drive a population to extinction.

Reid and Miller (1989) show that human exploitation, the effects of introduced species, and habitat disruption have been the major causes of recent historical extinctions of vertebrates (Table 5.3). King (1980), in a more detailed analysis for birds, showed

TABLE 5.2

Recorded Extinctions of Various Groups of Organisms Since 1600 A.D.

TAXON	CONTINENTAL	ISLAND	OCEANIC	TOTAL	PERCENT OF SPECIES
Mammals	30	51	2	83	2.1
Birds	21	90	2	113	1.3
Reptiles	1	20	0	21	0.3
Amphibians	2	0	0	2	0.01
Fish	22	1	0	23	0.1
Invertebrates	49	48	1	98	0.01
Vascular Plants	245	139	0	384	0.2

Data from W. V. Reid and K. R. Miller, *Keeping Options Alive: The Scientific Basis for Conserving Biodiversity*, 1989, World Resources Institute, Washington, D.C.; IUCN 1988 (fish and invertebrates); Nilsson 1983, supplemented by species listed as extinct in IUCN 1988 (vertebrates other than fish); Threatened Plants Unit, World Conservation Monitoring Centre, 15 March 1989, personal communication (plants); G. Vermeij, personal communication (marine invertebrates); Wilson 1988 (number of species).

TABLE 5.3

Major Causes of Vertebrate Extinctions

Group	PERCENT OF EXTINCTIONS				
	Human Exploitation	Introduced Species	Habitat Disruption	Other	Unknown
Mammals	24	20	19	1	36
Birds	11	22	20	2	37
Reptiles	32	42	5		21
Fish	3	25	29	3	40

Data from W. V. Reid and K. R. Miller, *Keeping Options Alive: The Scientific Basis for Conserving Biodiversity*, 1989, World Resources Institute, Washington, D.C.; Day 1981, Nilsson 1983, and Uno, et al. 1983.

that killing by humans and the effects of introduced predators have been the major causes of extinctions since about 1650 A.D. (Table 5.4). In North America, the Steller's sea cow and great auk were driven to extinction by their killing for oil and meat, respectively. Giant tortoises were eliminated from several islands in the Indian Ocean and from several of the Galápagos Islands off Ecuador by exploitation by early whaling crews (Honegger 1981). Killing by humans still threatens many species, such as African and Asian rhinoceroses (*See* Chapter 10).

Introduced predators, such as rats, cats, and mongooses, have eliminated many birds, mammals, lizards, and snakes from oceanic islands (*See* Chapter 12). The brown tree snake, native to New Guinea and neighboring areas, is a dramatic example. This snake was introduced accidentally to the island of Guam in the late 1940s or early 1950s, probably in shipments of fruit. An arboreal, nocturnal predator on eggs, and young and adult birds, it has nearly eliminated native forest bird species, driving six species to extinction and reducing the remaining four species to less than 100 individuals on the main island (Savidge 1987). Introduced herbivores have often decimated the native vegetation of islands, causing extinctions of both plants and animals. On San Clemente Island, one of the Channel Islands off southern California, goats introduced by early Spanish seafarers have eliminated a number of endemic plants.

TABLE 5.4

Causes of Historical Extinctions of Birds in Island and Continental Regions

CAUSE	CONTINENTAL REGIONS	OCEANIC ISLANDS
Hunting	61.5%	14.9%
Predation		41.8%
Competition		6.7%
Disease		5.6%
Genetic Swamping		0.7%
Weather		0.4%
Habitat Disturbance	23.1%	19.4%
Unknown	15.4%	10.4%
Number of Species	11	92
Number of Subspecies	2	83
TOTAL NUMBER OF TAXA	13	175

Source: Data from W. B. King, *Ecological Basis of Extinction in Birds*, 1980. Proc. XVII Int. Orn. Congr., 1978: 905–11.

Exotic plants and animals have caused extinctions in several other ways. Several island mammals and birds have been eliminated by introduced competitors. For example, in the Galápagos Islands, introduced black rats eliminated native rice rats from four of the six islands the latter originally occupied (Eckhardt 1972). Exotic plants are damaging competitors for native species in both continental and island areas (*See* Chapters 7 and 12). Introduced diseases have also contributed to extinction. Avian malaria (in the blood of introduced continental birds) and introduced mosquito vectors have combined to cause the extinction of several native Hawaiian birds (van Riper et al. 1986).

Genetic swamping has led to the extinction of 15 or more species of fish in North America (Miller et al. 1989). In New Zealand, the native gray duck has nearly been driven to extinction by hybridization with the introduced mallard duck (Gillespie 1985). In North America, the native black and Mexican ducks are facing a similar threat, due to the spread of the mallard duck into their ranges. Wild populations of the red wolf, which once occurred in the lower Mississippi Valley and along the Gulf Coast, have also disappeared due to genetic swamping by coyotes (Pimlott & Joslin 1968). The species now survives only in captivity, and on a North Carolina wildlife refuge to which it has recently been reintroduced.

Habitat disruption has long been an important cause of extinction (Table 5.3), and it is now a factor of increasing importance. Complete habitat conversion effectively eliminates even the populations of mobile animals that are able to escape immediate destruction. In all likelihood, the areas to which they flee are unsuitable or already fully occupied. Extinction of the migratory Bachman's warbler in North America, for example, was probably due to the combined destruction of breeding habitat in swamp forests of the southeastern United States and of wintering habitat in Cuban rain forests (*See* Chapter 21). In the tropics, the clearing of moist forests is likely to bring about the extinction of thousands, if not millions, of species in the coming century (Wilson 1992).

Humans have modified the habitats of native species almost everywhere by farming and ranching, change in the frequency of fire, and introduction of exotic plants and animals. The changes in habitat conditions can often be subtle, yet significant. The spread of crop and livestock farming over much of North America, for example, greatly extended the portion of the continent suitable for the brown-headed cowbird, a species that lays its eggs in the nests of other birds. Originally confined to the Great Plains, where it foraged in association with herds of bison, the cowbird is now found from coast to coast. Nest parasitism by this species now affects many other birds that are not adjusted to this threat through evolution. The extra pressure of cowbird parasitism may be a major factor in the decline of species such as the Kirtland's warbler in Michigan (Mayfield 1977) and the Bell's vireo in southern California (Fig. 5.5).

The areas of natural habitat that remain are being modified in many other ways. Fragmentation of habitat is a threat to many species. Agriculture and urbanization are rapidly transforming continuous areas of natural ecosystems into islands of natural habitat in a developed landscape sea. Fragmentation

FIGURE 5.5

This Kirtland's warbler, an endangered species that breeds only in a small area in the southern peninsula of Michigan, is feeding a young brown-headed cowbird that has displaced its own young.
(USDA Forest Service photo)

reduces the populations of many species to scattered groups of a few individuals, which possess a high risk of local extinction. In addition, small islands of natural habitat have a high edge-to-interior ratio, compared to larger areas, and thus are subject to stronger influences from surrounding agricultural and urban systems. Habitats of native species are also increasingly subject to chemical contamination by oil spills, pesticides, industrial chemicals, and urban waste discharges.

Deterministic processes may act simply by driving the population below a critical threshold. Exploitation, habitat conversion, fragmentation, and change may create a situation in which a species can no longer survive, even when areas of seemingly suitable habitat still exist. If the population of a species drops below a certain density, social interactions among individuals may become less frequent, and the benefits of these interactions lost. The extinction of the passenger pigeon (Fig. 5.6), once the most abundant bird in North America, may illustrate this process (*See* Reading 5.1). This mechanism may also have contributed to the decline of the California condor.

For birds (King 1980) and for amphibians and reptiles (Honegger 1981; Case et al. 1992), by far the majority of extinctions prior to 1900 were of island forms, which are especially vulnerable because of their small populations and lack of adaptation to man and introduced predators (Table 5.4). The risk of extinction on oceanic islands, such as the Galápagos and Hawaiian archipelagos, where some of the world's most remarkable biotas exist, is still very high. Since 1900, however, the rate of extinctions of animals in continental areas relative to island areas has been rising (Lomolino & Channell 1995).

On close examination, almost all recent extinctions are attributable to human activities. Among birds, King (1980) was able to recognize only a single extinction evidently due to natural causes—the disappearance of the Puerto Rican bullfinch on the West Indian island of St. Kitts at about the time that the island's forests were decimated by a hurricane.

FIGURE 5.6

The passenger pigeon was originally abundant in the eastern forested region of North America.

(Photo courtesy of the Bell Museum of Natural History, University of Minnesota)

Stochastic Processes and Extinction in Small Populations

In small populations, survival and reproduction of individuals become strongly influenced by **stochastic,** or random, factors. Chance events can directly affect individuals in the population itself, and random variation can occur in conditions of the habitat. **Demographic stochasticity** refers to chance variation in mortality and reproduction due to accidents that affect individuals. The few breeding females in a small population may fail to mate, produce young, or successfully rear young in a given year for any of several accidental reasons. Or, by chance, all the individuals of one sex might die in a given year. Small populations are also subject to **genetic stochasticity,** which can lead to loss of genetic variability or inbreeding problems. We shall examine this aspect of stochasticity in more detail in Chapter 25. Similarly, **environmental stochasticity,** or random variation in habitat conditions that affect survival and mortality, may cause unusual effects on small populations. Environmental stochasticity can range from the normal variations that occur around average conditions in most years to catastrophic variations, such as those caused by hurricanes or volcanic eruptions. For the same reason that flipping a coin just a few times may produce a run of all heads, random events like these in small populations can be the final cause of extinction.

WHAT MAKES SPECIES VULNERABLE TO EXTINCTION?

The features that make populations of particular species highly vulnerable to extinction have been investigated by several workers. Through an analysis of long-term censuses of land birds breeding on islands off the British Isles, Pimm et al. (1988) and Tracy and George (1992) showed that small population size was the strongest determinant of vulnerability, confirming theoretical

predictions by other workers. Overall, large-bodied species were more vulnerable to extinction because their populations tended to be smaller. In addition to body size, migratory behavior also made species more prone to extinction, apparently because extinction could occur not only because of failures of survival but also because of failure of surviving birds to return to an island.

Several other characteristics predispose species to extinction (Ricklefs & Cox 1972; Terborgh 1974). Species at the ends of long food chains tend to be low in abundance and often large in body size, making them vulnerable for reasons just noted. Species that are highly specialized for particular habitats, breeding sites, or food resources tend to be more at risk than do generalists. The occupation of insular habitats or ranges, where populations are often small and species may become specialized for highly distinctive environmental conditions, increases the risk of extinction when the environment is disturbed or changed. And, of course, species with characteristics of value that attract humans to exploit them are intrinsically vulnerable to extinction.

Several categories of species thus tend to be especially vulnerable to extinction:

1. *Large species with a low reproductive potential.*
   Mammals such as the great whales, rhinoceroses, and great apes, and birds such as the California condor and whooping crane are especially vulnerable because their populations are slow to recover, even when they receive protection. Once reduced in numbers, they remain at high risk for many years.

2. *Species with a high economic value.*
   Animals such as whales, sea turtles, elephants, rhinoceroses, and spotted cats are hunted legally and illegally for their meat and oil, tusks and horns, or pelts. The value of such products is sometimes enormous, and the incentive for killing even the last survivors in an area is very strong, especially in underdeveloped countries.

3. *Species at the ends of long food chains.*
   Animals such as hawks and owls, cats and canids, and many reptiles depend on food animals that are themselves not very abundant. Disruption that affects basic ecosystem productivity creates the greatest effect at these higher links in the food chain, making it more and more difficult for higher-level predators to meet their needs. In addition, a number of toxic materials can become concentrated along food chains, ultimately to a level toxic to species at their ends (*See* Chapter 22).

4. *Species restricted to local, insular habitats.*
   Species restricted to small islands or isolated habitats such as small lakes are at high risk because of their small total populations and limited geographical distribution. Only a small-scale disturbance is needed to eliminate such forms. In addition, the isolation of their habitats may have protected them against many of the biotic challenges of noninsular environments, and they may have lost their ability to cope with such challenges.

Originally the most abundant bird in North America, the passenger pigeon was estimated to have numbered about 3–5 billion birds, or about 25%–40% of all North American land birds. It was native to eastern North America, confining its activities mostly to areas of deciduous forest, breeding in the more northern portions of this region and wintering in southern areas. The birds moved from place to place in large flocks, and settled to breed where food supplies were abundant. Where migrations were concentrated, passing flocks would darken the skies for days. Near Frankfort, Kentucky, for example, Alexander Wilson counted flocks, estimated their size, and calculated that 2,230,272,000 birds had passed in migration over a period of days. In Ontario, another observer similarly calculated a mass migration of 3,717,120,000 birds. The nonbreeding birds occupied forest roosts hundreds to thousands of hectares in area.

The primary food of passenger pigeons throughout the year was mast—acorns, beechnuts, and chestnuts. For nesting, the birds required forest areas where bumper crops of mast had been produced the preceding fall, an event that occurs irregularly in the eastern forest region in any given place. In spring, as flocks of birds began moving north from their wintering range, social facilitation was important in locating areas of abundant mast. Flocks were attracted to birds feeding on the ground, and if the food source was abundant, more and more flocks would settle, and an enormous breeding colony might develop. Typical breeding colonies were about 16 kilometers long and 4–5 kilometers wide, with up to 90 nests per tree. Some were much bigger.

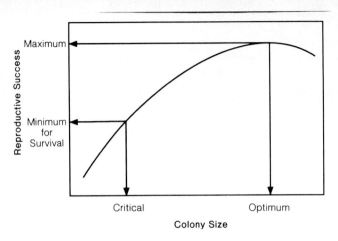

READING FIGURE 5.1

Hypothesized relationship of colony size and reproductive success for a colonial bird such as the passenger pigeon.

(Modified from Halliday 1980)

Several factors contributed to the decline of the species. Clearing of the eastern forests for farming in the 1700s and 1800s reduced and fragmented the area of suitable habitat. Farmers encouraged feeding on mast by pigs. And the birds themselves were subjected to heavy commercial hunting.

The final decline of the passenger pigeon may have been related to its dependence on social facilitation (Halliday 1980; Bucher 1992). Flocking and coloniality presumably confer benefits on members of the groups—greater survival or reproductive success (Reading Fig. 5.1). Finding the areas of abundant mast for nesting colonies may have depended on the number and size of flocks exploring the forested landscape. Individuals in small colonies may not have been able to detect or repel predators, or to locate good feeding areas as efficiently as those in large colonies. On the other hand, if numbers were too large, the birds may have suffered from competition for food or may have become vulnerable to disease. Thus, an optimal colony size probably existed, at which survival and reproduction were greatest. A critically small colony size must also have existed, at which survival and reproduction were barely able to maintain the population. From this, one can see that once the overall population of the species declined to the point that the benefits of flocking and coloniality became inadequate, the species would then automatically decline to extinction.

5. *Species specialized for habitats, breeding sites, or foods.*
   Species of old-growth forests, such as the spotted owl, are likewise vulnerable because even limited modification of their habitats may change conditions to their disfavor. Animals that breed in colonies are more subject to catastrophic accidents and the activities of predators, including humans. Animals such as the giant panda, a feeding specialist on bamboos, or the black-footed ferret, a specialist predator on prairie dogs, are especially vulnerable because they have few alternatives when something affects their food.

6. *Migratory species.*
   Migratory birds, mammals, fish, and even insects such as the monarch butterfly (Fig. 5.7), depend on habitat conditions in different geographical areas at particular parts of their seasonal cycle. Thus, they are vulnerable to habitat disruption in any of these regions.

GLOBAL RATES OF EXTINCTION

Present global rates of extinction can only be approximated. Rough estimates have been made by use of **species-area curves** (Fig. 5.8), which simply show the rate of accumulation of total

FIGURE 5.7

Monarch butterflies from temperate North America migrate to wintering sites in central Mexico and coastal California, where their extreme concentration makes them vulnerable to mass mortality from natural causes or human disturbance. This photo shows a wintering aggregation in Pacific Grove, California.

(Photo by G. W. Cox)

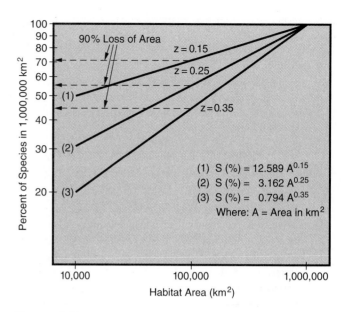

FIGURE 5.8

Species-area curves for island and continental areas, illustrating their use in estimating ultimate extinction values when habitat area is reduced.

TABLE 5.5

A Rough Estimate of Current Global Extinction Rates for Major Taxa	
	EXTINCTIONS/YEAR
Terrestrial Vertebrates	1–10
Aquatic Vertebrates	10–100
Plants	50–500
Aquatic Invertebrates	100–1,000
Terrestrial Invertebrates	1,000–10,000

number of species with increase in the area examined. Such a curve would be obtained by sampling the number of species in progressively larger land areas, either plots on continental areas or entire islands, from very small size (e.g., 1 hectare) up to very large size (e.g., 1 million hectares). The slope of this relationship is typically steeper for island areas than for continental areas (*See* Chapter 27). Using the equation for such a curve, with a slope value appropriate to the location, one can predict the decline in number of species if the area is reduced by an assumed amount. With a slope value of 0.25, this relationship suggests that a 90% reduction in habitat area will lead to about a 50% loss of species. For tropical forests, for example, Reid and Miller (1989) used this relationship to estimate that 4%–10% of species had been lost by deforestation in Latin America and the Caribbean by 1990, and that 42%–72% of all species would be lost if all forest except that now legally protected were lost. Pimm et al. (1995) inferred from this relationship that extinction rates were now 100 to 1,000 times their levels prior to human influence.

Use of the species-area curve to estimate future extinctions has been criticized on several grounds. In Brazil, for example, the area of the Atlantic coastal rain forest has been reduced by about 90%. This seems, however, to have led to almost no bird and butterfly extinctions, despite the richness of their faunas (Brown & Brown 1992). In this case, it is likely that extinction

does not follow instantaneously with reduction of habitat area, but occurs gradually over many decades.

Following a quite different approach, Smith et al. (1993) examined the change in status of species listed in databases of the World Conservation Centre in Cambridge, England. These databases list species of conservation concern, and characterize their rarity or endangerment along a scale from "not endangered" to "probably extinct." By analyzing species' changes along this ranking over a 4-year period, these authors were able to calculate the probable time for half the species in various taxonomic groups to become extinct. For birds and mammals, for which knowledge of the number of species and their status is best, they found that the time for extinction of half the world's species under current trends was 200–300 years. These estimates are close to those based on species-area curve analyses.

Current extinction rates for various groups of organisms differ greatly (Table 5.5). A reasonable estimate is that terrestrial vertebrates, the most carefully watched and guarded species, lose 1 to 10 species per year. Terrestrial invertebrates doubtlessly have the highest rate of extinction, which may be 1,000 to 10,000 species per year.

Summary

1. Extinction is the disappearance of an evolutionary line by demographic failure, conditions that reduce reproduction or increase mortality, or by genetic swamping, the absorption of one line in the gene pool of a related line by hybridization.

2. As early humans spread around the world in Pleistocene and post-Pleistocene time, they became agents of plant and animal extinction due to hunting, habitat conversion, and introduction of predators to new areas.

3. Extinctions result from deterministic pressures, such as direct killing, habitat disruption, the effects of introduced species, and pollution. Weakened social interactions in reduced populations can drive the species downward to extinction. In small populations, random processes affecting demography, genetics, or habitat conditions can be the final cause of extinction.

4. Species vulnerable to extinction include large-bodied forms with low reproductive potential, animals at the ends of long food chains, inhabitants of insular habitats, forms with high specialization for food or habitat, and forms with great economic value.

5. Global rates of extinction can only be estimated roughly, but appear to be 100 to 1,000 times those prior to human dominance of the biosphere.

Questions for Discussion

1. What are the implications of human-caused extinctions for the idea that pretechnological human populations often lived in harmony with their natural environment?

2. Consider the ecosystem type of greatest biodiversity in your geographical area. What are the deterministic and stochastic factors likely to be of greatest danger to endemic species of this ecosystem?

3. How will global climatic change likely influence rare or endangered species in your area? Consider the ways in which global change may affect temperature, rainfall, ultraviolet radiation, fire frequency, and similar environmental conditions.

Suggested Reading

Bucher, E. H. 1992. The causes of extinction of the passenger pigeon. *Current Ornithology* **9**:1–36. A synthesis of modern ideas about how various human impacts caused the extinction of the most abundant bird in North America.

Kaufman, L. and K. Mallory (Eds.). 1993. *The last extinction*. Second ed. MIT Press, Cambridge, MA. An examination of various aspects of the extinction process, including a comprehensive listing of North American extinctions.

Steadman, D. W. 1995. Prehistoric extinctions of Pacific island birds: Biodiversity meets zooarcheology. *Science* **267**:1123–31. A summary of evidence for humans as the agent of massive avian extinction throughout the Pacific region.

two

Terrestrial Ecosystems

I n this section we shall examine problems of
conservation biology in terrestrial ecosystems,
and identify principles leading toward their solution.
For this we must group terrestrial ecosystems into broad
ecological and geographic categories. First, we will
examine globally important types of terrestrial
ecosystems that are distinctive in their dominant
vegetation and animal life. These include temperate and
tropical forests, temperate grasslands, arctic and alpine
tundras, tropical savannas and woodlands, and deserts.
Second, we shall consider two distinctive terrestrial
ecosystems defined in a geographical fashion: coastal
environments and ocean islands.

chapter

6

Temperate Forests, Woodlands, and Shrublands

T emperate forests include deciduous, coniferous, and broad-leaved evergreen forest ecosystems. In North America, these forests span a wide range of climates, from the fringes of the arctic and the borders of grasslands to the cool, wet rain coasts of the Pacific Northwest. Temperate forests are also widespread in Eurasia, and occur in small parts of the southern hemisphere (Fig. 6.1). Temperate woodlands and shrublands include the evergreen communities of regions with Mediterranean climates—hot, dry summers and cool, moist winters (Fig. 6.1). In North America, these are the coastal sage scrub, chaparral, and oak woodlands of California. Similar ecosystems occur around the Mediterranean Sea, and in South Africa, central Chile, and Australia. In North America, distinctive pinyon-juniper woodlands and deciduous shrublands occur in the Rocky Mountains and Great Basin. Many of these ecosystems occur in heavily populated regions of the developed world, and thus present the challenge of sound management under conditions of intensive land use.

Temperate coniferous forests require 350-750 years to attain their distinctive characteristics.

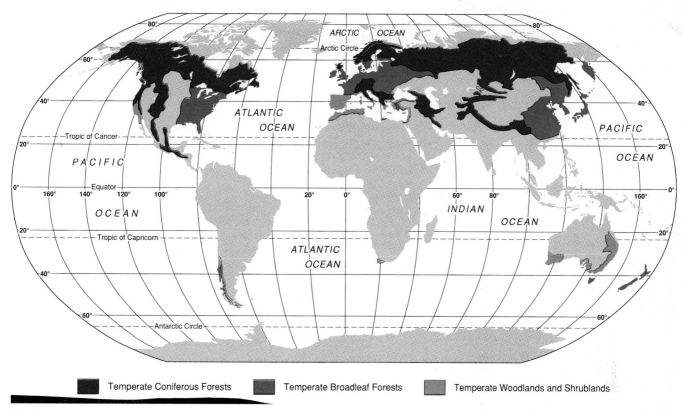

Temperate Coniferous Forests   Temperate Broadleaf Forests   Temperate Woodlands and Shrublands

FIGURE 6.1

Distribution of temperate forests and shrublands.

SUCCESSIONAL DYNAMICS

In these regions, biotic succession is a dominant ecological process (*See* Chapter 3). The pattern and time scale of succession, however, vary greatly from place to place (MacMahon 1981). On sites favorable for tree growth, succession may be a centuries-long process in which a series of woody plants replace each other, and the forest community changes greatly in structure. In the late stages, the major process may become plant replacement in gaps created by death or windfall of individual trees. At long intervals, catastrophes such as hurricane blowdowns or massive fires may reinitiate succession.

In other situations, succession may involve little replacement of species, with woody plants that appear in the pioneer phase maturing to form a late-successional community. Some of these communities are highly vulnerable to destruction by fire or insect outbreaks. Many pine and fir forests of dry or nutrient-poor habitats show this characteristic (Heinselman 1981), as do shrublands such as the California chaparral (Hanes 1977). The vegetation in such regions often consists of a mosaic of stands of different successional stages, only a fraction of which are late-successional in structure.

In either case, basic changes occur in ecosystem features during succession. During early stages of succession, especially following severe disturbance, losses of soil and nutrients may be high and the nutrient capital of the system may decline. Plant colonization and accumulation of biomass may be slow. Later, when a substantial vegetation cover has developed, biomass may begin to increase more rapidly and the ecosystem may begin to accumulate nutrients. The maximum rate of nutrient accumulation usually occurs in the phase of greatest biomass accumulation. Late in succession, increase in living biomass slows, but dead plant materials such as litter, fallen limbs and trunks, and standing dead trunks and stems may continue to increase. As storage of nutrients in biomass slows in these late stages, losses of nutrients from the ecosystem may increase. Species diversity changes through succession, but the greatest diversity is usually seen in the middle or late middle phase.

In some environments, where major disturbances occur rarely, or at very long intervals, **old-growth forests** may develop. These forests possess the greatest living biomass of any terrestrial ecosystem, and are home to many highly specialized animal species (*See* Reading 6.1).

Early and Late Successional Species

Where succession involves major changes in the characteristics of dominant species and can lead to long-lived, late successional communities, both plants and animals tend to be adapted to particular stages in the successional spectrum (Leopold 1966). For simplicity, we can consider most species to be either **early successional** or **late successional.** For these species, the basic patterns of adaptation, response to natural or human disturbance, and potential for exploitation of these kinds of species all vary with the stage of succession to which they are adjusted (Table 6.1).

OLD-GROWTH FORESTS

When succession proceeds for centuries, the forest acquires special ecological characteristics. In these old-growth forests the trees achieve enormous size (Reading Fig. 6.1A) and strata of herbaceous plants, shrubs, understory trees, and canopy trees become clearly developed. Patchiness is strong, due to succession in tree-fall gaps. Many large standing dead trees (snags) and decaying logs are present (Thomas et al. 1988; Feeney 1989). The forest floor develops a thick litter and humus layer, thickly penetrated by the roots of trees and shrubs. These old-growth forests are home to many specialized plants and animals. Thus, from an aesthetic and conservational point of view, old-growth forests are extraordinary and valuable. Commercial foresters sometimes consider them "over-mature," since no net increase in wood biomass is occurring. In the Pacific Northwest, old-growth redwood and Douglas-fir forests exhibit these qualities.

Most of the original old-growth forest has been cut. Only about 17% of the original old-growth Douglas-fir forest of the Pacific Northwest remains (Thomas et al. 1988), the bulk of this being in the United States National Forests. Bitter controversy has developed over how much old-growth Douglas-fir habitat should be preserved in the United States. Timber interests argue that their cutting and replacement by young stands is both economically justified and ecologically beneficial. Replacing old-growth forests by vigorous young stands, for example, has been claimed to be a way to counteract the greenhouse effect by removing CO_2 from the atmosphere. On the other hand, many species of birds, mammals, and amphibians are closely associated with this habitat, including the marbled murrelet, northern goshawk, northern spotted owl, Vaux's swift, silverhaired bat, red tree vole, and northern flying squirrel (Simberloff 1987; Marshall 1989).

Furthermore, analysis of the CO_2 releases and uptakes resulting from forest conversion shows that net storage does not result, and the greenhouse effect is not counteracted (Harmon et al. 1990).

Much of the recent controversy about old-growth forest in the Pacific Northwest has centered on the northern spotted owl (Reading Fig. 6.1B). The total population of the northern spotted owl is estimated to be 2,500 to 3,000 pairs (Simberloff 1987; Doak 1989). These birds occupy home ranges of 800 to 2,000 hectares in old-growth forests, and appear to feed heavily on flying squirrels. The value of old-growth timber is about $10,000 per hectare, so that an 800-hectare home range of a pair of owls contains about $8 million worth of timber. A decision about preserving habitat for a difference of 500 pairs of owls, by extension, involves $4 billion worth of timber. Modeling studies suggest, however, that continued harvesting of old-growth forest, as proposed by the United States Forest Service, will lead to increased fragmentation of old-growth stands and the severe decline of the species (Doak 1989). In this case, in June, 1990, under strong pressure from conservation groups, the United States Fish and Wildlife Service eventually listed the spotted owl as a threatened species, giving strong backing to the preservation of much of its old-growth habitat. Similar controversies exist for old-growth forests in coastal British Columbia and the southern panhandle of Alaska.

READING FIGURE 6.1A

Old-growth forests of the Pacific Northwest, like this coastal redwood stand in Muir Woods, northern California, require 350–750 years to grow. These stands are characterized by very large trees, distinct vertical strata of foliage, strong patchiness due to succession in tree-fall gaps, and numerous standing dead trees and decaying logs.

(U.S. National Park Service photo by George Grant)

READING FIGURE 6.1B

The northern spotted owl is largely restricted to old-growth coniferous forests of the Pacific Northwest.

(U.S. Fish and Wildlife Service photo by Randy Wilk)

TABLE 6.1

Characteristics of Early and Late Successional Species		
	EARLY SUCCESSIONAL	LATE SUCCESSIONAL
Habitat Requirements	Generalized	Specialized
Resource Requirements	Generalized	Specialized
Behavioral Plasticity	High	Low
Genetic Plasticity	High	Low
Reproductive Potential	High	Low
Impact of Disturbance	Often beneficial	Usually detrimental
Exploitation Potential	High	Low

Early successional habitats are usually variable and changing in their physical and biotic conditions. Extremes of physical conditions are greater, for example, on a bare rock outcrop than inside a mature forest. The specific kinds of plants and animals that exist in a weedy field, too, are more variable than the species making up the late successional forest. Thus, early successional species must be broadly tolerant of habitat conditions and generalized in their patterns of adjustment to their habitat. They must also be able to use a wide variety of foods, either plant or animal, since the biotic makeup of early successional situations is so variable.

Adaptation to early successional habitats and resources usually includes high behavioral and genetic adaptability—individuals can vary their responses in different situations. At the population level, species show high genetic plasticity; natural selection is able to adjust the local population to local environmental conditions.

Ecogenetic variation (*See* Chapter 4) typifies almost all widespread early successional species. An important population characteristic is usually high reproductive potential, which enables early successional species to colonize new areas rapidly, and also offsets the high mortality rates that usually occur under variable and unpredictable early successional conditions.

For such species, disturbance is often beneficial because it sets back the successional process that tends to eliminate the conditions to which they are adapted. Also, because of their plasticity and high reproductive potential, early successional species are often well suited to be exploited by humans—harvests of portions of their populations can be replaced quickly.

Late successional species, on the other hand, are adapted to more equable and predictable environments, where much more specialized habitat and resource requirements can be satisfied. Because such requirements can be counted on, a lesser premium exists on behavioral and genetic plasticity. High reproductive potential is less important than good parental care of a few offspring. When disturbance occurs, however, such species are usually unable to survive. Likewise, unregulated exploitation by humans usually cannot be tolerated because of the low reproductive potential, even when habitat conditions remain favorable.

Examples of Early and Late Successional Species

The mule deer, *Odocoileus hemionus* (Fig. 6.2a), is an excellent example of an early successional animal (Leopold 1966). In western North America, this species occurs from southern Alaska

a.

b.

FIGURE 6.2

A. A mule deer, a representative early-successional species. *B.* The woodland caribou, a representative late successional species.
(*a.* U.S. Fish and Wildlife Service photo by E. P. Haddon; *b.* Photo by H. R. Timmermann, Ontario Ministry of Natural Resources)

TABLE 6.2

Human Impacts on Temperate Forest, Woodland, and Shrubland Ecosystems		
	FORESTS	WOODLANDS AND SHRUBLANDS
Conversion	Variable	Moderate
Primarily to	Cropland, pasture, tree plantations	Urban land, pasture
Exploitation	Heavy	Variable
Primarily by	Timber harvest	Browsing by livestock
Fragmentation	Very extensive	Extensive
Successional Balance	Old growth depleted	Old growth increased
Species Diversity	Reduced	Unaffected
Exotic Invasions	Moderate	Variable
Types of exotics	Insects, plant diseases	Fire-adapted plants (South Africa)
Fire Relationships		
Frequency	Reduced	Reduced
Intensity	Increased	Increased
Soil Disturbance	Moderate to heavy	Light to moderate
Air Pollution Impacts	Moderate to extensive	Localized

far south into Mexico. It occupies habitats ranging from hot deserts to the borders of the cool, wet rain forests. Even in a single region, its food plants are varied, and over its total range the number of important food plants is far into the hundreds. Local populations differ in behavior; some are permanent residents, others make long seasonal migrations. Some 11 subspecies of mule deer are recognized, attesting to the high genetic adaptability of this species.

Taber and Dasmann (1957) found that disturbance can be a beneficial management tool for the Columbian black-tailed deer (*O. h. columbianus*), one subspecies of the mule deer. In northern California, they showed that dense chaparral could be opened up by prescribed burning to improve foraging conditions, reproduction, and hunting harvests. Thus, setting back succession by burning—clearly a "disturbance"—benefitted the population, and increased the hunting harvest.

The woodland caribou, in contrast, is a late successional ungulate similar in general characteristics to the mule deer (Fig. 6.2b). The woodland caribou includes the subspecies *R. t. caribou* and *R. t. sylvestris* of the caribou, *Rangifer tarandus*. In North America, woodland caribou have declined greatly in abundance, and are restricted to scattered locations from Alberta to Quebec in Canada, and to one small population in northern Idaho in the United States.

The key factor in the decline of the woodland caribou has been the change in structure of coniferous forests due to logging, and how this change affects the availability of critical foods (Cringan 1957). In winter, the critical food for woodland caribou is tree lichens—pale green, beardlike lichens that grow on tree branches. In a mature coniferous forest, the biomass of these slow-growing lichens is great, and every winter falling trees and branches make some available to caribou. In mature forests, tree lichens thus are a dependable food that the caribou cannot overexploit. Logging of the northern forests, however, not only destroys the old-growth trees and their lichen supplies, but creates a young, even-aged successional forest that lacks a dependable

supply of tree lichens. Thus, the decline of the woodland caribou seems to be largely the result of their high specialization for a food available only in late successional forests.

Other good examples of early and late successional species in North America are the ruffed grouse, *Bonasa umbellus,* and the spruce grouse, *Canachites canadensis,* respectively. The ruffed grouse, with 12 subspecies distributed from Alaska to Georgia, is a highly adaptable bird with a varied diet and an optimal habitat consisting of a mosaic of aspen stands of varying age—the products of succession after disturbance of mature forests (Gullion 1977). The spruce grouse, even though occurring from Alaska to Labrador and south beyond the Canadian border, shows less evolutionary differentiation (only 4 subspecies), and is restricted to mature coniferous forests, where it feeds almost exclusively on spruce buds and needles. The ruffed grouse is an important game species; the spruce grouse is not.

The widespread, behaviorally adaptable, prolific mourning dove, *Zenaida macroura,* is another good example of an early successional species, contrasting with the extinct passenger pigeon, *Ectopistes migratorius* (*See* Chapter 5, Reading 5.1). The mourning dove is a familiar bird of farmland, urban areas, and other disturbed habitats. It typically produces a clutch of two eggs, and usually renests two or more times each breeding season. The passenger pigeon, a bird of mature oak forests, was specialized in its breeding habits and feeding ecology. Its reproductive potential was low, each nesting pair producing only a single egg.

HUMAN IMPACTS ON FORESTS, WOODLANDS, AND SHRUBLANDS

Human impacts on temperate forests, woodlands, and shrublands are diverse, and vary greatly from place to place (Table 6.2).

In the United States, almost all of the virgin eastern deciduous forest has been cut (Noss et al. 1995) and only a little over one-third of this original ecosystem remains in forest (Klopatek et al. 1979). Much of these forests were cleared for farmland, because of the generally fertile soils on which they occur. Even greater deforestation has occurred in deciduous forest areas of Europe and Asia, and clearing of temperate forest now is proceeding rapidly in Chile.

In North America, most forests dominated by conifers have also been cut one or more times. In areas such as the southeastern United States, much of this forest has been replaced by intensively managed pine plantations. In more northern and western parts of North America, much of the original conifer and mixed conifer-hardwood forest remains as forest subject to commercial harvest and management. The exploitation and management of forest regrowth tend to reduce plant species diversity (e.g., Elliott & Swank 1994). In North America, major issues concern the small remaining areas of old-growth forest and policies concerning the extent and manner of timber harvest on public land. International timber companies are also promoting very rapid cutting of conifer forests in Siberia, where 4,100 square kilometers of forest are now being clear-cut annually (Rosencrantz & Scott 1992).

In areas of Mediterranean climate, growing human populations are bringing about the conversion of woodlands and shrublands to urban use. In California, for example, the coastal sage scrub ecosystem type has been 70%–90% destroyed (Noss et al. 1995). In much of the Mediterranean region of the Old World and Chile, livestock browsing and other human uses of shrublands over centuries or millennia have greatly modified the distribution and characteristics of these ecosystems.

The extensive conversion of these ecosystem types is correlated with their fragmentation, the effects of which we shall examine in detail in Chapter 13. Efforts to exclude fire have permitted the buildup of fuel and made many late successional stands vulnerable to catastrophic damage when fire does occur. Fire exclusion has also increased the areas of old-growth woodlands and shrublands, many of which are highly flammable.

Most forest, woodland, and shrubland ecosystems are relatively resistant to invasion by exotic plants. In eastern North America, disturbed deciduous forests have experienced invasion by trees such as Norway maple and ailanthus, and vines such as Japanese honeysuckle and kudzu. The Mediterranean shrublands of South Africa, however, are very prone to invasion by fire-adapted woody plants from several other regions of similar climate. Introduced insects and diseases, mostly from Europe and Asia, have created major problems in North American forests (Liebhold et al. 1995). Examples of introduced insects include the gypsy moth, *Lymantria dispar*, and the recently introduced pine shoot beetle, *Tomicus piniperda*. Diseases such as chestnut blight and Dutch elm disease have had devastating effects on deciduous forest trees, and the recently introduced poplar leaf rust has a similar potential.

Other impacts include disturbance by forest harvesting practices, overprotection of forest ungulates, and air pollution. Timber harvest by clear-cutting alters forest composition, and on mountainous land promotes erosion and landslides (*See* Chapter 24). Overprotection of populations of ungulates can lead to overbrowsing, change in tree composition, and altered soil nutrient dynamics (Alverson et al. 1988; Pastor et al. 1988). Air pollution impacts are now being felt over large areas of forests in eastern North America and western Europe, and even in areas of Mediterranean shrublands near large urban areas.

MANAGING SUCCESSION

To manage succession one must consider 1) the amount of habitat in different stages, 2) the size and location of areas of these stages, and 3) the nature and intensity of human use of different stages. The key strategy is to maintain the full range of successional stages. This strategy often implies the deliberate use of disturbance such as fire to create early successional habitats. It also implies protection of areas of late successional communities. A balance of areas of different stages must be maintained, such that species adapted to all successional stages are able to survive, in effect guaranteeing that biotic succession continues to be an active ecological process.

Today, some of our most endangered species require late successional forest ecosystems. Special efforts must be made to preserve the features that these species require. The red-cockaded woodpecker, *Picoides borealis*, for example, was once common in pine forests from Maryland to Texas. This species requires large pines, 75–95 years in age, that are infected by "red heart disease"—a fungal rot of the heartwood—for excavation of its large nest chambers, which serve as a communal nest for several females. Although pine forests are probably more abundant than ever within the woodpecker's range, intensive timber management has not encouraged the growth of old trees with heartwood disease. Red-cockaded woodpeckers can use large areas of younger forest for foraging. Thus, lengthening the harvest cycle to permit the growth of older trees, coupled with a harvesting design that leaves small clusters of old trees scattered through younger stands, may meet the needs of this bird without major loss of timber production (Seagle et al. 1987).

In some places, however, early or mid-successional stages of succession are disappearing. In eastern North America, land use tends to favor the extremes of forest succession: intensively farmed or grazed land, and protected forests and woodlots that are becoming more and more mature. In this landscape, successional shrublands and the youngest forest stages may be represented inadequately. As a result, many forest parks and preserves have initiated "controlled succession" plots, on which areas of mature forest are cut periodically and succession is allowed to proceed.

In many regions, animals are adapted to a landscape mosaic with a certain mix of communities of different successional status. The California black-tailed deer and ruffed grouse discussed earlier are good examples. In such ecosystems, too much or too little disturbance can both be detrimental, and management of succession must be planned very carefully.

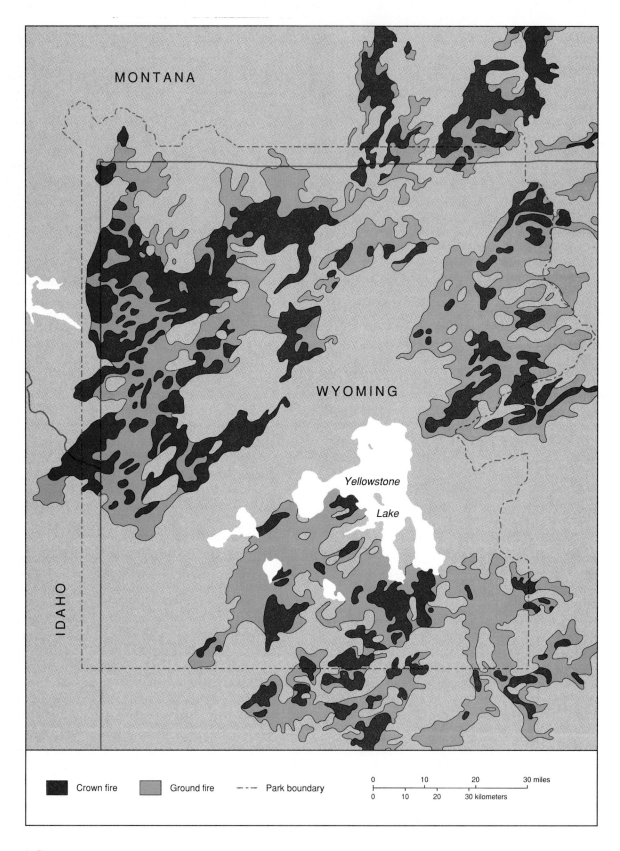

Crown fire   Ground fire   - - - Park boundary

0   10   20   30 miles
0   10   20   30 kilometers

FIGURE 6.3

The fires in Yellowstone National Park in the summer of 1988 burned about 35.7% of the park area.

(Courtesy of D. Despain, Yellowstone National Park, U.S. Park Service)

Managing Fire Relationships

Elsewhere, particularly in western North America, well-meaning protection of forests from fire and other disturbances has led to major changes in their characteristics through succession. In northern California, coast redwood stands on river floodplains were originally exposed to both flooding and fire. Redwoods are tolerant of both forms of disturbance, which essentially weed their groves of competitors and create favorable conditions for germination and growth of young redwoods (Stone 1965). Now protected from fire, many of these groves are being invaded by other tree species. In the Sierra Nevadas of California, protection of sequoia groves from fire has fostered a dense understory of shrubs and other trees, increasing the danger of severe damage when a fire inevitably does come (Biswell 1989). The same is true of many other forest and shrubland communities in western North America.

The recognition that fire is a natural, and often beneficial, ecological factor has led to policies of allowing natural fires to burn in some areas (the so-called "Let-Burn Policy"), as well as gradually reintroducing fire to forests by prescribed burning (Kilgore 1973). In sequoia groves, these efforts have resulted in the reduction of crown fire danger (Kilgore & Sando 1975).

In the summer of 1988, forest fires burned about 36% of Yellowstone National Park and surrounding areas of National Forest (Fig. 6.3). The stage for this event was set by a series of dry winters leading up to the driest summer since the Dust Bowl years of the 1930s. When fires of natural origin began, the park followed a policy designed to restore fire as a natural factor in the park's ecology. This policy called for natural fires that occurred within specified weather and fuel conditions to be classified as **prescribed natural fires** and allowed to run their course, unless they threatened buildings or other facilities. In 1988, however, fires that were initially classed as prescribed natural fires were converted by dry, windy weather into virtually uncontrollable fire storms. Although $125 million was spent in efforts to halt the fires, they were ultimately stopped only by autumn snows. Successional recovery of burned areas began in the following summer at lower elevations with a flush of growth of herbaceous plants and tree seedlings. At many higher elevations, however, recovery has proceeded very slowly. The fires caused little mortality of large animals, and in the long run have benefitted populations of elk and deer (Fig. 6.4), but their aesthetic impacts raised great controversy over fire policy in western parks.

Analysis of forest history of Yellowstone Park suggests that fires of comparable scale have occurred at intervals of 200 to 300 years (Romme & Despain 1989). Following fire, single-species stands of lodgepole pine tend to develop, due to adaptations of this species for post-fire seed dispersal and germination. In about 150 years, these stands reach maturity and some large trees begin

Figure 6.4

The Yellowstone National Park fires of 1988 caused little mortality of elk and other large animals, which began to forage in burned areas as soon as new plant growth appeared.

(U.S. National Park Service photo, courtesy of D. Despain)

to die, creating openings that are invaded by Engelmann spruce and subalpine fir. From then on, death of pines and growth of young spruces and firs creates a forest condition highly vulnerable to fire. Fallen dead trees create a heavy fuel load on the forest floor, and living trees of varying sizes provide a route for fire to climb into the canopy. In the 1980s, these conditions had developed over much of Yellowstone Park, largely as a result of natural successional processes.

Many ecologists argue, however, that although such fires may be part of the original natural order, they now represent an excessive disturbance of the small areas of natural landscape that remain in parks and national forests. Even in an area as large as the Greater Yellowstone Ecosystem, they feel that active management must be applied (Brussard 1991). A review of federal policies after the 1988 fires, furthermore, concluded that the fire management plan for Yellowstone National Park, developed to implement the policy of restoring fire as a natural force, was inadequate. In particular, its criteria for designating fires as prescribed natural fires that should be allowed to burn, as opposed to wildfires that should be controlled, were inadequate (Wakimoto 1990). Thus, efforts to use prescribed burning as a tool to manage the successional state of forests are sound, but must be designed to prevent large fractions of the landscape from becoming highly vulnerable to fires like those of Yellowstone National Park in 1988.

GLOBAL CHANGE IMPACTS ON FORESTS, WOODLANDS, AND SHRUBLANDS

Global warming scenarios based on a doubling of atmospheric CO_2 suggest that the climatic ranges of many temperate forest trees will shift northward by 500 to 1,000 km (Fig. 6.5; Davis & Zabinski 1992). This shift exceeds the natural dispersal ability of forest species. Disturbance regimes will also be altered, perhaps affecting forests more than the direct changes of climate (Franklin et al. 1992). The frequency and intensity of wildfires, severe storms, and outbreaks of pests and pathogens are likely to be altered, especially in areas suffering climatic stress due to warmer and drier weather. Exotic species are likely to be favored under such conditions, particularly in regions of Mediterranean climate, such as California (Westman & Malanson 1992). If climatic change of such magnitude occurs, the threat to biodiversity will be great, and preserves established under 1900s climate will be inadequate. Active programs to enable species to colonize areas that have become climatically favorable, and new strategies for biodiversity preservation, will be essential (Botkin & Nisbet 1992).

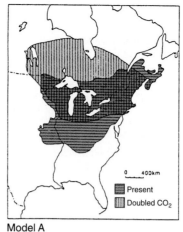
Model A

Model B

FIGURE 6.5

Predicted shifts in the climatic zone favorable to sugar maple in eastern North America by two global atmospheric circulation models, under an assumption of a doubling of atmospheric CO_2.

(From M. B. Davis and C. Zabinski, 1992, "Changes in Geographical Range Resulting from Greenhouse Warming: Effects on Biodiversity in Forests," 297–308 in R. L. Peters and T. E. Lovejoy, Eds., *Global Warming and Biological Diversity.* Copyright © 1992 Yale University Press, New Haven CT. Reprinted by permission.)

SUMMARY

1. Biotic succession is a key process in forest, woodland, and shrubland ecosystems. Where succession is an extended process, species tend to become specialized for early or late successional stages.

2. Humans have converted much of these ecosystems to agriculture, tree farming, and urban use. The remaining forest tends to be fragmented and lacking in representation of old-growth stands. Fire-exclusion has increased the vulnerability of these ecosystems to severe fire. Exploitation of these ecosystem types disturbs soil conditions and reduces species diversity. Air pollution and the effects of invasive exotic plants, insects, and diseases are additional human impacts.

3. Managing biotic succession is necessary for conservation in forest habitats. Maintaining the full range of successional and climax communities in an appropriate balance is a key strategy. Late successional ecosystems, particularly old-growth forests, contain highly specialized species that must be provided adequate protection from disturbance. Early successional ecosystems, however, contain many species with high exploitation potential.

4. A major challenge in management of these ecosystems is maintaining a normal disturbance regime, particularly involving fire. Even in the largest protected areas, such as Yellowstone National Park, active management is probably necessary to maintain a reasonable disturbance regime.

5. Global climatic change will likely modify the climatic zones suitable for forest species faster than these species can move on their own. Human assistance will be required to help these ecosystems adapt to changed conditions.

QUESTIONS FOR DISCUSSION

1. What specific ecological conditions do you think managers should strive to maintain in forests, woodlands, and shrublands on public lands?

2. Was the Yellowstone Ecosystem subject only to nonhuman ecological forces prior to European settlement of North America? Do you think that this ecosystem can now be managed by letting "natural" ecological processes operate without restraint?

3. What do you think the criteria should be for determining how much old-growth forest is preserved in western North America?

SUGGESTED READING

Keeley, J. E. and C. C. Swift. 1995. *Biodiversity and ecosystem functioning in Mediterranean-climate California.* Pp. 121–83 *in* G. W. Davis and D. M. Richardson (Eds.), Mediterranean-type ecosystems: The function of biodiversity. Springer-Verlag, New York, NY. A discussion of how species diversity is influenced by environmental conditions, particularly fire, in California shrublands and woodlands.

Maser, C. 1994. *Sustainable forestry: Philosophy, science, and economics.* St. Lucie Press, Delray Beach, FL. A comprehensive examination of the interaction of science and society in developing management policies for North American forest ecosystems.

Wilcove, D. S. 1994. Turning tangible goals into tangible results: The case of the spotted owl and old-growth forests. Pp. 313–29 *in* P. J. Edwards, R. M. May, and N. R. Webb (Eds.), *Large-scale ecology and conservation biology.* Blackwell Scientific Publications, London, England. A review of the issue of the northern spotted owl and the old-growth forests of the Pacific Northwest, and how this issue relates to the idea of ecosystem management.

chapter

7

Grasslands and Tundra

G rassland and tundra ecosystems lie between forests and deserts—broad-leafed forests and the arid mid-latitude deserts in the case of temperate grasslands, the coniferous forests and the cold polar or alpine deserts in the case of tundras. Both grasslands and tundras are dominated by grasses, or by grasslike plants such as sedges, together with broad-leaved herbaceous plants and low shrubs. The faunas of grasslands and tundras are also quite similar, and these ecosystem types share many of the same conservation problems.

In this chapter we shall first examine the distribution and some of the important ecological characteristics of these ecosystems. Then we shall consider how human activities have affected the plant and animal life of grasslands and tundras. Finally, we shall identify key strategies for the restoration and conservation management of these systems.

Reintroduction of grazing by bison is a key strategy in restoration of grassland ecosystems in central North America.

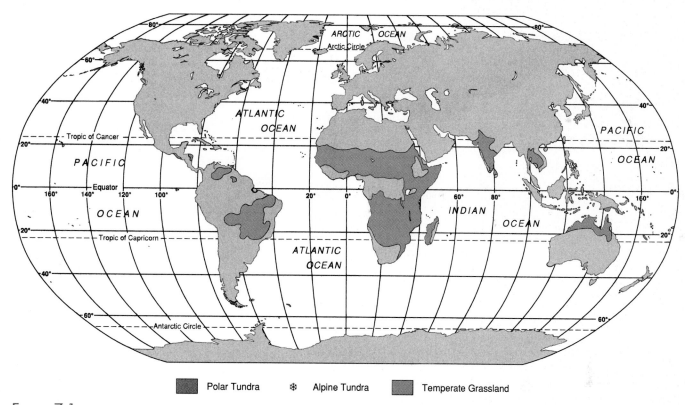

Polar Tundra   ❋ Alpine Tundra   Temperate Grassland

FIGURE 7.1

The distribution of temperate grasslands and arctic and alpine tundra in North America and other regions of the world.

DISTRIBUTION OF GRASSLANDS AND TUNDRA

Temperate grasslands occur in regions with strongly seasonal precipitation varying roughly between 25 and 75 centimeters per year (Fig. 7.1). Grassland climates nearly always exhibit one or more dry seasons severe enough to exclude most kinds of trees and shrubs. In some regions the dry season is summer, in others winter, and in still others spring and fall. These dry seasons also cause fire to be an influential environmental factor in many places. Frequent fire eliminates all but the most resilient species of trees and shrubs, and in many places prevents the replacement of grasslands by communities dominated by woody plants.

North America has several important grassland regions. In all of these areas the original grasslands were dominated by perennial **bunch grasses,** species that grow in dense clumps or large patches, rather than as a continuous turf. The most extensive grassland region occupies the Great Plains, where precipitation tends to occur both in summer and winter, but severe dry periods are frequent in spring and fall. In the Great Plains, total precipitation declines from east to west and from south to north. In the eastern and southern plains, tall grasses predominate, so that this region is known as the **tall-grass prairie** (Fig. 7.2). The leafy stems of many of these bunch grasses grow to heights of 1–2 m, and their flowering stalks even higher. The tall-grass prairie is a fire-dependent ecosystem, originally having been maintained and stimulated by frequent fire (Collins & Wallace 1990). Farther to the west and north lies the **mid-grass**

FIGURE 7.2

The tall-grass prairies of the eastern and southern Great Plains were dominated by bunch grasses with flowering stalks that reached more than 2 m in height.

(Photo by G. W. Cox)

prairie, where the dominant species tend to be of intermediate height. The dry western plains are home to the **short-grass prairie,** dominated by low-growing species of bunch grasses. Both the species of grasses and the luxuriance of growth of individual species change along this gradient.

Grasslands also occur on the Columbia Plateau of eastern Washington, Oregon, and western Idaho, and in the northern

FIGURE 7.3

The temperate grasslands of South America possessed a distinctive fauna of large herbivores, such as guanacos and rheas.

(Photo by Darla G. Cox)

FIGURE 7.4

The Afro-alpine vegetation of Mt. Elgon in western Kenya is dominated by tussock grasses and giant groundsels of the genus *Senecio,* of tropical affinity, rather than by arctic plants.

(Photo by G. W. Cox)

Great Basin and western Wyoming, a region known as the **palouse prairie** or **sagebrush steppe.** In California, **valley grassland** originally covered most of the central valley and the coastal zone in the south. In these regions, most precipitation comes in the winter. Finally, in southern Arizona, southern New Mexico, and northern Mexico lies a region of **desert grassland,** again with a combination of winter and summer rainfall, but dry conditions in spring and fall.

Similar grasslands exist on other continents. These include the steppes of Asia, the pampas of southern South America, the grassveld of southern Africa, and grasslands of southern Australia. These areas share basic features of grassland ecology, although the species that make up these grassland ecosystems are almost entirely different (Fig. 7.3).

Tundra ecosystems occur where severe cold, strong wind, and, in some cases, a permanently frozen subsoil zone known as **permafrost,** combine to exclude large woody plants (Billings 1973). **Arctic tundra** (polar tundra) occurs throughout the lowlands and hills of northernmost North America and Eurasia (*See* Fig. 7.1). Arctic tundra areas experience a very short summer with long days, and a long, cold, dark winter. Permafrost is a characteristic feature, contributing to cold, wet soil conditions and poor drainage. Ecosystems similar to those of arctic tundra also exist at high latitudes in some locations in the southern hemisphere. **Alpine tundra** occurs at high elevations in mountains at all latitudes. Day length and seasonal change in temperature and precipitation vary with latitude and other factors, and are often quite different from conditions of the arctic tundra. Permafrost is rare in alpine areas, and wind, rather than cold, is sometimes the critical factor that excludes larger woody plants.

Arctic tundra vegetation is dominated by perennial herbaceous plants such as sedges, low-growing forbs, and sprawling or creeping woody plants. In the northern hemisphere, the organisms inhabiting alpine tundra areas are often closely related to those of the arctic tundra, but in high mountain areas of the

tropics and southern hemisphere, they are derived mostly from lower elevation groups of organisms characteristic of those regions (Fig. 7.4).

ECOSYSTEM DYNAMICS IN GRASSLANDS AND TUNDRA

In the past, range ecologists have generally assumed a traditional model of grassland dynamics based on the process of biotic succession. This model assumed a "climax" state in equilibrium with climate and a grazing regime of native herbivores. Under this model, changes due to drought or overgrazing by livestock should be self-correcting by natural succession once the unusual stress is removed. The concept of grassland as equilibrium systems has been questioned by many recent ecologists, given the fact that these ecosystems are subject to great variability in rainfall, temperature, natural grazing pressure, and fire effects. Others (Westoby 1989; Laycock 1991) have shown that grassland ecosystems in many regions are better viewed as having **multiple stable states,** between which strong environmental influences may cause the ecosystem to shift. Overgrazing and fire exclusion, for example, may permit shrub invasion of grassland, leading to virtually complete replacement of grasses by woody plants. Changed, as well, are soil microbiota, insects and other invertebrates, and vertebrates. Under the new pattern of function the altered vegetation may be maintained even if grazing is eliminated and fire reintroduced.

In both grassland and tundra ecosystems, plant production usually is restricted to less than half the year by highly seasonal temperatures or rainfall. In arctic tundra areas, the growing season is often less than three months. In both ecosystems, grazing or browsing herbivores, ranging in size from meadow voles to bison and caribou, are prominent, and their feeding activities are year-round (Fig. 7.5), as are those of their important predators.

FIGURE 7.5

Barren-ground caribou are the most common large herbivore of the arctic tundra of North America.

(U.S. Fish and Wildlife Service photo by F. Mauer)

Climatic variation from year to year is also great. As Oksanen (1990) has shown, the strong seasonality of primary production predisposes these ecosystems to major fluctuations of herbivore populations.

One of the striking phenomena of tundra ecology, shown also to a significant degree in grasslands, is **population cycles** of many vertebrate herbivores and their predators. These fluctuations have attracted the attention of ecologists for more than a century, and many theories of their cause have been proposed. Two types of cycles are commonly distinguished, differing in the average interval between peak populations. In North America, cycles of 3–4 years are shown by various species of lemmings and meadow voles (a group of mammals collectively known as *microtine* rodents), together with their predators, such as the arctic fox and the snowy owl. Cycles of meadow voles of the genus *Microtus* occur both in the arctic tundra and temperate grasslands. In Eurasia, some kinds of grouse and ptarmigan are also reported to show cycles of 3–4 years duration. Cycles of 9–10 years are also shown in North America by northern hares, particularly the snowshoe hare, and several species of grouse and ptarmigan, as well as by their predators, particularly the Canadian lynx and red fox (Keith 1963). In northern grassland regions, species such as the sharp-tailed grouse and the black-tailed jackrabbit show 9–10-year cycles (Keith 1963; Anderson & Shumar 1986).

Larger tundra animals may show even longer cycles, although this is somewhat speculative. Caribou, for example, were abundant in Alaska in the mid-1800s, according to early prospectors and settlers. In the 1880s the populations of caribou declined. By the 1930s their numbers had recovered, but once again, in the 1970s, serious declines occurred. In western Alaska, for example, the caribou herd that migrates between a summer range on the arctic slope and a winter range south of the Brooks Range crashed from 242,000 in 1970 to 75,000 in 1976. In the 1980s Alaskan caribou populations showed strong recovery, and the western Alaska herd reached 230,000 by 1986 (Morgan 1988).

Whether or not caribou show true cycles is uncertain, however, due to the small number of peaks and lows that have been observed and the poor quality of information prior to the 1930s.

Some biologists doubt that fluctuations of smaller species are true cycles, with regularity of period and amplitude. Arctic lemmings, in fact, may show little more than chaotic fluctuations of numbers (Oksanen 1990). Nevertheless, these fluctuations are real, and are an important aspect of tundra and grassland ecology.

Many causal theories of 3–4 and 9–10-year cycles have been proposed and discarded by ecologists. Three major theories of 3–4 and 9–10-year cycles are now active. The first, the **herbivory hypothesis,** suggests that the interaction is between herbivores and their plant foods, and that populations of predators simply follow the abundance of their prey. Lemmings, northern hares, and caribou are active year-round, but their food plants grow for only about three months. In temperate grasslands, plants grow only during periods of warmth and moisture, again only a few months of the year. Thus, if herbivore populations reach high levels, they can exploit plants for a long time before plant recovery can even begin. In the arctic, nutrients may also become tied up in animal tissues and wastes, their recycling delayed by the cold, damp soil conditions that slow decomposition. Oksanen (1990) suggested that these mechanisms can account for fluctuations of herbivores with short generation times in habitats of low primary productivity.

Other biologists support the **food supply–predation hypothesis** as the cause of cycles. Keith (1981), for example, proposed that populations of snowshoe hares grow until they are halted by shortage of winter food. Populations of predators, which have also been growing because of the increased availability of snowshoe hares, are then able to catch up with the hare population. Starvation of hares and predation then reduce hare numbers to levels that permit plant recovery. To test this hypothesis, Keith et al. (1984) studied the ecology of snowshoe hares during the decline phase of a population cycle in Alberta, Canada. They showed that plant foods were in short supply, and that hares showed clear signs of malnutrition. Starvation of hares probably occurred only for a short time after the population had peaked, however, and the cause of death in most cases was predation. Keith et al. (1984) interpreted this to support the food supply–predation hypothesis, since both malnutrition and predation interacted to cause the decline of the hare population. Theoretical analyses by Oksanen (1990) suggest that truly cyclical fluctuations of herbivores may occur by this mechanism in productive, highly seasonal environments where specialized predator-prey interactions exist.

Still other biologists favor the **predation hypothesis,** which suggests that food shortage rarely affects herbivores such as lemmings and snowshoe hares, and that the basic cause of cycles is predation. In other words, without the action of predators, no cycles would occur. Trostel et al. (1987), for example, supported this hypothesis for snowshoe hare cycles. Long-term experimental studies in the Yukon Territory of Canada, designed to manipulate food availability and predation intensity, suggest that several relationships may be at work. These studies suggest that relative shortages of food may occur, predisposing

TABLE 7.1

Major Categories of Human Impact on Grassland and Tundra Ecosystems		
	GRASSLANDS	TUNDRAS
Conversion to Cropland	Extensive	Negligible
Grazing by Domestic Animals	Extensive	Slight
Physical Disturbance of Soil	Severe	Locally severe
Change in Fire Frequency	Reduced	Increased
Introduction of Exotic Plants	Extensive	Slight
Hunting and Animal Control	Extensive	Extensive

hares to predation by forcing them to move about more in search of food (Smith et al. 1988). Responses of shrub defenses to hare browsing do not seem to correlate with the population cycle, however, and crashes occur even when hare populations receive supplemental food, suggesting that predation is the key cause (Sinclair et al. 1988).

The fluctuations of populations of large animals, such as caribou, are likely to be the result of more complex interactions, but they may also be a natural phenomenon in tundra and other northern ecosystems. As suggested by the difference in cycle length for lemming and snowshoe hares, the size of the herbivore may be a strong determinant of the length of a population cycle. For mammals, it has been hypothesized that cycle length is approximately equal to 8.15 times the body mass (in kilograms) to the exponent 0.26 (Peterson et al. 1984). For an animal of 100 kilograms, about the size of a caribou, this suggests a cycle length of 27 years, or possibly longer.

HUMAN IMPACTS ON GRASSLAND ECOSYSTEMS

Humans have made an impact on grassland ecosystems in several different ways, most of them directly or indirectly related to ranching and farming (Table 7.1). In North America, most of the tall-grass prairie and much of the palouse prairie have been converted to rain-fed or irrigated farming. In Illinois, for example, only fragments of the original grasslands now exist. Surviving remnants tend to occur on the thin soils of bluffs (hill prairies), along railroad rights of way ("cinder prairies"), and in cemeteries. Elsewhere in North America, particularly in central California, most of the original grasslands have been replaced by irrigated farming operations. In North America, the conversion of grasslands to croplands began in the early 1800s. In some other world regions, this conversion has been less extensive and more recent. In the Argentinean pampas, for example, the conversion has accelerated greatly during the 1980s and 1990s.

In North America, livestock grazing preceded intensive farming in many grassland regions, such as the western Great Plains, the Columbia Plateau, the California grassland, and the desert grassland. The grazing patterns of cattle, sheep, and other livestock differ from those of native large herbivores, such as bison, pronghorn antelope, and elk. The overall grazing pressure under livestock tends to be more selective. As a result, many of the preferred species of plants decrease in abundance, and the

nonpreferred species increase. Range managers term these kinds of plants **decreasers** and **increasers.** Plants that increase under heavy grazing include woody species such as mesquite, juniper, and many species of shrubs and cacti. In some regions, such as the palouse prairies of the Columbia Plateau, hoofed animals were originally scarce, and the dominant grasses were not adapted to heavy trampling, as they appear to be in the Great Plains (Mack & Thompson 1982). With the introduction of livestock to the Columbia Plateau, the native perennial grasses have largely disappeared in many areas.

In addition, many kinds of weedy plants, especially species from the Old World that evolved in close association with early agriculture over the past 12,000 years or so, are able to invade disturbed and overgrazed grasslands. Range managers call these kinds of non-native species **invaders.** In the palouse prairie region, cheatgrass, *Bromus tectorum,* was introduced accidentally in the late 1800s, and spread through the region by 1928 (Mack 1981). This aggressive annual grass, native to Eurasia, has replaced native grassland species over vast areas. It forms a dense, highly flammable ground layer in areas with sagebrush and other shrubs that do not tolerate fire. When such a site burns, the result is a virtual monoculture of cheatgrass (Billings 1990).

In California, the original valley grassland was so quickly and so completely replaced by introduced Mediterranean annual grasses and forbs that its original composition is very uncertain. This is an example of the shift of a grassland ecosystem to an altered stable state.

Grassland areas are also subject to extended droughts. During such periods a major shift in the dominant species of grasses can occur, with species adapted to lower rainfall becoming more abundant. Heavy grazing under such conditions can severely damage rangeland, as demonstrated during the droughts of the 1930s and 1950s in the Great Plains (Albertson et al. 1957). During these dry years, heavily grazed ranges experienced a much higher decline in total plant cover and a much greater invasion by exotic annual weeds. Whereas well-managed ranges were able to recover from drought in about 5 years, heavily grazed areas required 20 years or more to recover.

The original role of fire has also been modified in most grassland regions. As the land was settled, the common Indian practice of burning grasslands was eliminated, and natural wildfires were deliberately controlled. In many locations, the reduced frequency of fire favored the invasion of grasslands by woody species, to the extent that many former grasslands have now become

FIGURE 7.6

Control programs for prairie dogs have greatly reduced their numbers, and some populations of the white-tailed prairie dog are now formally designated as threatened.

(Photo by Darla G. Cox)

FIGURE 7.7

The black-footed ferret, a specialist predator on prairie dogs, has been reduced to the brink of extinction by control of prairie dog populations on rangeland in western North America.

(U.S. Fish and Wildlife Service photo by Luther C. Goldman)

forests, shrublands, or desert scrub. In southern Texas, New Mexico, and Arizona, for example, the livestock grazing and the reduction of grassland fires have led to the invasion of much of the original grassland by mesquite, cholla cactus, and creosote bush (Humphrey & Mehrhoff 1958; Bock & Bock 1995).

Finally, grassland areas have been the sites of some of the most intensive hunting and deliberate control of animal populations. In the 1800s, the large native grazing animals of the Great Plains, such as bison, elk, and pronghorn, were almost completely eliminated by hunting. The original bison population of North America is estimated to have been 30 to 60 million; all but a few hundred were killed for their skins, meat, or simply to destroy the basic food on which groups of Plains Indians depended.

The past century has also seen massive campaigns to control populations of prairie dogs (Fig. 7.6) and other native animals. Prairie dog towns originally covered more than 40 million hectares of the mid- and short-grass prairies. The activities of these animals greatly modified the prairie vegetation, promoting broadleaf herbaceous plants at the expense of grasses. Intense grazing by prairie dogs tended to keep the vegetation in a young, actively growing, nutritious state. Although plant biomass within prairie dog towns often appeared sparse, primary production was, in fact, high. Both bison and pronghorn antelope, whose populations were also concentrated in the western Great Plains, were attracted to feed in these dog towns. Many other animals were associated with prairie dog towns, including the black-footed ferret, swift fox, burrowing owl, and mountain plover, and now are also declining (Miller et al. 1994). With the advent of cattle ranching, prairie dogs were viewed as competitors with cattle for grass. Organized campaigns to eliminate prairie dogs were thus undertaken.

Widespread control of prairie dogs was a major cause of the decline of the black-footed ferret (Fig. 7.7) to the edge of extinction. Near Meeteetse, Wyoming, where the last wild population lived in 1984, devastation of the local prairie dog population by plague caused the ferret population to crash.

The remaining animals were brought into captive breeding, and efforts are now being made to return them to the wild (*See* Chapter 26).

HUMAN IMPACTS ON TUNDRA ECOSYSTEMS

Tundra ecosystems have experienced some of the same kinds of impacts. Agricultural activities, however, have been a very minor cause of disturbance in tundra ecosystems. Crop cultivation is negligible in tundra regions. In northern Scandinavia, semi-domesticated reindeer are herded in tundra areas. Local efforts have been made to introduce reindeer herding to North America, and to develop ways of managing herds of musk oxen on tundra range. Some summer grazing of livestock also occurs in alpine tundra areas in the Temperate Zone.

Severe local impacts have resulted from disturbance of the tundra surface, both in arctic and alpine areas. Studies of the impacts of petroleum exploration activities show that the arctic tundra is sensitive to disturbance and very slow to heal. On the Alaskan north slope, recolonization of disturbed areas by native species is relatively rapid if the peaty surface layer of the soil remains intact, with 90%–100% plant cover often being achieved in 5–10 years (Van Cleave 1977). Where the organic surface layer is destroyed, succession may be much slower. In sites heavily disturbed by an exploratory drilling operation in 1949, effects were still severe 28 years later (Lawson et al. 1978). Where petroleum spills had occurred, or where increased thawing of the permafrost had been triggered by disturbance, very little recovery had occurred. In some places thawing of the permafrost led to surface subsidence and extensive erosion. In general, where disturbance of the tundra surface has led to permafrost melting, more than 30 years are required to restore surface stability (Lawson 1986). Even seismic surveys carried out in winter, when the tundra surface is frozen and snow-covered, crush and compact

FIGURE 7.8

The passage of vehicles conducting winter seismic surveys for evaluation of petroleum reserves on the Arctic National Wildlife Refuge creates trails that require many years for natural processes to heal.

(U.S. Fish and Wildlife Service photo)

the vegetated surface, often inducing permafrost melting and subsidence. Again, these effects are slow to heal (Fig. 7.8). These results suggest that major environmental damage may accompany exploitation of oil and gas reserves in areas such as the Arctic National Wildlife Refuge (*See* Reading 7.1).

Alpine tundra is likewise sensitive to physical impacts. Trampling by visitors on foot in alpine areas in the Rocky Mountains can cause severe impacts on the vegetation (Willard & Marr 1970; Billings 1973). Where foot traffic becomes concentrated along regular routes, tundra plants quickly die, and the soil surface begins to erode.

Fire in the arctic tundra seems to have been increased by human activities, contrary to the pattern in temperate grasslands. Although it might seem that cold, wet conditions in the arctic would make the vegetation hard to burn, the truth is that many tundra plants possess large quantities of resinous materials on their surfaces, making them highly flammable.

Hunting of animals such as caribou, and control of predators such as the wolf, are additional human impacts. The declines of caribou populations in the 1970s raised major conservation concerns. This was a period of cultural and economic change in many native villages of arctic Alaska and Canada. The winter hunting capabilities of Inuit and Indian peoples were increased considerably by the introduction of snowmobiles and better high-powered rifles, and some instances of wasteful killing occurred. Tundra fires, caused both by lightning and by human activity, were more frequent and extensive than usual, too. Although fire may be beneficial in the long term, by increasing habitat diversity along the border of forest and tundra, its short-term effect is the destruction of ground lichens and other caribou browse (Klein 1982). With decline in caribou numbers, as well, predation by wolves and other predators became relatively greater in importance (*See* Chapter 14). The high caribou populations of the 1930s through 1960s may also have overgrazed food plants and caused deterioration of range quality.

IMPLICATIONS OF GLOBAL CHANGE FOR GRASSLANDS AND TUNDRA

High-latitude ecosystems will likely experience the greatest degree of warming due to increases in greenhouse gases (*See* Chapter 23). Major ecological changes are thus likely to occur in the arctic tundra. Higher temperatures and a longer growing season will probably favor increased plant growth and the invasion of species from more southern regions. Increased depth of thaw into the permafrost layer may also destabilize large areas of wet coastal tundra, perhaps catastrophically (Billings & Peterson 1992). Warmer climate is likely to favor the northward advance of treeline in the arctic, and its upslope movement in mountain areas. Arctic tundra may thus be displaced in its southern areas, where populations of tundra mammals and birds are highest. The extent to which these losses will be offset by improved conditions in higher latitudes of the arctic is uncertain. In mountain areas at lower latitudes, alpine tundra may disappear from some areas.

Global change may cause positive feedback changes in the arctic. Some evidence suggests that climatic warming may lead to increased decomposition of organic matter stored in peaty tundra soils, thus increasing the rate of release of CO_2 into the atmosphere. Data from 1983 through 1990 suggest that the north slope tundra of Alaska has already become a net source of CO_2, rather than a CO_2 sink (Oechel et al. 1993).

GRASSLAND AND TUNDRA RESTORATION

Some of the earliest ecosystem restoration efforts were directed at tall-grass prairie in the midwestern United States, where these ecosystems had been almost completely destroyed (Kline & Howell 1987). In several locations, tall-grass prairies have been recreated on sites from which all native prairie species had been eliminated by farming. The techniques used in restoration have been derived largely from agronomy. Herbicides and mechanical cultivation have usually been used to eliminate existing nonprairie plants. Dominant prairie grasses and forbs are then seeded into the cultivated soil or transplanted individually onto the site.

Once a healthy stand of prairie dominants is created, efforts turn to adding typical, but less common, prairie species and to eliminating exotic herbaceous plants and native shrubs and trees that often tend to invade aggressively. Prescribed burning is often effective in reducing these nonprairie species (Fig. 7.9), but complete elimination of exotics is difficult. This is particularly true on small prairie plots where the use of fire is often complicated by restrictions on weather conditions under which burning can be done, and where seed sources for exotics may be abundant in neighboring areas. Most of these restoration efforts have been limited to recreation of the plant

THE ARCTIC NATIONAL WILDLIFE REFUGE

The Arctic National Wildlife Refuge, 7.7 million hectares in area, is located on the north slope of Alaska, bordering the Arctic Ocean. To the west lies the Prudhoe Bay oil field, and to the east the Canadian border. Much of the interior portion of the refuge is designated as wilderness, but 0.6 million hectares of the coastal plain, the "1002 area," is not so designated.

Environmentalists characterize the refuge as an "American Serengeti," because of its populations of caribou, musk ox, grizzly bear, gray wolf, arctic fox, waterfowl, shorebirds, and other animals. Others, including some U. S. legislators, consider the refuge

the most godforsaken land that you have ever looked at with the naked eye on the face of the Earth.

The refuge coastal plain is the summer calving ground for about 160,000 caribou belonging to the Porcupine Herd. This herd migrates between the refuge and wintering areas to the south and east, largely in Canada. Several thousand animals are harvested annually by Native American residents of the United States and Canada.

For a decade, oil companies have sought approval to drill for oil in the 1002 area of the coastal plain. Most experts estimate that about 3.2 billion barrels of oil could be extracted from this area, at a cost of about $10 per barrel less than the general world market price—$32 billion profit for the companies, or possibly savings for consumers. This would be about a third of the yield of the Prudhoe Bay oil field, now about two-thirds depleted. Oil from the 1002 area could be routed into the existing Alaska pipeline without much difficulty.

The wet coastal tundra is one of the most sensitive and least resilient ecosystems on earth. Seismic exploration of the refuge was carried out in the winter months of 1984 and 1985. Several kinds of vehicles, together with ski-mounted trailers, were used to survey 2,000 km of seismic lines.

This activity alone has created a major long-lasting impact on the tundra (Felix & Reynolds 1989; Felix et al. 1992). Even though the seismic work was done when the tundra surface was frozen, vegetational damage was often very great, especially on moist tundra dominated by sedges and dwarf shrubs. After four to five growing seasons little recovery was evident, and the effects on moist sedge-shrub tundra appeared to exceed the threshold beyond which recovery is impossible. Felix et al. (1992) concluded that continued exploration and development could lead to cumulative impacts that would change the basic character of the tundra ecosystem.

Environmentalists also fear that production activity and the increased road access to the refuge, eventually drawing visitors in vehicles, would modify the migration and calving activities of the caribou. One possibility is that the herd would be pushed into the interior hills during the calving period, where the risks of predation by grizzly bears and wolves are higher.

FIGURE 7.9

Prescribed burning is used to reestablish the natural role of fire in tall-grass prairie on the Konza Prairie Research Natural Area in the Flint Hills of northeastern Kansas. Fires are ignited around the periphery of a designated burn unit, and extinguish themselves when they reach the center.

(Photo courtesy of D. W. Kaufman)

community, with the assumption that success with plants will lead to colonization by the prairie animals that can exist on areas of the size in question.

At the University of Wisconsin, prairies 24 and 16 hectares in area have been created, the first efforts beginning in 1934 (Cottam 1987; Howell & Jordan 1991). These sites now contain over 350 species of native prairie plants. Periodic observations have shown that major changes in composition patterns occur as newly introduced native species find the microhabitats that are optimal for them, and as short-term climatic cycles influence microhabitat conditions within the landscape.

In a few locations, such as Grasslands National Park in Saskatchewan, Konza Prairie Research Natural Area in Kansas (Reichman 1987), the Niobrara Valley Preserve in Nebraska, and the Tallgrass Prairie Preserve under development by the Nature Conservancy in Oklahoma, larger areas of native grasslands are being restored. These areas have been used for livestock grazing but not extensively disturbed by cultivation. In these areas, the objectives are to establish prairie ecosystems containing the full complement of native animals, such as bison, pronghorn antelope, and elk (Fig. 7.10).

Little success has been achieved in restoration of tundra ecosystems. In the arctic, deliberate efforts to revegetate disturbed sites, such as the thick gravel pads on which roads,

FIGURE 7.10

Bison, the dominant original herbivore of Great Plains grasslands, are being reintroduced from areas such as the Wichita Mountains National Wildlife Refuge, the Tallgrass Prairie Preserve in the Osage region of Oklahoma, and Konza Prairie in the Flint Hills of Kansas.

(U.S. Fish and Wildlife Service photo)

camps, and work sites are placed in order to prevent melting of the underlying permafrost, have not been very successful. Seeding such areas with non-native perennial grasses can create a temporary cover that helps stabilize the disturbed soil surface (Van Cleave 1977), but colonization by native species is very slow (Bishop & Chapin 1989a). Current efforts involve planting native species, together with fertilization of the site and its surroundings, to encourage natural succession to occur as fast as possible (Bishop & Chapin 1989 a, b).

In alpine tundra areas restoration is somewhat more successful. A number of alpine grasses and a few forbs can be seeded onto disturbed areas, and many others can be reintroduced by transplantation (Brown et al. 1978). Seeding or transplanting is usually done in the late fall so that the species are able to begin growth as early the next spring as possible. Mulching and fertilization can aid the establishment of such species, as well. When fertilization is stopped, however, such plantings often tend to decline, indicating that restoration must consider ecological processes beyond those of initial establishment.

SUMMARY

1. Temperate grasslands occur in environments with low to moderate, highly seasonal rainfall. Tundra ecosystems are found at high latitudes, where long winters and wet soils underlain by permafrost prevent tree growth, and at high elevations, where severe conditions of temperature, wind, and soil exclude trees.

2. Many tundra and grassland mammals and birds show large population fluctuations or cycles, most commonly 3–4 or 9–10 years in their period. Whether the main cause of cycles is interaction of herbivore and plants, herbivore and predators, or all three groups, is still uncertain.

3. Human impacts on grasslands include conversion to cropland and livestock grazing land, reduction of fire frequency, killing of native animals, and introduction of exotic plants. These impacts have eliminated much of the tall-grass prairie and greatly changed the structure and function of other prairie types.

4. In tundra areas, human impacts largely have involved physical damage to the fragile vegetative cover, overexploitation of large animals, and increase in fire frequency. Global climatic warming may cause major changes in tundra environments.

5. Restoration of grassland ecosystems, including the reintroduction of large native ungulates, is under way in several locations in the Great Plains. Restoration of tundra ecosystems is more difficult, and has achieved only limited success.

QUESTIONS FOR DISCUSSION

1. Do you think productive and diverse native grassland ecosystems can be maintained under current grazing regimes of ranching in western North America?

2. Do you think that the low inertia and resilience of the wet coastal tundra of the Arctic National Wildlife Refuge are reason enough for prohibiting oil drilling and production in the refuge?

3. What are the implications of the great ecological changes imposed on grassland ecosystems by introduced exotic grasses and forbs for the strategy of managing preserves to recreate the conditions that existed prior to European settlement of these regions?

SUGGESTED READING

Felix, N. A., M. K. Reynolds, J. C. Jorgenson, and K. E. DuBois. 1992. Resistance and resilience of tundra plant communities to disturbance by winter seismic vehicles. *Arctic Alpine Res.* **22**:69–77. A scientific evaluation of the physical impacts of petroleum exploration activities on Alaskan wet tundra, with profound implications for management policy.

Joern, A. and K. H. Keeler (Eds.). 1995. *The changing prairie: North American grasslands.* Oxford Univ. Press, New York. A collection of 10 individually authored chapters on the ecology, cultural perception, and conservation biology of American grasslands.

Reichman, O. J. 1987. *Konza Prairie: A tall-grass natural history.* Univ. Press of Kansas, Lawrence, Kansas. A description of a native grassland in eastern Kansas, and its restoration and management as a tall-grass prairie research area.

Temperate and Tropical Deserts

Desert ecosystems exhibit perhaps the most challenging physical conditions in the terrestrial environment: extremes of both temperature and moisture. In many deserts, temperatures reach extremes at both ends of the scale, and although long periods of drought are frequent, cloudbursts and violent floods can occur. Despite these challenges, the diversity of plant and animal life in deserts is great, and patterns of adaptation to conditions of the physical environment are more striking than in any other terrestrial ecosystem.

Despite the intrinsic stability of desert ecosystems in the face of climatic variability—their ability to tolerate extremes of temperature, moisture, and wind—these systems are easily damaged by factors that disturb the soil surface. After such disturbances, furthermore, they show low resilience. Human impacts have degraded large areas of desert, and such areas may take decades or centuries to recover, even when protected (Webb et al. 1983). Restoration of disturbed areas is very difficult, and usually requires special techniques to create favorable microsites for plant growth and to provide supplemental moisture until perennial plants become fully established (Bainbridge & Virginia 1990). Because of the low productivity of desert environments and the sparse populations of plants and animals, large areas of land must be protected to guarantee the survival of desert forms (Louw & Seely 1982). Not surprisingly, some of the most serious problems of conservation ecology relate to desert ecosystems.

Desert plants and animals are extraordinarily diverse in their life forms and mechanisms of adaption to heat and aridity.

Outline

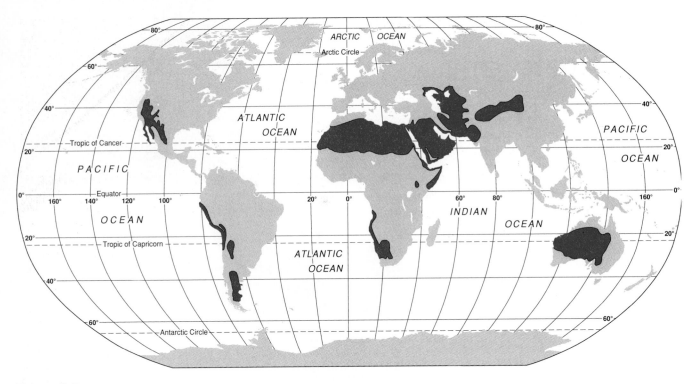

FIGURE 8.1

Distribution of deserts in North America and other world regions.

ORIGIN AND DISTRIBUTION OF DESERTS

Deserts have an annual rainfall of less than 25 cm, coupled with a potential evaporation from the soil surface and from plants well in excess of precipitation. Furthermore, rainfall is highly variable from year to year. As in grasslands, the precipitation that does occur tends to be highly seasonal. The wet season varies greatly in length and intensity in different desert areas. In some cases, rain falls mainly in winter, in other cases mostly in summer; still other deserts receive a little rain in both seasons. In some of the most extreme deserts, rains may fall only at intervals of several years, and seasonality is irrelevant. Temperature, too, varies among deserts, and ecologists commonly distinguish **hot deserts,** where freezing conditions are rare, from **cool deserts,** where winter temperatures regularly fall below freezing. **Cold deserts** also occur at very high latitudes (polar deserts) and at high altitudes, where precipitation is low and subfreezing conditions further limit water availability.

Three relationships contribute to the formation of deserts: 1) prevailing high atmospheric pressure, 2) rain shadows, and 3) cold coastal waters. At latitudes of 25°–30° N and S, subtropical high pressure patterns of the earth's atmospheric circulation system dominate most of the year. In these regions in the upper atmosphere, air flows converge from higher and lower latitudes and the air sinks downward toward the earth's surface. As it sinks, it warms by compression and declines in relative humidity—the actual amount of moisture in the air stays constant, but the capacity of the air to hold moisture increases. At the surface, the air is almost always hot and dry, and only very unusual

weather systems disturb this pressure belt and bring rain. An analogous system of high pressure dominates high arctic latitudes, creating desertlike patterns of low precipitation. The second mechanism operates where weather systems cross major mountain ranges. On the windward slope of the mountains, air is forced upward in elevation, and the expansion and cooling of the air leads to condensation of moisture, and thus heavy rains or snows. When the air has crossed the mountains, it descends in elevation, warming and decreasing in relative humidity. This **rain shadow** is usually severe enough to create desert conditions. Finally, where cold ocean currents exist along the western edge of continents, weather systems moving across these waters tend to cool and lose their moisture over the water. When they move onto the warmer land, however, the air warms and declines in relative humidity, thus creating dry conditions.

These mechanisms of aridity have produced deserts over about 12% of the continents (Fig. 8.1). In North America, the belt of subtropical high pressure contributes to the desert regions of northern Mexico and the southwestern United States—the more southern portions of the Chihuahuan and Sonoran deserts lie in this latitudinal zone. These are hot deserts, with the Chihuahuan having most of its rainfall in the summer, and the Sonoran Desert (Fig. 8.2) receiving either winter rain or a combination of winter and summer rain. The belt of subtropical high pressure is also responsible for the vast deserts of North Africa and the Middle East, together with the interior deserts of Australia.

The more northern parts of the Sonoran Desert, together with the Mojave and Great Basin Deserts, on the other hand, lie

FIGURE 8.2

The Sonoran Desert receives a combination of both summer and winter rain, and has one of the most diverse biotas of any desert region in the world. This scene is in spring in Organ Pipe Cactus National Monument, Arizona.

(U.S. National Park Service photo by Richard Frear)

FIGURE 8.3

Desert washes are stable in their general location, although individual channels are reworked by each flood.

(Photo by G. W. Cox)

largely in the rain shadow of the Sierra Nevadas, the Cascade Mountains, and the interior ranges of Nevada and Utah. Although the Mojave Desert is a hot desert and the Great Basin Desert is a cool desert, both receive their rain mostly in winter. The desert regions of Argentina and inner Asia are also examples of rain shadow deserts.

Cold upwelling oceans bordering the west coast of continents create the deserts of the Pacific Coast of Baja California in North America, the Atacama Desert of coastal Chile and Peru, and the Namib Desert of southern Africa. In these deserts, much of the moisture on which plant and animal life depends comes in the form of fog that blows inland, mostly at night, from the cold ocean waters. In these deserts one can find the seemingly contradictory association of cacti bearing epiphytic lichens and bromeliads, both largely dependent on moisture deposited by fog.

SUBSTRATE STABILITY IN DESERT ECOSYSTEMS

Substrate conditions, both physical and chemical, are a key determinant of the structure and function of desert ecosystems. Rocky hillslopes, eroded badlands, sandy washes, stony alluvial fans, fine-textured alluvial and lake bed deposits, and active sand dunes are all common desert substrates. Extreme conditions of salinity or alkalinity may develop in desert basins or poorly drained desert lowlands. Chemical leaching of carbonates may carry these minerals downward in the soil to a certain depth, where they are deposited to form a cemented horizon of material known as **caliche,** a formidable physical feature of the subsoil.

Despite the fact that thinly vegetated substrates of these varied types are exposed to fluctuating temperatures, frequent strong winds, and occasional violent rainstorms, desert landscapes undisturbed by humans develop a high degree of stability (Wilshire 1983). Major sand dune systems, although continually being reshaped in detail by wind, are stable. Their large-scale

structure is in equilibrium with the long-term pattern of prevailing winds and surrounding topography. Some dune plants and animals are dependent on stability of the major dune masses and interdunal flats, others depend on the ephemeral microhabitats created by the small-scale reshaping of the sands by wind. Desert washes are similar in their structure, varying in the pattern of small channels that change with each storm flood, but maintaining a stable position on the landscape. The vegetation of the floor of washes is often destroyed by the scouring of intense floods, but is quickly reestablished by plants whose seeds are stimulated to germinate by the physical abrasion that occurs during the flood itself (Fig. 8.3).

Many desert soils also tend to develop stabilized surface layers or crusts that reduce the vulnerability of the soil to wind and water erosion (Wilshire 1983). The fine roots of trees, shrubs, and herbaceous plants form a shallow, soil-holding network below the surface. In many places, removal of fine soil by wind erosion over thousands of years has concentrated pebbles and stones to form a **desert pavement**—a dense surface layer that protects the underlying soil from further erosion. More delicate **soil crusts** are also formed by several processes. Algae, fungi, and lichens can form a fine network of organic tissues at the soil surface, termed a **cryptobiotic crust.** Silt and clay particles can form a physically cemented surface layer in some soils. Precipitated salts may also form a protective crust over the surfaces of playas and alluvial lowlands. All of these layers and crusts are highly vulnerable to physical disruption by hoofed animals, motor vehicles, and human trampling. Once disturbed, cryptobiotic crusts probably require 200 years or so to reform (Belnap 1993).

HUMAN IMPACTS AND DESERTIFICATION

The low primary productivity and thin plant cover make arid and semiarid lands especially vulnerable to several human impacts: overgrazing, fuel-wood cutting, and physical disturbance of the soil surface. Heavy grazing by domestic animals such as cattle,

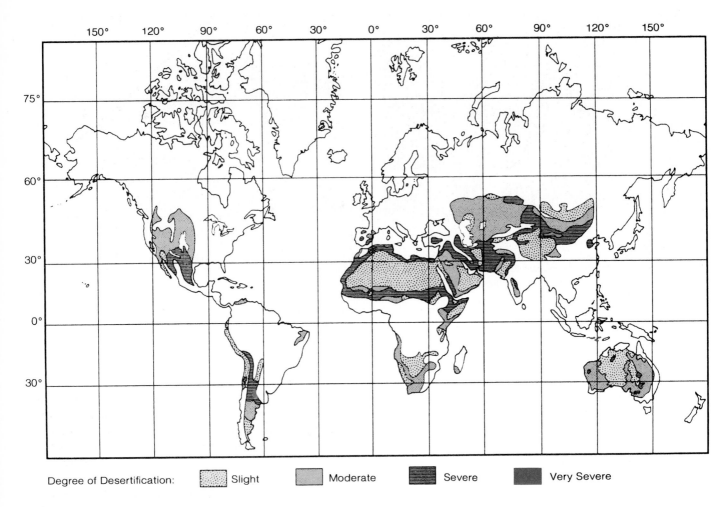

Degree of Desertification: [ ] Slight [ ] Moderate [ ] Severe [ ] Very Severe

FIGURE 8.4

Extent of desertification of world arid lands.

(Modified from Dregne 1983)

FIGURE 8.5

Desert bighorn sheep in many mountain ranges of the western United States have been decimated by diseases and competition from livestock and feral burros.

(U.S. Bureau of Land Management photo)

FIGURE 8.6
Feral burros are strong competitors with desert bighorn sheep because of their ability to browse woody plants heavily and their aggressive displacement of sheep from water holes.
(U.S. Bureau of Land Management photo)

sheep, goats, and camels, and by feral animals such as burros, can severely reduce the total cover and diversity of plants in desert ecosystems. Thinning of the plant cover and disturbance of the soil surface by trampling increase the vulnerability of the soil to erosion by wind and water. These changes also increase the patchiness of soil moisture and nutrients, which favors desert shrubs over grasses (Schlesinger et al. 1990). Once established, desert shrubs tend to reinforce this patchiness of resources and suppress the recovery of semiarid communities such as desert grasslands. In many developing areas, wood is the only fuel available, and wood-cutting is an additional pressure on arid-land vegetation. Other types of physical disturbance, such as cultivation, mining, and vehicular activity, cause even more serious degradation of desert landscapes.

Other major impacts result from irrigation. Desert soils, which lack a history of intense leaching by water, often contain large quantities of salts. Many desert regions also lack good drainage, and the water available for irrigation often contains substantial quantities of solutes. Consequently, when irrigation is undertaken without the construction of adequate drainage systems, the result is often waterlogged and salinized soils. In many areas, this has forced once-productive irrigated lands to be abandoned. Here, the original desert ecosystems and the productive agricultural systems that replaced them have both been lost.

Desertification is the degradation of arid, semiarid, and dry subhumid lands as a result of human activities and climatic variations. The effects of desertification are serious for both wildlife and human populations. In some severely desertified regions, such as the sub-Saharan zone of Africa known as the Sahel, famine has been a chronic problem since the late 1960s. Although the importance of human impacts relative to natural climatic cycles is still unclear in this region (Dodd 1994), the degree of human suffering is great. Desertification has thus become

a problem of major concern to international organizations, particularly the United Nations Environment Program. In 1994, an international **Convention to Combat Desertification, sponsored by the United Nations,** was signed by 90 countries (Kassas 1995). This agreement, which takes effect in 1996, establishes an organization to coordinate projects to protect and rehabilitate lands subject to desertification.

Desertification is not restricted to Africa, however, and surveys (Dregne 1983) indicate that almost 48% of the world's arid lands have suffered moderate to very severe desertification (Fig. 8.4). Slightly over 5% of this affected area consists of rainfed cropland, whereas almost 94% consists of rangeland. About 77% of the total rain-fed cropland in arid regions has experienced at least a 10% loss of crop productivity and 82% of rangeland is in fair-to-worse condition. About 675 million people inhabit regions that have become moderately to severely desertified; of these about 450 million have probably suffered substantial impairment of their livelihood (Dregne 1983). Unfortunately, global climatic warming is likely to exacerbate the desertification problem, expanding the area of land with desert climates perhaps 17% with a doubling of atmospheric CO_2 (Emanuel et al. 1985).

Desertification is extensive even in North America. About 80% of the arid rangeland and 50% of the arid, rain-fed cropland in the western United States are considered by Dregne (1983) to be at least moderately desertified. Along the edges of the warmer southwestern deserts, large areas of productive grasslands have been replaced by stands of desert shrubs (Schlesinger et al. 1990). In some of the cooler deserts, overgrazing has led to extensive erosion by wind and water. The Navajo Indian Reservation of northern Arizona and New Mexico, in fact, is considered the largest area of severely desertified land in the world (Dregne 1983).

CONSERVATION PROBLEMS IN NORTH AMERICAN DESERTS

In North America, the major conservation problems of deserts center on overgrazing by livestock or feral ungulates, off-road vehicle activity, and the invasion of exotic plants.

Overgrazing by Livestock and Feral Ungulates

Livestock grazing affects desert ecosystems in several ways. Large animals such as cattle and burros impact these systems by grazing and trampling, dispersing weedy shrubs and cacti. In North America, these effects may have contributed to the decline of several native desert animals. In the Mojave Desert of California, for example, cattle grazing may have contributed to the decline of populations of the desert tortoise (Campbell 1988) (*See* Reading 8.1).

The desert bighorn sheep is another species of concern in North America. Originally, 500,000–1 million desert bighorns probably occurred in the southwestern United States and northern Mexico (Cooperrider 1985). Today, total numbers are about 8,500–9,100, or about 1%–2% of the original population. Between 1850 and 1900, the populations of this species were hunted heavily, leading to their disappearance from many small mountain areas. Bighorn sheep are also very sensitive to several diseases of livestock, and some populations have disappeared in areas where domestic sheep and cattle have come into close contact with bighorns.

In Death Valley National Monument, as well as other areas, feral burros have contributed to the decline of populations of desert bighorn sheep (Fig. 8.5). Originally, bighorn populations were estimated to be about 5,000 animals; in the early 1970s they had declined to only about 600. The burro population of the monument was then estimated to be 1,500 animals, and showed a growth rate of about 18% per year (Norment & Douglas 1977). This suggested a loss of nearly three bighorns for every burro. Burros and bighorns feed on many of the same plants. Burros, however, can browse woody plants more heavily than can bighorns, because of a digestive system adapted to process woodier materials (Seegmiller & Ohmart 1981). Burros are also highly aggressive toward bighorns, and displace them from desert springs. Feral burros (Fig. 8.6) have created major

FIGURE 8.7

Off-road vehicle activity is destructive to both physical and biotic features of desert ecosystems.

(U.S. Bureau of Land Management photo)

management problems in several other locations in the North American deserts, particularly Grand Canyon National Park. Comparison of the vegetational characteristics and small mammal communities of areas accessible and inaccessible to burros (Table 8.1) shows just how extensive the impact of burro grazing can be (Carothers et al. 1976). Grazed areas had fewer plant species, less than a quarter the normal plant cover, and only about a quarter as many small mammals as ungrazed areas. Some mammal species that were common in ungrazed areas were absent in grazed sites, and vice versa.

Overgrazing has long been recognized as a problem in the Grand Canyon area. Prior to 1969, however, populations of these animals were periodically reduced by "burro hunts." In 1971, the killing of feral burros and horses on many federal lands was made a felony under the Wild Horse and Burro Act (Federal Law 92-195). The result was that feral burro and horse populations on federal land increased from about 17,000 in 1971 to 64,000 in 1985, and overgrazing has become a serious problem in many areas.

Recent policy has been to reduce burro populations by capturing them alive and putting them up for "adoption." Adopting a burro requires paying a fee of $75, and agreeing to conditions relating to ownership and care. This practice, which has resulted in the adoption of over 50,000 animals, has not

TABLE 8.1

Characteristics of Areas Inaccessible and Accessible to Burros in Grand Canyon National Park		
	INACCESSIBLE TO BURROS	ACCESSIBLE TO BURROS
Number of Plant Species	28	19
Total Plant Cover	80%	19%
Small Mammal Abundance (No./ha)	51.8	13.2
Canyon mouse	Absent	Common
Rock pocket mouse	Common	Absent

Data from S. W. Carothers, et al., 1976, "Feral Asses on Public Lands: An Analysis of Biotic Impact, Legal Considerations and Management Alternatives" in *North American Wildlife Conference*, 41: 396–405.

T he desert tortoise, *Gopherus agassizii* (Reading Fig. 8.1), one of four species of tortoises in North America, occurs in scattered areas of the Mojave and Sonoran deserts, including southeastern California, southwestern Arizona, southern Nevada, and extreme southwestern Utah, as well as much of Sonora and part of Sinaloa, Mexico. It is a long-lived animal, requiring 13 to 18 years to reach reproductive maturity and often living to an age of 50 or perhaps even 100 years. It is herbivorous, feeding on grasses and broadleaf herbaceous plants during the winter and spring and shrubs and cacti at other times. The home ranges of individuals vary in size, some as small as a few hectares, others more than 200 hectares in extent. In the warmer southern deserts, the animals use shallow burrows or depressions (pallets) built in shaded situations to avoid the extremes of heat and drought. In the cooler northern deserts, tortoises use shallow burrows in the summer, but construct deep winter dens in the banks of stream channels and similar sites. In the spring, mature females lay 2 to 14 eggs, which hatch after an incubation period of 90 to 120 days.

Although claims have been made that this tortoise once occurred in numbers greater than 1,000 per square kilometer over much of their range, and that their numbers have declined steadily since the early 1900s, few data support these claims (Bury & Corn 1995). Normal densities appear to be much less, with medium values being 8–20 per square kilometer. In the western Mojave Desert, however, much of their original range has been invaded over the past century by human activity and development, and a substantial decline in numbers have been documented since the 1980s (Corn 1994). This led to the Mojave Desert population of the species being listed as federally threatened.

Normally, desert tortoise populations show a sex ratio of about 150 females to 100 males, and about equal numbers of adults and juveniles. Declining populations, how-

READING FIGURE 8.1

The desert tortoise of the southwestern United States has been declining in abundance due to a variety of impacts related to human activity in deserts.

(U.S. Bureau of Land Management photo)

ever, are characterized by fewer females and fewer juveniles. These patterns indicate high female mortality, and poor reproduction—both likely the result of inadequate nutrition of adult females. In many parts of the Mojave Desert, cattle are trucked in during the spring, when new, tender shrub foliage and herbaceous plants stimulated by winter rains are most abundant. During these times, cattle and tortoises sometimes eat many of the same herbaceous plants, but how often cattle grazing seriously depletes the food of tortoises is unclear (Oldemeyer 1994). Tracy (1995) has estimated that normal tortoise populations require a density of about 16 grams per square meter of green herbaceous material to achieve full reproductive potential. Modeling of range condition suggests that forage availability may be driven below this level frequently in some locations, but uncommonly in others. He recommends that cattle should be removed from desert range when their grazing has reduced herba-

ceous forage to that level. Cattle may also cause mortality of tortoises by trampling them, or by defoliating shrubs and collapsing the shallow burrows that provide shelter from heat and drought.

Other contributing factors have been vehicle mortality, predation on young tortoises by ravens, the shooting and removal of animals (for pets) by humans, and, most recently, a viral respiratory disease that may have been introduced into wild populations by release of pet animals.

Recently, high levels of predation on young tortoises by ravens have been noted (Campbell 1988). Increased human activity in desert regions has led to increased numbers of ravens, which feed at garbage dumps and cattle ranches, and nest in trees or on utility poles. Predation by ravens may cause as much as 30% of total mortality of tortoises, leading to proposals for artificial reduction of raven populations as a means of increasing the survival rates of young tortoises.

TABLE 8.2

Effects of Motorcycle Passes on the Density and the Large Pore Volume of a Loamy Sand Soil in the Mojave Desert, California

NUMBER OF VEHICLE PASSES	SOIL DENSITY (g/cm³)	LARGE PORE VOLUME* (cm³/g)
0	1.52	0.21
1	1.60	0.19
10	1.68	0.17
100	1.77	0.15
200	1.78	0.14

*Pores > 0.045 mm in diameter.

Data from B. M. Iverson, et al., 1981, "Physical Effects of Vehicular Disturbances on Arid Landscapes" in *Science*, 212: 915–17, Washington, DC. American Association for the Advancement of Science.

TABLE 8.3

Effects of ORV Activity on the Abundance and Condition of the Creosote Bush, the Dominant Shrub Species, in the Mojave Desert of California

ORV ACTIVITY LEVEL	SHRUB DENSITY (No./ha)	CONDITION INDEX* (Per./ha)
None	240	216
Moderate	236	112
Heavy	145	56
Very Heavy ("Pit" Areas)	84	10

*Sum of decimal fraction of living branches for all shrubs present.

Data from R. B. Bury, et al., "Effects of Off-Road Vehicles on Vertebrates in the California Desert." U.S. Department of the Interior, Fish and Wildlife Service, *Wildl. Res. Rep.* No. 8, 1977.

been adequate to reduce feral animal populations in the wild. The "willingness to adopt" has largely been exhausted, and large numbers of burros and horses must be maintained in captivity at an expense of about $730 per animal-year. In 1985, the federal budget for the maintenance and adoption program was $16.7 million, based on the handling of 17,000 animals.

Physical and Biotic Impacts of Off-Road Vehicles

The use of **off-road vehicles (ORVs),** such as trail bikes, all-terrain vehicles (ATVs), dune buggies, and 4-wheel drive wagons and trucks for outdoor recreation has caused serious impacts on several types of ecosystems, especially sandy coastal ecosystems and deserts (Fig. 8.7). In the United States alone, there are probably 10 to 20 million motorcycles and ATVs, about half of which are used regularly for off-road recreation. A large fraction of the 5 million or so 4-wheel drive wagons and trucks are probably used for such activities, as well. These vehicles severely disturb desert soils and cause direct damage to plant and animal populations.

ORV activity disrupts the integrity of surface protective layers, compacts the soil (Adams et al. 1982; Webb 1983), and increases its vulnerability to erosion (Iverson 1980). Soil crusts are easily disrupted by the passage of vehicles. The weight of ORVs also tends to compact the soil (Iverson et al. 1981; Hinckley et al. 1983). The first few passes of an ORV do most of the compaction, however (Table 8.2), with the first pass causing more

compaction than any other. Loamy and gravelly soils with a mixture of textural classes (sand, silt, clay) are most susceptible to compaction, whereas dune sands and the fine alluvial soils of playas are least susceptible. Compaction is coupled with reduction in the pore space within the soil (Table 8.2), as well.

These effects promote erosion. Disruption of the soil crust exposes loose underlying particles to wind or water erosion. Compaction means that infiltration of water is slower, and that runoff, and the erosion that accompanies it, is greater. Experimental measurements of erosion caused by simulated rainfall on ORV recreational areas indicate that the general increase in erosion is 10–20-fold (Hinckley et al. 1983). On steep slopes, ORV trails promote severe gully erosion and occasionally cause massive **debris flows,** the mass slippage of destabilized, saturated soil.

Damage to vegetation is both direct and indirect. In the Mojave Desert, for example, the density of the dominant desert shrub, the creosote bush, declined by almost two-thirds on sites used as staging areas for vehicular recreation (Table 8.3). Moreover, when the fraction of the branches of individual bushes that were alive was taken into account, even moderate use areas experienced a decline of almost half in the amount of live shrub foliage, and staging areas showed only about 5% as much foliage as control (undisturbed) areas. Indirect effects, such as soil compaction, also tend to reduce the germination and growth of native annuals. Disturbance of the soil by ORV activity, on the other hand, favors the establishment of exotic weedy plants, such as the Russian thistle.

TABLE 8.4

Effects of ORV Activity on Vertebrate Animal Populations in the Mojave Desert of California

	ORV ACTIVITY LEVEL			
	None	Moderate	Heavy	Very Heavy
Reptiles				
Number of species	4–6	5	3–5	0–5
Density (per 2 ha)	29–75	30–42	8–33	0–16
Mammals				
Number of species	3–6	3–4	2–5	1–2
Catch per 100 traps	22.2	28.0	11.2	3.1
Birds				
Number of species	4	2	0	—

Data from R. B. Bury, et al., "Effects of Off-Road Vehicles on Vertebrates in the California Desert." U.S. Department of the Interior, Fish and Wildlife Service, *Wildl. Res. Rep.* No. 8, 1977.

ORV activity causes a marked decline in the abundance and diversity of animals (Table 8.4). Many small animals are crushed and their burrows collapsed by ORV passage. Among small vertebrates, ORV noise may cause hearing loss, reduced ability to detect predators, or unnatural behavior (Brattstrom & Bondello 1983). Roaring sounds that approximate thunder, for example, may stimulate spadefoot toads to emerge when water is unavailable for breeding. Moderate ORV activity may kill or frighten away important carnivores, such as coyotes, kit foxes, badgers, and larger snakes, with the result that small mammal populations may remain high, or even increase. Higher intensities of ORV activity, however, cause even these animals to decline. The diversity and density of birds are greatly reduced by ORV activity. In a Colorado Desert wash, for example, ORV activity reduced the number of breeding bird species by 90% and total density by more than 95% (Luckenback 1978).

Even on desert dune areas, where winds quickly obliterate the tracks of ORVs, populations of plants and animals are seriously reduced. Sand dune areas that are heavily used by ORVs, such as parts of the Algodones Dunes in southern California, become essentially denuded of vegetation. Many dune organisms are highly specialized forms that are endemic to individual dune systems. Most of the dune invertebrates and smaller vertebrates spend their inactive periods buried at a shallow depth, where they are vulnerable to being crushed by vehicle passage.

Recovery of deserts from ORV impacts is very slow. At Mojave Desert sites used for military exercises in the 1940s, for example, plant cover in 1978 was still less than 50% of normal (Lathrop 1983). Tracks formed by single passes of tanks still showed average penetrometer resistances 50% greater than undisturbed soil when examined in 1981 (Prose 1985). Based on observations at an abandoned mining town in Nevada, more than a century is needed to reestablish natural soil conditions and vegetation.

ORV activity on public lands is a good example of the "tragedy of the commons" (*See* Chapter 29). The ORV users themselves receive a certain amount of pleasure from their activities, but the widespread degradation of the desert environment reduces its values in many ways for the rest of society.

Invasion of Exotic Plants

Throughout western North America, exotic annual grasses and forbs from Europe and Asia have invaded grassland ecosystems. Some of these exotics appear to be evolving ecotypes adapted to desert habitats. Their effects are now being felt in winter rainfall areas such as the Mojave Desert and Great Basin Desert. The most serious of these species are annual grasses, particularly those of the genus *Bromus,* that germinate following the first winter rains and form a continuous ground layer. Later, when these plants have died and become dry, this layer can carry fire. Most desert shrubs cannot resprout after fire. Fire is thus being introduced to desert ecosystems, where its effects may cause great changes in vegetational structure.

Desert riparian areas throughout the southwestern United States are being invaded by the exotic salt cedar, *Tamarix ramosissima,* displacing native willow and cottonwood species (Busch & Smith 1993, 1995; Busch 1995). Salt cedar is a deep-rooted, salt-tolerant shrub that is capable of high transpiration rates but is also efficient in its water use when water is scarce. It thus contributes to the desiccation of desert watercourses. In the riparian zones of permanent streams it also forms dense stands that are highly flammable. Salt cedar resprouts vigorously after fire, thus tending to replace cottonwoods, which do not resprout, and willows, which resprout less vigorously. Cottonwoods are disappearing from many riparian communities as a result. Along major streams such as the Colorado River, salt cedar tends to form very dense thickets along the riverbank and on sandbars, stabilizing them to an unnatural degree. Deer and beaver do not feed on salt cedar, and den construction by beaver is impeded by the dense root network of salt cedar thickets (Kimball Harper, *pers. comm.*).

CONSERVATION PROBLEMS IN OTHER WORLD REGIONS

In desert regions of Africa and Asia, large mammals have been exterminated or reduced to widely scattered, small populations (Le Houerou & Gillet 1986). Most of this reduction has been the result of direct killing by humans, particularly since World War II. Some species, such as the Arabian oryx and Prezwalski's horse, have survived only in captivity. In northern Africa, the Arabian Peninsula, and the Middle East, the addax, scimitar-horned oryx, several species of gazelles, and several other desert ungulates have been reduced to the brink of extinction. The scimitar-horned oryx once occurred from Egypt to Senegal, but has been reduced by hunting, habitat destruction, and desertification to a single remnant population in Chad (Newby 1988). A large captive population of this species fortunately exists. Other large animals, which still occur in numbers in eastern and southern Africa, have largely disappeared from the dry savanna zones bordering the Sahara Desert.

Efforts are now being intensified to protect large desert mammals, and to reintroduce them to their former ranges. The Arabian oryx has been reintroduced to the Shaumari Wildlife Reserve in Jordan (Abu-Jafar & Hays-Shadin 1988) and to other reserves in Oman and Bahrain, on the Arabian Peninsula (Stanley Price 1989). The scimitar-horned oryx has been reintroduced to Bou-Hedma National Park in Tunisia (Bertram 1988). Unfortunately, some of the most depleted species of desert birds and mammals are still hunted in some areas of the Sahel, in some cases by wealthy members of Arab royal families (Newby 1990).

In recent decades, the growing populations of pastoral peoples and their herds of animals in regions bordering the African and Asian deserts have severely degraded the natural ecosystems of these regions (Le Houerou & Gillet 1986). The Sahelian zone of sub-Saharan Africa has more than twice the human population that can be supported on a long-term basis. Furthermore, the border regions of deserts in most of Africa and Asia have few natural preserves with any genuine protection of plant and animal life. Several desert regions of Africa and the Middle East are war zones, where conservation effort is presently impossible.

GLOBAL CHANGE AND DESERT ECOSYSTEMS

Global warming is likely to intensify desert climates in continental interior regions (Verstraete & Schwartz 1991), thus intensifying the desertification effects of human activity. Soil conditions in arid regions are likely to be affected in many ways (West et al. 1994). Cryptobiotic crusts are likely to become weaker, evaporation greater, wind and water erosion greater, soil salinities higher, and soil organic matter less. Desert soils are also a source of certain hydrocarbon compounds that act as greenhouse gases, so that expansion of desert environments could reinforce the global warming process.

DESERT MANAGEMENT PLANS

The impacts of ORV recreation and other unregulated use of desert land have forced state and federal agencies to develop more comprehensive desert management plans. The California Desert Conservation Area Plan, which applies to land administered by the U. S. Bureau of Land Management, took effect in 1980. This plan zones desert land into four categories: intensive use (drive anywhere), moderate use (drive only on existing roads and paths), limited use (drive only on designated roads and paths), and controlled use (closed to vehicles). In addition, the plan designates more than 75 areas of critical environmental concern, because of special ecological, archeological, or other significance. More than 65% of desert lands were zoned for limited and controlled use, but this leaves much of the California deserts open to relatively unrestricted vehicular activity.

Passage of the California Desert Protection Act in 1994 enlarged Death Valley and Joshua Tree National Monuments, and upgraded them to national parks. The act also created East Mojave National Preserve and designated other areas as wilderness, thus expanding the area of desert under strong protection.

SUMMARY

1. Deserts are formed a) at latitudes of about 25°–30° where prevailing high atmospheric pressure exists throughout most of the year, b) in the rain shadows of major mountains, and c) in coastal areas with cold oceans offshore.

2. Substrate conditions, both physical and chemical, are a key determinant of the structure and function of desert ecosystems. Physical, chemical, and cryptobiotic crusts are important features that help protect desert soils against severe erosion.

3. Desertification is the degradation of arid, semiarid, and dry subhumid lands as a result of human activities and climatic variations. Although perhaps most serious in sub-Saharan Africa, desertification affects almost half the world's arid lands to at least a moderate degree.

4. In North America, major detrimental impacts on desert ecosystems have resulted from overgrazing by livestock and feral ungulates, ORV activity, and introduced exotic annuals and riparian shrubs.

5. On a global scale, killing of desert wildlife continues to be a serious problem. In the future, global warming is likely to intensify desert climates and promote desertification.

Questions for Discussion

1. In what ways do desert ecosystems show high inertia to disturbance? Low inertia? How would you characterize the resilience of desert ecosystems?

2. Because of its destructive physical and biotic impacts on desert soils, should all ORV activity be prohibited on public lands?

3. How should problems with overpopulations of feral burros and horses be dealt with on public desert lands?

Suggested Reading

Bury, R. B., T. C. Esque, and P. S. Corn. 1994. Conservation of desert tortoises (*Gopherus agassizii*): genetics and protection of isolated populations. Pp. 59–66 in K. R. Beaman (Ed.), Proceedings of 1987–1991 symposia. The Desert Tortoise Council, Palm Springs, CA. Concise summary of the application of conservation biology principles to desert tortoise protection.

Glantz, M. H. (Ed.). 1994. *Drought follows the plow*. Cambridge University Press, Cambridge, England. Case studies of desertification as it has affected human food production in various world regions.

Zwinger, A. H. 1989. *The mysterious lands*. E. P. Dutton, New York, N.Y. A naturalist describes her exploration of the four major desert regions of North America.

chapter

9

Moist Tropical Forests

Outline

Moist tropical forests contain half or more of the earth's species of organisms, and are its most productive terrestrial ecosystem. These forests are concentrated in developing countries with rapidly growing human populations and rising economic aspirations. Tropical forests, and their plant and animal members, are an important resource, both for the countries in which they occur and for humanity at large.

The rate at which these forests are being destroyed is a matter of global alarm for several reasons. First, the resources of tropical forests must be managed on a long-term basis to enable forested tropical regions to achieve economic development, a fact fully recognized by scientists and developmental experts of the countries involved. The loss of mature tropical forests reduces the productivity of the land, and their recovery is extremely slow. Second, widespread deforestation contributes to regional and global climatic and hydrologic changes, many effects of which are likely to be detrimental. Third, extensive deforestation causes the extinction of numerous species, many of which have great potential economic value. One of the greatest challenges of conservation biology is to develop institutions that can maintain a major fraction of the richness and productivity of these forests for posterity.

Lowland tropical forests and their associated rivers are the richest continental ecosystems on earth.

DISTRIBUTION OF MOIST TROPICAL FORESTS

Moist tropical forests comprise many distinct forest types. Indeed, the richness of species in humid tropical regions is paralleled by the richness of the types of communities they form. In lowland areas, forest communities range from **tropical rain forests** (Fig. 9.1), in which abundant rain occurs in every month, to **tropical deciduous forests,** which have a dry season long enough that many or most of the species lose their leaves for several months. In major river basins, distinct forest types occur in areas that are seasonally inundated by the river flood. In Amazonia, these forest types include the **varzea forest**, bordering whitewater rivers that carry heavy sediment loads, such as the Amazon proper, and the **igapo forests,** bordering low-sediment, blackwater rivers such as the Rio Negro. These forests may be deeply submerged for several months, and their faunas are intimately adjusted to the seasonal cycle of inundation. At higher elevations, a variety of **tropical montane forests** can be recognized, culminating in high-elevation **cloud forests** that are bathed in mist much of the time. In a topographically varied tropical region such as Costa Rica or Peru, many more specific types of forest can be distinguished on the basis of the predominant tree species. All of these forest types fall into the general category of moist tropical forests.

FIGURE 9.1

Interior of a tropical rain forest in Costa Rica.
(Photo by G. W. Cox)

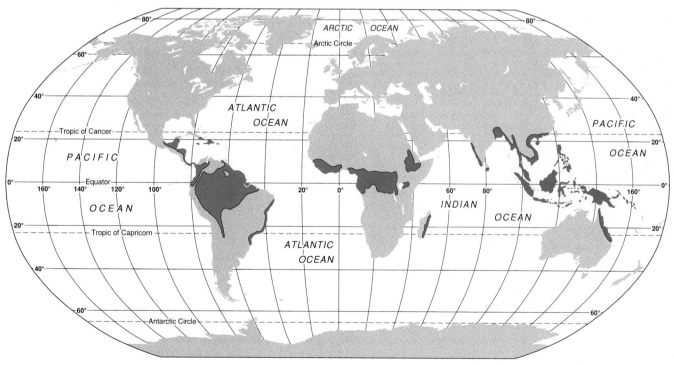

FIGURE 9.2

World distribution of original tropical rain forests.

TABLE 9.1

Original and Present-Day (1990) Moist Tropical Forests, and their Estimated Rates of Disappearance (Percent Per Year, 1981–1990) in Different Regions

Region Loss	ORIGINAL AREA		PRESENT-DAY FORESTS (1990)	
	(10⁶ km²)	(10⁶ km²)	% Lost	% Annual
New World	8.03	5.76	28.3	0.58
Africa	3.62	1.22	66.3	0.59
Asia/Pacific	4.35	2.24	48.5	1.09
TOTAL	**16.00**	**9.22**	**42.4**	**0.71**

Data based on FAO (1993), Forest Resources Assessment 1990. Tropical countries. FAO Forestry Paper 112. Moist tropical forest data taken as "tropical rain forest" and "upland formation" categories.

Tropical forests can also be characterized as primary or secondary. Mature stands that are free from signs of disturbance and have high biomass and diversity are termed **primary forests,** whereas stands recovering from disturbances such as fire, hurricane impacts, or cutting by humans are termed **secondary forests.** Secondary forests vary greatly in age, but by 60 to 80 years after disturbance they begin to approach primary forests in general appearance (Brown & Lugo 1990).

Moist tropical forests cover about 23% of the land area within the tropics, or about 7% of the total continental area of the earth. Major areas of moist tropical forests occur in the three main regions of the tropics: the New World, Africa, and the Southeast Asia/Pacific region (Fig. 9.2, Table 9.1). The largest area of such forests, both now and originally, is in the New World tropics, the smallest now in Asia. More significantly, about 80% of the remaining area of moist tropical forests occurs in only nine countries: Brazil (31% by itself), Indonesia, Zaire, Malaysia, Gabon, Venezuela, Colombia, Peru, and Bolivia (Myers 1980). This means that the decisions made by these few countries will ultimately determine the fate of most of the world's moist tropical forests.

CHARACTERISTICS OF MOIST TROPICAL FORESTS

Primary tropical forests are the richest and most productive of the earth's terrestrial ecosystems (Collins 1990). Of the estimated 12.5 million species of organisms living on earth (*See* Chapter 4), about half can probably be found in tropical forests. Because tropical forests are the least explored tropical environment, most of these species are as yet undescribed. Local endemism, or the restriction of species to relatively small geographical areas, is also high among tropical forest species (Gentry 1986).

The richness of life in tropical forests is overwhelming to anyone who tries to learn the species of even a small unit of forest. A single hectare of primary tropical forest commonly contains 100 to 250 or more species of trees, a richness much greater than that of temperate forests. In a single hectare Gentry (1988) found 283 tree species in wet, upper Amazonian forest. Others have identified 450 species in one hectare of the Atlantic

coastal forest of Brazil (Brooke 1993). In Malaysia, 835 species of trees were recorded on a 50-hectare plot—the world's record for tree diversity. Animal life is likewise diverse. In a 10-square kilometer area of lowland rain forest, about 150 species of butterflies, 60 species of amphibians, 100 species of reptiles, 125 species of mammals, and 400 species of birds can typically be found. Erwin (1983) estimated that more than 41,000 species of insects may occur on a single hectare of tropical lowland forest. This richness often forces biologists interested in ecological or evolutionary studies to do descriptive taxonomy before they can attempt other studies.

The biomass structure of primary tropical forests is likewise complex (Table 9.2). The total biomass of moist tropical forests is considerably greater than that of most temperate deciduous and coniferous forests. Maximum tree height is related to annual precipitation, reaching 70 to 80 meters in wet lowland forests. Wet lowland forests have several understory strata of smaller trees, and an abundance of **lianas,** or massive vines that expose their foliage in the forest canopy. The surfaces of leaves are often covered with **epiphylls,** a thin layer of algae, lichens, and mosses, which meet their nutrient needs by capturing nutrients from rainwater or fixing nitrogen from the air (Jordan 1985). Wet forests, especially the high mountain cloud forests, also possess a rich flora of **epiphytes** (Fig. 9.3), plants that grow on their trunks and branches. Epiphytes obtain their moisture from rain, and their nutrients from rainwater and the decomposition of organic matter. In many cases, the foliage of lianas and epiphytes exceeds that of the host tree. The canopy of the tropical forest is one of the world's ecological frontiers (*See* Reading 9.1).

Moist tropical forests are the most productive terrestrial ecosystem (Table 9.2), and are estimated to carry out about 29% of the total primary production of the biosphere. The productivity of nectar and fruits is especially high in these forests, and many birds and mammals depend on these resources. The biotic relationships that center on such producers can be very complex. Howe (1977) and Terborgh (1986) show that some trees function as **keystone mutualists** that are essential to the survival of many animals. A few species of figs and other fruiting trees, nut-bearing palms, and flowering plants that produce nectar are vitally important to many forest mammals and birds. These plants, in turn, often depend on animals that serve as **mobile links** by

THE LAST FRONTIER: THE TROPICAL FOREST CANOPY

Ecologists have long known that the canopy of the tropical forest was a zone of extraordinary richness and distinctive ecology. For decades, however, they had to rely on guns, binoculars, and fallen or felled trees to study the life of the canopy. An explosion of new sampling and access techniques in the 1980s opened tropical forest canopies to intensive study. Tower cranes (Reading Fig. 9.1), canopy dirigibles, aerial walkways, and climbing equipment adapted from mountaineering now give access for repeated observations and sampling of the canopy environment. A new branch of ecology has emerged: canopy science (Lowman & Nadkarni 1995).

The canopy environment is more like that of a tropical savanna than of the forest interior, with greater variations in solar radiation, humidity, wind, and the physiological processes these factors influence (Nadkarni 1988). The dense growth of epiphytes in rain forests and cloud forests leads to the accumulation of an organic soil mat on the larger limbs and upper trunks of trees. The tree branches themselves produce roots that penetrate these mats (Nadkarni 1981; Nadkarni & Matelson 1991). Nutrient cycling in these epiphytic communities thus is largely internal, although with gains from atmospheric deposition and losses by leaching.

The flora of the canopy of a lowland tropical forest includes the foliage of a hundred or more species of trees. Added to this are perhaps an equal number of lianas—vines that begin life on the forest floor and climb into the canopy, using trees as support, and sometimes engulfing and replacing them. Numerous species of epiphytes and parasitic plants attach to the canopy branches. Still other plants, known as **hemiepiphytes,** begin life in the canopy as epiphytes, but send roots downward, eventually tapping the forest floor and converting the plant into a liana.

Invertebrates and small vertebrates are numerous in the canopy, and especially in the epiphyte gardens. Stork (1991), in Bor-

READING FIGURE 9.1

The Smithsonian Tropical Research Institute's canopy crane in Panama.

(Photo by Antonio Montaner, courtesy of The Smithsonian Tropical Research Institute)

neo, sampled tree canopy arthropods, and identified over 3,000 species in a sample of nearly 24,000 individuals. A single tree, in fact, yielded over 1,000 species. Epiphytes are a rich microhabitat. In the Neotropics, epiphytic bromeliads hold water in their cupped leaf bases, creating tiny canopy reservoirs that are habitat for larval insects, snails, frogs, and even a species of crab (Moffett 1993).

A rich fauna of nonflying mammals occupies the canopy, primarily feeding on leaves and fruit, but also taking animal foods in an opportunistic manner. As many as 40 species of nonflying mammals coexist in the tropical forest canopy (Emmons 1995). These animals range upward in size to bears and gorillas that make only limited forays into the canopy, but include many large species such as orangutans, gibbons, and many species of monkeys that rarely visit the forest floor. Some of these species are accomplished leapers, enabling them to move between trees whose branches are not in

contact. Some smaller forms have evolved gliding capabilities to get from one tree to the next. Bats, which reach their greatest diversity in tropical forests, are the most mobile mammals, and have become the most specialized in feeding, utilizing fruit, nectar, insect, and blood foods.

Birds are more diversified in their feeding ecology. The largest predators of the canopy are birds. Some of the largest raptors are able to take monkeys, and many smaller species concentrate on birds and reptiles. Others are specialized for nectar, fruit, or insects. In the American and Asian tropics, the resident bird fauna is joined in the northern hemisphere winter by many species of temperate zone migrants, some of which are canopy foragers. The ecology of canopy birds, as well as of other small vertebrates, is still poorly known, but studies in Panama and Costa Rica have shown, for example, that bird activity may vary seasonally, and that many species forage extensively in epiphytes (Munn & Loiselle 1995).

carrying pollen from individual to individual, or seeds from mature tree to potential establishment sites (Gilbert 1980). As for other forests, however, the bulk of the net primary production enters detritus food chains on the forest floor. A rich invertebrate and decomposer biota, operating under warm, moist conditions, rapidly breaks down litter, so that litter biomass is typically small. Detritus food chains culminate in a rich community of forest floor insects, amphibians, and lizards that, in turn, sup-

port a diverse array of predators. These predators range from army ants, and the ant birds that follow them and feed on insects they flush out, to larger reptiles, mammals, and birds.

From the diversity and productivity of the forests, one might conclude that their soils must be highly fertile, and that the potential for converting the land to permanent agriculture is great. Although some tropical soils are indeed very fertile, especially those of volcanic highlands, vast areas of the forested tropics

TABLE 9.2

	TROPICAL RAIN FOREST	TEMPERATE DECIDUOUS FOREST	TEMPERATE CONIFEROUS FOREST
Biomass, Net Primary Productivity, and Quantities of Nitrogen and Calcium in Various Ecosystem Compartments for Tropical Rain Forest and Temperate Forest Ecosystems*			
CHARACTERISTIC			
Biomass (Mg/ha)	410.0	195.0	200.0
Canopy tree leaves	9.0	3.0	10.0
Canopy tree trunks	360.0	142.0	114.0
Understory	4.0	3.0	2.0
Roots	33.0	37.0	38.0
Litter	4.0	10.0	36.0
Net Production (Mg/ha/yr)	50.0	15.0	10.0
Total Nitrogen (Mg/ha)	8.0	8.0	6.0
Live biomass	1.8	1.0	0.7
Litter	0.2	0.5	1.8
Soil	6.0 (75%)	6.5 (81%)	3.5 (58%)
Exchangeable Ca (kg/ha)			
Live biomass	745	1000	640
Litter	5	10	15
Soil	40 (5%)	1100 (52%)	575 (47%)

*Values are rough approximations derived from several sources.

FIGURE 9.3

Epiphytes are abundant in cloud forests, sometimes covering the limbs of trees with a dense layer of bromeliads, ferns, orchids, and other plants.

(Photo by G. W. Cox)

have very poor soils. Many of these soils are ancient and highly weathered. Their low organic content (due to high decomposition rates), the dominance of highly weathered clay minerals with poor nutrient-holding capacities, and the abundant rainfall make such soils prone to loss of nutrients by leaching. In some locations, such as the **caatinga forest** of the northern Amazon Basin, soil is little more than highly leached, quartzitic sand that does not retain appreciable quantities of nutrients in any form (Herrera 1985). Leaching is particularly severe for mineral cations such as calcium (Table 9.2), magnesium, potassium, and sodium. For these nutrients, the vast bulk of the active nutrient pool is in the living biomass of the forest, contrary to the case in temperate forests (Jordan 1985). For nitrogen and phosphorus, concentrations of available nutrients in the soil may not be much less than for temperate forests, except in caatinga forests, but more of the active nutrient capital still tends to occur in living material than is true for temperate forests because of the greater biomass of the mature forest (Table 9.2).

The high productivity and biotic richness of the forests in areas of poor soils result from an efficient mechanism of retention and recycling of nutrients by the mature forest ecosystem (Jordan 1985). This mechanism is primarily a high concentration of mycorrhizal roots of trees in the surface soil, where nutrients released by decomposition can be taken up immediately (Fig. 9.4). **Mycorrhizae** are symbiotic associations between fungi and the fine roots of higher plants. In these associations, the filaments of the fungi extend outward into the soil or litter from the rootlets themselves. In some mycorrhizae the fungal filaments simply form a sheath around the rootlet; in others they penetrate into the interior of the root. In both cases, the fungal associate of the root assists in the uptake of water, nutrients, or both. The intimate association of some mycorrhizae with decomposing organic matter results in **direct nutrient cycling,** in which nutrients pass directly from the dead matter into fungal filaments and then into plant rootlets without entering the mineral soil.

Mycorrhizal root mats are highly effective in uptake of nutrients released by decomposition of litter. In forests on leached quartzitic sands at San Carlos de Rio Negro, Venezuela, where these mats are 15–40 centimeters thick, Stark and Jordan (1978) found that 100% of the calcium and phosphorus in leaves placed on the ground surface was taken up by the root mat; none passed into deeper soil layers by leaching. Because of the high concentration of absorbing roots close to the soil surface, most trees of mature tropical forests have few deep roots. Many of these trees have **plank buttresses,** flanged extensions of the

FIGURE 9.4

Mycorrhizal root mat in a tropical rain forest in Costa Rica.
(Photo by G. W. Cox)

FIGURE 9.5

Plank buttresses of trees in tropical rain forests are believed to be a
mechanism to brace these shallow-rooted trees against toppling by wind.
(Photo by G. W. Cox)

trunk base that connect with lateral roots (Fig. 9.5). Plank buttresses are thought to provide support for the trunk in trees that
do not have deep roots to anchor them.

When mature tropical forests are cleared, and the root systems of the trees killed, this system of nutrient retention and recycling is destroyed. Nutrients then are leached into deeper soil
layers, or flushed out of the soil and carried into streams. This
dispersion of nutrients means that successional plants must seek
nutrients in deeper soil layers, and gradually reconcentrate them
over many years. Successional trees in tropical forests thus tend
to have widely spreading, deeply penetrating roots, rather than
shallow root systems with surface root mats.

Leaching of nutrients is one of the factors that forces indigenous farming peoples throughout the tropics to practice
shifting cultivation. In these farming systems, which vary
greatly in detail, a plot is cleared by felling and burning the
woody vegetation. Burning releases large quantities of nutrients
in the ash. Crops are grown for two to several seasons, and the
plot is then abandoned to succession. Loss of fertility by leaching
from the surface soil, together with weed and insect pest buildup
and, often, an increase in soil acidity that causes aluminum ions
to exert toxic effects, lead to a rapid decline in yields of crops in
successive growing seasons. Without heavy artificial fertilization
it soon becomes unprofitable to farm the plot.

Shifting cultivation is practiced by about 240 million people
in tropical forest regions (National Research Council 1982).
When the interval between farming activity is 70 or more years,
shifting cultivation is probably an ecologically stable strategy that
does not degrade the tropical forest. By initiating succession at
scattered locations, in fact, shifting cultivation may promote biodiversity by increasing the frequency of stands of different age and
successional status. If disturbance from shifting cultivation is small
in scale and short in duration, nutrients are retained in deeper lay-

ers of the soil, and recovery by succession is rapid because of the
rich source of propagules in the surrounding forest. In addition,
some shifting cultivators, such as the Kayapo Indians of Brazil, actively manage their farm plots to encourage successional recovery
(Hecht & Cockburn 1989). As human populations grow, however, the number of subsistence farmers in forests increases. The
period of cultivation lengthens, the fallow interval between cropping periods shortens, and farm plots are spaced closer together.
With such changes, the nutrient capital of the landscape at large is
depleted by leaching and export of harvested materials.

Recovery of moist tropical forests to a mature state is very
slow. Many of the trees and other plants of the mature forest
have specialized seed dispersal mechanisms, usually involving
fruit-eating animals. How quickly such species reinvade an area
depends on the distance from a source of seeds and the abundance of the animal dispersal species. In Costa Rica, Opler et al.
(1977) estimated that about 1,000 years is required for full successional recovery of primary lowland forest, even when sources
of seed were close at hand.

TROPICAL DEFORESTATION

Moist tropical forests are being converted to other vegetation
types at an alarming rate, although actual conversion rates are
still poorly known. Current estimates are based largely on

TABLE 9.3

Estimated Percentages of Tropical Forest Species that have Disappeared or are Projected to Disappear Due to Decline in Total Forest Area						
WORLD REGION	LOST PRIOR TO 1990	AT CURRENT DEFORESTATION RATE			AT 2× CURRENT RATE	WITH LOSS OF ALL BUT CURRENTLY PROTECTED AREAS
		1990–2000	1990–2010	1990–2020	1990–2020	
Africa	10–22	1–2	2–4	3–7	6–14	35–63
Asia/Pacific	12–26	2–5	4–10	7–17	22–44	29–56
New World	4–10	1–3	3–6	4–9	9–21	42–72

Data from W. V. Reid and K. R. Miller, *Keeping Options Alive: The Scientific Basis for Conserving Biodiversity,* 1989 World Resources Institute, Washington, DC.

surveys by the United Nations Food and Agriculture Organization (FAO) (1993). These indicate that from 1981 to 1990 an annual average of 70,600 square kilometers of moist tropical forest, or about 0.71% of the total forest remaining (Table 9.1), was cleared permanently. Myers (1989), however, estimated that 142,200 square kilometers of tropical forest disappeared in 1989. Data from satellite imagery show that the pattern of deforestation, as it follows roads and streams, also fragments once continuous forest, and greatly increases the portion of the forest subject to edge influences (Skole & Tucker 1993).

Primary forest is also being converted to secondary forest, and the fraction of secondary forest is thus increasing as a result of disturbance. More than 31% of the remaining moist tropical forest is now estimated to be secondary forest (Brown & Lugo 1990). As the remaining forest area shrinks, and as human populations in the tropics grow, the rate of forest conversion is likely to increase.

The total loss of moist tropical forest has been least in the New World, where the original area was greatest (Table 9.1). In Asia, about half of the original forests have been lost and the continuing annual rate of loss is greatest. In Madagascar, one of the world's tropical forest hotspots of biodiversity, about 66% of the original moist forest has disappeared (Green & Sussman 1990).

The ultimate causes of tropical deforestation are complex, and are deeply rooted in international trade and third-world debt (Kahn & McDonald 1995). Chief among them are growth of human populations and efforts for economic development in tropical forest countries, combined with the appetite for tropical products such as timber and beef in developed countries. These forces drive several deforestation processes. The change from shifting cultivation to permanent farming due to the growth of human populations is a major process, particularly in Africa and Latin America. Clearing of forest for cattle ranching is a major process in Latin America. In some cases, the beef produced by these ranching operations is exported to developed countries for use in fast foods; this relationship has been termed the **"hamburger connection"** (Myers 1981). Often, clearing of forest for ranching is promoted by international developmental organizations (Fearnside 1987) and by governmental incentives and subsidies, based on the legal definition of such action as "improvement" of the land (Hecht 1989). In Southeast Asia, cutting of tropical hardwood forests for timber export has been the major cause of forest destruction. The rapid depletion of timber in this region is now causing timber companies to move to other parts of the tropics.

The thinly populated Amazon Basin, where the largest area of unspoiled tropical forest remains, is viewed by several nations as their last frontier. Brazil, in particular, considers Amazonia a vast underdeveloped area that must be integrated into the national economy and culture (Lovejoy 1985). Brazil has encouraged development of its portion of the basin by the construction of highways, by agricultural settlement schemes, and by encouraging major corporations to undertake large-scale developments (Hecht & Cockburn 1989). The Trans-Amazon Highway, crossing the Amazon Basin from east to west, opened up much of the southern part of the basin to settlement. In the eastern State of Para, the Grande Carajas Program promoted mining and smelting of metal ores, using charcoal derived from the tropical forest as a principal fuel (Oren 1987). In the western state of Rondonia, the Polonoroeste Project promoted settlement on small farm plots, giving this region the most rapid deforestation rate in Brazil. The unsuitability of this land for permanent farming has kept the settlers in a state of poverty and forced many to abandon their original land and clear still more forest.

Another project, the Jari River Forest and Agricultural Enterprize, was initiated in the late 1960s on a 1 million hectare site on a tributary of the Amazon River 350 kilometers west of the port of Belem, Brazil. The original forest was cleared and replaced with pulpwood plantations, rice fields, and pastures for beef cattle. By 1981 about $1 billion had been invested in development. Pulpwood production fell well below expectations, however, and in 1982 the entire operation was sold to a consortium of Brazilian companies for only $280 million. Having destroyed an extensive area of tropical forest, the project still operates, but at a low level of profit (Russell 1987).

IMPACTS OF DEFORESTATION

Destruction of tropical forests threatens enormous numbers of species with extinction. Lovejoy (1980) estimated that 15%–20% of forest species would disappear between A.D. 1980 and 2000 due to forest loss. Simberloff (1986) estimated that 12% of plant species and 15% of bird species in the Amazon Basin would disappear by A.D. 2000 due to deforestation. Raven (1988) projected a loss of 25% of tropical forest species by A.D. 2015. The most comprehensive estimate (Table 9.3) is that 5%–15% of

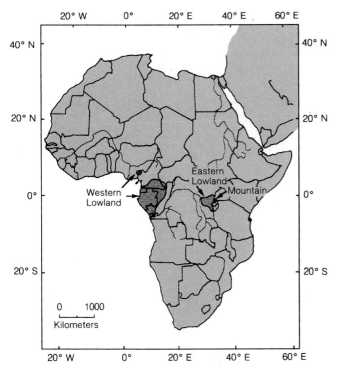

FIGURE 9.6

Ranges of the three subspecies of gorilla in Africa.

FIGURE 9.7

Mountain gorillas are almost entirely restricted to high elevations of the Virunga Mountains on the border of Zaire, Uganda, and Rwanda in East Africa.

(WWF photo by R. Wiederkehr)

tropical forest species will be lost between A.D. 1990 and 2020. (Reid & Miller 1989). If about 6.25 million species still exist in tropical forests, this translates into losses of 10,000 to 31,000 species per year. Even higher losses are projected if rates of forest destruction increase, and a catastrophic loss if the only areas to survive are those that are now protected. From these estimates it is obvious that many species must already have disappeared in tropical forest regions: 12%–26% in Asia, 10%–22% in Africa, and 4%–10% in the New World. Extinction of these species is an immense scientific, aesthetic, and economic loss.

Most extinctions due to continued tropical deforestation will be of plants and invertebrates, particularly insects. Nevertheless, the vertebrates of tropical forest regions are increasingly under threat. Killing of large animals for food and commerce has eliminated many species from areas that are still forested (Redford 1992). Deforestation and hunting by humans are major threats to forest primates throughout the tropics. In the Amazon Basin, 6 of 18 species of mammals listed as rare or endangered are monkeys (Barrett 1980). In the Atlantic forest region of eastern Brazil, where the original moist tropical forest has been reduced to about 5% of its original area, 16 endemic species or subspecies of primates occur, all of which must be considered as endangered (Quintela 1990).

In Africa and Asia, deforestation threatens wild populations of all of the great apes, including the African gorilla. Three subspecies of gorillas exist: western lowland, eastern lowland, and mountain. The lowland gorillas occur in two relatively extensive areas of west-central and central Africa (Fig. 9.6), with the population of the western form estimated to be about 40,000 individuals and that of the eastern form about 5,000 (Harcourt et al. 1989). With the rapid clearing of tropical forests, however, these numbers are declining rapidly.

The mountain gorilla is much less abundant, with slightly more than 400 individuals surviving in the wild. Mountain gorillas are restricted to the high, volcanic Virunga Mountains on the border of Zaire, Rwanda, and Uganda (Fig. 9.7), where they number about 290, and to a smaller area in the Bwindi Forest, Uganda, where about 115 live. In the Virungas, the gorillas occur in wet forests from about 2,100 to 4,000 meters in elevation. All three countries have established parks or preserves, which are designed to form a continuous sanctuary for the gorillas. In the wake of the Rwandan civil war, however, chaotic conditions exist throughout the region. Poaching of adult gorillas, capture of infants for sale as pets, and the accidental capture of gorillas in snares set for other animals continue to affect the Virunga population. Clearing of the forest inside the preserves by squatters and war refugees is also a major problem.

Wild populations of several forest pachyderms are also threatened by habitat loss and poaching. In Africa, populations of forest elephants in five countries in the Congo Basin have probably been reduced by 44%–56% by poaching (Michelmore et al. 1994). Asian elephants have declined by nearly 75% since 1940, especially in Malaysia, Cambodia, and Laos. The wild population is now much smaller than that of the African elephant, probably no more than 40,000 animals (Cohn 1990). About 10,000–13,000 also live in captivity, mostly as work animals. Asian elephants have suffered less than African elephants from ivory poaching, since females and many males lack tusks. Habitat destruction, capture of animals for work use, killing of animals responsible for depredations or human deaths, and the impacts of wars and revolutions have all contributed to decline of the species (Sukumar 1989).

The Sumatran and Javan rhinoceroses formerly occurred in forests from eastern India throughout most of southeastern Asia. The Sumatran rhino also occurred on Sumatra and Borneo, the Javan on Sumatra and Java. Poaching for horn (See Chapter 10) and habitat destruction have greatly reduced both species. The Sumatran rhino now occurs only in scattered parts of its original range, and the Javan rhinoceros only in western Java and southern Vietnam. Only about 500 Sumatran rhinos survive in the wild, and the Javan rhinoceros numbers only about 50 individuals on Java and in mainland Vietnam (Khan & Foose 1994).

DEFORESTATION, CLIMATE CHANGE, AND GLOBAL WARMING

Deforestation will likely produce major changes in regional hydrology and climate. Deforestation tends to increase surface runoff into streams during rainy periods. Widespread deforestation in the upper Amazon Basin has been implicated in an increase in the mean high water level of the Amazon River (Gentry & Lopez-Parodi 1980). For the period from 1970–1978, high water level averaged more than two meters higher than during the period 1962–1969, whereas low water levels were essentially the same.

Tropical deforestation may lead to a decline in local precipitation (Myers 1988), a change noted at Barro Colorado Island, a Smithsonian research area in Gatun Lake, Panama. Deforestation also reduces the rate of return of moisture to the atmosphere by transpiration and increases the landscape albedo, or ratio of solar energy reflected to that absorbed (Myers 1988). Modeling of these relationships for the Amazon Basin suggests that rainfall will be reduced even more than total evaporation from the landscape, so that runoff into streams will also decline (Shukla et al. 1990). The result may be a shift to a new climatic regime, with longer dry seasons and lower total rainfall, so that even with protection from cutting, forests may be unable to recover.

Deforestation effects may interact with increased atmospheric CO_2 and global warming. Models of global climate predict less warming in tropical regions than at high latitudes. Moderate warming, combined with deforestation, however, might lead to greater seasonal differences in rainfall over the humid tropics (Hartshorn 1992), as well as more frequent droughts (Bawa & Markham 1995). As a result, the frequency of fire in tropical forests might increase. Other possible changes include a greater incidence of destructive storms, such as hurricanes, and loss of the cloud forest climate in some mountain regions. The observation that tropical forests might already be experiencing an increased rate of tree death and replacement (Phillips & Gentry 1994) suggests that some of these effects are now occurring.

CONSERVATION OF TROPICAL FOREST DIVERSITY

International conservation groups have been cooperating with the governments of several nations to develop plans for biotic reserves to preserve the diversity of the forested tropics. In Amazo-

nia, a plan has been developed for a system of reserves covering 5%–6% of the basin. At least 48 proposed reserves are to be distributed systematically in eight biogeographic subdivisions of the basin. Half of these are to be larger than 5,000 square kilometers, a size thought to be adequate to reduce the rate of extinction of large mammals to less than 1% per century. These reserves are to be centered on areas of high biotic diversity, which may have been refugia of tropical forests during dry periods of the Pleistocene, corresponding to periods of continental glaciation at high latitudes. In 1990, a conference was held in Manaus, Brazil to identify priority areas for preserves based on biotic diversity, landscape features, and climate.

To evaluate sites for their preserve potential, a **Rapid Assessment Program (RAP)** has been established. This program relies on the expertise of individuals with intimate familiarity with major taxonomic groups, such as flowering plants, birds, and mammals. In a week or so, a team of such experts can obtain an inventory of a site's biodiversity adequate to evaluate its potential as a preserve. The danger of such an approach was revealed in 1994, when a small airplane carrying an RAP team crashed in remote forest, killing three of the world's foremost experts on tropical biodiversity.

Development of a network of biotic preserves for the Amazon Basin has proceeded well on paper, but many designated reserves lack effective protection from illegal exploitation and settlement (Johns 1988). The Brazilian government remains committed to agricultural and industrial development of the Amazon region. Unfortunately, the weak economies of Brazil and neighboring countries are likely to favor projects that involve rapid and poorly regulated exploitation of mineral, timber, and land resources, with little consideration of long-term ecological effects (Johns 1988).

SUSTAINED FOREST RESOURCE MANAGEMENT EFFORTS

Preserving tropical forest biodiversity, however, requires much more than a set of preserves. Ways of managing forested environments for sustained production of food and fiber for domestic consumption and export must be devised (Gradwohl & Greenberg 1988). Without such management strategies, the growing populations of tropical countries will be forced into destructive, short-term exploitation of forest resources simply to meet their immediate needs.

As one approach, Brazil has pioneered **extractive reserves,** in which the forest is protected from logging, but tree products such as rubber, oils, fruits, nuts, tubers, and other substances can be harvested on a sustained-yield basis (Fearnside 1989). The first extractive reserve, covering 4,000 square kilometers, was approved in 1989 in the state of Acre. In 1990, four additional reserves, totaling 21,600 square kilometers, were established and 10 additional forest areas were being considered. A related approach, known as **agroforestry,** is the development of multi-species, tree-dominated communities of plants that yield timber, fuelwood, basic food products, and industrial products such as rubber and oils. These ecosystems are intensively managed and exploited, but mimic the basic structure and function of a natural forest. A successful, small-scale example of agroforestry has been developed at Tome-Assu in the state of Para, Brazil (Jordan 1987). This project includes managed successional sequences in which annual crops such as corn, beans, and cotton are followed by perennials such as ginger and black pepper, and later by tree crops such as rubber, cacao, and various kinds of palms.

Several schemes for sustained-yield timber harvest from tropical forests have been suggested. Jordan (1982), for example, suggested a system of contour strip-harvesting, in which narrow belts of forest, following the contour of hillsides, are harvested at intervals of several years. Harvesting would also be designed to minimize the impact on the forest floor, so that root damage is small and nutrients can be captured within the cut area by trees capable of resprouting or in the unharvested zone downslope. In essence, the system simulates a small-scale pattern of shifting cultivation. A model project of this sort was begun in 1985 in the Palcazu Valley of eastern Peru (Hartshorn 1990), a wet montane area with nutrient deficient, acidic, clay soils poorly suited to permanent agriculture. Under management by a cooperative of local Amuesha Indians, income from the cutting of timber was projected to be about $3,500 per hectare harvested. Unfortunately, political instability in this region has hindered full development and evaluation of this project.

International conservation groups and the governments of countries such as Brazil are now pursuing ways to develop systems of preservation and sustained management of moist tropical forests. In January, 1990, a conference in Manaus, Brazil explored ways to manage development of 80% of the basin, emphasizing tourism, park development, and multiple-use systems of forest exploitation. In 1990, the International Tropical Timber Organization, with representatives of 23 producer countries, adopted A.D. 2000 as a target date for attaining sustainable production levels of timber. Most conservation experts believe that if effective plans for protection and sustained use are not implemented by this date, the world's moist tropical forests will be irreparably damaged.

SUMMARY

1. Moist tropical forests cover about a quarter of the land area of the tropics, and are concentrated in the New World, Africa, and Southeast Asia. These forests include lowland rain forests, floodplain forests, and deciduous forests, together with montane and cloud forests.

2. Moist tropical forests have the highest primary productivity and biodiversity of terrestrial ecosystems, and contain about half of all living species. Their productivity depends on biotic mechanisms of nutrient retention and recycling, particularly mycorrhizal roots of trees.

3. Over 142,000 square kilometers of tropical forest, or roughly 1.5% of the remaining forest, are probably being destroyed annually. The total loss has been least in the New World and greatest in Africa, but the continuing annual rate of loss is greatest in Asia.

4. Deforestation will lead to extinction of many species. Between 5% and 15% of tropical forest species probably will be lost between A.D. 1990 and 2020, with forest primates being one of the large animal groups most severely affected. Deforestation may also promote drier climatic conditions.

5. Innovative systems of sustainable food and timber production are needed to meet the needs of human populations and reduce the demands for exploitation and conversion of remaining undisturbed forests.

QUESTIONS FOR DISCUSSION

1. Considering the ecological risks associated with tropical deforestation, what sorts of scientific efforts, involving both ground-level and remote sensing technologies, do you think should be implemented?

2. Is developing ways of protection and sustainable use of moist tropical forests in the national interests of the developed temperate zone nations?

3. Do you think the promotion of commercial products derived from rain forest trees will help preserve tropical forests or hasten their conversion to plantations of selected species?

SUGGESTED READING

Collins, M. (Ed.). 1990. *The last rain forests. A world conservation atlas.* Oxford University Press, New York, N.Y. A richly illustrated survey of the remaining rain forests, their biota, and their indigenous peoples.

Moffett, M. W. 1993. *The high frontier. Exploring the tropical rain forest canopy.* Harvard University Press, Cambridge, MA. An illustrated summary of recent canopy investigations by an accomplished photographer and journalist.

Redford, K. H. 1992. The empty forest. *BioScience* **42**:412–22. An evaluation of the impacts of human activities on the animals of neotropical forests, and the implications of these impacts for forest ecology.

Tropical Savannas and Woodlands

Tropical savannas are grasslands with a scattering of shrubs or trees. **Tropical woodlands,** into which savannas grade, show a higher density of trees, but they do not form a continuous canopy, and grasses and herbs still form a dense ground layer. Tropical savannas and woodlands cover about 20% of the earth's land area (Fig. 10.1). They occupy about 65% of Africa, 60% of Australia, 45% of South America, and smaller areas in India, Southeast Asia, Central America, and other locations (Sarmiento 1984; Cole 1986).

Tropical savannas and woodlands are some of the world's most important wildlife ecosystems. African and Asian savannas are home to a rich diversity of ungulates, mega-herbivores such as elephants and rhinoceroses, and predators such as lions, hyenas, and canids. The savannas of East Africa contain the greatest biomass and diversity of large herbivores and their predators on earth. The Australian savannas support an endemic, diverse fauna of kangaroos, wallabies, and smaller marsupials. South American savannas, in contrast, have few large herbivores, aside from three species of deer and the capybara, a large rodent (Ojasti 1983). They do have a highly endemic fauna of smaller rodents, anteaters, armadillos, and canids. Savannas and woodlands in all areas except perhaps Australia face the challenge of rapidly growing human populations, and are thus becoming endangered ecosystems.

Plant and animal diversity in savanna ecosystems of East Africa are interested in a complex and intimate fashion.

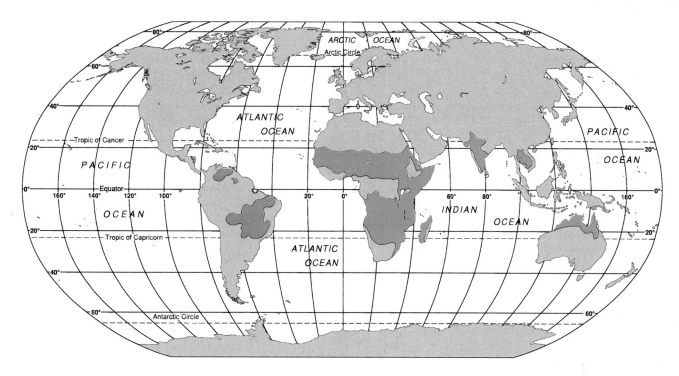

FIGURE 10.1

World distribution of tropical savannas.

THE SAVANNA AND WOODLAND ENVIRONMENT

Savannas and woodlands experience a highly seasonal climate with one or two dry and wet seasons annually. Total annual rainfall ranges from about 30 to 160 centimeters or more, but is often patchy and highly variable from year to year. Temporal and spatial heterogeneity thus play a major role in the ecology of these ecosystems (Sinclair 1995).

The pattern of wet and dry seasons results from the seasonal shift of the **intertropical convergence,** the belt of most intense solar heating of the land, and the zone into which the trade winds of the northern and southern hemispheres flow (Fig. 10.2). The intertropical convergence at any given time is close to the latitude where the sun is directly overhead, and thus most intense in its radiation. This overhead position migrates seasonally, passing northward over the equator on March 21, reaching the Tropic of Cancer on June 21, moving southward and passing over the equator again on September 22, and reaching the Tropic of Capricorn on December 22.

The zone of maximum solar heating is also a zone of convectional rainfall—thunderstorms created by the rising masses of heated, moist air. Proximity of the intertropical convergence thus means a rainy season. Near the equator, therefore, rainy seasons tend to occur around March and September, and dry seasons around December and June (although secondary influences tend to shift these periods somewhat). At the limits of the tropics, a single, short rainy season occurs when the intertropical convergence is nearby, and the rest of the year is dry. Thus, the sun's seasonal movement defines a complex pattern of dry and wet seasons within the tropics.

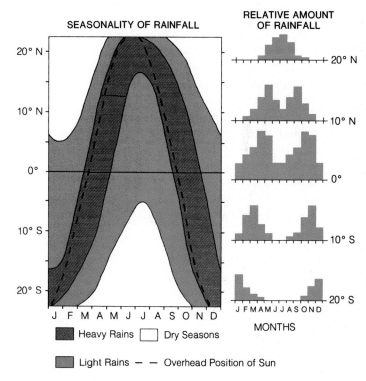

FIGURE 10.2

Wet and dry seasons within the tropics are determined by the seasonal movement in the overhead position of the sun, which determines the latitude of most intense heating of the earth's surface. Periods of heaviest rainfall, corresponding to the location of the intertropical convergence, tend to occur shortly after the direct overhead position of the sun has occurred at a given location.

(Modified from S. Niewolt. 1977. *Tropical Climatology.* John Wiley and Sons, London.)

Seasonal changes in moisture, combined with soil nutrient conditions, directly influence the growth of the vegetation. Grasses and forbs grow actively during the wet seasons, and their shoots die during the dry seasons, creating conditions favorable to fire. Fire is thus an important factor in almost all savanna and woodland areas. Frequent fire tends to kill woody plants and stimulate perennial bunch grasses. The chance of fire is, in turn, influenced by how much plant material has been harvested by herbivores. Thus, moisture supply, nutrient availability, herbivore activity, and fire are the major determinants of savanna and woodland ecology. Savannas in some areas, including parts of northern Australia, Central America, and northern South America, may actually be of human origin, the result of fires set by aboriginal peoples (Bourliere & Hadley 1983; Cole 1986).

The seasonality and variability of rainfall make migration one of the key patterns of adaptation to savannas by both wildlife and humans. The life cycles of migratory ungulates are closely keyed to the seasonal conditions of different areas. During the wet seasons, the plants that grow in the drier portions of the migrational range are typically higher in protein than those in wetter areas. They are also higher in many mineral nutrients (McNaughton 1990). These areas are thus optimal for feeding by animals in later stages of pregnancy, as well as for growth of their young. The drier areas are also more open, making it difficult for predators to stalk and kill ungulates. During the dry season, of course, the wetter areas of the migrational range form a drought refuge, with forage adequate for adult survival.

The native human peoples of savanna regions in Africa and Asia largely lived by **pastoralism,** the herding of animals such as cattle, sheep, goats, and camels. These livestock provided humans with most of their food and clothing, in the form of milk, blood, meat, and skins. Like migratory ungulates, pastoralists moved their herds seasonally to exploit grazing habitats that were optimal at different seasons. Many pastoralist peoples still retain a basic subsistence lifestyle, and are thus an intimate part of the ecology of savannas.

BASIC TYPES OF SAVANNAS AND WOODLANDS

The dominant plants of savanna ecosystems are varied in their life forms. Trees, shrubs, succulents, grasses, and broad-leafed herbaceous plants are all important in the vegetation. Often, communities dominated by plants of different life forms exist as a mosaic resulting from the complex interaction of climate, soils, fire, and animal influences (See Chapter 3). In these interactions, some of the large herbivores, such as elephants, giraffes, and hippopotamuses, play keystone roles (McNaughton & Sabuni 1988; Owen-Smith 1988, 1989). By their intense feeding on woody plants, for example, elephants can create and maintain the more open savanna and savanna woodland habitats that are optimal for many other large herbivores. Elephants are capable of pushing over trees up to half a meter in diameter to obtain foliage. They also rip off strips of bark with their tusks, eventually killing trees by girdling them. In dry years, even the large, soft-wooded

baobab trees can be killed when deep cavities are gouged into their trunks by elephants to obtain the moist xylem wood. Heavy elephant browsing can keep large areas of some trees, such as the mopane of southern Africa, in a resprouting, shrub-like condition, effectively preventing their maturation as normal trees (Owen-Smith 1988).

Based on fertility, savannas can be classified as **eutrophic,** with rich soil nutrient supplies, or **dystrophic,** with low nutrient availability (Scholes & Walker 1993). Combining fertility with moisture, four basic types of savanna ecosystems can be recognized. These types are thus defined by basic influences of climate and soils. They differ strikingly in their biotas, and especially the role of herbivores in nutrient recycling (Ruess 1987). Even more importantly, they differ in their inertia and resilience in the face of disturbance and stress imposed by human activities.

Dry eutrophic savannas are characterized by low rainfall, 30–70 centimeters annually, but possess soils rich in nutrients. These savannas have soils that have not been depleted of nutrients by leaching over a long period of geological time. Often, they occur on young volcanic soils in which weathering releases large quantities of mineral nutrients. In other cases, they exist on alluvial soils that receive nutrient inputs by occasional flooding. Although rainfall is low, the richness of the soil means that during wet seasons plants grow rapidly and produce forage of high nutrient quality. This type of savanna supports the greatest biomass and diversity of large animals. The high quality of the plant forage means that many kinds of vertebrates with specialized grazing and browsing diets can satisfy their food needs. The frequency of fire in this type of savanna depends on the intensity of herbivory, but often grazing is heavy enough that fuel for fires does not remain into the dry seasons. Likewise, consumption of most of the plant material by large herbivores means that the dead matter available to termites is relatively small. The high nutrient quality of plant material also promotes decomposer activity, so that decomposition is rapid during the wet season, and nutrients are quickly recycled.

Dry eutrophic savannas do not show the greatest inertia, as they are subject to droughts that may cause heavy mortality of plant and animal life. They are, however, highly resilient as a result of their fertility and of the adjustment of the vegetation to intense grazing. Savannas of this type are among the most important wildlife ecosystems on earth, and are exemplified by the Serengeti-Mara Ecosystem of Kenya and Tanzania (See Reading 10.1).

Moist dystrophic savannas contrast sharply with those described earlier. These savannas receive 60–160 centimeters of rain annually, but occur on soils that are ancient and have been leached of their nutrients over millions of years. The high rainfall permits a high level of primary production, but the bulk of the plant tissue produced is low in nutrient quality and high in fiber. Specialist herbivores consequently are often unable to meet their food needs, and the large herbivore fauna tends to be dominated by animals such as elephants and cape buffalo, which are able to process large quantities of plant material and extract the small fraction of digestible material present. In this type of savanna, therefore, the principal herbivores are frequently

THE SERENGETI-MARA ECOSYSTEM: A CASE STUDY

L ying just south of the equator in East Africa (Reading Fig. 10.1A), the Serengeti-Mara ecosystem contains the world's greatest concentration of large mammals (Sinclair 1979; Sinclair & Arcese 1995). Covering more than 25,000 square kilometers, this ecosystem centers on the Serengeti National Park in Tanzania and the neighboring Masai Mara National Reserve in Kenya. Climate and soils are key factors in the region's ecology. Northern and western areas, where total annual rainfall averages up to 120 centimeters, are covered by savanna and woodland. The southeastern portion of the region is plains grassland, with an annual rainfall of about 50 centimeters. Rainfall is strongly seasonal, coming mainly from December through April. Fertile volcanic ash soils permit high plant production when moisture is available.

The Serengeti-Mara ecosystem shows dramatically many of the major relationships that are important in ecosystems (Reading Fig. 10.1B). Among these are spatial differences in climate and soils, stochastic disturbances by physical and biotic factors, plant-herbivore interactions, and interactions of herbivores with the physical environment. The phenomenon of multiple stable states also is clearly shown. Increasingly, too, this ecosystem presents severe challenges of wildlife conservation in a region of rapidly growing human population.

More than 3 million ungulates of 28 species inhabit the Serengeti-Mara region (Reading Fig. 10.1C), the most abundant being the wildebeest (1.3 million), the Thomson's gazelle (440,000), the plains zebra (200,000), the topi (100,000), and the Cape buffalo (55,000). Of these, the wildebeest, Thomson's gazelle, plains zebra, and eland are mostly migratory; most individuals of the remaining species live in the same location throughout the year (Reading Fig. 10.1D). Together they consume an astounding 66% of the aboveground net primary production of the ecosystem—a fraction probably greater than for any other terrestrial ecosystem (McNaughton 1985). This intense grazing, rather than degrading the productivity of the system, may actually stimulate it. Areas subjected to natural grazing show about twice the net primary production of protected areas (McNaughton

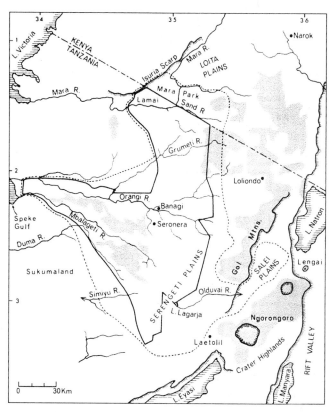

READING FIGURE 10.1A

The Serengeti-Mara ecosystem. The solid line indicates the boundary of Serengeti National Park, and the dashed line shows the total area used by migratory ungulates.

(From A. R. E. Sinclair, "Dynamics of the Serengeti Ecosystem" in A. R. E. Sinclair and M. Norton Griffiths, Eds., *Serengeti: Dynamics of an Ecosystem.* Copyright © 1979 The University of Chicago Press, Chicago IL. Reprinted by permission.)

1985). In part, this reflects the fact that grazing, with return of all elements to the soil of the system in urine, feces, and decomposed carcasses, stimulates rapid recycling of nutrients. Animals tend, in fact, to be concentrated in "hot spots," where certain nutrients are more abundant in the soils. In these places, intense grazing creates **grazing lawns** (Reading Fig. 10.1E), which are low, dense stands of grasses that are kept in an actively growing, juvenile state (McNaughton 1985).

The migrations follow a gradient of forage abundance and nutritional quality. The wildebeest, plains zebras, and Thomson's gazelles spend the wettest portion of

the year, typically from December through May, on the southern Serengeti plains. Here the calves are born, and the forage is most nutritious. As the forage begins to dry and disappear in May and June, the usual pattern has been for these species to move northwest into the western corridor, where they tend to remain until August or September. Following this, they gradually concentrate in the northern area, near the Kenya border. When the rains resume, the animals return south to the plains. In recent years, however, migration patterns have shifted somewhat, and many animals now move directly north and south between the plains and the Mara region.

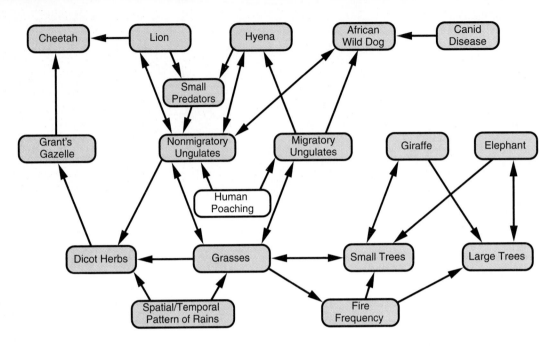

READING FIGURE 10.1B

Major interactions between climate, vegetation, fire, and animals in the Serengeti-Mara ecosystem.

(Modified from Sinclair 1979, 1995)

READING FIGURE 10.1C

Plains zebras are the migratory ungulates in the Serengeti-Mara ecosystem.

(Photo by G. W. Cox)

READING FIGURE 10.1D

About 100,000 topi, one of several nonmigratory ungulates, occupy the Serengeti-Mara ecosystem.

(Photo by G. W. Cox)

READING FIGURE 10.1E

Intense grazing by animals such as wildebeest can maintain grasses in a low, actively growing state known as a grazing lawn.

(Photo by Darla G. Cox)

Grazing by the migratory species also appears to be complementary. Zebras are usually the first animals to move into new areas. They feed in the tallest grasses, and consume the coarsest components of the vegetation. The wildebeest follow, foraging in the shortened stands of grasses and selecting primarily the more tender leaves. Finally, when the stands of grasses have been decimated and trampled by these species, the Thomson's gazelles appear to feed on re-sprouting grasses and the leaves and fruits of low broad-leafed herbs exposed by removal of the grasses. Whether this sequence results from the fact that grazing by each species benefits the next, or simply reflects a tendency for the species to avoid each other as forage begins to become scarce in the dry season is argued by different workers.

An equally impressive group of carnivores—lions, leopards, cheetahs, spotted hyenas, wild dogs, several jackals, and others—prey on these herbivores. The most abundant large predators are the spotted hyena, which numbers about 7,500 individuals, and the lion, numbering about 2,800. Clans of hyenas maintain permanent territories, but the clan members also commute long distances to reach migrating herds of ungulates when these prey are not available locally. Most of the other carnivores are permanent residents, and their numbers seem to be limited by the abundance of resident, rather than migrant, prey. One of the benefits of migration, in fact, may be that it enables these species to escape population limitation by resident carnivores (Fryxell et al. 1988). Hyenas and lions appear to be major agents of population regulation for other carnivores, including cheetahs, jackals, and smaller species. The African wild dog, long a species of concern because of its small numbers and declining population, essentially disappeared from the ecosystem in 1991. An epizootic of rabies and possibly distemper is likely to have been the major factor in this population crash (Ginsberg et al. 1995). Wild dogs are present in the surrounding region, and may recolonize the park.

Interactions of herbivores and plants are conspicuous (Sinclair 1979). Feeding by the large grass-eating herbivores, such as wildebeest, reduces the biomass of grasses, and permits an increase in the biomass of broad-leafed herbs, which are the preferred foods of gazelles, particularly the Grant's gazelle (Reading Fig. 10.1F). As the numbers of large grass-eaters have grown, so have the numbers of Grant's gazelles and their specialist predator, the cheetah (Reading Fig. 10.1G).

READING FIGURE 10.1F

The Grant's gazelle is a specialist feeder on broadleaf herbaceous plants.
(Photo by G. W. Cox)

READING FIGURE 10.1G

The cheetah is a specialist predator on the Grant's gazelle.
(Photo by Darla G. Cox)

Plant-animal interactions also involve the savanna trees (Sinclair 1979). Giraffes browse on small acacias, keeping them hedged and shrublike. Heavy giraffe browsing thus delays or prevents young acacias from maturing into trees (Reading Fig. 10.1H). This in turn favors the giraffe, since all of the foliage of small trees can be browsed by these animals. Elephants, now confined mostly to the Mara area in Kenya, browse trees and shrubs heavily, killing some and stunting others. In the 1950s and 1960s, elephants were forced into Serengeti Park from neighboring areas by farming developments. Their impact on acacia woodlands was great, reducing tree canopy cover by up to 50% (Pellew 1983). The greatest change was in the northern part of the park where acacia woodlands were most extensive. Elephants also browse heavily on tree seedlings and saplings, a factor preventing reproduction of trees in this part of the ecosystem. In this region, a transition from one ecosystem state, woodland, to a quite different state, open grassland, seems to be in progress.

Interaction of disease and animal hosts has exerted one of the most long-enduring influences on the Serengeti-Mara ecosystem. In the 1890s, a virus disease of cattle, known as **rinderpest,** reached sub-Saharan Africa. In a few years this disease decimated the native African cattle, causing famine among the people who raised them. It also greatly reduced the populations of large ungulates and, under the reduced grazing pressure, permitted trees to invade what were formerly open grasslands, creating the woodland areas of the present ecosystem. Major rinderpest outbreaks recurred until the early 1940s, but vaccination of domestic cattle has now largely

READING FIGURE 10.1H

Heavy browsing by giraffes can cause severe hedging of preferred woody plants, greatly slowing their growth to tree size.
(Photo by G. W. Cox)

eliminated the disease. After rinderpest control, the numbers of large herbivores climbed steadily. In 1961, wildebeest in the Serengeti-Mara area numbered only 250,000; now there are 1,300,000.

More change seems likely, as the growing human population of East Africa is now encroaching on this remarkable natural ecosystem. Over 1 million people now live along the ecosystem's western border. Poaching has nearly extirpated the black rhinoceros, and has reduced Cape buffalo numbers in the northwestern part of the park. Nearly 200,000 large animals are taken illegally by poachers annually. Most are wildebeest, for which the illegal harvest appears to have reached a maximum sustainable level. Resolving this human influence is the key to the survival of this ecosystem.

termites, which create distinctive landscapes dotted with their large nests. Termites may consume a third or more of the total primary productivity in these savannas (Josens 1983). Many termites process plant material by using it as a substrate for growth of certain fungi, which they then consume. The lower abundance of large herbivores means that much plant material remains in the dry season, making fire a nearly annual event in savannas of this type. Because of the low nutrient quality of the plant material, decomposition is slow, and nutrient recycling is much slower than in dry eutrophic savannas.

Moist dystrophic savannas and woodlands show greater inertia than the drier eutrophic savannas and woodlands. Ecosystems of this type are widespread in central and western Africa in the climatic belt immediately north of the forests of the Congo Basin and West African coastline. The extensive *miombo* woodlands of central and southern Africa exemplify this type of ecosystem. Many of the savannas of Brazil and northern Australia also fall in this category.

Dry dystrophic savannas combine the less desirable features of the first two savanna types. These savannas occur where shallow soils and less than average rainfall severely limit moisture availability to plants during the dry seasons. They have a low, highly variable productivity and a low biomass of vertebrate herbivores. Like moist dystrophic savannas, termites are major herbivores and fire is frequent. Both inertia and resilience are low, and the stresses of drought or overgrazing are likely to convert these systems to desertlike ecosystems. Savannas of this type occur in parts of Africa and South America.

Moist eutrophic savannas occur where basic climatic and soil conditions might permit forest or dense woodland types to exist, but where fire, animal, or possibly human influences have resulted in the conversion of these vegetation types to open woodland and savanna. Plant productivity and nutrient quality are high, and thus diversity and abundance of large herbivores are great.

Decomposition and nutrient cycling also are rapid. Inertia of this type of savanna is high, but resilience is limited. Excesses of fire or animal influence can transform the plant community to one dominated by a different life form. In fact, the diversity of animal and plant life in this type of system often depends on a delicate balance of soils-animal-fire influences that create a mosaic of communities dominated by plants of different life forms. Savannas of this sort also occur in parts of East Africa.

HUMAN POPULATIONS AND SAVANNA PRESERVES

Rapidly growing human populations, especially in Africa, Asia, and South America, are placing greater pressures on savanna areas that were once the exclusive domain of wildlife and subsistence pastoralists. Increasing conflict between wildlife and human interests is inevitable.

One of the most serious effects is increasing constraints on the ability of migratory species to move between their dry- and wet-season ranges. Nairobi National Park, for example, borders the city of Nairobi, Kenya. The park itself, only 114 square kilo-

meters in area, is continuous to the south with drier plains that are utilized and administered by the Masai tribe, whose traditional way of life is pastoralism, their livestock mostly cattle. Like the Serengeti-Mara ecosystem, the Nairobi Park region possesses migratory ungulates. During the wet seasons, plains zebra and wildebeest disperse from the park onto the plains to the south, and during the dry seasons they return to the park, where grazing conditions remain more favorable. Growth of the city of Nairobi, population growth of the Masai people themselves, and the gradual transition of subsistence pastoralism to more intensive ranching and farming are placing increased pressure on this wildlife ecosystem. These changes are tending to interrupt migration corridors and to confine wildlife more fully to the park itself.

Populations of pastoralists and their herds of livestock are increasing in and near many African wildlife areas. Traditionally, pastoralists have lived in relative harmony with wildlife, except for predators such as lions, since they did not hunt wild animals for food. Some ecologists have suggested that pastoralists' herds often increase to levels beyond those that can be sustained in dry years, resulting in range degradation (Lamprey 1983), which also affects wildlife. Others (Ellis & Swift 1988) doubt that pastoralism substantially contributes to range degradation. Nevertheless, increasing numbers of pastoralists and their livestock, together with other socioeconomic changes, will require careful management if savanna wildlife is to survive in many parts of Africa.

A major goal of conservation biologists in such areas is to find ways to involve local peoples in park management and wildlife conservation, so that they share both responsibility for and economic benefit from these activities. In Kenya, for example, ways are being explored to enable Masai pastoralists and wildlife to utilize the entire geographical area, both inside and outside parks, on which migratory animals depend (D. Western, *pers. comm.*). Specific policies include the sharing of park income with people living in areas used by wildlife outside the parks, promoting ecotourism in areas around parks, and even permitting limited use of park lands for livestock grazing. These policies are being implemented with considerable success in the Amboseli National Park region on the border of Tanzania.

Fencing of the borders of wildlife areas has sometimes had disastrous effects on migratory savanna animals. In Botswana, for example, long fences have been constructed to separate cattle ranching and wildlife areas, the objective being to reduce transmission of diseases from wild ungulates to cattle. These fences have diverted dry-season migrations of wildebeest from the extensive wetlands of the Okavango delta to Lake Xau, an area that itself is being used with increasing intensity by cattle ranchers (Williamson et al. 1988). Restricted access to permanent water during the dry season has contributed to a severe decline in numbers of wildebeest.

In Kenya, several game parks lie close to major cities and centers of intensive farming. Fencing has become necessary to separate wildlife from both farming and residential areas. Lake Nakuru National Park, established in 1960, lies in an area of the Rift Valley that was settled by Europeans in the early 1900s. Killing of wildlife by these settlers eliminated several species of large mammals, including lions, cheetahs, and hyenas. The park,

TABLE 10.1

PARK	PARK AREA KM²	NOW	NUMBER OF SPECIES AFTER ISOLATION
Serengeti NP, Tanzania	14,504	31	30
Mara NP, Kenya	1,813	29	22
Meru NP, Kenya	1,021	26	20
Amboseli NP, Kenya	388	24	18
Samburu NP, Kenya	298	25	17
Nairobi NP, Kenya	114	21	11

The Numbers of Large Herbivores Expected to Survive in Some East African National Parks if the Effective Area of Wildlife Habitat is Reduced to the Legally Protected Minimum Area

Data from D. Western and J. Ssemakula, 1981. "The Future of the Savannah Ecosystems: Ecological Islands or Faunal Enclaves?" Oxford, England: Blackwell Scientific Publications Ltd.

with a total area of 200 square kilometers, 38 square kilometers of which are covered by the lake itself, lies adjacent to the city of Nakuru, now a major commercial and industrial center. This park has become the first East African park to be completely fenced. Freed from predators, populations of waterbuck grew to about 6,000 animals and impala to about 4,000 individuals, and severe overbrowsing of the woody vegetation became apparent. More recently, hyenas and lions have been reintroduced. This park now faces the challenge of maintaining a balanced system of predators, prey, and vegetation in a relatively small, fully enclosed environment.

FRAGMENTATION OF SAVANNA WILDLIFE HABITAT

Like forests, savanna ecosystems are being fragmented, and the remaining parks are becoming islands in agricultural and urban landscapes. How serious this isolation is for the survival of the wildlife of these parks is uncertain. Miller and Harris (1977) noted that Mkomazi Game Reserve in Tanzania has lost several species of large mammals, perhaps in part because of its isolation. When it was established in 1952, the reserve had 43 species of large mammals; by 1977 it had only 39 (Miller 1978). Projecting this rate of loss, Miller suggested that in about 114 years, the game reserve would only contain 22 species of large mammals.

Soulé et al. (1979) offered an even grimmer scenario of probable losses from isolated preserves, based on a projection of losses at a rate like that seen on newly isolated oceanic islands. They predicted that losses of all species over 500 years might range from 34%, for the largest parks, to 65%, for the smallest preserves. Western and Ssemakula (1981), on the other hand, noted that the relation of number of species to park area was very weak, and that projection of losses based on faunal relaxation theory (See Chapters 12, 27) was not well-justified. They also suggested that long coevolution of savanna animals greatly reduces the chance of competitive exclusion. They concluded that even small parks, such as Nairobi National Park, could be viable "faunal enclaves" as long as wildlife could continue to make reasonable use of areas outside the park boundaries. Nevertheless, Western and Ssemakula (1981) concluded that if parks were reduced from the effective ecological units now

being used by animals to the legally protected areas alone, significant losses would likely occur (Table 10.1). For Nairobi National Park in Kenya, for example, only 11 of the 21 species now present were predicted to survive.

East (1981, 1983) noted that many of the smaller savanna parks have very small populations of many large mammals—often less than 25 individuals. In these cases, local extinction can easily occur due to any of several factors, including drought, poaching, fire, disease, and competition from livestock. Problems of inbreeding may also arise in such small populations (See Chapter 25). Thus, active management on a regional basis will be required to maintain species diversity of many savanna parks.

As for desert regions, global warming is likely to intensify the problems faced by savanna preserves. Many are likely to experience hotter and drier climates, which are likely to reduce their carrying capacity for savanna wildlife.

POACHING THREATS TO ELEPHANTS AND RHINOCEROSES

African elephants have declined greatly in numbers in recent years, largely due to poaching for ivory. African elephants occur in 35 countries, but their populations are declining in at least 28 of these. In 1980, the population of African elephants (Fig. 10.3) was estimated at about 1.3 million animals. In East Africa as a whole, the center of wildlife tourism, numbers declined from 353,000 in 1977 to 88,650 in 1989. By 1991, leading elephant biologists estimated that the population had declined to about 471,500 animals (Table 10.2). These estimates are rather uncertain, however, and the numbers of animals remaining in the forested regions of the Congo Basin, in particular, are poorly known. Estimates of forest elephants in Cameroon, Central African Republic, Congo, Equatorial Guinea, Gabon, and Zaire in 1989 suggested that they may total only 140,000 (Michaelmore et al. 1994). Thus, the total estimate of forest and savanna elephants of 274,910 for these countries in 1989 might well have been two times too high. The 1995 population of African elephants might well be little more than 300,000.

Although poaching has been the main cause of the decline of African elephants, several other factors have contributed. In

FIGURE 10.3

In Kenya, African elephants have declined in numbers from more than 130,000 in 1973 to about 16,000 in 1989 due to drought and poaching.

(Photo by Darla G. Cox)

FIGURE 10.4

During the 1980s, the black rhinoceros suffered heavy poaching for its horn in most of central and eastern Africa.

(Photo by G. W. Cox)

TABLE 10.2

Estimated Populations of the African Elephant in 1991*	
REGION	ESTIMATED NUMBERS
Western	14,575
Central	174,588
Eastern	77,290
Southern	205,000
TOTAL	**471,453**

Data from Morrell, 1990.

Uganda, the chaos that prevailed in the years during and after the dictatorship of Idi Amin led to the decimation of elephant populations in all of the national parks by poachers and military personnel. In Kenya, drought and poaching caused the loss of two-thirds of the elephant population during the 1970s and 1980s. The population is recovering slowly in the 1990s.

In late 1989, the Convention on International Trade in Endangered Species (*See* Chapter 26) designated the African elephant as endangered and banned all trade in animals and ivory. Several nations of southern Africa have opposed this ban, arguing that they are managing their elephant populations well, and should be able to cull animals and sell their products to earn money for wildlife conservation.

The African black rhinoceros (Fig. 10.4) was originally distributed throughout most of the savanna region of the continent. The African white rhinoceros was formerly widespread in the savannas of southern Africa and in a small area from Uganda west into the Central African Republic (Owen-Smith 1988). The range of the black rhinoceros has been reduced to isolated pockets in central, eastern, and southern Africa. Natural populations of the white rhinoceros survive only in Garamba National Park, Zaire, and in game reserves in Natal Province, South Africa (Owen-Smith 1988). The Indian rhinoceros formerly

occurred along the foothills of the Himalayas from Pakistan to easternmost India; it now survives in isolated preserves in India, Bhutan, and Nepal.

Rhinoceroses are poached primarily to obtain the horn, which has purported medicinal values in Asia, and is carved into elaborate dagger handles in countries of the Arabian Peninsula. In Asia, powdered rhino horn is a common component of medicines for skin diseases, muscle aches and sprains, and colds, as well as being considered an aphrodisiac. In 1989, the retail value of powdered horn for these uses in Taiwan was about $42,880 per kilogram for Asian horn and $3,347 per kilogram for African horn (Martin & Martin 1989). In various countries of the Arabian Peninsula, elaborately carved daggers are traditionally given by fathers to their sons at puberty, and the most prestigious of these have handles made from rhinoceros horn. African rhino horn entering the dagger trade has a wholesale value of about $500 per kilogram, with the carved dagger handles retailing for up to $12,000 each.

The various uses of African and Asian rhino horn resulted in the killing of about 2,580 animals per year in the late 1970s and early 1980s. During this period, populations of the black rhinoceros and the northern white rhinoceros were decimated in several countries of central and eastern Africa (Table 10.3). The total population of the black rhinoceros declined from almost 15,000 in 1979 to little more than 2,500 in 1993 (Brooks 1994), and the northern white rhinoceros was extirpated except in Garamba National Park, Zaire. In recent years, heavy poaching has also spread into more southern countries of Africa.

In Zimbabwe and Namibia, dehorning of wild black rhinoceroses has been tested in an effort to discourage poaching. In these regions, the rhinoceroses were believed to use their horns mainly in territorial fighting with other animals, rather than in feeding or predator defense, so that loss of the horns would not be detrimental. The cut horns regenerate rather quickly, however, and conflicting evidence exists about the reproductive success of normal versus dehorned females.

TABLE 10.3

Populations of Rhinoceroses in the Wild*			
SPECIES AND COUNTRY	1979	1988	1993
African Black	14,875	3,780	2,550
Ctr. Afr. Republic	3,000	50	0
Kenya	1,500	350	417
Tanzania	3,795	400	132
Zambia	2,750	200	33
Zimbabwe	1,400	1,750	381
South Africa	630	630	897
Other	1,710	400	690
African White	3,841	?	6,784
Northern			
Zaire	400	22	32
Sudan	400	0	0
Other	21	0	0
Southern			
Zimbabwe	180	?	134
South Africa	2,500	?	6,376
Other	340	?	242
Indian		1,200–1,500	1,900
Sumatran		500–900	500
Javan		65–70	50

*Compiled from various sources

Stable populations of both black and white rhinoceroses exist in parts of southern Africa, particularly South Africa. In Natal Province of South Africa, Hluhluwe and Umfolozi Game Reserves hold large numbers of both species. These populations are a source of animals for reintroduction elsewhere. A coordinated plan has been developed for the conservation of the black rhinoceros in South Africa and Namibia (Brooks 1989). Under this plan, the population, representing three subspecies, could be increased from about 1,000 animals to nearly 5,000.

In Kenya, efforts are being made to aggregate black rhinoceroses from scattered small populations into larger groups in relatively secure areas. Lake Nakuru National Park, which is now completely fenced, has been chosen as one rhinoceros sanctuary (Lever 1990). Seventeen rhinos from a private ranch have been moved to the park, where only two animals had existed. Other fenced or carefully protected sanctuaries are being established in Tsavo, Aberdare, and Nairobi National Parks, as well as on certain private ranches (Cohn 1988).

RESTORING THE ECOLOGICAL BALANCE IN SAVANNA PRESERVES

In spite of the wild character that still invests many of the regions where savanna parks exist, the ecosystems of many of these parks are missing some of their original members. In other cases, the confinement of large animals within park boundaries has resulted in exaggerated impacts of these species on the park environment. Efforts are now being made to restore natural relationships in several locations.

In Natal Province, South Africa, elephants were eliminated from the region now occupied by Hluhluwe ("Shush-SLOO-wee") and Umfolozi National Parks in the late 1800s.

Large predators such as lions, cheetahs, and wild dogs were apparently also eliminated. These two parks, connected by a broad corridor of wild land through which large animals move freely, have seen major vegetational change since the 1930s, when they were established. This change is apparently due to absence of the elephant, a keystone herbivore, and of the several predators. Protected populations of many savanna herbivores reached high densities, exerting heavy pressure on the vegetation. Woody plants invaded open grassland, in spite of controlled burning. In turn, major declines of black rhinoceros and several savanna ungulates occurred, with local extinction of at least three species (Owen-Smith 1989).

To restore the biotic balance of Hluhluwe and Umfolozi Parks cheetahs and African wild dogs were reintroduced in the 1970s. This effort was aided by the natural recolonization of the area by lions. In the early 1980s, elephants were reintroduced to the two parks, and by the late 1980s their population had grown to about 100 animals (Owen-Smith 1989).

In Amboseli National Park, Kenya, in contrast, poaching in surrounding areas led to a high, year-round concentration of elephants within the park. This dense population led to widespread destruction of trees and shrubs, converting woodlands and thickets to open grasslands (Western 1989). Meanwhile, outside the park, areas without elephants experienced brush invasion, which reduced forage for cattle and increased the potential for livestock disease transmitted by the tsetse fly, which flourishes in dense brushland. A secondary effect of the vegetational change has been to enable tourist vehicles to penetrate into almost every corner of the park. The impacts of tourist vehicles on park landscapes in East Africa have now become a serious problem in themselves. Ways to encourage elephant use of areas outside the park are now being sought, in the interest of improving both environments.

SUMMARY

1. Tropical savannas, grasslands with scattered shrubs or trees, grade into tropical woodlands with a higher density of trees, but not a continuous canopy. These ecosystems, rich in vertebrate wildlife, are widespread in Africa, South America, and Australia.

2. Savannas can be classified according to high or low soil fertility and high or low moisture availability, with those richest in vertebrate wildlife having high fertility.

3. The Serengeti–Mara ecosystem of East Africa illustrates the complex interactions of soils, climate, fire, historical factors, and plant-herbivore and predator-prey interactions. Poaching by humans is now becoming a major factor in the dynamics of the ecosystem.

4. Savanna preserves are tending to become islands of wildlife habitat in landscapes dominated by human activity. Conservation strategies that encourage people to preserve wildlife because of its value to them are being sought in several countries.

5. Poaching has seriously threatened survival of elephant and rhinoceros populations in much of Africa and Asia.

QUESTIONS FOR DISCUSSION

1. In what ways can activities of animals, both invertebrates and vertebrates, affect abiotic and vegetational conditions of the landscape? To what extent are patterns of diversity of landscape conditions, vegetation, and animal life mutually interdependent in savanna ecosystems?

2. If the Serengeti population of wildebeest crashed due to excessive killing of animals by humans, how do you think the rest of the ecosystem (plants, animals, soils) would respond?

3. Do you think that developing ways for human populations to obtain economic value from the wildlife ecosystems with which they live in close association is a recipe for protecting these ecosystems, or one for destroying them?

SUGGESTED READING

Sinclair, A. R. E. and P. Arcese (Eds.). 1995. *Serengeti II: Dynamics, management, and conservation of an ecosystem.* Univ. of Chicago Press, Chicago. A comprehensive summary of recent research on the world's most remarkable savanna ecosystem.

Solbrig, O. T. and M. D. Young. 1992. Toward a sustainable and equitable future for savannas. *Environment* **34**(3):6–15, 32–35. A nontechnical analysis of human impacts on savanna ecosystems and a discussion of policies for their sustainable management.

Werner, P. A. (Ed.). 1990. Savanna ecology and management. *J. Biogeogr.* **17**:341–557. A collection of 23 articles, most fairly technical, on aspects of ecological function in savannas throughout the tropics.

Coastal Ecosystems

chapter 11

S ome of the richest terrestrial ecosystems are those that border lakes and oceans. Beaches and coastal dunes, both on mainland areas and on barrier islands, together with sea cliffs, are the most important of these coastal systems. These environments are ecosystems in their own right, but they are also breeding or nonbreeding habitats for many animals that are also members of other ecosystems. These include waterfowl and other freshwater birds, many raptors and seabirds, and all species of marine turtles. Still another coastal ecosystem of great importance is salt marsh. We shall consider salt marshes as integral parts of estuarine marine ecosystems (*See* Chapter 18).

The immediate coastline also attracts human development, both commercial and residential. Because of their highly dynamic geological nature, however, many coastal environments that are vital to wildlife are poorly suited to intensive development. Sandy coastal environments, including barrier islands, fall into this category. Severe storms can reshape these areas overnight, destroying property worth millions of dollars. Unfortunately, many such areas have been developed, often with the aid of government subsidies for road and bridge construction. Following storm damage, many such areas have received still other subsidies in the form of emergency relief funds and low-interest disaster loans to repair damage. Understanding the dynamics of coastal environments, and planning appropriate human uses that are compatible with their wildlife values are major conservation challenges.

Cormorants and other water birds, mammals, and reptiles depend on specialized coastal habitats.

Outline

FIGURE 11.1

Marram grass and other colonists of sandy beaches spread by rhizomes that connect individual shoots and anchor them in the loose substrate.

(Photo by G. W. Cox)

FIGURE 11.2

Colonization of inland edge of the beach at Cape Hatteras, North Carolina, by marram grass results in the deposition of windblown sand and the formation of a foredune.

(Photo by G. W. Cox)

ECOSYSTEMS OF SANDY COASTLINES

Beaches and coastal sand dunes occur along most ocean coastlines, as well as along the shores of many large lakes. They are best developed where rivers discharge large volumes of sandy sediment into shallow coastal waters, or where rising sea levels have caused the erosion of headlands, creating large volumes of sandy sediment along the coastline itself. The combined influences of a sandy substrate and marine (or lake-influenced) climate usually make sandy coastal ecosystems sharply distinct in habitat conditions from the ecosystems lying inland. Often, too, these systems are discontinuous, being separated by stretches of rocky coastline. The distinctiveness and discontinuity of these ecosystems mean that they often contain isolated, endemic populations of plants and animals.

Sandy coastal ecosystems are highly dynamic. The beaches themselves change in form seasonally. During periods of frequent storms, intense wave action erodes the beach, carrying sand into deeper water and producing a long, flat profile across which the waves expend their energy. During storm-free periods, gentle waves carry sand toward the shore, building up a short, steep beach leading up to a flat berm or beach terrace. These changes can usually be seen on lake and ocean beaches in the temperate zone from winter to summer.

Sandy areas inland from the beach are also highly dynamic. The first plant colonists of newly deposited sand are often perennial grasses that spread by rhizomes (Fig. 11.1). These grasses trap windblown sand and cause the formation of a **foredune ridge** (Fig. 11.2). They also begin to stabilize the sand surface enough to permit other plants to become established, facilitating biotic succession. Succession often leads to a dense vegetation of grasses, shrubs, and trees that cover and stabilize the sandy substrate.

As long as the vegetation cover is undisturbed the landscape will be stable. Where this cover is disturbed, however, an

FIGURE 11.3

Disturbance of the plant cover in Indiana Dunes State Park, at the south end of Lake Michigan, has led to the formation of a blowout. Inland from the blowout, sand is deposited in an advancing dune front.

(Photo by G. W. Cox)

even more dramatic example of the potential instability of sandy coastal environments may occur. Wind action may initiate a **blowout**—an erosion basin that gradually enlarges as wind sweeps out sand and carries it inland, where it is deposited as an advancing dune front (Fig. 11.3). The advancing dune may bury mature forests, highways, and residential developments. Blowout formation is a normal phenomenon of coastal dunelands, but, of course, a process that can also be triggered by disturbance of vegetation by human activity.

In North America, coastal sand dune systems occur in many locations along the Atlantic and Pacific Coasts, as well as along the shores of the Great Lakes. In Oregon, Washington, and

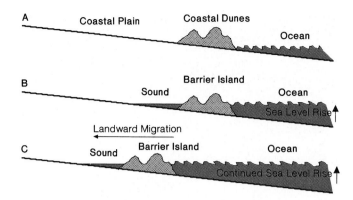

FIGURE 11.4

Barrier islands are believed to have originated by the isolation of beach and dune complexes by rising sea levels as continental glaciers melted. *A.* Formation of coastal dunes and beach ridges at a low interglacial sea level. *B.* Rising sea level caused by glacial melting allows water to flood in behind coastal dunes and ridges, isolating them as barrier islands. *C.* Continued sea level rise, coupled with the dynamics of transport of sand from the front to the back of barrier islands, causes their landward migration.

FIGURE 11.5

At Canaveral National Seashore, Florida, wind transport, storm overwash, and inlet formation have all carried sand landward from the ocean side of the barrier island, at right, to create a mosaic of peninsulas, islets, and marshes on the inner side of the barrier. (U.S. National Park Service photo)

northern California, coastal dunelands reach perhaps their greatest extent and size. Here, these ecosystems occupy about 42% of the coastline, often extending inland for 2–4 kilometers (Wiedemann 1984). These dune ecosystems have developed since the last glacial maximum, and are often perched on nonsandy marine terraces below sea level. Much of this coastal duneland has been covered and stabilized by vegetation, but spectacular, active blowouts and high dunes exist in many locations.

Barrier islands and beaches are also important types of sandy coastal ecosystems. **Barrier islands** are long, narrow islands formed from beach sediments and lying more or less parallel to the seacoast. **Barrier beaches** are similar, differing only in that they are connected to the mainland at one end. Barrier beaches and islands lie offshore from about 10% of the world's seacoasts. These landforms are well developed along the Atlantic and Gulf coasts of the United States, where about 295 barrier islands total 6,500 square kilometers in area.

Barrier beaches and islands are one of the most dynamic landforms on earth. Several mechanisms of origin of barrier islands have been suggested, but the most accepted is the **mainland beach detachment hypothesis** (Fig. 11.4). This hypothesis suggests that barrier islands originate at the end of periods of continental glaciation, when ocean levels are beginning to rise as a result of the melting of continental ice sheets (Dolan et al. 1980). At the height of the last Pleistocene glaciation, for example, sea level was about 120 meters below its present level, and the coastline of North Carolina and other regions now bordered by barrier islands was far oceanward. As sea level rose, headlands were eroded, producing large volumes of sediment that contributed to the formation of extensive coastal beach and dune systems. The rising seas also flooded inland in river lowlands and embayments, spreading parallel to the coast behind the coastal dunes. Eventually, the rising sea completely cut the beach and

dune complex from the mainland, forming a barrier island that shelters a **sound** on its inner side. Some barrier beaches and islands also form by the growth of sand spits parallel to the coast, beginning from a headland. Breaching of a long spit by one or more inlets then may create a barrier island system.

After barrier islands were formed in late glacial time, they began to migrate coastward. This migration results from the transport of sediment from the ocean side to the sound side of the island. Sediment transport occurs in several ways (Fig. 11.5). Wind carries sand inland from the beach, and blowout formation creates active dunes that advance toward the inland side of the island. During winter storms or hurricanes, heavy waves on top of a storm surge may break through the dune system and overflow the island, washing large quantities of sand to the back side of the island, and extending the island into the sound. Storm waves may also cut through the island, forming a new **inlet** or **pass** through which currents may carry sediment into the sound, depositing it as an underwater delta. Landward migration has enabled barrier islands to keep up with rising sea level, which would have inundated them in their original location.

Historical records document major changes in the position of island shorelines due to landward migration. Along much of the coastline of Virginia and North Carolina, the average rate of migration is about 1.2 meters per year, but this rate is as great as 8.0 meters per year in a few spots (Inman & Dolan 1989). The shells and other biotic relics that one finds on the ocean side of a barrier island also attest to island migration. These are often the shells of mollusks that live in the sheltered sounds on the opposite side of the island. The island has simply migrated across the site where these shells were deposited in sound sediments, and they have been exposed on the beach. In a number of locations tree stumps—relics of a forest once growing on the inland side of the island—are exposed on barrier island beaches.

The fate of barrier islands is thus eventually to fuse with the coastline. "Fossil" barrier islands from the last interglacial, such as the Ingleside Barrier along the Texas coast between Corpus Christi and Galveston, demonstrate this fate. The Ingleside Barrier is a sandy coastal formation fused with the continental coastline, but bordered oceanward by Matagorda Island and other modern barrier islands.

Another dynamic feature of barrier islands is the migration of the passes or inlets between islands. Longshore currents tend to deposit sediment on the up-flow side of passes and remove sediment on the down-flow side. Again, large changes in the measured position of passes are documented by historical records.

The dynamic nature of beach and dune systems, and especially of those that develop into barrier beaches and islands, means that these environments are poorly suited to urban development. Most parts of the United States Gulf and South Atlantic coastline have a 5%–15% chance of being hit by a hurricane in a given year. Nevertheless, summer home developments and resorts exist on—in some cases completely cover—70 barrier islands from New Jersey to Texas. Almost annually, urbanized barrier islands somewhere experience massive damage as a result of the natural processes that reshape beaches and dunes, and slowly carry the islands landward. In 1995, for example, Hurricane Opal caused over $1 billion in damage to property on barrier islands and beaches along the Florida Panhandle. Damage of this nature and scale has almost always led to massive expenditures of public funds to replace roads, bridges, and other public facilities that many feel should never have been built in the first place.

PLANT AND ANIMAL LIFE OF SANDY COASTAL ECOSYSTEMS

Plant Communities

From the beach inland, a transect of successional development of plant communities is usually evident. Grasses that invade the beach and initiate foredune growth include marram grass (*Aimophila breviligulata*) and sea oats (*Uniola paniculata*). On tropical and subtropical beaches, beach morning-glory (*Ipomoea pes-caprae*) is a sprawling vine that invades beaches. On the Pacific coast of North America, a variety of prostrate shrubs and vines form a distinctive coastal strand community. Farther inland, other grasses, broad-leafed herbs, vines, and shrubs invade the stands of the pioneers. Along ocean coastlines, tolerance of salt spray is a requisite of all of these early successional species.

Succession often leads to forest communities with a distinctive composition and nutrient cycling regime. On Fire Island National Seashore, off the coast of Long Island, New York, for example, a 200- to 300-year-old forest of American holly, white sassafras, and shadbush has developed on the protected inner side

FIGURE 11.6

The piping plover, a nesting shorebird of lake and ocean beaches, is designated as endangered or threatened in all of its North American range.

(U.S. Fish and Wildlife Service photo by J. P. Mattsson)

of the barrier island (Art 1976). Known as the Sunken Forest, this community has a nutrient budget dependent almost entirely on atmospheric inputs.

At the south end of Lake Michigan, succession leads through a complex sequence of cottonwood, pine, oak, and mixed hardwood forests. Between dune ridges paralleling the shore, pond and marsh communities also occur, and in drier locations tall-grass prairie occupies dune sands. These communities occur on beach ridges of Lake Michigan that were formed at different stages in glacial retreat, and are of differing age. These dunes, in fact, were the site at which the concept of plant succession was recognized. H. C. Cowles, who began his studies of dune succession in 1896, was the first of several prominent ecologists to study successional processes at this location. Indiana Dunes State Park, established in 1923, and Indiana Dunes National Lakeshore, created in 1972, preserve the full range of this remarkable successional sequence.

In the Pacific Northwest, the humid climate and complex microtopography of the coastal dunelands likewise result in a complex of communities, some 21 in all (Wiedemann 1984). The isolation and the distinctive nature of coastal dunes mean that they often contain endemic species. The Nipomo Dunes, in coastal Santa Barbara County, California, for example, contain 18 species of rare or endangered plants, several of them highly adapted to the dune environment (Smith 1976).

Terrestrial Vertebrates

Sandy coastal ecosystems, especially barrier islands, are breeding grounds or winter homes to many water birds. Many species of gulls, terns, plovers, herons, egrets, ibis, rails, and other water birds nest on barrier islands or in the salt marshes bordering the sounds on their landward sides. Migratory shorebirds and waterfowl use many of the same areas as stopover points on migration,

FIGURE 11.7

The key deer, restricted to a few islands in the Florida Keys, is a small subspecies of the white-tailed deer with a total population of a few hundred individuals.

(U.S. Fish and Wildlife Service photo by John Oberheu)

or as wintering habitats. Several species of raptors, including the bald eagle, osprey, and peregrine falcon, are closely associated with coastal ecosystems, especially in winter.

Most immediately threatened by development and disturbance of sandy coasts are nesting species such as the least tern and the piping, snowy, and Wilson's plovers that nest on open sandy beaches (Fig. 11.6). Least terns require long stretches of bare sand for nesting, their protection against predators being isolation and camouflage of their nests and young. Increasing human use of beaches has led to the disappearance of these species from much of their original ranges.

Under the geographic isolation provided by barrier islands, many endemic forms of terrestrial vertebrates have evolved. Because of destruction of vegetation, many of these have become endangered. Along the coast of Georgia, the Cumberland Island pocket gopher, *Geomys cumberlandius,* a barrier island endemic species, is now considered endangered. Several endemic subspecies of beach mice, cotton rats, and rice rats on barrier islands along the Florida and Alabama coasts are also listed as threatened or endangered because of habitat disturbance.

The Florida Keys, a chain of islands of tropical reef origin, harbor many endemic reptiles and mammals, several of which are threatened or endangered (Schomer & Drew 1982). These include the endangered key deer, *Odocoileus virginianus clavium,* a diminutive subspecies of the white-tailed deer (Fig. 11.7). Reduced to less than 30 individuals in the late 1940s, this subspecies has recovered, under protection, to several hundred animals. Other endangered forms include endemic subspecies of the raccoon, marsh rabbit, and several rodents. Several tropical West Indian birds and reptiles also reach their northernmost distribution in the Florida Keys.

FIGURE 11.8

The Kemp's ridley turtle nests exclusively on a barrier beach at Rancho Nuevo, Tamaulipas, Mexico. Because of the rapid decline of this species, eggs are collected for incubation in protected enclosures.

(U.S. Fish and Wildlife Service photo by David Bowman)

Sea Turtles

Sea turtles are intimately tied to sandy coastal ecosystems. Of the seven species of sea turtles, six are formally classified as threatened or endangered (Table 11.1). These species are very diverse in their habits, some being herbivores, others carnivores (Hendrickson 1980). Five species are widely distributed in tropical and subtropical oceans, but two are much more restricted. The Kemp's ridley is confined largely to the Caribbean and Gulf of Mexico, and the Australian flatback is limited to the northern coastal waters of Australia (Zangerl et al. 1988).

Sea turtles are tied to sandy coastal ecosystems by their nesting habits. The females of all species come ashore to deposit their eggs in nests excavated in beach sands (Fig. 11.8). From 50 to 250 eggs are laid at a time, with some females nesting several times in one season. The eggs incubate for 45 to 75 days by the heat of the sun-warmed sand. When the young hatch, they crawl immediately to the water and begin a juvenile life that may be 10 to 50 years in duration, during which they may travel

TABLE 11.1

Sea Turtles and their Ecological Characteristics. Endangered (E) and Threatened (T) Species are Designated			
SPECIES	HABITAT	MAJOR FOOD	NESTING
1. **Leathery Turtle (E)** *Dermochelys coriacea*	Tropical open ocean	Jellyfish	Scattered tropical beaches (colonial)
2. **Hawksbill Turtle (E)** *Eremochelys imbricata*	Tropical oceans near coral reefs	Sponges	Coral reef beaches (solitary)
3. **Kemp's Ridley Turtle (E)** *Lepidochelys kempi*	Caribbean coastal waters	Crustaceans	Rancho Nuevo beach, Tamaulipas, Mexico (colonial)
4. **Olive Ridley (T)** *Lepidochelys olivacea*	Tropical coastal waters except Caribbean	Crustaceans	Scattered tropical beaches (colonial)
5. **Loggerhead Turtle (T)** *Caretta caretta*	Subtropical to temperate coastal waters	Mollusks and crustaceans	Scattered beaches (colonial)
6. **Green Sea Turtle (T)** *Chelonia mydas*	Tropical and subtropical coastal waters	Sea grass, algae	Scattered beaches (colonial)
7. **Flatback Turtle** *Natator depressus*	Northern Australian coastal waters	Reef and benthic invertebrates	Tropical beaches (colonial)

thousands of kilometers from their home beach. Recent observations, for example, indicate that over half of the juvenile loggerhead turtles in the Mediterranean Sea were hatched on North American beaches. During their first year, turtles of several species may live a pelagic life in association with masses of *Sargassum* seaweed accumulating in the centers of oceanic gyres and in drift lines formed where surface water masses converge and sink (Carr 1987). In the North Atlantic, the Sargasso Sea is such an area.

Eventually, sea turtles return as adults to mate and lay eggs on or near their natal beaches. Some species nest at widely scattered locations, but others concentrate their nesting on certain beaches year after year. At Tortuguero Beach on the Caribbean coast of Costa Rica, for example, green, hawksbill, and leatherback turtles nest in large numbers (Carr 1967). Adults of these species congregate offshore over a period of days or weeks, and then come ashore together in what is termed an **arribada** (arrival), concentrating their egg-laying into a period of a few nights.

Commercial exploitation of sea turtles for a variety of body parts and products has been a major cause of decline of all the endangered and threatened species (King 1981; Ross 1981). All are hunted for their meat, and the nests of all species are robbed of eggs for human consumption, especially in parts of Latin America where the eggs are believed to impart sexual vigor. The green turtle is hunted heavily for its **calipee,** the cartilaginous material lining the inner side of the ventral shell that is the essential ingredient of green turtle soup. Green turtle oil is used in certain cosmetics, and the skin and shell of the ridleys and the hawksbill are used for manufacture of leather goods and trinkets. Stuffed juveniles, mainly green and hawksbill turtles, are also a common tourist item. A 1987 analysis of turtle products sold in Japan indicated that this trade involved about 2 million animals annually, almost a million of which were stuffed juveniles (Gregg 1988).

Sea turtles have been hunted for centuries. Indeed, one of the earliest laws relating to wildlife exploitation in the New World was made to protect young sea turtles in the Bahamas (Carr 1967). Often, hunting is concentrated on or offshore from nesting beaches. This hunting, together with the raiding of nests by animals such as dogs and pigs that forage on the beaches, is certainly one of the factors responsible for the decline of sea turtle populations.

Recently, another major cause of mortality has been recognized: the capture of young turtles at sea in the nets of shrimp trawlers and in other fishing gear. Since turtles are air-breathing reptiles, being trapped in nets often results in drowning if the animals are held submerged for more than an hour or two. Analyses of population dynamics of animals with a long juvenile life suggest that this cause of mortality may be more serious than the more obvious mortality of adults and eggs at nesting beaches (Crouse et al. 1987). By the late 1980s, annual shrimp trawler mortality of loggerhead turtles along the Atlantic and Gulf Coasts of the United States was estimated to be at least 5,000 animals, and possibly as many as 50,000 individuals, with the mortality of Kemp's ridleys being somewhere between 500 and 5,000 (National Research Council 1990). Several kinds of **turtle excluder devices (TEDs),** essentially a trap-door arrangement that allows turtles and other large animals to escape trawl nets (Fig. 11.9), have been developed and tested. TEDs are now required on all United States shrimp trawlers. The use of TEDs is still resisted bitterly by the fishing industry, which claims that they cause major losses of shrimp. In 1993 and 1994, reports suggested that many TEDs were being disabled, and that high mortality of sea turtles was occurring once again. Mexico also requires the use of TEDs on shrimp trawlers.

Special efforts are being made to encourage recovery of the Kemp's ridley in the Gulf of Mexico, the most endangered of the sea turtles. The Kemp's ridley of the Caribbean and Gulf of Mexico nests almost entirely on a beach at Rancho Nuevo, Mexico. This population has declined catastrophically in recent decades. In 1947, an estimated 40,000 nesting female turtles were seen on the beach on one night. The estimated number of nesting females declined to about 350 in 1990 (National

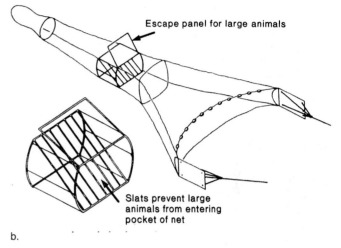

Escape panel for large animals

Slats prevent large
animals from entering
pocket of net

a.   b.

FIGURE 11.9

a. The turtle excluder device (TED) enables animals the size of young sea turtles to push open a trap door and escape from trawl nets designed for shrimp. *b.* Diagram of TED structure and placement.

(*a.* U.S. National Marine Fisheries Service photo) (*b.* Modified from National Research Council 1990)

Research Council 1990). During the egg-laying season, the beaches at Rancho Nuevo, Tamaulipas, are now guarded against predators and poachers. Clutches of newly laid eggs are collected (Fig. 11.8) and placed inside predator-proof enclosures or incubated artificially. Over 65,000 newly hatched turtles were released at Rancho Nuevo in 1979 through this effort. The nesting population of Kemp's ridleys now appears to be increasing, and over 1,800 nests were recorded in 1995.

A second approach that has been tried is **headstarting,** or raising young turtles to the age of about a year and a weight of about 600 grams, before their release. These turtles are then released in areas believed to be optimal habitat for animals of this age. Although headstarting has been used with some species for about 30 years, there is little evidence that headstarted animals survive even as well as those entering the ocean at hatching (National Research Council 1990).

Efforts have also been made to introduce certain species to new, more protected nesting beaches. In 1978, an effort was begun to establish a new nesting colony of Kemp's ridleys on Padre Island National Seashore, Texas. Eggs taken from turtles at Rancho Nuevo have been incubated in Padre Island sand, and the young headstarted for 9 to 11 months. Over 13,400 headstarted young were released into the Gulf of Mexico (Shaver 1992). This program has been discontinued, since none of the headstarted turtles has appeared to nest at Padre Island. A number of nestings of Kemp's ridleys have been recorded along the Texas coast, however, so that the species might colonize this area by a natural process.

PROTECTING BARRIER ISLAND ECOSYSTEMS

Sandy coastal ecosystems are particularly vulnerable to physical impacts on the vegetation or the soil surface. Trampling by humans is a serious problem in some areas, tending to promote the expansion of foredune vegetation into areas farther back from the beach, where other types of dune plants normally dominate (McDonnel 1981). Off-road vehicles can exert heavy damage to any type of dune vegetation, as well as causing mortality of animal life (Godfrey et al. 1980). At Sand Lake, along the central Oregon coast, for example, the activities of up to 5,000 ORVs on a single weekend have destroyed some of the finest examples of duneland vegetation (Wiedemann 1984). Even when ORV activity is confined to nonvegetated areas of the beach, repeated disruption of the sand surface promotes erosion and wind transport of sand inland, where excessive deposition can kill vegetation and lead to blowout development (Hosier & Eaton 1980).

The Outer Banks are a series of barrier beaches and islands fronting the Atlantic Coast for about 290 kilometers from southern Virginia to central North Carolina. The barriers enclose sounds up to 45 kilometers wide. Much of the northern portion of the barrier system is privately owned, but most of the southern portion has been preserved as Cape Hatteras and Cape Lookout National Seashores. Five National Wildlife Refuges border the sounds enclosed by the Outer Banks.

The Outer Banks show the dynamic nature of barrier island systems. Many inlets have formed, closed, reformed, and shifted. The tiny town of Currituck, lying inside Currituck Banks, was a bustling fishing port in the 1700s and early 1800s. Closing of the inlet serving the port in 1828 cut off access to the sea, freshening waters of that part of the sound and destroying the estuarine fisheries that were the port's mainstay.

Oregon Inlet separates Bodie and Pea Islands a short distance south of the community of Nags Head. Although an inlet has periodically existed at about this location at several times since settlement of the eastern seaboard, the present Oregon Inlet was formed during a hurricane in 1846. Since its formation, the inlet has migrated southward more than 2.8 kilometers, shifting about 23 meters per year (Inman & Dolan 1989). A bridge 5 kilometers long was constructed across this inlet at a cost of several million dollars. The bridge has been an expensive victim of the shifting inlet, with several more millions of dollars being spent to protect and extend its southern end, while its northern end spans an increasing length of dry land.

Hurricanes and winter storms have struck the Outer Banks many times, destroying property in beach communities such as Kill Devil Hills and Nags Head. On Hatteras Island, the chance of a tropical storm in a given year is 18%, the chance of a hurricane 11%, and the chance of a high intensity hurricane 8%. Intense storms create a storm surge of 2 meters, topped by storm waves that often breach the foredunes. Water flows across the islands, carrying beach sand and debris to be deposited as overwash fans. In severe storms, much of the islands are awash from the combination of surge and high waves. With each storm the beachline retreats a little and the inner side of the island advances landward.

The shoreward migration of the Outer Banks is evidenced by the Cape Hatteras lighthouse (Reading Fig. 11.1). Built at a location about 1,000 meters from the water in the late 1860s (Dolan et al. 1973), it now stands at the edge of the beach. To save this lighthouse by moving it inland will cost about $8.7 million. If it is not moved, costly revetments or groins must be built to protect it as long as possible, an exercise in delaying the inevitable loss of the site to the sea.

The Outer Banks are immensely important to wildlife. Pea Island National Wildlife Refuge, south of Oregon Inlet, was established in 1938. About 265 species of birds breed in or regularly visit the refuge. It is a major wintering area for 25 species of ducks, Canada and snow geese, and tundra swans, as well as a breeding site for herons, ibis, rails, terns, and many other water and land birds. It is also an important stopover area for many migrating shorebirds.

Human occupation of the Outer Banks began in the mid-1700s. On Shackleford Island, originally largely forested, settlers cleared much of the forest. Heavy livestock grazing disturbed the vegetation, fostering blowouts and actively migrating dunes that buried the remaining forest. The island was abandoned after a hurricane in 1889. Farther north, resort development

READING FIGURE 11.1

The Cape Hatteras lighthouse, built about a kilometer from the ocean in the 1860s, now stands in imminent danger of destruction, testifying to the retreat of the barrier island shoreline.

(Photo by G. W. Cox)

began at Nags Head in the 1830s. With little concern about island dynamics, the foredunes were flattened and houses built immediately behind the beach line. This pattern continues to the present, in spite of several disastrous hurricanes. Elsewhere, attempts were made in the 1930s to stabilize the shorelines by planting marram grass behind the beaches. This effort resulted in an artificially high foredune system that some scientists believe has increased beach erosion by deflecting storm waves seaward. Again, this effort reflects failure to recognize the dynamics of barrier islands.

Today, the protected areas of the Outer Banks illustrate the valuable use to which these coastal environments can be put. More than 1 million people visit the national seashores and wildlife refuges of the Outer Banks annually.

Other impacts have resulted from the introduction of plants to coastal dunes, ostensibly to stabilize them. European beachgrass (*Ammophila arenaria*) and gorse (*Ulex europeus*), among other species, were introduced to dunes along the Pacific Coast (Wiedemann 1984). European beachgrass has spread widely, crowding out many of the native strand plants that fostered a distinctive hummocky foredune topography. It also creates a much stronger foredune, unnatural for the region, that traps sand that formerly moved inland in greater quantity, feeding other duneland landscapes. Gorse, a shrub, has covered thousands of hectares of dunelands in southern Oregon.

The environmental unsuitability of barrier islands for urban development alone does not deter many developers, who nevertheless demand that public funds be used to provide roads, bridges, and other basic services, and to repair these services when they are damaged. When these funds are

FIGURE 11.10

Thick-billed murres, one of the most abundant seabirds of the auk family, utilize narrow ledges of sea cliffs for nesting.

(U.S. Fish and Wildlife Service photo by Karl W. Kenyon)

withdrawn, development often becomes impractical. In 1982, the United States Congress passed the **Coastal Barrier Resources Act,** which designated 186 barrier island units as a **Coastal Barrier Resource System.** These barrier islands extend along 751 miles of Atlantic and Gulf of Mexico shoreline. The key feature of this act is that these areas are ineligible for federal flood insurance or for federal road and bridge funds. Thus, public subsidization of development has been eliminated.

The importance of barrier islands to wildlife is evidenced by the location of national wildlife refuges and other wildlife preserves. About 35 national wildlife refuges and 10 national seashores lie on or inside barrier islands of the United States Atlantic and Gulf coasts. These refuges and parks represent appropriate uses of dynamic sandy coastal environments: wildlife conservation and recreation.

OTHER COASTAL ENVIRONMENTS

Cliffs, offshore rocks, and islands are major nesting areas for coastal water birds and certain raptors. More than 60 species of colonial water birds, including petrels, pelicans, cormorants, herons, egrets, ibis, gulls, terns, auks, and other birds use such habitats in the United States (Fig. 11.10). In California, for example, some 23 species of seabirds nest at 260 locations along the coast. The total populations of these birds are about 700,000 individuals. Of these, about 89% belong to four cliff-nesting species: the common murre, Cassin's auklet, Brandt's cormorant, and western gull.

Colonial nesting makes many of these species vulnerable to exploitation. In the late 1800s and early 1900s, nesting colonies of many species, particularly egrets and ibis, were killed for their feathers, which were used for decorating women's hats. Until early in this century, sea bird colonies were also exploited for eggs almost everywhere (*See* Chapter 18). This practice is still pursued in many parts of Latin America, Africa, and the Middle East (Croxall et al. 1984). Often, egg collectors destroy all the eggs present on an initial visit to a colony, and then return a few days later to harvest freshly laid eggs for sale in markets.

Many colonies of coastal water birds have also been decimated by the introduction of rats, cats, foxes, and other predatory animals to nesting islands (Croxall et al. 1984, Bertram 1995). Increased numbers of opportunistic predators, such as ravens and gulls, their populations supported by human food sources such as garbage dumps and fish processing operations, have also contributed to the decline of some coastal birds.

Increasingly, the impacts of human visitation of nesting colonies is a serious cause of nest failure and mortality (Anderson & Keith 1980; Croxall et al. 1984). Recreational activities, tourism, and even scientific study can disturb colonies. Off-road vehicles or humans on foot can cause birds to flee from their nests, often spilling the eggs or young and leaving the nests open to predation by gulls, ravens, or other predators.

GLOBAL CHANGE AND COASTAL ECOSYSTEMS

The terrestrial and wetland ecosystems of ocean coastlines are among the environments most threatened by global warming. Coastal wetlands, particularly along the Gulf Coast of the United States, are disappearing because of reduced river sediment deposition and land subsidence due to ground-water withdrawals (White & Tremblay 1995). Accelerated sea level rise due to global warming will exacerbate present losses. Efforts to protect developed portions of the coastal landscape by dikes and seawalls will tend to cause natural wetlands to be squeezed out of existence in the zone between protected areas and the ocean. Rising sea levels and shifts in ocean currents will likely increase the rates of migration and change in shape of barrier island. In sounds behind barrier islands, rising water levels will probably exceed the rate of sediment deposition, leading to deeper sounds and the loss of much of the mid-sound marshlands (Ray et al. 1992).

SUMMARY

1. Sandy coastal ecosystems are dynamic and ecologically fragile. Beaches change seasonally and are reshaped frequently by storm events. Disturbance of duneland vegetation can lead to the formation of blowouts and active dune migration.

2. Barrier islands are coastal beach and dune systems that were isolated by rising sea levels as Pleistocene continental glaciers began melting, and are now moving landward as sea level continues to rise gradually. Barrier islands shelter wetlands that are major nesting and wintering habitat for many groups of water birds.

3. Coastal dunelands are rich in endemic species and in species with ecological affinities to desert and grassland ecosystems from which they are often greatly distant. These coastal environments are of critical importance to sea turtles, which use them for nesting in tropical and subtropical regions worldwide.

4. Land use planning should recognize the high public costs of urban development on sandy coasts, especially barrier islands, and the appropriate use of these areas for recreation and wildlife conservation.

5. Many coastal areas, particularly cliffs and offshore islets, are rookeries for marine birds and mammals, and need special protection against disturbance and overexploitation.

QUESTIONS FOR DISCUSSION

1. Consider the landscape features of a sandy coastal area near where you live, or which you have visited. How many of these features might have been the product of human activity and disturbance?

2. What are some recent examples of the use of public funds for projects such as beach protection or restoration, road and bridge access, disaster relief, and the like on barrier islands and sandy coastal regions? Do you think these expenditures are justified?

3. If laws are changed to discourage building on privately owned coastal lands that are subject to the effects of storms and rising sea level, should the landowners be compensated for decline in their land value?

SUGGESTED READING

Bjorndal, K. A. (Ed.). 1995. *Biology and conservation of sea turtles.* Revised edition. Smithsonian Institution Press, Washington, DC. Fourteen accounts covering recent developments in sea turtle biology and conservation.

Eisma, D. (Ed.). 1995. *Climate change: Impact on coastal habitation.* Lewis Publishers, Boca Raton, FL. Eleven contributed chapters on global warming, sea level rise, and the implications of these trends for coastal ecosystems and human activities.

Pilkey, O. H. 1990. Barrier islands; formed by fury, they roam and fade. *Sea Frontiers* **36**(6):30–36. An illustrated description of the mainland beach detachment process of barrier island formation and migration.

Islands

chapter 12

T he plants and animals of ocean islands have suffered some of the most serious impacts of human disturbance. More species have become extinct on islands than in the vastly more extensive continental environment (*See* Chapter 5). At the same time, island ecosystems contain some of the most remarkable plants and animals that exist on earth (Carlquist 1965). Preservation of the members of these island ecosystems is thus one of the major challenges of conservation ecology.

Islands are of two major types: oceanic and continental. **Oceanic islands** are those that have never been connected to a continent, good examples being the Hawaiian Islands and the Galápagos Islands (Fig. 12.1). Most oceanic islands are of volcanic origin, and many show extensive areas of recent volcanic activity (Fig. 12.2). The flora and fauna of such islands is thus derived exclusively by overwater transport (including recent introductions by man). The biotas of oceanic islands are often a very biased sample of the kinds of forms that exist on the nearest continents. **Continental islands** are those that once possessed a continental connection. Trinidad, for example, lying off the northern coast of South America, was connected with Venezuela during the last Pleistocene glacial period, when sea level was about 125 meters lower than at present. Continental islands may possess a flora and fauna much like that of the continental areas to which they were once connected.

The giant tortoises of the Galápagos are one of the many unique and threatened species restricted to oceanic islands.

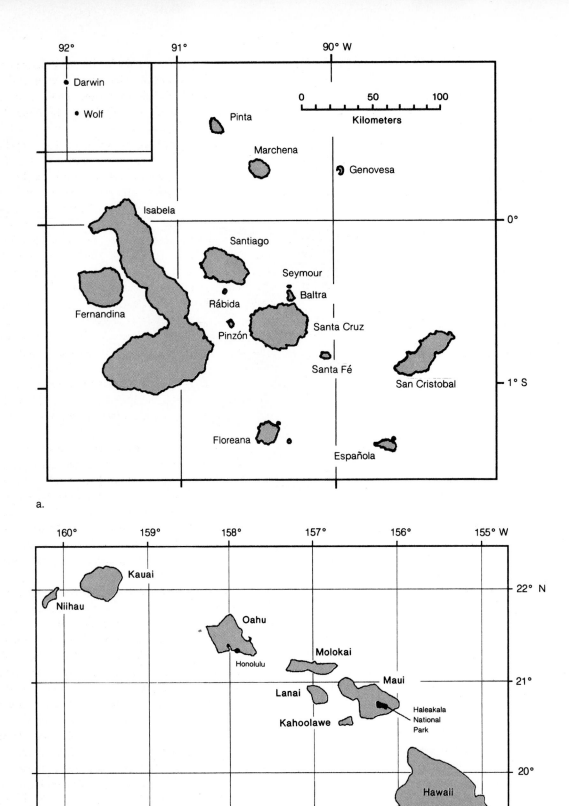

FIGURE 12.1

The Galápagos (a) and Hawaiian (b) archipelagos.

Figure 12.2

The Galápagos Islands are oceanic islands of volcanic origin, as evident in this view of the eastern coast of Santa Cruz from the neighboring small island of Bartolome.

(Photo by Darla G. Cox)

CHARACTERISTICS OF ISLAND BIOTAS

The isolation of island ecosystems profoundly influences biogeographic, ecological, and evolutionary processes (Mueller-Dombois et al. 1981). As a result, the organisms of islands, and the communities they form, possess many distinctive characteristics. These characteristics are the product of the biogeographic hurdles that organisms have to surmount to reach islands, and of the distinctive evolutionary environment they encounter once they arrive on islands.

Species diversity is low in island ecosystems. In the case of small oceanic islands, the number of species present is largely the outcome of two processes: colonization of the island by species from other areas, and extinction of the populations of species inhabiting the island. Isolation causes colonization rates to be low, but at the same time, the risk of extinction may be high because of the small area of the island and the small size of the population of an individual species. For archipelagos such as the Hawaiian Islands and the Galápagos Islands, speciation may be a second source of new species. Speciation requires that populations of an ancestral form colonize two or more islands, where the different populations can eventually diverge to become distinct daughter species. It is doubtful that speciation is accelerated on islands compared to continental areas, and, of course, the products of insular speciation are also affected by high extinction rates. The low diversity of island biotas is illustrated by the native land birds of the Galápagos Islands, lying 1,000 kilometers west of South America. The Galápagos Islands are one of the few archipelagos of substantial size and diversity for which we have an idea of the full biota prior to disturbance by humans. Only 28 land bird species occur on the Galápagos, the product of perhaps as few as 13 colonization events. Some of these species are undifferentiated from their relatives on the coast of Ecuador. Others, such as 13 species of finches of the

subfamily Geospizinae, are forms that were derived by speciation within the archipelago (Grant 1986). An area of equivalent size and vegetational diversity on the South American continent, however, would probably have a bird fauna 10 or 20 times as rich.

Even continental islands exhibit a reduced level of species diversity. When the continental connection existed, their biotas were presumably as rich as those of similar neighboring areas. But when the land connection is severed, the rate of colonization of the area drops, and the rate of extinction increases due to the fact that an isolated land area with populations of small size has been created. For some time, the rate of extinction exceeds the rate of colonization, and the number of species declines, a process known as **biotic relaxation.** In the Gulf of California, for example, lizards occur on islands that were connected to the mainland during the height of Pleistocene glaciation. At that time, the future island areas probably had faunas identical to those of the mainland at large. With isolation, however, the number of species on the islands has declined, and the longer the isolation, the greater this loss has been (Wilcox 1978).

A dramatic example of biotic relaxation is provided by forest birds on Barro Colorado Island, Panama (Karr 1982). This island, about 15.6 square kilometers in area, was created in 1913 by the flooding of Gatun Lake as part of the construction of the Panama Canal. Since the first faunal surveys of the island, some 23 species of forest birds are known to have disappeared. Comparisons of the present bird fauna of the island with that of comparable forests in nearby Parque Nacional Soberania, on the mainland, however, suggest that as many as 50 additional species were probably present on Barro Colorado Island at the time of isolation, and have since disappeared from the island. This loss amounts to over 27% of the probable original breeding bird fauna of the island.

A second characteristic of islands is high endemicity. **Endemic species** are those that have evolved in, and are restricted to, a particular area. All of the native land birds of the Hawaiian Islands, for example, are endemic. Over 40% of the plant species on well-isolated oceanic islands are commonly endemic (Table 12.1). The isolation of the islands means that once a species becomes established, its evolution is likely to be independent of that occurring in its parental population, thus leading to divergence. Endemic groups of plants and animals on islands often show striking examples of **adaptive radiation,** the evolutionary diversification of species for different ways of life. The Hawaiian honeycreepers, birds of the endemic subfamily Drepanidinae, show remarkable patterns of specialization of beaks for feeding on insects, seeds, and nectar (Scott et al. 1988). The Galápagos finches, which form the endemic subfamily Geospizinae, have diversified from an ancestral seed-eater to become fruit-eaters, wood-excavators, bark-scalers, and foliage-gleaners (Grant 1986).

A third characteristic of island biotas, especially those of oceanic islands, is absence or poor representation of certain ecological groups. Large predators tend to be absent or poorly represented on islands compared to continental areas. In the Hawaiian Islands, for example, the Hawaiian hawk and the short-eared owl are the only predatory land birds; the hawk is restricted to

Table 12.1

Native and Endemic Plant Species on Various Islands and Archipelagos, and their Conservation Status

	NUMBER OF SPECIES		PERCENT OF SPECIES	
Location	Native	Endemic	Endemic	Threatened Endemic
North Atlantic				
Azores	600	55	9	42
Canary Islands		500		75
South Atlantic				
Ascension	25	11	44	82
St. Helena	60	50	83	8
North Pacific				
Hawaiian Islands	1200–1300	1140–1235	95	50
South Pacific				
Galápagos	541	229	42	
New Calidonia	3250	2474	76	6
Norfolk Island	174	48	28	94
New Zealand	2000	1620	81	8
Juan Fernandez	147	118	80	79
Indian Ocean				
Mauritius	850	280	33	42

Data from W. V. Reid and K. R. Miller, *Keeping Options Alive: The Scientific Basis for Conserving Biodiversity.* 1989 World Resources Institute, Washington, DC.; W. L. Wagner, et al., "Status of the Native Flowering Plants of the Hawaiian Islands" in C. P. Stone and J. M. Scott, (Editors), *Hawaii's Terrestrial Ecosystems: Preservation and Management,* 1985, University of Hawaii, Honolulu; and data from Davis, et al., 1986.

one island, Hawaii. Flightless mammals, including browsing and grazing mammals as well as carnivores, tend to be absent. No native terrestrial mammals, for example, occur in the Hawaiian Islands. Early successional plants that tend to invade areas disturbed by animals and even other types of disturbance are often poorly represented. In the absence of large grazing animals that can disturb the vegetation, of large rivers that can cause flooding and erosion, and often of causes of fire, evolution has not favored adaptation of plants for disturbed sites.

A fourth characteristic is modified ecology and behavior. Flightlessness among birds (Fig. 12.3) and insects, and fearlessness of humans and other large animals are common features of island animals. Presumably, these characteristics reflect the absence of predators for these species. Sensitivity to grazing or browsing is a comparable characteristic of many island plants. The physical or chemical defenses, such as thorns or bad-tasting chemicals, by which continental plants defend themselves against herbivores, have often been lost. Where grazing or browsing animals are absent, natural selection does not maintain energetically costly defenses of these sorts, favoring instead the diversion of energy to some adaptation that has an important function in the island situation.

Finally, a fifth characteristic is **ecological release:** the expansion of the ecological niche by occupation of more habitats and the use of more kinds of resources. In a sense, island species make up for the shortage of species by becoming generalists. In a comparison of bird communities in Panama and on the small Caribbean island of St. Kitts, for example, Cox and Ricklefs (1977) found that island species both occupied more kinds of habitats and were more abundant within each habitat than were birds in a continental area with a similar range of habitats. In

Figure 12.3

The flightless Galápagos cormorant is endemic to the Galápagos Islands.
(Photo by G. W. Cox)

some cases, species that show ecological release may evolve ecological races or sexual differences that are like the differences shown by different species in continental regions.

Human Impacts on Island Ecosystems

The characteristics of island biotas have combined to make them highly vulnerable to the impacts of human activity (Mueller-Dombois & Loope 1990). The most serious human impacts have been due to hunting and harvesting of particular species, destruction of native vegetation, and the introduction of alien plants and animals.

Hunting has affected only a few selected species, but the impact in these instances has often been heavy because of the fearlessness of the animals, or their physical inability to escape humans. The dodo, a large, flightless pigeon that inhabited the island of Mauritius, in the Indian Ocean, was hunted to extinction in the seventeenth century. Because of their fearlessness, dodos were captured for meat by the crews of ships that visited the islands. On many oceanic islands, such as in the Mascarene Islands and Seychelles in the Indian Ocean, populations of giant tortoises were hunted to extinction, or severely reduced in numbers, for the same reason. Polynesian colonists of the Hawaiian Islands probably contributed to elimination of many endemic flightless birds by hunting (*See* Chapter 5).

Where man settles, destruction of the native vegetation follows. On islands, this can quickly lead to disappearance of distinctive vegetation types that evolved in isolation. Clearing and burning of island vegetation is an obvious cause of damage to native vegetation. Overall, however, introduced plants and mammalian herbivores have probably been responsible for the greatest damage to island vegetation.

Many plants, brought to islands as crop, forages, or ornamentals, have become problem weeds (Smith 1989). Many species invade natural communities, especially after disturbance, and crowd out or overgrow native species. Others, particularly perennial grasses that become dormant seasonally, increase the incidence of fire. Some exotics also cause major changes in soil water and nutrient status, which may be detrimental to native communities. Examples of these effects are given in the case studies later in the chapter.

Many kinds of herbivorous mammals have been introduced to islands. Early ships often introduced goats or other mammals to islands to create a meat supply for future use. Other large herbivores, such as pigs and donkeys, brought to the islands as domestic species, gave rise to feral populations. Feral populations of these species, lacking control by predators and exploiting vegetation unadapted to grazing or browsing pressures, often virtually strip islands of their plant cover. The tussock grasslands of the islands of the southern oceans were extremely vulnerable to overgrazing (Holgate & Wace 1961).

Damage to island vegetation by grazing and browsing mammals has been responsible for many extinctions of native plants and animals as well (Holgate & Wace 1961; Brockie et al. 1988). Tristan da Cunha, in the South Atlantic, was originally covered by a rich forest and tussock grassland. Settlers in the early 1800s brought many kinds of domestic animals that decimated the native vegetation. The tussock grass itself is now extinct on the island, together with several endemic birds (Holgate & Wace 1961). Predators such as dogs, cats, mongooses, rats, and mice have been introduced deliberately or accidentally to many islands. Introduction of mongooses to several islands in the West Indies may have led to extinctions of populations of some native lizards and snakes, although Corke (1992) suggests that other human impacts may have been involved. The introduc-

tion of the brown tree snake to Guam has resulted in the loss of almost all species of native birds (*See* Chapter 5). Rats and mice, ubiquitous followers of man, reached many islands in shipments of cargo, or by shipwrecks. One or more species of *Rattus* have been introduced to about 80% of oceanic islands, where they have contributed to extinction of at least 31 species of birds. Rats appear to be responsible for the loss of the tuatara, *Sphenodon punctatus,* the only living representative of an order of lizardlike reptiles, from islands of northern New Zealand (Cree et al. 1995). Together with cats, they have also caused enormous reductions in the populations of seabirds on many islands (Moors & Atkinson 1985).

Introduced competitors have also been a cause of extinctions on islands. Introduced plants are often competitors for native species, as are introduced rodents for native species. Introduction of land birds to Hawaii, New Zealand, and other island regions has also contributed to the decline of native bird faunas.

KEYSTONE EXOTICS AND THE STABILITY OF ISLAND ECOSYSTEMS

The examples described earlier suggest that certain types of species have a profound impact on island ecosystems. In general, these are species that differ substantially in way of life from those of the islands in question. Such species can be considered as keystone exotics—species capable of restructuring almost completely the ecosystems they invade. Good examples of keystone exotics include goats and donkeys (browsers) on the Galápagos Islands, the brown tree snake (an arboreal predator) in Guam, and pigs and the banana poka in Hawaii. These species all differ strikingly in their way of life from the animals or plants native to the islands, and native forms are simply not adapted to resisting their detrimental impacts.

This suggests that not all introductions to islands are likely to be deleterious. One illustration of this comes from New Zealand, to which hundreds of exotic animals and plants have been introduced. Some 15 species of ungulates have been introduced to, or spread as feral animals into, forested areas of New Zealand. In addition, the arboreal brush-tailed opossum, also a herbivore, has been introduced from Australia. These animals have caused severe, widespread damage to the native forests. In addition, many exotic birds have also been introduced to New Zealand, and these have replaced native forest species in many areas.

In the Hauraki Gulf, near Auckland on North Island, however, several forested islands have remained free of introduced mammalian herbivores (Diamond & Veitch 1981). Although many of the introduced birds have reached these islands, they do not enter the forests and displace native forest birds. This suggests that birds from continental areas are not inherently superior to native island species. Thus, the introduced herbivores are keystone species. If the effects of these exotics can be controlled, the various introduced birds are not themselves likely to cause massive disruption of the New Zealand avifauna. We shall examine strategies for control of exotic species in detail in Chapter 20.

The Galápagos Islands (See Fig. 12.1) are an oceanic archipelago lying on the equator about 1,000 km west of South America. The islands, part of Ecuador, are volcanic, and range in age from about 1 to 5 million years. The Galápagos Archipelago is one of the few oceanic island groups that was not colonized by humans in prehistoric time. Thus, the biota present when the islands were first visited by European voyagers represented the result of evolution unaffected by human activity. Island environments range from arid lowlands to humid highland forests. Recent volcanic activity has created landscapes of lava and ash in many places.

The biota of the Galápagos has been derived largely from South America, and is strongly endemic. Over 42% of the 541 native plants, many of the native birds, and all of the native mammals and reptiles are endemic (Reading Fig. 12.1A). Several genera of plants and animals have shown elaborate patterns of adaptive radiation in the archipelago, making it an important natural laboratory of evolution (Grant 1986; Cox 1990).

Human impacts date from the sixteenth century. The islands were discovered in 1535 by Spanish voyagers from Panama (Jackson 1985). From the 1500s through the 1700s they were visited by whalers and buccaneers. Settlement began in the early 1800s. Only four islands—Isabela, Santa Cruz, San Cristobal, and Floreana—have been settled, and the total human population is now about 9,000. In 1959, the uninhabited regions of the archipelago, about 90% of their total area, were declared a national park. In 1964, the Charles Darwin Research Station, located on Santa Cruz Island, was established to coordinate research and conservation activities in the archipelago. Tourism is now the major economic activity in the islands, with more than 60–70,000 visitors a year (Powell & Gibbs 1995) being attracted by the natural history of the archipelago (Reading Fig. 12.1B).

Hunting and killing have reduced or eliminated populations of several Galápagos species. Many populations of giant tortoises (Reading Fig. 12.1C) were hunted to extinction or severely reduced in numbers.

Tortoises were taken primarily by crews of ships, because the animals could be kept alive as a source of meat for months. Commonly, a single ship would take aboard 100 to 400 tortoises at one time. The total number of animals removed from the Galápagos probably numbers in the hundreds of thousands (Steadman 1986). At least three races of the Galápagos tortoises were hunted to extinction, and a fourth survives as only a single individual. Recovery and protection of populations on six other islands is one of the major conservation goals of the Charles Darwin Research Station. A recent upsurge in poaching of tortoises has led to increased concern about their survival. Other species have probably been eliminated by killing by humans. On Floreana, killing of the native Galápagos hawk and barn owl by settlers was probably the final cause of local extinction of these species, since both were very tame (Steadman 1986).

Settlement, concentrated in the higher elevations where moisture permitted farming, resulted in the direct destruction of unique vegetation types. Forests of *Scalesia*

READING FIGURE 12.1A

The marine iguana is one of the ecologically unique, endemic reptiles of the Galápagos Islands.

(Photo by Darla G. Cox)

READING FIGURE 12.1B

Tourist visitors to the Galápagos Islands, now numbering about 60,000 to 70,000 annually, must be accompanied by a licensed guide and must remain on designated trails.

(Photo by G. W. Cox)

READING FIGURE 12.1C

Giant tortoises similar to this individual on Santa Cruz Island in the Galápagos originally occurred on oceanic islands in several parts of the world.

(Photo by G. W. Cox)

trees, an endemic genus of the sunflower family, were severely affected by land clearing, as well as by other impacts. Only fragments remain of these original highland forests (Reading Fig. 12.1D).

The Galápagos Islands provide a good case study of the effects of both animal and plant introductions. Most of the larger islands, especially those with human settlements, have 6 to 10 species of introduced mammals (Reading Fig 12.1E). Goats have been introduced to several of the smaller islands. In many cases, these introductions led to population explosions (Eckhardt 1972). On the island of Pinta, for example, one male and two female goats were introduced in 1959. By 1970 the population on this island of only 60 square kilometers was estimated to be between 5,000 and 10,000 goats. On Santiago Island, goats introduced in the early 1800s increased to 80,000 to 100,000 animals in the 1970s and 1980s (Schofield 1989). Goats consume a wide variety of plants, including ferns, herbaceous plants, shrubs, trees, and cacti (Schofield 1989). On several islands goat browsing decimated the lowland vegetation and severely damaged the forest understory at higher elevations. On Floreana, Eckhardt (1972) estimated that over 70% of the native plant species were reduced or eliminated by goat browsing. On Isabela, goats have recently invaded northern portions of the island that were formerly goat-free.

Cattle, pigs, horses, and donkeys have also become feral on some of the inhabited islands. Cattle grazing in highland areas of Santa Cruz has severely disturbed the native shrublands that occur above the *Scalesia* forests. In overgrazed areas of these shrublands exotic grasses tend to invade.

Many introduced plants have also become problems (Schofield 1989). Guava, *Psidium guajava,* has become the dominant shrub over large areas of the highlands in the wake of overbrowsing by goats and other animals. Recovery of the original highland vegetation is now inhibited in many places by this weedy shrub. The quinine tree, *Cinchona succirubra,* a tree that spreads by windblown seeds, has invaded much of the high-elevation forest and shrubland of Santa Cruz. Lantana, *Lantana camara,* a weedy tropical shrub, is another highly invasive exotic on Santa Cruz and San Cristobal.

Vegetation destruction has secondarily affected many animals (Steadman 1986). Damage and destruction of the *Scalesia* forests has probably caused the extinction of populations of the sharp-beaked ground finch, which has disappeared from all four islands that are inhabited by humans. Browsing by goats, donkeys, and other large mammals has probably been responsible for several extinctions of birds. On Floreana Island, for example, destruction of the forests of tree prickly pears by these animals probably led to the extinction of the native mockingbird and the large-billed ground finch, both of which depended on this plant.

In the Galápagos Islands, introduced black and Norway rats have probably led to the extinction of native rice rats on seven islands (Eckhardt 1972). Rice rats survive only on islands free of these species.

Major efforts are being made to control the spread of invasive plants and to eradicate introduced animals. Goats have been eliminated from five small islands, and have been greatly reduced or perhaps eliminated on Pinta, a medium-sized island. Substantial recovery of the vegetation has occurred on these islands.

READING FIGURE 12.1D

Forests of *Scalesia* trees are now restricted to small areas of the highlands of Santa Cruz in the Galápagos.

(Photo by G. W. Cox)

READING FIGURE 12.1E

Introduced mammals of many kinds occur on most of the Galápagos Islands, but populations of some species have recently been eradicated as part of efforts to restore the habitats of the islands.

(Photo by G. W. Cox)

The Hawaiian archipelago, a set of volcanic islands, atolls, and reefs (*See* Fig. 12.1), spreads along almost 2,450 kilometers of tropical ocean, the youngest islands at the eastern end of the chain (Wagner et al. 1985). Hawaii, the youngest of the main islands, is probably less than a million years in age; Kauai, the westernmost large island, is 3.8–5.6 million years old (Carlquist 1980). The rocks and atolls farther west are eroded remnants of still older islands, so the archipelago as a whole has been in existence for a much longer time, perhaps as long as 70 million years (Wagner et al. 1985).

The Hawaiian Islands are even richer than the Galápagos in biological diversity. Large portions of the islands have a humid tropical environment favorable to plant and animal life. Most of the plants and animals native to the islands are endemic (Reading Fig. 12.2A), perhaps the highest level of endemicity for any area of comparable size (Carlquist 1980). Of the 1,200 to 1,300 species of native flowering plants, 95% are endemic (*See* Table 12.1), the result of colonization of the islands by about 270 ancestral forms (Fosberg 1948). Among invertebrates, 22 to 24 colonizations by land snails, followed by speciation, have resulted in an extraordinary fauna of about 1,000 species (Gagne & Christensen 1985). Similarly, 300 to 400 arthropod colonists have undergone speciation to produce more than 6,000 descendent species. Among land birds, a single ancestral finch radiated to produce 47 species and subspecies of songbirds specialized for diverse feeding niches. Many of these disappeared during the Polynesian period, and most of the survivors are now endangered (Scott et al. 1988).

When visited by Captain Cook in 1778, the Hawaiian Islands had already sustained major biotic losses due to changes resulting from Polynesian settlement (*See* Chapter 2). The Polynesian human population grew to about 200,000 to 250,000. About 80% of the native lowland forests were cleared for farming during the Polynesian period (Kirch 1982; Stone 1985). The Polynesians also hunted many of the native birds for food and feathers and introduced several predatory mammals and a number of plants. These effects evidently led to the extinction of many birds, and probably a variety of other plants and animals.

Rates of environmental disturbance and extinction of species have increased

READING FIGURE 12.2A

The Haleakala silversword, a member of an endemic genus of the Sunflower Family, was almost driven to extinction by goat browsing before establishment of Haleakala National Park on Maui, Hawaii.

(Photo by G. W. Cox)

following European settlement. Large areas of native highland forests have been logged and replaced by cattle ranches and plantations of pines and eucalyptus. Koa, *Acacia koa,* one of the dominant native trees, is a valuable wood for furniture and cabinet making. 'Ohi'a, *Metrosideros polymorpha,* on the other hand, is less valuable for its wood, and has been harvested for fuel for electric power generation, even after native forests were reduced to less than a quarter of their original extent (Holden 1985). The greatest disturbance of native vegetation, however, has resulted from the introduction of exotic plants and animals. Over 600 exotic plants have become established in the wild, and about 86 of these have become serious pests (Smith 1985). Some, such as guava and lantana, are the same species that are serious pests on the Galápagos. Several other trees, shrubs, vines, and grasses are serious invaders of native vegetation, however.

Firetree, *Myrica faya,* native to the Canary Islands, has invaded wet and moist sites, including open volcanic ash areas, on all of the main islands. This tree, which has nitrogen-fixing root nodules, forms pure stands that outcompete native plants (Vitousek & Walker 1989). This exotic thus modifies basic nutrient cycling processes of

READING FIGURE 12.2B

The banana poka, a species of passion flower native to South America, has invaded native highland forests in the Hawaiian Islands.

(Photo by G. W. Cox)

ash-soil ecosystems. Koa haole, *Leucaena leucocephala,* a nitrogen-fixing leguminous shrub, is ubiquitous in dry to moist lowland areas throughout the islands. It also forms dense stands that crowd out all other plants.

The banana poka, *Passiflora mollissima,* a South American vine of the passion flower family, has caused serious damage to native upland forests. Banana poka (Reading Fig. 12.2B), its seeds spread by pigs that feed on its fruits, grows into the crowns of the native forest trees and smothers their foliage (Reading Fig. 12.2C). About 500 square kilometers of forest are being attacked by this vine (Mueller-Dombois & Loope 1990).

Broomsedge, *Andropogon virginicus,* a perennial grass from the southeastern United States has spread throughout large areas of former wet mountain forest. Although it is a perennial, broomsedge becomes dormant during the temperate zone winter. At this time broomsedge stands transpire little or no water, so that the soil can become saturated with water for long periods, leading to a serious landslide problem (Mueller-Dombois 1973). During the dormant season, broomsedge also becomes prone to fire, and is responsible for a major increase in fire frequency in areas where it occurs (Smith 1985).

Efforts have been made to find means of biological control of the most serious plant invaders. Partial control of lantana has been achieved by introduced insect enemies. Controlling problem grasses in this fashion is

Reading Figure 12.2C

In the Hawaiian Islands, banana poka vines climb into the crowns of native forest and smother the trees.
(Photo by G. W. Cox)

difficult, however, because the importation of possible enemies is often opposed by the sugar cane industry (because sugar cane, a grass, might also be attacked).

Introduced animals add to the destruction of the native vegetation. Pigs are a major cause of damage to the understory vegetation of native forests, reducing, as well, the regeneration of the native canopy trees. Dispersal of seeds of both the banana poka and guava is aided by pigs, which feed heavily on their fruits. Goats are a serious problem in dry to moist, more open areas. Goat browsing has seriously reduced a number of native species, as shown by the recovery of these forms in areas fenced to exclude them (Loope & Scowcroft 1985). Eradication of these animals is virtually impossible over large areas of rugged mountain topography. Many resident people also hunt pigs and goats, and regard them as a desirable resource. Efforts to control these animals have therefore centered on trying to reduce their numbers to less destructive levels by hunting, and by excluding them from specific areas by fencing.

Rats, mice, mongooses, and feral cats are other problem exotics. Black rats damage the flowers, fruit, and bark of many shrubs and trees in the native forests. These smaller animals are also predators on native invertebrates and birds. The black rat led to the extinction of the Laysan rail on Midway Island, after the species had been translocated there from Laysan Island, where its habitat had been decimated by the grazing of introduced rabbits.

Many kinds of land birds have been introduced to the Hawaiian Islands, and few native birds survive in the lowland areas of the islands. Most of these species are confined to cities, farmland, and vegetation dominated by exotic plants. Nevertheless, competition from some of these introductions may be a factor in decline of native land birds (Mountainspring & Scott 1985). The introduction of avian pox and avian malaria to Hawaii, the latter resulting from the combined introduction of mosquito vectors and exotic birds carrying the malaria parasites, has contributed to the extinction of native birds. These diseases are now one

of the major threats to the remaining endemic birds (Scott & Sincock 1985; van Riper et al. 1986; Warner 1968).

As a result, the Hawaiian Islands contain more than a quarter of the threatened and endangered species in the United States. About 177 native plants are known or believed to have become extinct (Wagner et al. 1985), and 199 were listed as endangered or threatened in 1995. Over 30% of the species in the United States being considered for federal listing as endangered or threatened were from Hawaii. Many extinctions have occurred. On Lana'i, for example, about 20% of the native plants have disappeared, including several species endemic to that island (Hobdy 1993). Since European discovery, 13 species of Hawaiian land birds have become extinct and several others have been reduced to less than 100 individuals (Scott et al. 1988). Most recently, in late 1989, the last Kauai o'o', a male that had responded to taped recordings of its song, could no longer be found. About 30 Hawaiian birds are now listed as threatened or endangered.

Summary

1. Oceanic islands, which have never been connected to continental areas, have derived their biotas by aerial or over-water dispersal, whereas continental islands have biotas similar to those of continental areas to which they were once connected.

2. Oceanic islands tend to have low species diversity and high endemism, to be deficient in nonflying vertebrates and predators, and their species to be generalized and often lacking in defenses against herbivorous mammals and large predators.

3. Humans have influenced the biota of the Galápagos Islands by the direct killing of certain species, such as tortoises, and by the introduction of herbivorous and predatory mammals and weedy shrubs and trees.

4. In the Hawaiian Islands, Polynesians deforested lowland areas, hunted certain native birds, and introduced several predators, leading to extinctions of many birds. Since European colonization, these influences have intensified, and hundreds of exotic plants and animals have been introduced, many with detrimental impact.

5. Certain introduced species act as keystone exotics that trigger the restructuring of island plant and animal communities; such species are typically different in basic life-form or pattern of resource use from species native to an island.

Questions for Discussion

1. How do you think terrestrial vertebrates such as land iguanas and rice rats, whose ancestors lived in South America, colonized the Galápagos Islands?

2. When the Polynesians colonized Hawaii, they introduced dogs, pigs, rats, and chickens. How do you think each of these animals affected native species of the islands?

3. What are the implications of increased human travel and commerce for the biotas of oceanic islands? Is there any effective way of preventing the occasional invasion of keystone exotics to island areas such as Hawaii and the Galápagos?

Suggested Reading

Cox, G. W. 1995. Galápagos Islands. Pp. 167–179 in *Encyclopedia of environmental biology, Vol. 2.* Academic Press, Orlando, FL. A summary account of the geology, biology, human history, and conservation status of the Galápagos Islands.

Royte, E. and C. Jones. 1995. Hawaii's vanishing species. *Nat. Geogr.* **188(3)**:3–37. Dramatically illustrated account of past and present challenges to the survival of Hawaii's native flora and fauna.

Scott, J. M., C. B. Kepler, C. van Riper III, and S. I. Fefer. 1988. Conservation of Hawaii's vanishing avifauna. *BioScience* **38**:238–53. A concise, well-illustrated description of the native land and sea birds of Hawaii and of efforts to protect them.

part

three

Special Problems of Terrestrial Ecosystems

Certain problems of conservation ecology cut across many terrestrial ecosystems. Fragmentation of natural ecosystems is a serious threat to almost all types of forest, shrubland, and grassland ecosystems. The management of large predators presents unique problems because of their relations with game animals and livestock, and in some cases because of their danger to human safety.

chapter

13

Habitat Fragmentation

A s human populations grow, and human transformation of the landscape intensifies, terrestrial ecosystems of all types are being converted into island remnants. Once-continuous expanses of nature are being fragmented into patches isolated by agricultural, urban, or disturbed habitat. Fragmentation of the eastern forests of North America, leaving scattered woodlots surrounded by farmland and suburban residential areas (Fig. 13.1) first drew the attention of conservation biologists to this phenomenon. Now, however, fragmentation is recognized as a serious problem for shrublands, grasslands, savannas, and other terrestrial ecosystems.

The effects of fragmentation are related primarily to 1) creation of isolated patches of habitat, and 2) increase in importance of edge effects. First, we shall examine theory relating to the dynamics of populations in insular habitats, and then consider edge effects and their influence in habitat patches.

Tropical forests and other natural ecosystems are being reduced to islands in landscapes transformed by human activity.

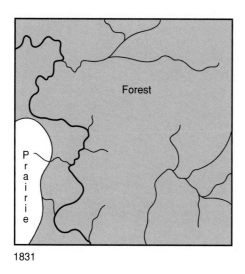

1831

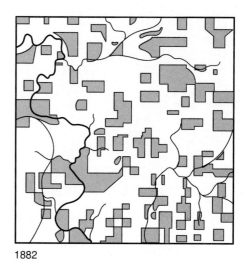

1882

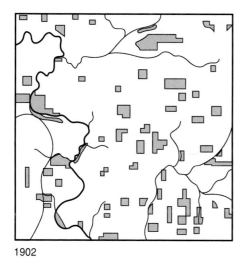

1902

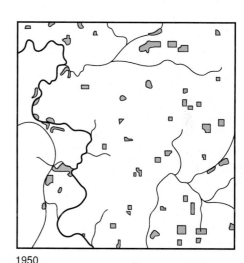

1950

FIGURE 13.1

From the time of settlement of Cadiz Township, Green County, southern Wisconsin to the mid-1900s, the forest was reduced from a nearly continuous cover to isolated woodlots covering less than 1% of the original area.

(From J. T. Curtis, "The Modification of Mid-Latitude Grasslands and Forests by Man" in W.L. Thomas, *Man's Role in Changing the Face of the Earth*, Copyright © 1956 The University of Chicago Press, Chicago, Il. Reprinted by permission.)

EFFECTS OF PATCH ISOLATION

Many of the effects of patch isolation can be understood by considering basic processes of island biogeography. The basic principles of island biogeography were outlined by MacArthur and Wilson (1967). They were applied quickly to issues of habitat fragmentation and preserve design (*See* Chapter 27).

Island Biogeographic Theory

In small island areas where speciation is unimportant in the addition of new species to the biota, the number of species is determined by the processes of colonization and local extinction. This relationship can be illustrated by considering how the rates of colonization of an island and of extinction of species on the same island are related to the number of species present (Fig. 13.2). If an island is stripped of its plant and animal life by, say, a

volcanic explosion, but nearby islands or continental areas are unaffected, we expect that recovery will soon begin, as dispersal brings potential colonists to the island. If conditions are relatively favorable, many of these will probably become established. Species that are well adapted for over-water dispersal and for occupation of severely disturbed habitats will probably be the first colonists. Later, species with weaker dispersal mechanisms and specialized habitat requirements will arrive. Eventually, as the number of species on the island approaches that on the nearby continent, the rate of colonization necessarily declines. Thus, the rate of colonization should tend to be high in early stages of recovery, and should theoretically decline to zero when all species from the source areas have colonized the island.

Once species have colonized, however, some of them may then suffer extinction. When only a few hardy, abundant species are present, the rate of extinction will obviously be low. As the number of species increases, the abundance of each tends

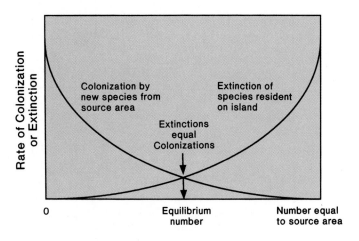

Number of Species on Island

FIGURE 13.2

The rate of colonization by new species from a source area, the extinction rate of resident species, and the equilibrium number of species on an island as predicted by basic island biogeographic theory.

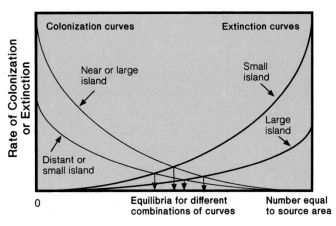

Number of Species on Island

FIGURE 13.3

The effect of island size and isolation from a source area of colonists on the equilibrium number of species on an island.

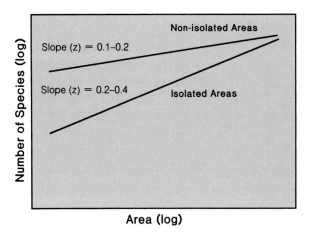

FIGURE 13.4

Species-area curves for biotas in which the areas examined for numbers of species are isolated have a steeper slope than those for areas of nonisolated, continuous habitat of similar nature.

to become smaller, each tends to become more specialized in its habitat and niche, and each tends to encounter stronger biotic challenges such as competition, predation, and disease. As a result, the rate of extinction increases. On our hypothetical island, colonization rate—high in the early stages of recovery—declines through time, and extinction rate—low in the early stages—increases. Eventually, these two rates become equal, so that the number of species on the island reaches a steady state value. In actuality, at this steady state, the biota fluctuates randomly about an average number of species defined by average rates of colonization and extinction. In addition, even though the number of species remains relatively constant, the composition of the biota is constantly changing, as certain species disappear by extinction, and others arrive and colonize the island.

These **colonization and extinction curves,** and the equilibrium biotas they define, vary with factors such as island size and distance from the sources of potential colonists (Fig. 13.3). Islands that are very small, thus presenting a small "target" for dispersal propagules, or very far from a dispersal source will have colonization curves beginning at much lower rates than those for islands that are large or near source areas. Extinction curves for very small islands, where the individuals of any species are few in number and confined in space, will rise at a much faster rate than those for large islands. A small, far island, therefore, will have a low colonization rate, a high extinction rate, and a low equilibrium number of species. A large, near island will have just the opposite.

A major corollary of this relationship is the principle of **biotic relaxation.** If an island is suddenly created from an area that was once part of a much larger region of similar habitat, it probably contains more species than can be maintained by the dynamics of colonization and extinction. As a result, extinction will run ahead of colonization, and the number of species will decline toward the steady state number. Case

(1975) showed that relaxation of the number of species of lizards had apparently occurred on islands created in the Gulf of California by rising post-Pleistocene sea levels. In Panama, Karr (1982) estimated that 50–60 species of forest birds have disappeared from Barro Colorado Island, an area of 15 square kilometers of upland tropical forest made into an island by the impoundment of Gatun Lake during construction of the Panama Canal in 1915. This represents about 19%–22% of the 268 or so species that were likely to have been resident on the area before it became an island. As we shall see, biotic relaxation has major implications for the design of biotic reserves.

The combined effects of area and isolation on the steady-state number of species can be seen by examining **species-area curves** (Fig. 13.4). These are simply curves of the number of species found in relation to size of the area examined. When both values are plotted as logarithms, the result is usually a straight line, the slope, z, which is a measure of how fast the number of species changes with a unit change in area. Species-area curves for island situations, where each point represents an island of a certain area, tend to be steeper that those for continental situations, where each point represents an arbitrarily bounded area of certain size. The z-value for species area curves tends to range from 0.20 to 0.40, with z-values for continental areas from 0.10 to 0.20. The steeper slope of the line for island areas means simply that as one goes to smaller and smaller islands the number of species drops off faster than that expected for a continental situation. The reason for this difference is that when an extinction occurs on an area of certain size on a continent, it is easily offset by recolonization, since no habitat barrier has to be crossed. On an island, the presence of a water barrier increases the difficulty of recolonization, and an extinction is not offset as quickly. This means that on small islands, where extinctions are frequent, fewer species are present, on average, than on a mainland area the same size.

Island Biogeography and Habitat Islands

The fragmentation of continuous ecosystems into patches surrounded by ecosystems of very different types creates "habitat islands" that are similar in many ways to oceanic islands. Island effects have been noted commonly in forest woodlots, but are also seen in fragmented grasslands (Herkert 1994), shrublands (Knick & Rotenberry 1995), clearcut openings in forest (Rudnicky et al. 1993) and other ecosystems. In these habitat islands local extinction and recolonization create a **turnover of species.** In fragmented forests, for example, isolation of woodlots reduces the interchange of forest species, and increases the likelihood of extinction of populations in individual woodlots. Small, isolated woodlots often lack the resources necessary to support viable populations of forest species, and make it difficult for these species to recolonize if they do accidentally disappear from areas that are minimally adequate. Thus, even though the number of species may remain about the same, the particular combination may change considerably.

Turnover of species can be seen in long-term census data for bird populations in isolated woodlots in eastern North America. In Trelease Woods, a 22-hectare area of forest near Champaign, Illinois, censuses of breeding birds have been conducted nearly every year since 1927 (Kendeigh 1982). Through 1980, some 62 species of breeding birds were recorded, but only 9 species were present every year. On the average, the species composition showed about a 13.6% change from one year to the next. Each year, some species disappeared, perhaps for only a year or two, and new species appeared. In England, censuses of woodlots showed that turnover from one year to the next ranged from a mean of over 40% for woodlots less than 0.25 hectare to nearly zero for woodlots of 20 hectares or larger (Hinsley et al. 1995). The species responsible for most turnovers were the scarcer species, which tended to come and go against a background of more abundant, regularly present species.

Biotic relaxation can also be seen in habitat islands. In tropical forest remnants in Brazil, Willis (1979) determined that woodlands 250 and 21 hectares in area suffered losses of 41% and 62%, respectively, of their breeding birds. An area as large as 1,400 hectares suffered a 14% loss of species.

Many North American national parks qualify as habitat islands. Major parks of montane western North America have experienced an excess of extinctions over colonizations since their establishment in the late 1800s or early 1900s (Newmark 1995). The rate of extinction has also been greatest in the smaller parks. Mount Rainier National Park in Washington, for example, falls within the original ranges of 68 species of native mammals. With settlement and development of land surrounding this park, many species disappeared (Newmark 1986). By 1920, only 50 species remained, 73.5% of the original forms. By 1976, only 37 species, 54.4% of the original mammals, remained. Some of these species, such as the wolf, may have disappeared because of direct persecution by humans, but many have become extinct because the area is now effectively a small habitat island with a high extinction rate, and because areas from which recolonization might occur are now quite distant.

Extinctions of species in fragmented habitat thus may not occur until many generations later, so that fragmentation creates an "extinction debt" that is paid long into the future (Tilman et al. 1994). This debt may involve not only rare species and stochastic final causes of extinction (See Chapter 5), but also dominant species that are gradually driven to extinction in an almost deterministic manner. Many dominant plant species achieve their high competitive success at the expense of reduced dispersal and colonization abilities. In Minnesota oak savannas, for example, bur oak is the dominant plant. Fragmentation of savanna habitat can easily create patches too far apart for dispersal of acorns of this species. The individual patches of savanna that remain are subject to edge effects, including the invasion of weedy species and exotics that reduce bur oak reproduction and increase mortality, as well as stochastic processes. Patches, and ultimately the entire set of patches in the region, thus eventually may lose their bur oaks.

Island effects are quite evident in species-area curves for habitat islands of different size. In eastern Illinois, Blake and Karr (1984) censused breeding bird communities in woodlots varying in size from 1.6 ha to 600 ha. For forest interior bird species, they found a species-area slope (z) of 0.57, and for long-distance migrants a slope of 0.30, an indication that these groups had declined in species number in a manner typical of oceanic islands. Observations like these have led to the recognition of **area-sensitive species,** or species that are likely to disappear when habitat patches reach some critical minimum size, even though they may still be much larger than the activity area of individuals of the species (Robbins et al. 1989). In forest areas ranging from 0.1 to over 3,000 hectares in Maryland and neighboring states, for example, 26 species showed significant decreases in relative abundance with decrease in forest area (Robbins et al. 1989). These included permanent residents such as the pileated woodpecker, short-distance migrants such as the white-breasted nuthatch, and long-distance migrants such as the scarlet tanager.

NEOTROPICAL MIGRANT BIRDS IN MIDWESTERN UNITED STATES FORESTS

I n southern Wisconsin, Illinois, and northern Missouri, deciduous forests have become heavily fragmented. In these areas, forest interior and edge birds, especially those that winter in neotropical areas (*See* Chapter 21), are suffering heavy nest parasitism and predation. The brown-headed cowbird, a species originally associated with migratory herds of bison, is a nest parasite, laying its eggs in the nests of many songbirds. Animals such as blue jays, crows, skunks, raccoons, and snakes are the agents of nest predation. Cowbirds and many of the nest predators are most common in open, agricultural, or forest-edge habitats, and thus are benefitted by forest fragmentation. In central Illinois, Robinson (1992) examined nesting success of forest birds in woodlots ranging from 14 to 65 hectares in size. He

found that about 80% of open-cup nests suffered predation, and about 76% of the nests of neotropical migrants were parasitized by brown-headed cowbirds. Nests of the wood thrush held, on average, four times as many cowbird eggs as wood thrush eggs. From the 15 wood thrush nests he examined, only one young thrush was fledged.

Robinson and several colleagues (1995) extended their studies of nesting success and forest fragmentation to a four-state region of the midwestern United States. They monitored more than 5,000 nests on study areas in Wisconsin, Indiana, Illinois, and Missouri, and were able to test the relationship of fragmentation to nesting success for nine species. Nest parasitism by cowbirds increased as forest cover decreased for all nine species. In study areas with less than

55% forest cover, most wood thrush nests contained cowbird eggs. Nest predation rates also increased as forest cover decreased, with ground-nesting birds showing the highest levels of predation.

The study areas in Illinois, southern Wisconsin, and northern Missouri were most severely affected by these fragmentation influences. Robinson et al. (1995) suggest that populations of some of the forest birds, especially the neotropical migrants, in this large region are maintained by dispersal from surrounding regions with more extensive forests. Such regions include the Missouri Ozarks, northern Wisconsin, and parts of southern Indiana. A conservation strategy for the migratory birds of this portion of the Midwest obviously requires planning at an interstate level.

Metapopulation Dynamics in Fragmented Habitats

Habitat fragmentation may create a metapopulation—a set of partially isolated subpopulations (e.g., in isolated woodlots) that can still exchange individuals by dispersal—from a continuous regional population (*See* Chapter 3). These subpopulations are likely to vary in size and permanence. Extinction and colonization dynamics then become dominant processes. The smallest subpopulations will tend to go extinct most often, and their occurrence will thus depend on recolonization. Some of these small subpopulations may never be able to achieve a reproductive success adequate to offset their general mortality rate, and thus are **sink populations** that are maintained almost entirely by dispersal from **source populations** that do exhibit a positive population growth rate.

The survival of a regional population thus depends on a relationship involving the entire set of subpopulations. If subpopulations become too isolated and too many become population sinks, the entire regional population will tend to decline gradually toward extinction. This pattern was observed by Hanski et al. (1995) in metapopulations of a meadow butterfly in Finland. They found that networks of meadow habitats tended either to be healthy, with populations in most meadows, or unhealthy, lacking butterflies or losing them from too many meadows in the network.

Source and sink relationships may exist over very large regions. The populations of many neotropical migrant birds, for example, may be maintained in parts of the midwestern United States with severely fragmented forests only by dispersal from source regions (*See* Reading 13.1). In fact, sink habitat units may be "attractive traps" for emigrants from demographically

productive populations, and thus may reduce the ability of other metapopulation units to sustain themselves. On the other hand, sinks may contribute to genetic diversity and long-term survival of a regional metapopulation if occasional dispersal from sinks to sources occurs.

The need to understand complex relationships of metapopulations in patchy environments, often with the patch pattern changing through time by disturbance and succession, has helped foster a new generation of ecological models. **Spatially explicit models** are mathematical simulations that combine a population or ecosystem simulator with a landscape map (Dunning et al. 1995). This type of model, still in its infancy, might describe, for example, the dynamics of each subpopulation and the change in the habitats of these subpopulations due to natural disturbance, biotic succession, and conservation management. The precise spacing of the subpopulations on the landscape would influence the pattern of dispersal among subpopulations, and the viability of the metapopulation as a whole. With such a model, management alternatives could be explored to determine the conditions necessary for regional survival of endangered species in a patchy or fragmented landscape.

EDGE EFFECTS IN HABITAT PATCHES

Transformation of continuous forest into scattered woodlots has a major impact on the forest environment (Saunders et al. 1991). The microclimate of the forest is greatly modified, especially near the woodlot edge. Temperature, humidity, and wind patterns along woodlot edges differ greatly from conditions in interior portions of woodlots. These differences, in turn, affect the soil environment and favor different plants and animals. Newly created woodlots may experience high rates of tree mortality (Chen et al. 1992; Esseen 1994). The composition of understory plants may be affected. In North Carolina, Fraver (1994) found changes in the understory community to distances of 10–50 meters into woodlots, depending on the directional exposure of the woodlot edge. In Oregon, Mills (1995) suspected that negative microclimatic edge effects on soil fungi, the major food of red-backed voles, reduced the numbers of this species in small woodlots.

Biotic edge effects may have even greater impacts on forest interior animals. Wilcove (1985) and others (See Paton 1994) have concluded that nests of birds in small woodlots and near the woodlot edges in the eastern United States are subject to much heavier predation than those in larger forest tracts and interior locations. Predators from adjacent open habitats, such as skunks, crows, jays, and grackles, penetrate into the edges of forest areas. The nest-parasitic brown-headed cowbird is also a species that frequents the edge zone of woodlots. In the case of small woodlots, the entire area may become accessible to such species. Haskell (1995) notes, however, that many experiments on nest predation have used eggs that are not vulnerable to small mammalian predators such as chipmunks and mice, which may be abundant in interior areas of woodlots.

Correlated with these predation and nest parasitism observations, many studies have shown that the number of species of breeding birds declines much faster as one goes from large to small woodlots than from large to small areas within a region of continuous forest (Galli et al. 1976; Ambuel & Temple 1983; Lynch & Whigham 1984; Blake & Karr 1984). This decline is greatest among species typical of forest interior habitats, among insectivorous and predatory species, and among species that are long distance migrants than among other kinds of birds (Foreman et al. 1976; Whitcomb 1977; Whitcomb et al. 1981; Blake 1983, 1986).

In California, Soulé et al. (1988) showed a similar pattern for native chaparral communities in canyons isolated in an urban development. The number of chaparral bird species declined with decreasing canyon size and with increasing time since isolation. In addition, canyons that were still visited by coyotes retained more native chaparral birds than those that were not. Apparently, coyotes control the abundance of small predators such as foxes, skunks, and domestic cats, favoring the survival of species such as the California quail that are vulnerable to these animals. When coyotes disappear, these "mesopredators" increase in numbers, increasing predation on birds.

EVALUATING THE ADEQUACY OF PRESERVES

Examination of species-area relationships for preserves can reveal how strong the island effect is for various groups of species. Kitchener and his colleagues (1980a, 1980b, 1982) examined these relationships for mammals, birds, and lizards in 23 preserves in the wheatbelt region of Western Australia. The preserves, varying in size from 34 to 5,119 hectares, exist in an intensively farmed landscape. For birds, the species-area slope (z) was 0.18, indicating that the faunas of the various preserves were

FIGURE 13.5

The western gray kangaroo, *Macropus fuliginosus*, is absent from Australian wheatbelt preserves smaller than about 200 hectares. (Photo by G. W. Cox)

not strongly affected by their isolation. Many of these birds are adapted to relatively open vegetation types, and are perhaps able to use shrubby growth and trees along roads, fence lines and streams, or surrounding farmsteads, enabling them to recolonize preserves where extinctions have occurred. For lizards the z value was 0.25, indicating that preserves were showing a slight island effect. For nonflying mammals, however, the z value was 0.39, indicating that these preserves were suffering very severely from isolation in a fashion similar to oceanic islands (Fig. 13.5). This high z value means that many species had, in fact, become extinct on the smaller preserves since they were isolated.

Other workers have used species-area curves to examine how conversion of reserves into habitat islands affects the net rate of species loss due to the excess of extinction over colonization. Miller (1978), for example, estimated that the Mkomazi Game Reserve in northern Tanzania, which now contains 39 species of large mammals, would suffer severe biotic relaxation if it became isolated from neighboring reserves. Although it is 3,276 square kilometers in area, Miller estimated that it would likely lose 17 species over the next 300 years. Soulé et al. (1979) and Burkey (1995) also applied species-area curves derived from oceanic islands to survival of species in East African parks. Their models likewise predicted substantial losses, even in some of the largest preserves, such as the Mara-Serengeti Park.

The species-area relation can also be used to estimate the size of the area required to support a specified fraction of the regional biota. Care must be used in formulating such estimates, however, since the slopes of species-area relations may be influenced greatly by the position of a few data points (Boecklen & Gotelli 1984). For mammals of the Australian wheatbelt, however, Kitchener et al. (1980b) estimated that a preserve of 43,000 hectares would be necessary to support all 25 species, or a preserve of 32,700 hectares to support 90% of them. This suggests that the existing preserves, the largest being 5,119 hectares, are inadequate to preserve the mammal fauna of this region. In the

FIGURE 13.6

In the Biological Dynamics of Forest Fragments Project, tropical forest areas, such as these woodlots 1 and 10 hectares in area, were created and are being censused regularly to determine the changes induced by isolation.

(Photo by Richard O. Bierregaard, Jr.)

serves in areas such as the Amazon Basin.

Experimental studies are now in progress in the central Amazon Basin to obtain a more exact picture of the effects of creating habitat islands. These studies, the **Biological Dynamics of Forest Fragments Project,** rely on the Brazilian law that 50% of land under private ownership in Amazonia must be left in forest. With the cooperation of the operators of new ranches in a region north of Manaus, Brazil, 20 forest patches of 1, 10, 100, 1,000, and 10,000 hectares have been established (Fig. 13.6). The biotas of these areas were censused prior to the isolation of certain patches by land clearing. Censuses will be continued over at least a 20-year period to reveal how isolation affects the numbers and kinds of species that survive in them. The data already available (Lovejoy et al. 1986) indicate that newly isolated forest fragments experience severe climatic and biotic change at their edges. This study has also revealed that different animal groups respond quite differently to fragmentation. Hummingbirds, for example, persisted well in fragments as small as 1 and 10 hectares, whereas insectivorous birds of the forest understory declined in diversity (Stouffer & Bierregaard 1995).

United States, Robbins et al. (1989) similarly found that forest areas 3,000 hectares in area would be needed to retain all of the forest interior birds of the Middle Atlantic States. Similar analyses of savanna preserves in Africa suggest that preserves must be at least 10,000 square kilometers in area to maintain the large mammal faunas of this region (East 1981). Fortunately, a number of the existing parks and game preserves are larger than this minimum value. Maintaining these large park areas in the face of human population growth will be very difficult, however, and it is possible that lands will be removed from some of the large parks, reducing their effectiveness as preserves for the large mammals.

Terborgh (1975) used data on the apparent loss of bird species on forested tropical islands isolated from the adjacent continent by rising sea level to estimate extinction rates for tropical islands or preserves of different size. The islands in this analysis ranged from Trinidad, off the coast of Venezuela, to small continental islands in the Gulf of Panama. These islands were all connected to the mainland of South or Central America about 10,000 years ago during the low sea level periods of the Pleistocene. Terborgh assumed that the Pleistocene faunas of these islands were equal to the present faunas of areas of equal size on the mainland. From these very rough data, Terborgh calculated that an area of about 2,500 square kilometers was needed to reduce extinction rates to less than 1% per century. Based on this estimate for birds, other workers suggested that a doubling of this area would achieve a similarly low extinction rate for large or wide-ranging tropical forest animals in general. For lack of any better estimate, this value, 5,000 square kilometers, has become a standard minimum size for major tropical forest pre-

SUMMARY

1. In insular environments, species diversity is determined largely by the balance of colonizations, which tend to increase the number, and extinctions, which tend to reduce it.

2. Continuous versus fragmented environments differ in their species-area curves. Fragmented environments have a steeper slope because extinctions in small fragments are not as quickly offset by recolonization as those in areas of nonfragmented habitat.

3. Areas of habitat that become fragments may experience biotic relaxation, the decline in number of species due to excess of extinctions over colonizations. They may also experience microclimatic and biotic edge effects that further reduce the area of habitat with interior conditions.

4. Habitat fragmentation may create regional metapopulations that consist of many partially isolated subpopulations that are subject to exchanges of individuals by dispersal. Some of these metapopulation units may be sources of emigrants and other sinks that depend on immigrants for survival.

5. Island biogeographic theory, coupled with field experiments, can suggest the size and arrangement of preserves necessary to allow the survival of species in fragmented habitats.

1. Do concepts of insular biogeography apply only to terrestrial environments? Do metapopulations exist only for terrestrial species? Can you think of cases in which fragmentation is significant in wetlands or aquatic habitats?

2. Do problems associated with excessive fragmentation of habitats mean that a condition of minimum possible fragmentation is most desirable from a conservation standpoint?

3. Should housing developments adjacent to areas of natural habitat be subject to special regulations, such as a prohibition on domestic cat ownership, in order to minimize urban edge impacts?

SUGGESTED READING

Harris, L. D. and G. Silva-Lopez. 1992. Forest fragmentation and the conservation of biological diversity. Pp. 197–237 in P. L. Fiedler and S. Jain (Eds.), *Conservation biology. The theory and practice of nature conservation, preservation, and management.* Chapman and Hall, New York. Patterns and impacts of forest fragmentation, with emphasis on northern Florida.

Saunders, D. A., R. J. Hobbs, and C. R. Margules. 1991. Biological consequences of ecosystem fragmentation. A review. *Cons. Biol.* **5**:18–32. A summary of the varied consequences of habitat fragmentation on survival of species.

Whitney, G. G. 1994. *From coastal wilderness to fruited plain: A history of environmental change in temperate North America 1500 to the present.* Cambridge Univ. Press, Cambridge, England. How settlement of eastern North America changed the landscape, with particular emphasis on forest conversion and fragmentation.

chapter

14

Predator Ecology and Management

H ow predators should be managed in wildlife ecosystems is one of the most controversial topics in conservation ecology. Predators are diverse in nature, including members of the carnivore and top carnivore trophic levels of the ecosystem. Tiger beetles and insectivorous songbirds are thus predators, along with killer whales and tigers. The activities of some predators may be the dominant factor in structuring the composition of some biotic communities. Predation by sea otters, for example, appears to control virtually the entire composition of the kelp bed ecosystem (*See* Chapter 18).

Controversy about predators, however, has centered on large mammalian carnivores whose prey include wild ungulates, occasionally livestock, and sometimes humans. Survival of many of these large carnivores in the wild is one major concern. Tigers, leopards, lions, cheetahs, jaguars, wolves, African wild dogs, and other such species are now extinct or endangered over much of their original range, and present urgent problems of conservation management. The effects of large carnivores on their prey is another concern. For game animals, a major question is whether predation does or does not limit prey populations at levels well below their carrying capacity, so that predator control might increase substantially the numbers of animals available to hunters. For predators that sometimes kill domestic animals, the basic question is whether losses are great enough to justify the costs of control and, if so, how control should be carried out. Finally, concern about large predators involves their danger to humans visiting park and wilderness areas.

As we shall see, the predator-prey relationship is complex, even when we consider only large mammalian predators. Studies of predation are continually revealing new aspects of the relationship, so that our understanding of the phenomenon is constantly changing. In any case, we should begin by noting that predation may range from being an unimportant factor in the dynamics of the prey species to the opposite extreme—limitation of prey at a low density. The latter relationship is employed to human benefit in the biological control of agricultural pests.

Management of vertebrate predators involves many scientific, economic, and ethical issues.

ECOLOGY AND BEHAVIOR OF PREDATORS

Vertebrate predators show distinctive features of behavior and ecology. These include systems of territoriality, basic patterns of response to changes in prey abundance, and patterns of selectivity in the capture of prey.

Territoriality is the defense of an area of habitat by an individual, a pair, or a group of individuals against the intrusion of others of the same species, or sometimes those of related or ecologically competing species. Individual mountain lions, for example, seem to defend at least the core areas of their overall ranges, although these ranges tend to overlap considerably, especially for males and females (Hornocker 1970; Neal et al. 1987). Pairs of many species of raptors, such as golden eagles and peregrine falcons, defend territories. Packs of gray wolves (Pimlott 1967) and prides of lions (Schaller 1972) defend group territories (Fig. 14.1). In all cases, direct aggressive challenge of invaders is the ultimate defense. In many vertebrates, however, vocalizations serve to define territorial boundaries: the howling of wolves, the roaring of lions, and the singing of birds. Scent-marking is a frequent way by which many mammals mark territorial boundaries. Most animals use more than one territorial defense mechanism; wolves scent-mark territory boundaries, proclaim ownership by howling, and attack trespassing animals.

Predators vary greatly in their response to changing prey abundance. Their responses are of two basic types: functional and numerical (Holling 1959). A **functional response** is a change in the effort exerted in hunting a particular type of prey. For example, as the numbers of lemmings increase in an arctic tundra area, arctic foxes and snowy owls may shift more of their hunting effort to these animals. When lemmings decline, these predators may switch back to alternate prey, such as arctic hares. A **numerical response** is a change in the density of predators, due to reproduction and mortality. Increased lemming numbers may allow arctic foxes and snowy owls to produce more young, thus increasing their population size. When lemmings crash, however, predator reproduction may fail, and many predators may die, reducing their numbers. In a given situation, the overall response of predators is thus a combination of their functional and numerical responses. Given the widely differing characteristics of predators, this overall response can be quite variable. In general, however, the smaller the predator and the higher its reproductive potential, the greater is the importance of the numerical response. The larger the predator and the lower its reproductive potential, the more important its functional response becomes. Nevertheless, most predators show both numerical and functional responses to changes in prey abundance.

A third important feature of predation is **predator selectivity,** the extent to which a predator concentrates on substandard prey—individuals that are very young or old, weak, diseased, or injured. High selectivity, in this sense, means that the predator is taking prey that are likely to contribute little to prey population growth, and that are likely to die soon in any case. Low selectivity means that many of the prey taken would probably contribute significantly to prey population growth.

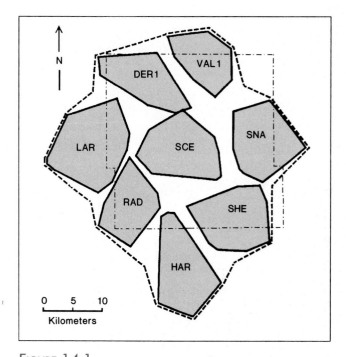

FIGURE 14.1

Winter territories of gray wolf packs in northeastern Minnesota in 1982–83. These areas were revealed by tracking radio-collared animals on the 839 square kilometer Bearville Study Area (dot-dash line) about 50 kilometers north of Grand Rapids, Minnesota. The dashed line enclosed the annual census area for wolf packs.
(Modified from Fuller 1989)

Information on selectivity of predators is difficult to obtain. Temple (1987), however, studied selectivity in hunting by a tame red-tailed hawk for various kinds of prey. He compared the condition of prey captured by the hawk with that of a random sample of individuals from the prey populations. Selectivity was high when prey were hard to capture and the success rate of attempted captures was low. In effect, capture attempts constituted a "test" of the condition of the prey, in which any substandard quality of the prey increased the chance of capture. Temple found that gray squirrel prey, which were difficult to capture, were lower in body fat, more heavily parasitized, and had more physical defects than animals in the population at large. These differences were not seen for chipmunk prey, which were more easily captured. Reviewing the literature, Temple found that high selectivity for prey difficult to capture was shown by many other predators (Table 14.1). Pursuit predators such as lions, cheetahs, African wild dogs, and wolves tend to test the condition of their prey, and thus often show strong selectivity (Fig. 14.2).

PREDATION IN AN ECOSYSTEM CONTEXT

Predation is influenced strongly by many physical and biotic factors of the environment, such as weather conditions, fire influences, structure of the vegetation, alternate prey availability, and the other kinds of predators present. These conditions may allow

TABLE 14.1

Difficulty of Prey Capture in Relation to Degree of Selectivity for Substandard Individuals by Large Mammal Predators

PREDATOR	PREY	DIFFICULTY OF CAPTURE	SELECTION FOR SUBSTANDARD INDIVIDUALS
Tiger	Deer	Difficult	Yes
	Pig	Easy	No
Lion	Wildebeest	Difficult	Yes
	Zebra	Difficult	Yes
	Cape buffalo	Difficult	Yes
	Gazelle	Easy	No
Cheetah	Wildebeest	Difficult	Yes
	Gazelle	Easy	No
Puma	Elk	Difficult	Yes
	Deer	Easy	No
Wolf	Moose	Difficult	Yes
	Caribou	Difficult	Yes
	Deer	Easy	Sometimes
Wild Dog	Wildebeest	Difficult	Yes
	Zebra	Difficult	Yes
	Gazelle	Easy	No
Coyote	Deer	Difficult	Yes
	Pronghorn	Difficult	Yes
	Sheep	Easy	No
Hyena	Wildebeest	Difficult	Yes
	Gazelle	Easy	No

Source: Data from S. A. Temple, 1987, "Do Predators Always Capture Substandard Individuals Disproportionately from Prey Populations?" in *Ecology,* **68:**669–74, Tempe, AZ. Ecological Society of America.

FIGURE 14.2

Pursuit predators such as African lions tend to show strong selectivity in their predation on animals such as wildebeest.

(Photo by G. W. Cox)

predation to range from being of minor importance to having a severe depressing influence on prey populations. Management efforts may also compound the complexity of the relationship. Small areas of protected, high-quality habitat, for example, may become predation traps by attracting predators to locations where prey have become concentrated and are easily found (Bergerud 1987).

Coyote Predation on Mule Deer

Little doubt exists that coyotes prey on deer, particularly fawns, and that predation has some impact on deer populations. In Colorado, Bartmann et al. (1992), for example, found that coyote removals from an area of 140 square kilometers led to reduced predation rates on deer fawns. But it is also clear that coyote predation is influenced by many factors.

A good example of the complex influence of ecosystem conditions is provided by predation on mule deer in California. Mule deer herds in many parts of the state have declined in recent decades, and predation is commonly cited as a major cause. Most wildlife biologists believe that the main cause is habitat loss and deterioration (McCullough et al. 1990). The mechanisms are complex, however, and may involve basic changes in several aspects of wildlands ecology.

The North Kings herd, a migratory population in the Sierra Nevadas east of Fresno, declined from 17,000 animals in 1952 to 2,000 animals in 1986 (Neal et al. 1987), for example. This herd winters in the foothills bordering the San Joaquin Valley. In summer, the herd moves to higher elevations in the Sierras, where mild, moist conditions favor forage production. The fawns are born in June and July at these higher elevations. The decline of the herd is due to low recruitment of young animals, yet fecundity is very high. Females produce an average of 1.5 fawns annually. Many of these fawns, however, do not survive to enter the adult population.

In the summer range of the North Kings herd in the 1970s, rodents and rabbits were the primary prey of coyotes for most of the year. From mid-June to the end of July, however, the coyote diet consisted mostly of fawns (Salwasser 1974). This finding did not indicate whether coyotes were preying on healthy fawns, were culling poorly nourished fawns that would have died soon in any case, or were simply scavenging carcasses of stillborn or dead fawns.

Further studies revealed that the summer range of the North Kings herd was deficient in food and cover for deer, due primarily to overprotection from fire. Fire scars on large trees indicated that between A.D. 1580 and 1920 fires occurred every 8 to 9 years. By the 1970s, however, no major fire had occurred in over 60 years. The result was growth of high brush and dense forest stands that provided little food for does late in pregnancy or for the fawns after they were born. Thus, coyotes found easy prey in weakened fawns. Coyote predation, in this instance, appeared to be largely an indication of poor range condition.

Efforts in the late 1970s and 1980s to improve range conditions of the North Kings herd, using techniques such as prescribed burning to open up overly mature brushlands, did not lead to the recovery of the herd, however, and it has continued to decline. Some investigators now suggest that the deer population has fallen below a critical ratio with coyotes and mountain lions, the latter having been protected for many years (Bertram 1984; Neal et al. 1987). Mountain lions in the range of the North Kings herd may have increased from about 12 individuals in the early 1970s to about 40 individuals in the late 1980s (Neal et al. 1987). Studies with radio-collared fawns and does suggest that predators kill as much as 48% of fawns and 22% of does annually. In 1985–1986 Smith (1990) found that coyotes were feeding very little on deer, even during the period of fawn drop, suggesting that much of this predation was due to mountain lions. Thus, recovery of this deer herd, even with improved range conditions, may now be inhibited by a high ratio of mountain lions to deer.

Dealing with this situation has now become complicated by California politics. In 1990, a voter initiative proposition protected the mountain lion from hunting (Fig. 14.3). This action prevented culling or controlled hunting to be used to aid recovery of the deer population. By 1995, the protected status of the mountain lion had become a political issue due to attacks on humans by mountain lions.

Gray Wolf Predation on Ungulates

Perhaps because of its notoriety, the North American gray wolf has become one of the best-studied large predators (Pimlott 1967; Mech 1979; Allen 1979; Fuller 1989). Yet even in this case, a full understanding of the predator-prey relationship is far from being obtained.

Isle Royale, an island of 570 square kilometers in area lying near the north shore of Lake Superior, is the site of one of the longest studies of wolf ecology (Fig. 14.4). Isle Royale was lumbered and depopulated of large game in the early 1900s. In 1912, however, moose reached the island by walking over the

FIGURE 14.3

The mountain lion has been protected against hunting in California by a voter initiative proposition.

(U.S. Forest Service photo)

frozen lake surface in winter. Moose found an early successional forest habitat that was ideal in quality, and their population exploded to between 1,000 and 3,000 individuals in the early 1940s. At this time severe overbrowsing was evident.

Wolves reached the island in 1949, crossing the ice in similar fashion. Soon an apparent "steady state" relationship of about 20 to 22 wolves and about 600 moose developed (Fig. 14.5). These populations corresponded to a moose:wolf ratio of about 30:1. Under this regime, wolves harvested about 142 to 150 moose per year—essentially a quarter of the population—or about 6.8 to 7.5 moose per wolf. Under these conditions, wolf predation appeared to be limiting the moose population well below the levels set by food availability in the 1930s and 1940s (Pimlott 1967).

Other studies suggested that the wolf density on Isle Royale was typical of the maximum density that wolf populations could achieve, about one wolf per 26 square kilometers. This limit seemed to be set by the behavior of the wolf. Pack territoriality, combined with dominance behavior that limits the size and structure of packs, appeared to set an upper limit to the density of wolves in a region. This was interpreted to be an intrinsic population control, and even more interestingly, one that appeared to be tuned to a fairly low density of large prey, such as that probably existing in mature forests of North America prior to European settlement (Pimlott 1967). Before European settlement, in fact, large predators may have limited the densities of prey such as moose, deer, and caribou in much of North America (Ballard et al. 1987).

Since the late 1960s, however, the wolf and moose populations of Isle Royale have fluctuated widely (Peterson & Page 1988). Moose have increased to over 2,400 animals and fallen to less than 600. Wolf numbers have oscillated between 12 and 50

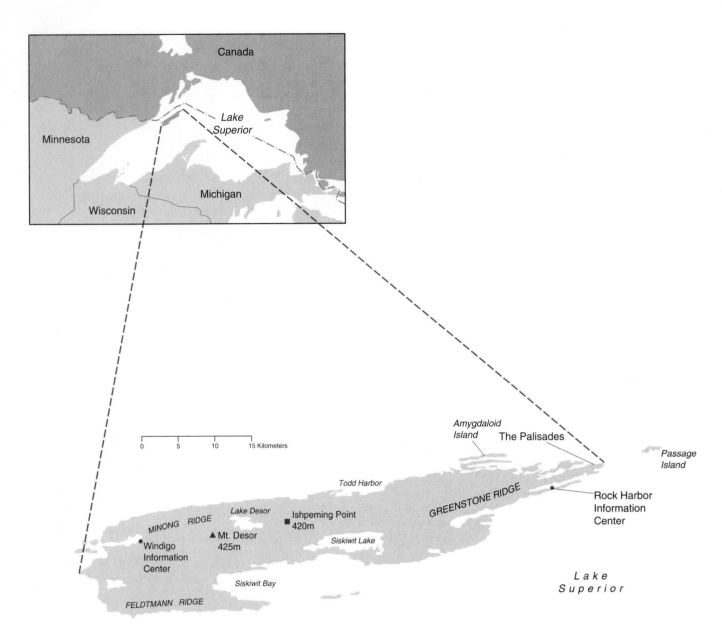

Isle Royale National Park

FIGURE 14.4

Isle Royale, an island of 570 square kilometers lying near the north shore of Lake Superior, is the site of one of the longest studies of wolf ecology.

FIGURE 14.5

A large pack of gray wolves travels through deep snow on Isle Royale, Michigan, the site of long-term studies of wolf predation.

(Photo by Rolf O. Peterson)

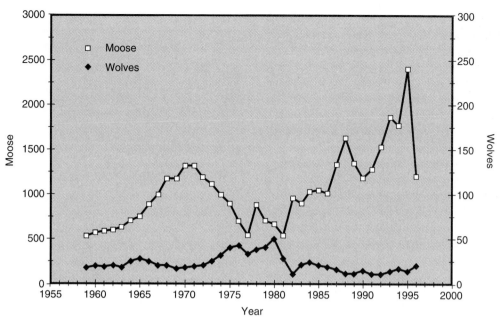

FIGURE 14.6

Estimated numbers of moose and wolves on Isle Royale since 1958.

(Data courtesy of Rolf O. Peterson)

TABLE 14.2

Populations of Moose, Caribou, and Wolves on the Tanana Flats, Alaska Before and After Artificial Reduction of the Wolf Population					
				RATIOS	
YEAR	MOOSE	CARIBOU	WOLF	MOOSE/WOLF	CARIBOU/WOLF
Before					
1965	23,000	5,000	200	115	25
1975	2,800	1,800	240	12	8
After					
1978	3,500	3,100	80	44	39

Data from W. C. Gasaway, et al., 1983, "Interrelationships of Wolves, Prey, and Man in Interior Alaska," in *Wildlife Monographs*, No. **84**:1–50, Bethesda, MD: Wildlife Society.

(Fig. 14.6). These fluctuations seem to be related partly to changes in quality of food supply for moose, and thus the health and vigor of individual moose, which are able to defend themselves effectively when they are in good condition. From 1987 through 1992, however, the wolf population fell to about 12 animals, perhaps the result of a disease, canine distemper, introduced to the island. The wolves at this time also showed no genetic variability, suggesting that problems might also have resulted from inbreeding (*See* Chapter 25). In 1994 and 1995, wolves showed some recovery, while moose numbers reached record levels then crashed. Thus, it now appears that even on Isle Royale, wolves may limit the size of the moose population at certain times, and not at others.

Studies in Alaska give another perspective on this relationship (Gasaway et al. 1983; Bergerud 1988). On the Tanana Flats, an area south of Fairbanks, moose and caribou populations crashed between 1965 and 1975 (Table 14.2). The de-

cline was attributed to combined effects of severe winters with deep snow, hunting harvest by humans, and wolf predation. In 1965, the total ratio of large ungulates to wolves was about 140:1, far above the 30:1 ratio at which wolves seemed to be controlling moose on Isle Royale. By 1975, however, this ratio had fallen to 20:1, a level at which limitation of prey populations by wolves was likely. From 1976 to 1979, wolf numbers were reduced experimentally, the intent being to increase the ratio of prey to wolves. By 1978, this experiment seemed to show success, and moose and caribou had reached a combined ratio of 83 ungulates per wolf.

Further studies (Gasaway et al. 1992), including a second experimental reduction of wolves in eastern Alaska, suggested that combined predation by wolves and grizzly bears often limits moose and caribou at levels well below those set by food. These investigators contended that where wolf and bear populations had been lightly hunted, moose populations tended to be low and predation on newborn calves heavy. They concluded that if ungulates are reduced to low densities by severe winters, they can become limited at this low level by predation. This analysis provided the scientific basis for proposals by the Alaska Fish and Game Department in 1992 and 1993 to reduce wolf and bear populations in parts of eastern Alaska. The conclusions of this study were questioned by a number of scientists, and the predator reduction efforts were strongly opposed by environmentalists and animal rights groups, leading to their cancellation.

Nevertheless, high wolf:prey ratios may limit or depress the abundance of moose and other ungulates. Caribou numbers may show a similar relationship, being depressed by wolf predation in forested areas where wolf populations are high, due to the presence of moose as alternate prey, but being independent of wolf predation in tundra areas where wolf numbers are low (Bergerud 1988).

In Minnesota, radio-tracking of wolves has been used to obtain data on movements, territories, and predation on white-tailed deer, their primary prey there (Fuller 1989). Wolf densities in the study area were near the maximum of one animal per 26 square kilometers. The ratio of white-tailed deer to wolves was about 158:1, and a small population of moose was also present. On average, individual wolves killed about 19 deer annually, 11 of these being fawns.

These studies suggest that critical ratios of prey to wolf exist for the maintenance of stable ungulate populations, both with and without hunting harvest of the ungulate prey (Keith 1983). In the absence of hunting, where moose are the primary prey, a ratio of 35:1 appears to be minimal, and where the smaller white-tailed deer are the primary prey, the minimal ratio is about 90:1. Ratios necessary to sustain these ungulate populations with additional hunting harvest would be larger.

Thus, even for individual predator species, the effect of predation may vary from controlling to noncontrolling. Favorable range conditions, however, seem to permit prey species such as moose, mule deer, and other early successional ungulates to exceed the critical ratios of prey to wolf below which predation is a controlling factor. A major strategy for management in these cases is thus improvement of range condition for ungulates, and direct predator control may be necessary only to correct imbalances that have developed because of unusual events.

PREDATORS AND LIVESTOCK DEPREDATIONS

A second focus of public concern is predation on livestock by large carnivores such as wolves, mountain lions, and coyotes. In North America, this concern now centers on the coyote, the predator most abundant on rangeland, and on sheep, the domestic animal most often taken. Once again, this relationship is far from being fully understood.

The most comprehensive evaluation of sheep losses to predators is a 1974 USDA study (Table 14.3) in which losses to coyotes, other predators, and other causes were evaluated (Gee et al. 1977). This study, covering 78% of the United States sheep

TABLE 14.3

Mortality of Adult Sheep and Lambs from Various Causes in 15 Western States as Determined for 1974

| | PERCENT MORTALITY | |
Cause of Death	Lambs	Adults
Coyote Predation	8.1	2.5
Other Predators	3.3	0.9
Subtotal	11.4	3.4
Other Causes	8.2	5.1
Unknown	3.6	1.9
Subtotal	11.8	7.0
TOTAL	**23.2**	**10.4**

Data from J. H. Schrivener, et al., 1985, "Sheep Losses to Predators on a California Range, 1973–1983" in *Journal of Wildlife Management*, **38**:418–21. Bethesda, MD: Wildlife Society.

ranching area, showed that heaviest losses were for lambs, but that substantial losses of adult sheep also occurred. What was also evident is that predation losses for both lambs and adult sheep are less than half of all mortality, and that losses to coyotes amount to only about a third for lambs, or a quarter for adults, of total losses. A very similar picture (Table 14.4) was given by a study of sheep losses in central California from 1973 to 1983. Thus, predation by coyotes is only one of the problems of sheep ranching, an industry in decline because of marginal production economics.

Large sums have been spent on coyote control, with little evaluation of the benefits. In 1974, $6,883,000 was spent for coyote control, mostly to hire professional predator control personnel. The result was the killing of 88,092 coyotes, at an average cost of $78 per animal. The cheapest methods were use of the **M-44 device,** a metal tube that uses a shotgun shell to shoot cyanide into the mouth of any animal that pulls on the baited end, and poisoning carcasses of dead livestock with **Compound 1080** (sodium fluoroacetate). These latter methods are also the least selective, and frequently kill other predators and scavengers.

No real evaluation has been attempted of the benefits of coyote control on an industry-wide basis. In 1974, the total economic loss to coyotes was estimated as $27 million, based on the deaths of about 1.08 million lambs and adult sheep. Each sheep lost to predation thus had a value of about $25. At a cost of $78 per coyote killed, control would have to save slightly over three animals per killed coyote to break even in cost effectiveness. Whether such a benefit is realized is highly uncertain. At best, this analysis suggests that coyote control efforts are probably only marginally cost effective.

Several alternatives to continuing predator control can reduce predation on domestic animals (Robel et al. 1981). Sheep losses can be reduced by burial or removal of carcasses of dead

FIGURE 14.7

Recovery of populations of the American alligator in the southern United States has led to an increase in human injury by these animals. (USDA Forest Service photo)

Data from J. H. Shrivener et al. 1985, "Sheep Losses to Predators on a California Range, 1973–1983" in *Journal of Wildlife Management*, 38:418–21. Bethesda Md: Wildlife Society.

TABLE 14.4

Mortality of Adult Sheep and Lambs from Various Causes in Central California from 1973 to 1983

Cause of Death	PERCENT MORTALITY	
	Lambs	Adults
Coyote Predation	2.4	1.3
Other Predators	0.3	0.2
Subtotal	2.7	1.5
Other Causes	0.8	1.6
Unknown	2.5	2.6
Subtotal	3.3	4.2
TOTAL	**6.0**	**5.7**

animals, which reduces the food supply for coyotes and lessens the tendency of coyotes to become conditioned to sheep as food. The use of trained sheep dogs or llamas to protect herds from predators is a second effective technique. Confinement of sheep in corrals, especially lighted corrals, also reduces losses. Managing herd reproduction so that the lambs are born in fall, rather than spring, reduces predation on lambs. In the fall, coyotes are usually better fed and have more alternate prey available than after a long winter when natural prey are scarce. Finally, the selective killing of individual coyotes that have become predators on sheep, as opposed to a general coyote suppression program, is an effective, and sometimes necessary, technique.

PREDATORS DANGEROUS TO HUMANS

Large carnivores in parks or wilderness areas sometimes present risks to humans visiting or living in these locations. In North America, this problem has centered on bears—the grizzly bear (*See* Reading 14.1), polar bear, and black bear—and the mountain lion. In parks where grizzly and black bears are protected, they often become habituated to sources of food associated with man, but are still highly aggressive, leading to many incidents of human injury or death. Through 1980, 41 deaths from grizzly bear attacks and 20 by black bears occurred in North America (Herrero 1985). Additional deaths have occurred since then, 4 in 1995, for example. In Canada, 6 persons were killed by polar bears between 1965 and 1985 (Feazel 1990). At Churchill, Manitoba, the concentration of wintering polar bears along the shore of Hudson Bay leads to frequent encounters between bears and people. In California, recovery of the mountain lion population has been accompanied by several attacks and two human deaths in recent years.

Other animals are occasionally a problem in North America as well. With the recovery of American alligator populations in the southern United States, injuries from these animals have become more frequent (Fig. 14.7). In California, cases of injuries to children by coyotes in suburban residential areas have been reported luridly in newspapers. Large ungulates such as bison, moose, elk, and deer have also caused injuries and deaths of unwary park visitors that unwisely treat them as tame animals.

Elsewhere in the world, these sorts of problems are even greater. In India, where efforts have been made in recent years to protect the Bengal tiger, human deaths are frequent near tiger preserves. At the Sunderbans Preserve, for example, about 45 human deaths per year were recorded prior to 1983. Most of these deaths were of humans entering the preserve to collect firewood, honey, or other items. An interesting experiment was undertaken at this preserve to condition tigers against attacking humans by the use of electrified human mannequins. These mannequins, placed in the forest, give a strong shock to any animal that comes into contact with them. Seemingly, this negative conditioning has been effective, and only 12 deaths occurred in 1984. Encouraging residents of the park vicinity to wear face masks on the back of their head when engaged in activity in bush areas near the park has also deterred tiger attacks. However, for India as a whole, about 50 tiger deaths per year still occur. Elephants are also a danger to humans in Africa and Southeast Asia, causing about 100–150 deaths annually in India (Sukumar 1989).

REINTRODUCTION OF PREDATORS TO PARKS AND PRESERVES

Large predators have been eliminated from much of their original ranges by human killing, often with the aid of governmental eradication programs. Restoration of such species to major parks and preserves is now being encouraged by many conservation groups. Black bears have been restored to national forest areas in

In North America, public concern about risks from grizzly bears was aroused in 1967 by a rash of maulings and the deaths of two girls in Glacier National Park, Montana. These events led a prominent biologist, Gardner Moment (1968, 1969), to suggest that the presence of grizzly bears in parks such as Yellowstone and Glacier was incompatible with human use. Specifically, Moment (1969) suggested that

> grizzly bears should be removed from parks such as Yellowstone and Glacier, which are unsurpassed in eminence as part of our heritage of natural wonders, and are ideal for the family hiking and camping that are traditional in them.

The grizzly bear is the North American representative of a species that ranges across much of Eurasia and North America. In North America, grizzlies originally occurred west of the Mississippi from the arctic tundra regions of Alaska and Canada southward to northern Mexico. This species includes the so-called brown bears of parts of Alaska and Eurasia. Grizzlies are large, females weighing up to 1,200 kilograms and males 1,800 kilograms. Despite their size, they are agile, and can run at a speed of 48 kilometers per hour for short distances. Grizzlies are highly aggressive, and maintain a security zone between themselves and other large animals by threats and chases. When another large animal, such as a human, enters this zone, a grizzly reacts in reflex manner by a threat or charge. For adult females with cubs, the security zone may be more than a hundred meters in radius. Animals feeding on the carcass of a prey also maintain a large security zone.

In Alaska and Canada, grizzly bear populations are estimated to number 10,000–15,000 individuals. In the contiguous United States, however, grizzlies occur only in several areas of national parks and forests (Reading Table 14.1), with an estimated total population of 650–900 animals. Populations in the Yellowstone National Park area had been declining until about 1989, largely due to illegal shooting, but have begun to increase slightly since then. In 1988, about 170–180 animals inhabited the park and surrounding national forest areas.

Past park practices contributed to some degree to the grizzly bear danger to visitors. Prior to 1930, for example, bears in Yellowstone Park were fed to attract them to sites for visitor viewing. After this was discontinued, grizzlies became attracted to garbage dumps that each year received the refuse of a growing number of visitors. At the dumps, the bears acquired familiarity with human foods. Realization that the bears were becoming conditioned to such foods, and that this probably reinforced their raiding of campground areas, led to the closing of these dumps in 1970. Garbage is now trucked to disposal sites outside Yellowstone and Glacier Parks.

Grizzly bears have caused 13 deaths and numerous injuries in United States and Canadian national parks over the years. These cases, however, are few relative to deaths and injuries in the same parks from other natural hazards. In Glacier National Park, for example, only 3 deaths by grizzly bears occurred between 1916 and 1977 (others have occurred since 1977). During this same period there were 36 deaths from drowning, 16 from hiking falls, 6 from falling rocks or trees, and 11 from other natural accidents (Cauble 1977). Nevertheless, deaths from grizzly bear attacks receive great publicity.

Several patterns can be seen in grizzly bear attacks on humans. Through 1970, more than 80% of attacks involved sows with cubs (Herrero 1970). Over half of these were in campgrounds in developed areas of parks (at a time when bears had access to garbage dumps that were not far away). Over 30% involved back-country hikers, however, and in almost all of these cases a sow with cubs was the culprit. In

Arkansas (Smith et al. 1991) and final plans are being made for reintroduction in eastern Kentucky and Tennessee (McDade & Emmott 1995). Canada lynx have recently been reintroduced to Algonquin State Park in New York.

The most controversial reintroduction programs have involved wolves. In 1986, the United States Fish and Wildlife Service reintroduced red wolves to the Alligator River National Wildlife Refuge in North Carolina. By 1990, 21 captive-bred animals had been released, and two young had been reared successfully in the wild (Phillips 1990).

The gray wolf is now being reintroduced to the Greater Yellowstone Ecosystem in Wyoming and neighboring states. Wolves were eliminated from this region in about 1926. Reintroduction of the species is consistent with the objective of creating in Yellowstone National Park a complete, naturally functioning ecosystem like that existing prior to European arrival in North America. The Greater Yellowstone Ecosystem should be able to support about 11 to 15 packs, or a total of about 110 to 150 animals (Singer 1990). Models of the interaction of the wolves and prey suggest that wolf predation might reduce the size of elk herds by about 15%–25%, with lesser or very minor

Populations of Grizzly Bears in North America in 1988, and as Proposed Under the Recovery Plan for Endangered Portions of the Population*		
LOCATION	1988 POPULATION	RECOVERY TARGET
Canada and Alaska	10,000–15,000	———
Glacier National Park (Area), MT	440–680	560
Yellowstone National Park, WY	170–180	301
Cabinet Wilderness, MT	12	70
Selkirk Mountains, ID	?	———
North Cascades, WA	?	———

*Compiled from various sources

these cases, hikers likely came upon the bear at a blind point along a trail, finding themselves suddenly within the security zone of a sow with cubs. Since the mid-1970s, injuries in Glacier National Park have involved more single adult grizzlies, perhaps as the result of increase in numbers of bears and reduction in their fear of humans in this protected population (Martinka 1988). The rate of injuries by grizzlies has increased in recent years from 0.7 to 2.1 per million visitors in this park as well, indicating that the problem is still significant.

The response of the U.S. National Park Service to this problem has not been to remove the animals from Yellowstone and Glacier National Parks. In fact, the grizzly bear is the object of a recovery program intended to establish stable populations in at least four areas of the contiguous United States (See Reading Table 14.1). In the Yellowstone National Park area the goal is to increase numbers to about 301 animals, and to restore natural patterns of behavior of the species (Yellowstone National Park 1982). At the same time, efforts are being made to minimize the chance of close encounters of visitors with grizzlies, especially sows with cubs. Garbage dumps inside the park have been closed, bear-proof garbage containers provided, and camping regulations adopted to prevent bears access to foods and food wastes. Bears that become problem animals in visitor-use areas are removed to remote parts of the park. Certain areas of heavy bear use are permanently or temporarily closed to hiking. Educational leaflets and films are used to inform individuals of the danger from bears, and of ways to minimize risks. These include hiking in groups, avoiding streamside paths and game trails that bears frequent, making noise when hiking in bear country, and keeping a food-free camp by cooking and storing food more than 100 meters downwind from the sleeping area (Herrero 1985).

effects on other species (Boyce 1990). The total population of elk, the most abundant large ungulate in this ecosystem, is now about 40,000 to 50,000 animals.

Opposition to wolf reintroduction in Yellowstone centered on possible depredations on livestock in areas outside the park and wilderness areas. Some livestock have been killed in the vicinity of Glacier National Park, by the small wolf population there. Reintroduction to the Greater Yellowstone Ecosystem is coupled with a provision for allowing animals that wander outside park and wilderness areas to be killed. In addition, a fund for reimbursement of damages caused by wolves to live-stock and other human property has been established. In addition, many hunters opposed reintroduction of wolves on the grounds that elk and deer summering inside the park migrate to areas outside the park, where they can be hunted in autumn. Reduction of herds with summer ranges in the park might therefore reduce hunting harvests.

Plans are also being made for reintroduction of the Mexican wolf, a subspecies of the gray wolf, to White Sands Missile Range, New Mexico, and national forest areas in Arizona (See Reading 14.2).

The Mexican wolf, *Canis lupus baileyi,* is one of the five currently recognized subspecies of the North American gray wolf, and is the one most seriously endangered (Reading Fig. 14.2). The total population in 1995 consisted of only 137 individuals in captivity. These individuals were scattered among 24 zoos and animal parks in the United States and Mexico. This population is derived from only seven individuals that were brought into captivity and became successful breeders. A relict population may exist in northern Mexico, where about 50 animals were estimated to exist in the late 1970s. More recent efforts to locate wild Mexican wolves have been unsuccessful.

The range of the Mexican wolf extended from southeastern Arizona, southern New Mexico, and southwestern Texas, south through the interior of Mexico to the vicinity of Mexico City. Mexican wolves lived mostly in woodland habitats mixed with open grasslands at elevations about 1,350 meters. The wolf was extirpated throughout its range largely due to killing by ranchers, and, in the United States, by federal animal control efforts between 1915 and 1960.

Because the Mexican wolf has been designated an endangered species, the United States Fish and Wildlife Service developed a recovery plan, which recommended reintroduction to the wild. A draft environmental impact statement (EIS) for reintroduction was released for public comment in 1995. This EIS considers several reintroduction alternatives, the proposed alternative being to release animals into one or both of two areas: the San Andres Mountains in the White Sands Missile Range in New Mexico, and the Blue Range portion of the Apache National Forest in Arizona. The release areas lie within larger recovery zones throughout which the reestablished population is expected to disperse. The total recovery area for the San Andres Mountain animals includes the White Sands Missile Range and

READING FIGURE 14.2

The Mexican wolf, a race of the gray wolf, is being considered for reintroduction to locations in the southwestern United States.

(Photo by Russell Lampertz, Wild Canid Survival and Research Center, Eureka, Missouri)

adjacent federal land. For the Blue Range animals, the total recovery area includes the entire Apache and Gila National Forest areas in Arizona and New Mexico.

If reintroduction is approved, releases will be spread over several years, beginning in 1997, and will use captive-bred animals whose genetic characteristics are already well represented in the captive population. Animals will be gradually acclimated to living and breeding in the wild, a process that may last several months. The target recovery population is about 120 animals ranging over about 15,360 square kilometers. The White Sands recovery area is projected to be capable of supporting about 20 wolves, and the Blue Range recovery area about 100 animals. The cost of this program is expected to be about $7.2 million.

A reestablished Mexican wolf population is expected to prey primarily on various ungulates, such as deer and elk. Javelina, feral horses, and even exotic oryx in the White Sands Missile Range could be secondary prey species. Predation is expected to reduce densities of principal prey species somewhat below what they would be in the absence of wolves. Some predation by wolves on domestic animals, especially cattle, might also occur; however, fewer than 40 cattle are predicted to be taken by wolves annually. The Defenders of Wildlife has developed a plan to reimburse ranchers for confirmed losses. The wolves would also be designated a **nonessential experimental population,** allowing greater flexibility for controlling animals that prey on livestock.

On the positive side, enhancement of the wilderness nature of the Apache and Gila National Forests might attract more back-country visitors, bolstering the recreational economy of the region. Visitors might also obtain greater satisfaction from back-country visits, especially if they hear or see Mexican wolves. Furthermore, if a viable population can be established, expenditures for maintaining captive populations might be reduced, and the long-term survival of the subspecies assured.

Surveys have shown that most people in the region support the Mexican wolf recovery program. But as for wolf reintroduction efforts elsewhere, this proposal has been opposed by some ranchers and opponents of federal land management policies in the West. The New Mexico Cattle Growers Association has opposed the action, claiming that, in spite of the plan for reimbursing predation losses, it will likely cause them economic harm. Supporters of the County Rights Movement in Arizona and New Mexico, who favor return of federal lands to local control, oppose the action as another example of federal land management that ignores the interests and wishes of local residents.

Summary

1. Predators vary in degree of intrinsic control of their own populations by territoriality and other social behaviors. Predators show both functional and numerical responses to changes in prey density, and may show selectivity for substandard prey if prey capture is difficult.

2. The influence of a particular predator on a particular prey depends on many factors in the ecosystem, so that predation can vary from being a strong regulatory control on a prey species to being a factor of minor significance.

3. Predator depredations are a significant cause of mortality for some domestic animals, such as sheep on open rangeland, but several techniques of herd management, together with selective predator control, usually can provide adequate protection.

4. Several large predators, such as bears and large cats, pose danger to humans living near or visiting wildland areas, thus requiring special educational and management programs to achieve public safety.

5. Reintroductions of predators are being conducted in many locations to restore original ecosystem relationships and to effect the recovery of endangered species.

Questions for Discussion

1. Does the demonstration that a predator is limiting the density of a game species at a level below that set by its food supply justify a program of occasional or regular predator control?

2. Should a predator such as the coyote be controlled on public lands specifically to reduce depredation losses by ranchers leasing grazing rights for sheep or other livestock on these lands?

3. What criteria should be used in considering the reintroduction or population recovery of large carnivores in areas near human populations or in wildlands with heavy visitor use?

Suggested Reading

Carbyn, L. N., S. H. Fritts, and D. R. Seip (Eds.). 1996. *Ecology and conservation of wolves in a changing world*. Canadian Circumpolar Press Institute, University of Alberta, Canada.

Craighead, J. J., J. S. Sumner, and J. A. Mitchell. 1995. *The grizzly bears of Yellowstone: Their ecology in the Yellowstone Ecosystem, 1959–1992*. Island Press, Washington, DC. An account of the bear management controversies and ecological studies of grizzlies in Yellowstone, focusing on the efforts of John and Frank Craighead to protect the species.

Gasaway, W. C., R. D. Boertje, D. V. Grangaard, D. G. Kelleyhouse, R. O. Stephenson, and D. G. Larsen. 1992. The role of predation in limiting moose at low densities in Alaska and Yukon and implications for conservation. *Wildl. Monogr.* **120**:1–59. A controversial study that formed the scientific basis for proposals for reduction of wolf and grizzly bear populations in Alaska in 1992 and 1993.

part

four

Aquatic Ecosystems

A quatic ecosystems make up more than 70% of the biosphere. The biotic resources of these ecosystems have long been attractive to humans, and many have been exploited heavily since ancient times. Aquatic ecosystems have been the source of food, water, energy, and even minerals. They have served as routes of exploration, travel, and commerce. They have provided recreation. Not least of all, they have served as sewers to dilute and flush away the wastes of human activity. Not surprisingly, almost all major human settlements are situated on the banks or shores of streams, lakes, or oceans. Aquatic ecosystems thus have received all degrees of impact by humans. Some are so changed that we can only guess at their pristine state. Others are only now beginning to feel the impact of human activity.

chapter

15

Lakes, Ponds, and Marshes

L ake, pond, and marsh ecosystems are as diverse as the full range of terrestrial biomes. They range from large, deep, and geologically ancient lakes, such as Baikal in the former Soviet Union, to shallow seasonal marshes and tiny temporary pools. Their waters likewise range from fresh and nutrient-deficient, as in Lake Tahoe, to fresh and rich in nutrients, as for Lake Erie, and to hypersaline, as for the Great Salt Lake.

Few types of ecosystems have been abused as badly as lakes. The impacts of humans have been direct and indirect, intentional and inadvertent, biotic and abiotic. That these systems retain as much wildlife as they do is testimony to considerable resilience. In the developed nations, diversion of freshwater inflow for agricultural or urban use, and the filling or drainage of ponds and marshes to create land for urban development or farming, are destroying many of these ecosystems. Throughout the world, the watersheds of lakes, ponds, and marshes have been modified by forest clearing, agricultural activity, and urban growth. Their animal life has been exploited carelessly, and their basins have served as convenient disposal sites for agricultural, urban, and industrial wastes. Intentionally or by accident, exotic plants and animals have been introduced. Often, these impacts have distorted basic patterns of nutrient cycling, usually by flooding the ecosystem with nutrients from sewage discharges, fertilizer runoff, or accelerated erosion of the watershed.

In this chapter we shall first examine these impacts in detail, giving special attention to enrichment of the ecosystem by increased nutrient inflow. We shall defer fuller discussion of some impacts, such as those of acid deposition and chemical pollution, to later chapters. Next, we shall consider some of the special lake, pond, and marsh ecosystems of major conservation importance. Finally, we shall consider national and international efforts to conserve wetlands ecosystems.

Outline

Waterfowl are among the many valuable wildlife resources of freshwater lakes and wetlands.

FIGURE 15.1

About 900,000 eared grebes use Mono Lake as a summer and autumn staging area prior to southward migration to wintering areas at the Salton Sea in California and in the Gulf of California.

(Photo by J. R. Jehl, Jr.)

FIGURE 15.2

Wilson's phalaropes spend several weeks at Mono Lake, where they deposit fat that serves as fuel for migration south to wintering areas on salt lakes of the South American altiplano.

(Photo by J. R. Jehl, Jr.)

DIVERSION AND DRAINAGE

Diversion of water for agricultural and urban use has had a major impact on lakes in semiarid regions such as western North America and central Asia. Many such lakes are saline and fish-free, but possess distinctive invertebrate faunas that are heavily exploited by water birds. In North America, Mono Lake, Pyramid Lake, and other lakes in the Great Basin have lost much of their water inflow by diversion. In the former Soviet Union, the Aral Sea has shrunk in surface area by 40% since 1960, due to the diversion of water for irrigation.

Mono Lake, in eastern California, has been a focus of great controversy in North America (Mono Basin Ecosystem Study Committee 1987). This lake, with a solute content nearly three times that of the ocean, has an aquatic biota of algae, brine shrimp, and the larvae of brine flies. The abundant brine shrimp and brine fly larvae make the lake an important stopover area for several migratory birds (Jehl 1988). Up to 900,000 eared grebes spend 3 to 8 months at the lake in late summer and autumn to molt and store fat for southward migration (Fig. 15.1). About 100,000 Wilson's phalaropes spend 3 to 5 weeks at the lake for the same purpose (Fig. 15.2), and about 54,000 red-necked phalaropes visit the lake on brief migratory stopovers. In addition, 61,000 or so California gulls nest on islands that are free of terrestrial predators such as coyotes.

By the mid-1930s, Los Angeles had acquired water rights for much of the watershed of Mono Lake, and in 1941 began diverting water into the Owens Valley Aqueduct leading to Los Angeles. Until 1970, about 56,000 acre-feet of water were diverted annually. In 1970, with growth in Los Angeles water demand, the aqueduct tunnels were enlarged and diversions in- creased to about 90,000 acre-feet annually. Increased diversion caused a drop of over 12 meters in lake level, a 50% decrease in surface area, and more than a doubling of salinity. Dropping water level created land connections to several islands on which California gulls were nesting, making them accessible to coyotes, so that the gulls were forced to move to other islands. It also exposed the striking tufa formations—tower-like calcium carbonate deposits formed in the mineral-rich water. Studies revealed that if the diversions were continued, the lake level would continue to fall, and the water would become saline enough to kill the lake invertebrates and to eliminate most nesting islands for California Gulls.

In 1988, the United States Forest Service, which manages Mono Lake and its watershed, concluded that a reduction of 50%–75% in diversion was needed. In 1994, the City of Los Angeles agreed to this plan. As a result, the surface level of the lake is expected to rise about 5 meters, sufficiently high to maintain the isolation of all the major islands and cover the exposed tufa formations.

Ponds and marshes have also suffered destruction. In the upper Great Plains of the United States and Canada, drainage of prairie ponds and marshes, collectively termed **potholes,** has been especially widespread (Weller 1981). Ten million or more potholes originally occupied the undulating topography left by retreat of Pleistocene glaciers. Perhaps 50%–60% of these have been drained and converted to farmland (Mitsch & Gosselink 1986). Iowa has lost about 99% of its original wetlands and North Dakota about 60%, for example (Tiner 1984).

Major wetland losses have occurred in other parts of the continent. Coastal Louisiana, with about 40% of the coastal wetlands of the coterminous 48 states, is losing wetlands at a rate of 100 to 150 square kilometers annually (Baldwin et al. 1990). Land subsidence following petroleum removal, construction of canals, and other types of development are apparently contributing to these losses. In the coterminous states, only about 40 million hectares of coastal and inland wetlands now remain, roughly 46% of those originally present (Tiner 1984).

Loss of wetlands, particularly in the prairie pothole region, has contributed to a major reduction in waterfowl populations from their original levels in North America (Bethke & Nudds 1995). Disappearance of the whooping crane from most of its former breeding range (*See* Chapter 26) was partly the result of early habitat destruction. Since the 1920s, however, waterfowl numbers have fluctuated widely, largely due to wet and dry cycles that affect breeding populations in the prairie pothole region. Severe population declines were recorded in the 1930s, early 1960s, and late 1980s. Wetter years in the 1990s are now helping these waterfowl populations to recover. The dependence of waterfowl numbers on favorable habitat, however, means that any further losses will certainly translate into permanent reduction of populations.

In California, vernal pools (*See* Chapter 4) were originally widespread in the Central Valley and in coastal areas from Santa Barbara south. Agricultural and urban developments have eliminated about 90% of the original vernal pool habitat, and several species of plants are now endangered.

WATERSHED MODIFICATION

Even where the water inflow to lakes is not directly impaired, changes in watershed vegetation and land use affect lakes. Clearing or thinning of watershed vegetation typically leads to increased runoff and erosion, and thus to greater transport of dissolved materials and silt into lakes. The seasonal pattern of flow is changed as well, and wet season flows increase while dry season flows decrease. Where a forest cover is removed, temperatures of streams are also altered, typically becoming warmer in summer due to increased exposure of the stream surface to sunlight. These changes modify the biota of watershed streams in a complex fashion, which in turn influences the biota of the lakes into which they flow.

OVEREXPLOITATION AND INTRODUCTION OF EXOTICS

Commercial fisheries in large lakes have been overexploited nearly everywhere, as illustrated by salmonid fisheries in lakes of northern Eurasia and North America. The fisheries of most of these lakes were exploited heavily before modern techniques of assessing fish stocks were available. Large catches in the early years of fishing nearly always led to harvests, and even to scientific estimates of sustainable harvest, that exceeded the long-term potential of the ecosystem (Regier & Loftus 1972).

The decline in yield that follows overexploitation favors other biotic changes, such as the establishment of exotic species. Some exotics arrive by natural dispersal or are introduced accidentally. Species purported to be better game or commercial fish, however, are often introduced deliberately. Few lakes have escaped such meddling. Introduced species may completely restructure the lake ecosystem, often to the detriment of valuable, endemic species.

The introduction of the peacock bass, *Cichla ocelaris,* to Gatun Lake in Panama illustrates the extent of ecosystem restructuring that can occur. A large, predatory fish native to the Amazon River, the peacock bass was introduced as a sport fish to ponds in the upper drainage of the Chagres River in 1967. Floods carried it into Gatun Lake, where it gradually spread, eliminating or reducing the populations of native fish (Zaret & Paine 1973). Decline of the smaller fish led to the disappearance of several fish-eating water birds. Loss of these fish permitted increase in zooplankton and insect larvae, including those of *Anopheles* mosquitos, which are carriers of malaria.

In similar fashion, introduction of the large opossum shrimp, *Mysis relicta,* to lakes in the western United States has sometimes decimated smaller zooplankton populations. In turn, this has led to a decline of fish the opossum shrimp were supposed to increase, and sometimes even to a decline of fish-eating birds such as the bald eagle (Spencer et al. 1991). Introduction of fish to previously fish-free lakes has also disrupted these ecosystems, often reducing or eliminating invertebrates that are intolerant of fish predation.

EUTROPHICATION

Eutrophication, the increase in fertility and productivity of an ecosystem due to an increased rate of nutrient input, is perhaps the most pervasive impact of human activities on lake ecosystems. Eutrophication may be natural, due to the slow increase in fertility of a lake's watershed through geological time. Almost all cases of recent eutrophication, however, are due to human activity. Sewage discharges, organic waste discharges by industries, runoff of fertilizers from agricultural lands, and increased leaching and erosion of watersheds disturbed by farming and construction all increase nutrient inflows to lakes. For most fresh waters, phosphorus is the limiting nutrient for primary production, and excess inputs of phosphorus the primary culprit in eutrophication due to human activity (Schindler 1974). The continuing growth of human populations and the exponential use of phosphorus fertilizers in human food production are the underlying causes of most of the eutrophication of aquatic ecosystems (Forsburg 1994).

Many of the impacts of eutrophication result from the fact that deep lakes in the temperate zone become thermally stratified in summer. Solar heating creates a surface zone of warm, low-density water—the **epilimnion**—overlying the cold, high-density bottom water—the **hypolimnion** (Fig. 15.3). These layers are separated by the **thermocline,** a relatively thin zone where temperature and density change rapidly with depth. Summer stratification prevents exchanges between the epilimnion and hypolimnion. Wind-induced currents circulate only within the epilimnion, and oxygen is not carried downward, nor are nutrients released by decomposition carried upward. In cooler periods of autumn and spring, this stratification breaks down, and currents circulate through the entire lake profile.

Lakes vary greatly in their fertility and productivity. A lake with low concentrations of nutrients and a low rate of nutrient input is termed an **oligotrophic** lake. Oligotrophic lakes have a

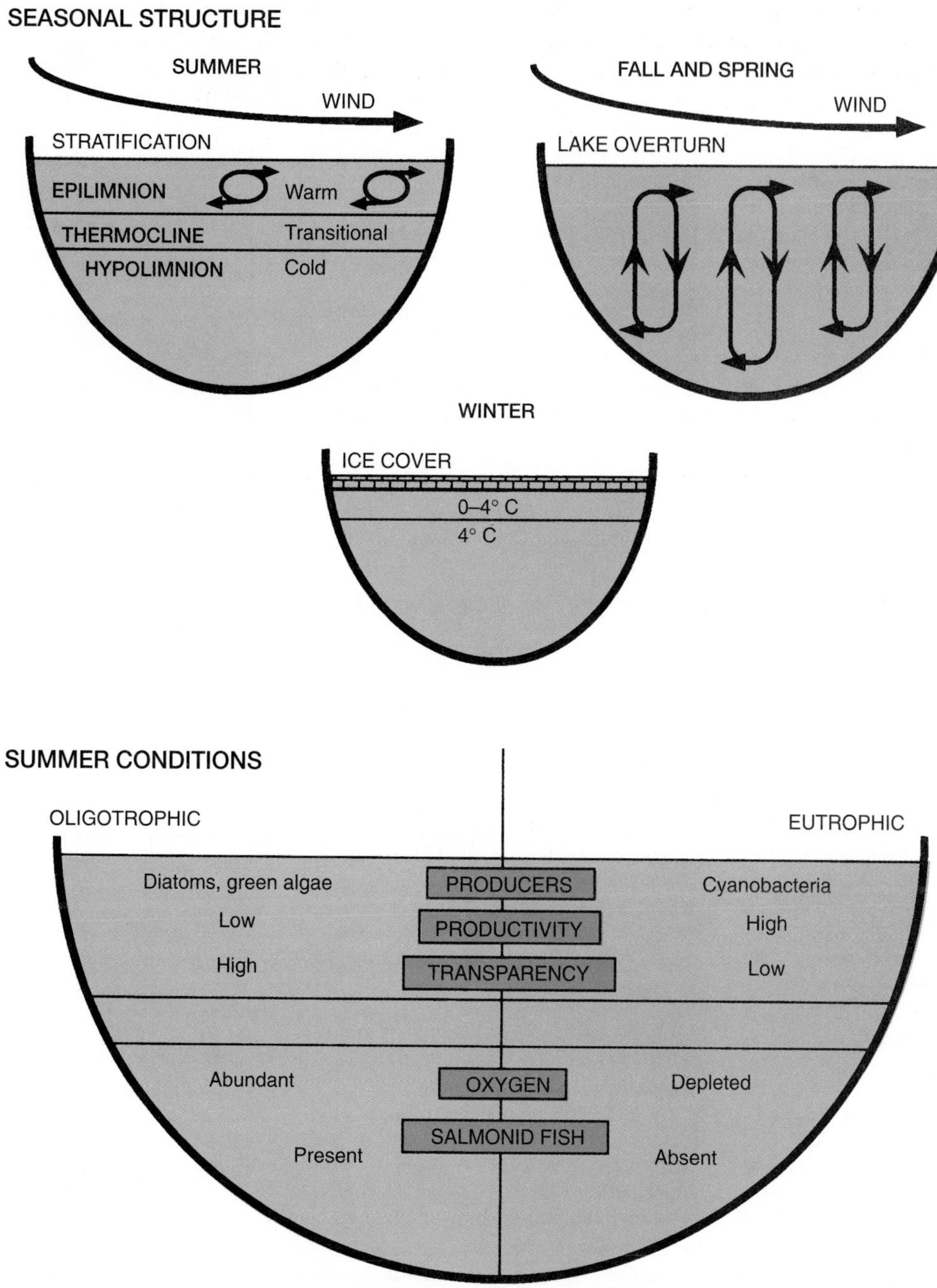

SEASONAL STRUCTURE

SUMMER

WIND

STRATIFICATION

EPILIMNION — Warm
THERMOCLINE — Transitional
HYPOLIMNION — Cold

FALL AND SPRING

WIND

LAKE OVERTURN

WINTER

ICE COVER

0–4° C
4° C

SUMMER CONDITIONS

OLIGOTROPHIC

EUTROPHIC

Diatoms, green algae	PRODUCERS	Cyanobacteria
Low	PRODUCTIVITY	High
High	TRANSPARENCY	Low
Abundant	OXYGEN	Depleted
Present	SALMONID FISH	Absent

FIGURE 15.3

The vertical structure and seasonal cycle of oligotrophic and eutrophic lakes in the Temperate Zone.

low phytoplankton density in the epilimnion, and thus tend to be clear—the abundance of phytoplankton largely determines the clarity of lakes free of massive sediment pollution (Fig. 15.3). The low abundance of producers—mainly diatoms and green algae—is coupled with a low primary productivity, and this, in turn, means that the amount of dead organic matter sinking into the hypolimnion is small. Thus, in summer, bacterial decomposition of organic matter is rarely great enough to deplete the oxygen supply of the hypolimnion. Oligotrophic lakes are thus favorable to salmonid fish such as lake trout, whitefish, and other

deep-water fish that demand cold, high-oxygen waters, and, of course, to invertebrates such as mayfly larvae and freshwater shrimp, on which these fish feed. The dominant fish of oligotrophic lakes are high-quality game fish and table fish, so that even though their productivity is not great, their value is.

With an increase in fertility and productivity, a lake becomes **eutrophic.** As phytoplankton become more abundant, the clarity of the water declines. The producer community changes in composition, often to dominance by blue-green algae, many of which are incompletely harvested by herbivores.

THE ST. LAWRENCE GREAT LAKES: A CASE STUDY

T he North American Great Lakes were originally oligotrophic lakes with faunas of salmonid fish such as lake trout, Atlantic salmon (Lake Ontario), and various whitefish, such as the ciscos and the lake herring (Regier 1979). Lakes Ontario, Michigan, Huron, and Superior had 10 to 13 salmonids, and Lake Erie, the shallowest of the lakes, 4 species. All of the lakes have been profoundly changed by the impacts of human activity.

Overfishing contributed to the decline of several whitefish, as well as other prize species such as the lake sturgeon, in the 19th and early 20th centuries. In 1879, for example, 1,780 tons of sturgeon were harvested in Lake Michigan. In only 20 years, this harvest had declined to less than 10 tons.

With settlement, the watersheds of the lakes were cleared of most of their forests. Deforestation led to higher temperatures of the streams flowing into the lakes, making them less favorable for spawning by native sturgeon and salmonids, and more favorable for spawning of introduced species, particularly the sea lamprey. Several attempts to reestablish Atlantic salmon in Lake Ontario have failed because streams flowing into the lake are too warm and shallow to provide the well-oxygenated gravel bottoms required for spawning (Moss 1988). Drainage of the Great Black Swamp at the western end of Lake Erie led to massive siltation in the western end of the lake by the Maumee River, contributing to major ecological change (Egerton 1987).

Overfishing and watershed changes, together with canal construction, enabled several exotic species to invade the lakes. The Erie Canal, linking the Hudson River and Lake Ontario, allowed the sea lamprey to invade Lake Ontario in the 1860s. There, the lamprey found suitably warm spawning streams in its watershed. Later, with reduction in the number of locks in the Welland Canal that allowed shipping to pass around Niagara Falls, the lamprey reached the upper lakes, colonizing them between 1921 and 1946. The sea lamprey is a predator on other fish, feeding on them by attaching to the body surface with a sucker-like mouth, rasping through the body wall with horny teeth, and draining the body fluids of the prey (Reading Fig. 15.1A). The lamprey had severe impacts on the native salmonids in Lakes Ontario, Michigan, Huron, and Superior. The lamprey first attacked the largest deep water salmonids, such as the lake trout. When their populations declined, the lamprey shifted to the next largest species. The lake trout and Atlantic salmon were driven to extinction in Lake Ontario before 1900, due to overfishing, watershed change, and sea lamprey predation. In the upper lakes, sea lamprey predation caused catastrophic declines in salmonid populations in the 1930s and 1940s, driving several species to extinction.

Decline of salmonid fish was correlated with invasion of the lakes by the alewife and rainbow smelt. The alewife followed the lamprey into Lake Ontario in the 1870s, and into the upper lakes between 1931 and 1953, filling the plankton-feeding niche of the lake herring, one of the lake whitefish. The smelt first entered Lake Michigan, escaping from a reservoir on a tributary stream into which it had been stocked. Smelt spread throughout the lake system between 1923 and 1935.

Lake Erie suffered the most severe impacts of eutrophication. Because it is the shallowest of the lakes, eutrophication caused the greatest changes in the hypolimnion (Regier & Hartman 1973). In the

READING FIGURE 15.1A

The sea lamprey, introduced to the Great Lakes, caused the decline of many species of native salmonid fish. These lake trout were severely wounded by lamprey attacks.

(Photo courtesy of Marquette Biological Station, U.S. Fish and Wildlife Service)

1960s, oxygen concentrations below 3 ppm occurred throughout more than 70% of the lake during summer stratification. Eutrophic conditions led to changes in the communities of phytoplankton, zooplankton, and invertebrates. Mayflies disappeared from large areas of the lake bottom, and were replaced by midge larvae and tubificid worms tolerant of low oxygen. These invertebrates, however, are poor food for salmonid fish, which were also excluded from deep water areas by lack of oxygen. The blue pike, lake

---

Scums of these algae may accumulate on the surface, blow to shore, and pile up in windrows, where they rot and create a vile stench. Rotting blue-green algae often taint the water with foul chemicals that are difficult to remove by municipal water treatment. With high productivity, large quantities of dead organic matter also sink to the bottom waters of the lake. There, bacterial decomposition may exhaust the supply of oxygen during summer stratification, killing invertebrates and fish. Ultimately, these physical and chemical changes cause the replacement of animals typical of oligotrophic lakes with invertebrates and fish tolerant of low oxygen. The fish fauna of eutrophic lakes in the temperate zone is dominated by sunfish, bass, yellow perch,

carp, catfish, and other forms, some of which are neither good sport fish nor good table fish. Although the fish harvest from such lakes may be much higher than that of the preexisting oligotrophic lake, its value may be primarily for production of fish meal and fertilizer.

RECOVERY OF LAKES FROM EUTROPHICATION

Eutrophication has been likened to the "aging" of lakes, based on the idea that it is a natural trend that occurs over the geological

Reading Figure 15.1B

Electrical barriers are used to prevent adult lampreys from entering tributary streams of the Great Lakes to spawn.

(Photo courtesy of Marquette Biological Station, U.S. Fish and Wildlife Service)

trout, and longjaw cisco became extinct, and the walleye fishery collapsed. The original fishery, which yielded about 22,700 tons of high quality table fish annually, was replaced by a fishery of roughly equal volume, but one that yielded mostly trash species used for production of fish meal.

Beginning in the 1970s, efforts have been made to counteract some of these impacts. Control of sea lampreys was attempted in several ways. Dams and electrical barriers were installed at the mouths of tributary streams in which lampreys spawned (Reading Fig. 15.1B). A **lampricide,** TFM, that would kill lamprey larvae during their early life in streams was discovered and put into use. Some control was achieved initially in the watersheds of Lakes Michigan, Huron, and Superior, leading to increases in the harvest of lake whitefishes (Dahl & McDonald 1980; Smith & Tibbles 1980). In 1993, however, lamprey populations were still out of control in northern Lake Huron.

Eutrophication has been fought by reductions in phosphorus discharges from municipal sewage treatment plants. By the late 1980s, more than a two-thirds reduction in phosphorus input to Lake Erie had been achieved (Makarewicz & Bertram 1991). The phytoplankton and zooplankton communities of Lake Erie have become less dominated by species typical of eutrophic conditions, and summer oxygen depletion has become less severe in deep water. Most dramatically, the walleye has once again become a major sport fish in Lake Erie.

A somewhat different response—some might say a form of ecological "tinkering"—has been the introduction of exotic salmonids to the Great Lakes system. Coho,

chinook, pink, and kokanee salmon from the Pacific Coast have been introduced to the upper lakes in an effort to reestablish a salmonid sport fishery. Initially, these populations depended entirely on introduced fry. Chinook and pink salmon, however, now spawn in the upper lakes, and appear to have established self-sustaining populations.

Invasion of the Great Lakes by exotics continues without effective control (Mills et al. 1994). From 1960 to 1990, more species invaded than in any previous 30-year period, and, altogether, 139 exotics have become established. The most serious of the recent invaders are the zebra and quagga mussels, the spiny water flea, and the Eurasian ruffe, a small perchlike fish with an enormous reproductive potential. Most recent introductions seem to have been brought from the Old World in ballast water of cargo ships. The zebra and quagga mussels form dense beds on both hard and soft substrates, and are very effective in filtering particulate matter from the water. Zebra mussels in western Lake Erie, for example, may consume 26% of the primary production of this region of the lake (Madenjian 1995). Dense populations of these mussels are likely to reduce or eliminate native freshwater clams and restructure the entire benthic community (Mills et al. 1994; Schloesser & Nalepa 1994). These clams are expected to spread to most freshwater ecosystems in North America, causing billions of dollars of fouling and ecological damage. The spiny water flea, a large zooplankton that many fish are unable to eat, is causing major restructuring of open-water zooplankton communities (Lehman & Caceres 1993). The ruffe could cause millions of dollars of damage annually to yellow perch and walleye fisheries in the Great Lakes.

lifetime of a lake basin. By this analogy, recovery from eutrophication—"de-aging"—is not possible. Fortunately, cultural eutrophication does not constitute an irreversible change in lake ecology, and many lakes that have become eutrophic from human activities can be restored to an oligotrophic condition. Recovery is most rapid for lakes that have major inflows and outflows of water, so that their water volume is replaced in a few years.

Lake Washington, in the Seattle, Washington, metropolitan area, provides a good example of such recovery (Edmondson & Lehman 1981). During and after World War II, Seattle and its inland suburbs grew rapidly. The sewage systems built to serve these communities used the most expeditious approach to

disposal of secondary effluent—discharge into Lake Washington. By 1955, the lake was receiving 24,200 cubic meters of secondary effluent from 10 systems. At this point, sewage contributed 56% of total phosphorus input to the lake. The result was severe eutrophication of a large lake in the center of a major metropolitan area. Clarity of the water, which in the mid-1930s enabled a black-and-white Secchi disk to be seen to a depth of over 6 meters, declined to about 1 meter. Massive blooms of blue-green algae made the lake a nuisance to nearby residents.

Fortunately, studies by limnologists at the University of Washington not only documented eutrophication of the lake, but showed the cause—phosphorus input in secondary sewage

T he Everglades, originally a flowing wetland nearly 10,200 square kilometers in area covering much of the southern tip of Florida, face an uncertain future. Will their ecological degradation and transformation continue—even accelerate—or will major steps to restore this rich ecosystem be begun?

The Everglades lie in a broad, shallow trough formed by the higher topography along the eastern and western coastlines of the Florida peninsula. The Everglades are part of a wetlands complex that originally extended from near Orlando south through the lakeland of the central peninsula to the sawgrass marshes, cypress swamps, and mangrove forests of Collier, Monroe, Broward, and Dade Counties. In the north, water was collected by the meandering Kissimmee River and channelled into Lake Okeechobee. Overflow from Lake Okeechobee, augmented by rainfall along the way, spread over the low, gradually sloping flatlands to the south, eventually to drain into the ocean from Biscayne Bay to Florida Bay and the mangrove swamps that front the Gulf of Mexico from Cape Sable to Everglades City. The flow was slow—36 or so meters per hour—and sheetlike, varying in depth from a few centimeters to slightly over a meter. It was highly seasonal, greatest in summer, the wet season in South Florida. Year-to-year variation was also great. These variations, together with the influences of fire and hurricanes, were integral aspects of Everglades ecology. The heart of the Everglades consisted of wetlands ranging from dense sawgrass marsh to open sawgrass marsh studded with tree islands and hardwood hammocks (Reading Fig. 15.2A).

The Everglades ecosystem was originally oligotrophic, with most nutrients entering the system through precipitation. Nevertheless, the subtropical climate and diversity of habitats supported a distinctive plant and animal biota that combined temperate and tropical elements. Aquatic food chains beginning with algae supported

READING FIGURE 15.2A

The Florida Everglades are a vast wetland consisting of open marsh, wooded hammocks, cypress swamps, and coastal mangroves.

(Photo by G. W. Cox)

enormous populations of invertebrates and fish that, in turn, were life support for reptiles, birds, and mammals. The seasonal changes in water depth and distribution provided, in wet periods, for a vast area of production and, in dry periods, for concentration of aquatic animals to the benefit of their predators. The American alligator was a keystone species, creating deep "alligator holes" and maintaining water channels on which many aquatic animals depended. Dry-season fires contributed to landscape heterogeneity, as well, burning the peaty soil deeply in places, lightly or not at all in other places.

Bordering the Everglades proper were still other wetland systems, which were an integral part of the southern Florida wetlands system. These included cypress swamps that flanked the eastern edge of the Everglades as a narrow strip and covered a vast area, the Big Cypress Swamp, to the west. Mangrove swamps spread along the coastlines of the Gulf of Mexico and Florida Bay, where the fresh water of the Everglades met the salt water of the ocean.

Perhaps half of the original Everglades survives untransformed. South Florida has become a major winter agricultural area, a mecca for tourists and retirees, and a center of commerce between the United States and countries of the Caribbean and Latin America. Agricultural and urban developments have spread over vast areas of the northern and central Everglades and along both coasts. Much of the Kissimmee River has been channelized. Vast areas south of Lake Okeechobee have been converted to agriculture. About 2,250 kilometers of canals have been dug to drain agricultural lands and protect developed areas from flooding. Large, diked water conservation areas have been created west of Fort Lauderdale and Miami to store water and assure the recharge of aquifers that provide the municipal water for these and other Atlantic coastal cities. Pressure for continued urban growth is great. The region's human population of 6 million is projected to triple in the next 50 years.

Even the untransformed areas, however, have suffered human impacts. Canals,

effluent. Based on this knowledge, Seattle undertook the construction of a metropolitan sewage disposal system that now discharges effluent into Puget Sound. Diversion of effluent from Lake Washington was accomplished in stages, and was almost complete in 1968. Recovery of the lake was rapid, and by 1978 algal blooms had ceased to occur, water clarity was again at mid-1930s levels, and the phosphorus concentration in the lake water was essentially equal to that measured in 1933 (Fig. 15.4).

ULTRAOLIGOTROPHIC LAKES: A SPECIAL CASE

Several highly oligotrophic lakes have exceptional aesthetic or biotic values. Among these are Lake Tahoe in the western United States and Lake Baikal in Russia (*See* Reading 4.1).

READING FIGURE 15.2B

Since the 1930s, breeding water birds like these great egrets have declined in numbers by 90% in the Florida Everglades.

(Photo by G. W. Cox)

levees, and highway embankments have constrained and diverted the original pattern of water flow. Water movements are now controlled primarily to benefit agriculture, prevent flooding, and recharge the Atlantic coastal aquifer. Park and preserve lands are last to receive water in dry years and are the recipients of overflow in wet years. Fertilizers from farming operations and sewage disposal have enriched the remaining water flows, favoring cattails over sawgrass. Mercury, released by accelerated decomposition of peaty agricultural soils and introduced by fertilizers and urban pollution sources, has become a major pollutant of the aquatic ecosystem. The inflow of water that nourished mangroves and maintained estuarine salinities in Florida Bay has been reduced greatly, and the fisheries of this bay have plummeted. Australian pines and paperbark trees, together with Brazilian pepper trees, have invaded native pinelands and wetlands. Breeding water birds have declined, their colonies have shifted in location, and their nesting has become later in

the year (Bildstein et al. 1991). Numbering 300,000 in the 1930s, fewer than 15,000 now breed in the Everglades (Reading Fig. 15.2B). Some 56 species of plants and animals, including the Florida panther, West Indian manatee, wood stork, and snail kite, are designated as federally endangered or threatened.

Much of the southern Florida wetlands that survives in near-natural form is now incorporated into federal, state, and private preserves. Federal preserves include Everglades National Park, Big Cypress Preserve, and the Loxahatchie and Florida Panther National Wildlife Refuges. State preserves include Collier Seminole State Park and Fakahatchee State Preserve. The National Audubon Society maintains Corkscrew Swamp Sanctuary. These protected areas make up only a small fraction of the functional Everglades ecosystem, however, and their survival depends on maintenance of this total ecosystem.

A long-term plan of restoration steps aimed at alleviating some of these distress

signs has been drawn up by the Army Corps of Engineers, working in cooperation with state and regional agencies. If implemented, this plan will require years and billions of dollars to complete. Several major actions are envisioned. The Kissimmee River will be de-channelized and returned to a more natural, meandering pattern of flow. Diversions of water into the Caloosahatchee River, leading to the west coast, and the St. Lucie Canal, leading to the east coast, will be reduced and more water sent southward. Some 16,000 hectares of filtration marshes will be created to receive agricultural runoff waters and extract nutrients before the water passes into the Everglades system. The quantities, routes, and timing patterns of southward water delivery will be modified to approach their original patterns more closely. Finally, the sloughs that delivered water into Florida Bay will be revitalized and the inflow of water to this estuarine system restored.

These efforts can only be the first steps toward fuller restoration, a process that must be an ongoing interplay of research and refined management. The pattern of sheet flow must be reestablished over areas that experienced this pattern. Recreating seasonal and year-to-year variability in water conditions, some of which might seem like catastrophes in the short term, is also essential. Preventing massive invasion of the wetlands by exotics will be a constant battle. Fire and other techniques must be used to maintain the local mosaic of sloughs, tree islands, and other habitat features. Endangered species must be given their needed special protection. All of this must be done with coordination of the various federal, state, and local agencies that are involved. All of this must be done in a region subject, ultimately, to the effects of sea level that is now rising fast enough, 30 centimeters a century, that coastal mangrove ecosystems do not appear to be keeping up with the rate of change!

Because of its setting and water clarity, Lake Tahoe is one of the most beautiful lakes in the world. The Secchi disk depth of Lake Tahoe, once about 40 meters, is declining about 0.4 meter per year (Goldman 1989), and was 20.4 meters in 1995. Growth of green algae in shallow waters is conspicuous, and the primary productivity level of the lake has nearly trebled since 1958. The development of resort communities around the lake

is the main cause of this eutrophication. Even though sewage effluent from communities in the Tahoe Basin is exported, disturbance of the land surface by construction activity has increased nutrient inflow, particularly that of phosphorus, by erosion and leaching. Runoff from paved surfaces and rooftops also carries nutrients deposited by atmospheric fallout into the lake. Eutrophication in Lake Tahoe cannot be reversed as quickly as in

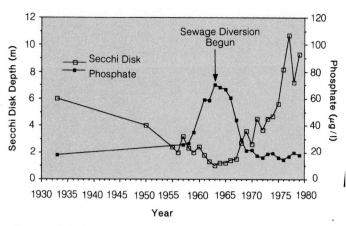

FIGURE 15.4

Changes in phosphate (annual mean concentration) and maximum Secchi disk depth in July and August in Lake Washington from 1933 to 1979.

(Data from W. T. Edmondson and J. T. Lehman, 1981. "The Effect of Changes in the Nutrient Income on the Condition of Lake Washington" in *Limnology and Oceanography*, **26**:1–29. Seattle, WA: University of Washington.)

Lake Washington, since Tahoe's great depth and water volume are such that 700 years are required for its water to be replaced by inflow and outflow.

Introductions of kokanee salmon, *Onchorhynchus nerka,* and opossum shrimp to improve sport fishing have led to the decline of the native zooplankton of Lake Tahoe. The loss of these zooplankters, and their grazing on open-water phytoplankton, may also be contributing to increase in algal populations in the lake waters.

AFRICAN RIFT VALLEY LAKES

Several of the Rift Valley lakes of East Africa possess rich, highly endemic faunas of great significance. Of particular importance are Lakes Victoria, Tanganyika, and Malawi. These lakes are ancient and the last two very deep. Through climatic cycles that have isolated and reconnected portions of their basins, speciation and adaptive radiation have produced rich endemic biotas. These species now face the varied challenges of growing human exploitation and impact.

Lake Victoria, bordering Kenya, Tanzania, and Uganda, has a rich, highly endemic fish fauna. More than 350 species of small cichlids, known as haplochromines, occur in the lake. Many haplochromines are brightly colored fish prized in the aquarium trade. Haplochromines, native and introduced tilapias, and native catfish supported local fisheries that used traditional techniques and preserved fish by sun-drying.

In the late 1950s, the Nile perch, *Lates niloticus,* a large predatory fish, was introduced to Lake Victoria. The Nile perch, which reaches a weight of 250 kilograms, has almost completely restructured the fish fauna of the lake, and severely affected the traditional fisheries (Payne 1987). About 80% of the fish biomass of the lake is now made up of Nile perch. The perch cannot be sun-dried, and must either be dried over fires, using hard-to-get firewood, or frozen. The Nile perch has become a major commercial fish, with the Kenyan harvest alone being about 100,000 tons annually, but about 40% of the haplochromines have become extinct (Kaufman 1991).

Lake Tanganyika, 1,470 meters deep, has existed for between 1.6 and 6.0 million years, and completely lacks oxygen in its deepest layers (Burgis & Morris 1987). The lake has a highly endemic fauna, including 60 species of aquatic snails, 30 of them endemic. The fish fauna is dominated by a group of endemic herrings—a typically marine group of fish—commonly known as "Tanganyika sardines." Lake Malawi, 704 meters deep and lacking in oxygen in its bottom waters, is also several million years old (Burgis & Morris 1987). The fish fauna—almost completely distinct from that of Lake Tanganyika—is richer than that of any other lake in the world. Some 245 species are known from Lake Malawi, 219 of them endemic. Of the endemics, 190 are cichlids, which have undergone an astounding evolutionary radiation. The cichlids have diversified greatly in feeding ecology, to the point that up to 30 species sometimes occur together. The biotas of Lakes Tanganyika and Malawi have not been damaged by introduced species such as the Nile perch, and must be rigorously protected against such an action.

LAKE ECOSYSTEMS AND GLOBAL WARMING

Global warming will likely have great impacts on lake, pond, and marsh ecosystems, although the nature of these impacts is still difficult to predict. Based on long-term studies of lakes in the boreal forest region of Ontario, Canada, Schindler et al. (1990) suspect that northern lakes will become ice-free for a greater portion of the year, increase in solute content, and develop a deeper layer of warm epilimnetic water in summer. Other studies suggest that lake productivity in general is likely to increase, with a corresponding increase in summer oxygen depletion in stratified lakes (Carpenter et al. 1992). Thus, the suitability of these lakes for species typical of cold, oligotrophic waters may be reduced considerably.

Freshwater wetlands cover about 15% of arctic and subarctic lands. Many of these high-latitude ecosystems contain large quantities of peat, and are the source of almost a quarter of all natural emissions of methane, one of the most active greenhouse gases, into the atmosphere (Carpenter et al. 1992; Bridgeham et al. 1995). Thus, how these systems respond to warmer temperatures and longer summers by releases of CO_2 and methane will determine whether they will intensify or oppose greenhouse effects.

Protection of Wetlands

In the United States, the **Clean Water Act (Federal Water Pollution Control Act),** as reauthorized in 1987, gives lakes, ponds, marshes, and other wetlands special protection. This act requires that filling or dredging actions minimize detrimental impacts, or, if such impacts are unavoidable, mitigate them by restoration of degraded wetlands or creation of artificial wetlands (Barton 1987). This act has been subject to intense controversy because of its regulation of development on private property that meets the criteria for a wetland.

The United States and Canada cooperate in protection of the Great Lakes through activities of the **International Joint Commission,** established by the **Boundary Waters Treaty** of 1909 to regulate matters such as lake levels and water quality. The **Great Lakes Fishery Commission,** established in 1955 by the **Convention on Great Lakes Fisheries** controls the sea lamprey and coordinates other aspects of fisheries research and management. Only since about 1976 have these complementary commissions been actively coordinating efforts, and an integrated approach to regulation of the Great Lakes Ecosystem is far from being implemented. Inadequate funding of these commissions in the 1990s has seriously limited their activities.

Several international efforts have been mounted to protect wetlands (Bildstein et al. 1991). The **Convention on Wetlands of International Importance,** developed at Ramsar, Iran in 1971, provides a framework for the protection of wetlands of international importance to waterfowl. Commonly termed the **Ramsar Convention,** this agreement permits member nations to designate areas for inclusion on a *List of Wetlands of International Importance.* The convention provides for cooperation in research and protection of listed wetlands. By 1989, some 54 nations, including the United States, were parties to the Ramsar Convention, and more than 400 sites had been placed on the list, including Everglades National Park, Florida. A priority for this program is the addition of southern hemisphere members such as Botswana, Argentina, and Brazil. The wetlands of the Okavango Delta of Botswana constitute one of the most important wildlife areas of southern Africa (*See* Chapter 16). Those of Argentina and Brazil are important both as breeding areas for resident wildlife and as wintering areas for North American shorebirds.

In the Western Hemisphere, **Wetlands for the Americas,** a nongovernmental organization, was organized in 1985 to promote a system of protected wetland areas of major importance to migratory waterbirds and other wetland species. This system, complementary to the Ramsar program, now includes more than 24 sites in seven countries.

Summary

1. Lake, pond, and marsh ecosystems have suffered from drainage, water diversion, watershed modification, eutrophication, fisheries overexploitation, and the introduction of exotic species.

2. Ultraoligotrophic lakes such as Lake Baikal in Russia, one of the earth's freshwater hotspots of biodiversity, and Lake Tahoe in the western United States, are at high risk from toxic chemical and nutrient pollution.

3. Introduced exotic species, especially predatory fish, are a major threat to the rich, endemic fish faunas of Lake Victoria and other African Rift Valley Lakes.

4. The ecosystems of the Everglades and Florida Bay have suffered because of massive disruption of the hydrology of southern Florida by drainage, diversion, and construction of levees and highway embankments. Eutrophication and exotic plant invasion are affecting many parts of the Everglades, as well.

5. Global change is likely to lead to increase the productivity of all types of wetland ecosystems, reducing the suitability of many lakes for species requiring oligotrophic conditions. Increased release of CO_2 and methane from wetlands may also exacerbate global warming.

Questions for Discussion

1. How is eutrophication in aquatic ecosystems linked to the nutrient dynamics of watershed ecosystems?

2. Considering the massive biotic changes in the Great Lakes that have resulted from invasion of exotic species, is full restoration of original conditions possible? What should the objectives of ecological management of these lakes be?

3. To an extent perhaps greater than for any other national park, the values under protection in Everglades National Park are biological. Will this make it easier or more difficult to achieve long-term protection in the face of current and future threats to the park?

Suggested Reading

Lodge, T. E. 1994. *Everglades handbook. Understanding the ecosystem.* St. Lucie Press, Delray Beach, FL. The origin, ecology, human impact history, and proposed first restoration efforts for North America's most remarkable wetland ecosystem.

Mills, E. L., J. H. Leach, J. T. Carlton, and C. L. Secor. 1994. Exotic species and the integrity of the Great Lakes. *BioScience* **44:**666–76. A description of the accumulation of exotic species' impacts from the days of the Erie Canal to the present.

Worthington, E. B., and R. Lowe-McConnell. 1994. African lakes reviewed: creation and destruction of biodiversity. *Env. Cons.* **21:**199–213. Origin of the biotas of African lakes, and a history of their colonial and postcolonial management and mismanagement.

chapter

16

Rivers and Streams

Outline

R ivers and streams hold only about 1.3% of the water in all freshwater ecosystems combined (Berner & Berner 1987). Nevertheless, they play an important role in the flux of water through the biosphere, and are an important habitat for many types of organisms. Rivers also transport large quantities of dissolved and suspended organic and inorganic materials to the oceans. Annually, river transport of inorganic sediment is estimated to be about 152 tons for each square kilometer of watershed. Human-induced changes in watersheds, primarily deforestation, livestock grazing, and farming, have about doubled the natural sediment discharge of rivers to the ocean.

The major rivers of the world are concentrated in two latitudinal zones: the moist temperate to subarctic belt and the humid tropics. Most rivers at high latitudes in the Northern Hemisphere were heavily disturbed in recent geological time by Pleistocene glaciation. The fish faunas of these river systems are not very rich, but they show distinctive patterns of adaptation, foremost among which is migration between fresh and salt water. Many species are **anadromous,** entering these rivers only to spawn. Anadromous fish include salmon, shad, striped bass, sea lampreys, and others. Other species, such as the American eel, are **catadromous,** living in fresh water but returning to the sea to spawn.

The surface beauty of many rivers and streams often hides serious ecological impacts on the subsurface ecosystem.

RIVERS AND THEIR FLOODPLAINS

Rivers are integrally tied to the riparian ecosystems of their banks and floodplains (Hunt 1988; Bayley 1995; Sparks 1995). Riparian ecosystems are among the world's most threatened environments, and many riparian species are listed as endangered. In the Sacramento Valley of California, for example, where riparian woodlands originally covered more than 3,130 square kilometers, only about 1.6% remain (Hunt 1988).

The interrelation of river and floodplain is clearly illustrated by the disastrous flooding along the Mississippi River in 1993. Drainage of wetlands in the Mississippi watershed, together with construction of levees that prevented flood flow from spreading over the floodplain, were responsible for the severity of flooding. Restoration of about half the area of wetlands that have been drained in the upper Mississippi basin would likely have alleviated flood damage considerably (Hey & Philippi 1995).

HUMAN IMPACTS ON STREAM ECOSYSTEMS

Rivers and streams have always been of great economic importance to humans. These ecosystems have provided humans with food and water, irrigated their crops and watered their livestock, served as routes of travel and transportation, provided power for individuals and industries, and flushed away the wastes of human activity. In some regions, such as western Europe, the long history of human impact has almost totally transformed river ecosystems (Petts et al. 1989). Elsewhere, however, rivers still retain elements of their original condition, even where human impacts have been severe.

Given these impacts, it is not surprising that the biotas of most rivers and streams have suffered greatly. Rivers are resilient ecosystems, and much of their wildlife has hung on tenaciously. The opportunity still exists to restore wildlife in abundance to most rivers and streams. In this chapter, we shall first consider some of the major impacts of human activity on rivers and streams, and then examine management strategies for these ecosystems.

Water Diversion

Growth of human population and economic activity increases demands for water for agricultural, urban, and industrial uses. Often, this use does not return water to the channels from which it was diverted. Many coastal cities, such as Los Angeles and New York City, take water from rivers, but discharge wastewater directly into the ocean. Water is also diverted for irrigation, where much of it is dissipated by evaporation. The use of water in cooling systems of power plants also leads to its evaporation.

The result of this consumptive use is often the near destruction of the lower portion of the river system. The Colorado River of western North America and the Nile River of Egypt (White 1988) now discharge little or no water into the ocean. Loss of freshwater flows, with their sediment and nutrients, has major impacts on coastal marine environments (*See* Chapter 18).

FIGURE 16.1

The birdwing pearly mussel of the upper Tennessee River is one of 100 species of freshwater mussels that have become extinct or endangered in eastern North America due to siltation, acid mine drainage, and habitat destruction.

(U.S. Fish and Wildlife Service photo)

Watershed Modification

Deforestation of watersheds leads to changes in stream physical and chemical characteristics (*See* Chapter 15). These include streambed structure, volume and seasonal pattern of flow, water temperature, silt load, detritus and nutrient concentration, and pattern of biological productivity.

In mountain regions, including most of western North America, timber harvesting by clear-cutting has had catastrophic impacts on the structure of stream channels. Channels in these areas are often structured by **large woody debris,** such as fallen trees that span the channel (Cummins et al. 1995). Debris dams form pools and trap gravel, creating the special environments that are required for the spawning of adult salmon and the early life of young salmon (Maser & Sedell 1994). When these forests are clearcut, the source of large woody debris is removed. Many of the fallen logs themselves may be removed because of their timber value. As a result, the pools disappear and the stream channel tends to become clogged with medium-sized stones, producing a wider, shallower bed. In steep headwater areas, landslides may occur on clearcut slopes, leading to debris torrents that scour the downstream channel, destroying pools by removing logs and other natural dams, removing gravel beds used by spawning salmon, and leaving a bare, bedrock channel.

Siltation results from watershed activities, such as clearcutting, farming, or construction, that promote erosion. Siltation is a growing problem in many regions of the world as a result of deforestation. Even in North America, siltation is a growing problem in many farming regions, due to cultivation of high-erosion crops such as cotton, maize, sorghum, and soybeans (Smith et al. 1987). Siltation has been a major factor in the endangerment of freshwater mussels in the Mississippi, Ohio, and Tennessee River systems (Fig. 16.1). Mussels feed by pumping water through the gill chambers that filter out fine particles of organic matter. Nonorganic sediment overwhelms this filtration system, and may also interfere with oxygen exchange between animal tissues and the water, suffocating the animal. Other factors in the decline of freshwater mussels include acid mine drainage (*See* Mining Impacts), destruction of stream habitat by damming and channelization, and the effects of exotic species such as the zebra mussel (*See* Chapter 15). Of 297 freshwater mussels, the American Fisheries Society considers 21 to already be extinct, 77 to be endangered, 43 to be threatened, and 72 to be of special concern (Williams et al. 1993). By early 1993, zebra mussels had dispersed from the Great Lakes into the main portions of the Ohio, Tennessee, and Mississippi Rivers, placing still greater stress on populations of native mussels.

Overexploitation of Fisheries

Many rivers are important sources of fish and other animal food for humans. The Mekong River of southeastern Asia yields about 500,000 tons of fish annually, the Zaire River of Africa about 210,000 tons (Davies & Walker 1986), and the Amazon River of South America about 210,000 tons (Junk 1984). Indeed, these rivers are the most important sources of protein for the surrounding human populations.

River fisheries, like those of lakes, are easily over-exploited. In the late 1800s, for example, the Delaware River in the eastern United States yielded over 11,300 tons of shad and other finfish annually (Thomann 1972). Overfishing and various forms of pollution have reduced this harvest to less than 50 tons annually.

Pollution

Pollution of streams is much more varied than that affecting lakes. Organic wastes, salts, toxic wastes, pesticides, and thermal effluent are major categories of stream pollutants, each with its distinctive impact on stream ecosystems.

Pollution of streams by organic wastes, primarily sewage but also wastes from industries such as canneries and paper pulp plants, has long been a major problem in streams. Organic pollution affects stream ecosystems because bacterial decomposition of the organic matter consumes oxygen, often creating anoxic conditions for long distances downstream from discharge points. The standard measure of such pollution, in fact, is known as **biological oxygen demand (BOD),** and is measured as the quantity of oxygen needed to decompose the organic matter in one liter of water.

Municipal sewage systems that discharge untreated sewage or provide only primary treatment, which is designed to kill pathogenic organisms but not digest organic materials, are the most frequent causes of organic pollution. Development of modern secondary treatment plants that do digest organic components of sewage is leading to a gradual reduction of this type of pollution in the United States and other developed countries.

Chemical pollutants such as toxic industrial wastes, pesticides, and salts enter streams in a variety of ways. Waste disposal, agricultural runoff, and the precipitation and fallout of products of combustion and smelting introduce heavy metals, industrial chemicals, and pesticides into streams. Drainage waters from irrigated lands often carry large quantities of salts and toxic substances such as arsenic and selenium. Salts and toxic elements such as arsenic, cadmium, and selenium are among the pollutants on the increase in stream waters in the United States in recent years (Smith et al. 1987). Some rivers have received heavy pollution by synthetic industrial chemicals. The Hudson River has experienced perhaps the most severe pollution by polychlorinated biphenyls (PCBs) in the world due to discharges from industrial plants (Limburg 1986). From the late 1940s through the mid-1970s more than 200 tons of PCBs were discharged into the river, causing massive contamination of the biota (*See* Chapter 22). The Buffalo and Niagara Rivers are severely contaminated by toxic chemicals entering the river from chemical dumps and landfills. In the Amazon Basin, gold-mining activities are contaminating several major rivers with mercury, used to concentrate gold particles (Barbosa et al. 1995).

Discharge of heated water, or **thermal pollution,** results from the use of water in the cooling systems of fossil fuel or nuclear power plants and certain industries. Temperatures of the water leaving such systems are commonly 38° to 46° C, well above the tolerance limits for most aquatic organisms.

Mining Impacts

In regions where coal or metal ores are mined, spoil material is washed into streams, causing heavy siltation. Sulfur compounds in the spoil are oxidized to produce sulfuric acid, and leaching carries large quantities of this acid, together with heavy metals, into streams, creating a form of pollution known as **acid mine drainage.** The most extensive problems of acid mine drainage occur in areas where strip mining for coal is practiced. In coal-mining regions of Pennsylvania, West Virginia, Kentucky, and Tennessee almost 7% of all streams were acidic due to acid mine drainage, based on surveys done in 1986 (Herlihy et al. 1990). In the northern Appalachians, acid mine drainage was over twice as frequent a cause of stream acidification as was acid deposition from the atmosphere (*See* Chapter 23).

Strip mining is practiced by two main techniques. **Area strip mining,** practiced on level lands, involves the removal of the soil and rock (overburden) in an elongate trench overlying coal or ore deposits. This overburden is then dumped into the adjacent trench that has just been mined. With the overburden removed, the coal or ore can be excavated. **Contour strip mining** is practiced on hilly land (Fig. 16.2). In this case, overburden is removed above an outcrop of coal or ore, forming a cliff known as a **highwall.** The depth of overburden and the resulting highwall height depend on the thickness of the deposit to be mined. For coal, the highwall:coal seam thickness ratio is about 15:1, so that cliffs to heights of 15 or more meters are common. In the early days of strip mining, little or no reclamation of the mined landscape was required. Today, most states

FIGURE 16.2

Contour strip mining and mountaintop removal on Bolt Mountain in West Virginia has left hillsides devasted by highwalls and tailings, as well as streams polluted by acid drainage.

(Photo by R.L. Smith)

In California, mining of zinc and copper ores in the upper Sacramento River basin has led to heavy metal pollution of tributaries of the Sacramento, and to the mortality of steelhead and chinook salmon. The most notorious of these operations, the Iron Mountain Mine northwest of Redding, left a mountain honeycombed with shafts through which rainwater percolates, reacting with pyrite and metal ores to yield an outflow of intensely acidic water rich in toxic metals. This outflow is trapped by a reservoir built specifically to prevent it from entering the Sacramento River. Nevertheless, the capacity of this reservoir often has been exceeded, leading to 33 major fish kills in the Sacramento River since 1940. Efforts to contain this pollution source have already cost $30 million, and $100 million in expanded control facilities have been recommended (Paddock 1993).

Damming

The construction of dams on the world's rivers has increased exponentially since the mid-1800s. More than 730 dams are now inaugurated annually (Fig. 16.3). Although the most suitable dam sites have already been utilized in the developed countries, numerous potential sites exist in developing nations. It seems likely that electric power and irrigation water needs, coupled with the disadvantages of fossil fuel power production, will continue to favor dam construction for several decades.

Construction of dams for hydroelectric power production and water storage causes profound change in stream ecology both below and above the dam. Above dams, the entire stream and riparian zone is destroyed by inundation. At the extreme, the river is converted into a set of warm, quiet pools between which upstream movement of organisms is impossible. Damming thus interferes with fish that migrate between the ocean and fresh water, such as species of salmon, which spawn in fresh water, and certain eels that spawn in the ocean. Many

require that the landscape be restored to a natural contour, covered with topsoil, and revegetated. Regardless of the extent of reclamation efforts to recreate a naturally contoured landscape, strip mining replaces solid bedrock with highly porous, unweathered sedimentary rubble that is subject to leaching.

Acid mine drainage converts streams with a rich plant, invertebrate, and fish fauna into systems occupied by only a few kinds of acid-tolerant bacteria, algae, and invertebrates. In the Beaver Creek area of eastern Kentucky, for example (Collier et al. 1970), strip mining in the late 1950s reduced stream pH from 6.0–7.0 to 3.3–3.9, eliminating all eight species of fish that were originally present, and reducing the abundance of benthic invertebrates by 6- to 7-fold (Table 16.1).

TABLE 16.1

Changes in Stream pH and Biota Due to the Influence of Acid Mine Drainage in the Beaver Creek Watershed, Eastern Kentucky		
	CANE BRANCH (ACID IMPACTED)	HELTON BRANCH (UNIMPACTED)
pH[1]	3.5	6.9
Sulfate Concentration[1] (ppm)	242	0.4
Benthic Insects[2] (No./m²)		
Ephemeroptera	0	365.5
Plecoptera	0	367.6
Trichoptera	0	185.0
Megaloptera	19.3	8.6
Coleoptera	2.1	107.5
Diptera	301.0	814.8
Odonata	tr	37.6
TOTAL	322.4	1886.6
Fish Biomass[1] (kg/ha)	0	18.4

[1]Mean for 1956–1966. [2]Mean for 1959–1965.

Source: Data from C. R. Collier, et.al., 1970 "Influences of Strip Mining on the Hydrologic Environment of Parts of Beaver Creek Basin, Kentucky, 1955–66" in *Geological Survey Professional Paper 427-C, viii + 80.* Washington, DC: U.S. Geological Survey.

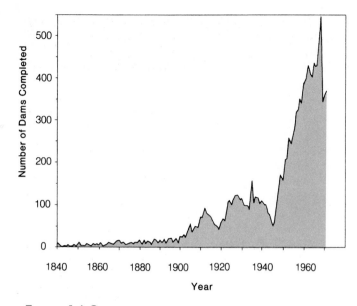

FIGURE 16.3

Rate of inauguration of new dams on the world's rivers since 1840.
(Modified from Berner & Berner 1987)

other strictly freshwater fish also migrate extensively within major river systems, such as the Amazon and Parana Rivers of South America and the Mekong River of southeastern Asia (Davies & Walker 1986; Barthem et al. 1991), and damming thus constrains their movements.

Dams, as they are designed to do, also modify the seasonal pattern of flow volume in particular, reducing flood flows downstream. This alone changes the character of the river system. On the Platte River in Nebraska, for example, the floods that formerly scoured the channel, clearing it of woody vegetation and maintaining a broad complex of open sandbars and channels, no longer occur. The river channel has become more confined, and much of the original channel has grown up in riparian cottonwood woodland. Likewise, below Glen Canyon Dam on the Colorado River, woody vegetation and freshwater marshes have invaded areas formerly scoured by flood flows (Stevens et al. 1995).

Stream temperatures, sediment loads, flow volume, and salinity are all changed dramatically by damming, as well. The general result of damming is to increase evaporation from reservoir and irrigated land surfaces, increase the concentration of dissolved materials, trap sediments in reservoir basins, and decrease terminal water and sediment flows into the ocean.

Channelization

In some situations, such as highly urbanized or intensively farmed areas, humans sometimes view rivers primarily as channels to carry flood runoff. In such places, the idea of straightening a meandering stream channel to speed flood water outflow, or **channelization,** often arises. From the standpoint of stream and riparian wildlife, and often from a broad aesthetic viewpoint, channelization can be disastrous (Simpson et al. 1982).

When a stream channel is straightened, it is shortened and its downhill gradient is steepened. The result is swift outflow of water, which affects not only the channelized section, but areas both upstream and downstream. Within the channelized section erosion is usually accelerated, widening and deepening the channel. Channelization of the Blackwater River in Missouri (Emerson 1971), for example, led to a deepening of the channel by about 3.5 meters. Widening of the channel has necessitated replacement of bridges 15 to 30 meters long by structures 60 to 124 meters in length. Often, channel widening increases the total area occupied by the channel, in spite of the shortening that occurred. From the standpoint of aquatic wildlife, the microhabitats important to many organisms are eliminated by channelization, so that the biomass of life per unit area of stream declines. In the Blackwater River, fish biomass declined from 632 to 126 kilograms per hectare of stream area as a result of channelization (Emerson 1971).

Channel stabilization on the Missouri River by damming, dredging, and various forms of bank protection illustrate the extent of impacts (Hunt 1988). The Missouri River Bank Stabilization and Navigation Project has resulted in large losses of island and sandbar habitat, riparian woodland and wetland, and total water surface area of the river itself. These changes have led to the decline of riparian wildlife such as river otter, mink, and many other species.

Combined Impacts: Major North American River Systems

The multiple impacts of human activity affect the wildlife of streams, especially migratory and wide-ranging species. The magnitude and economic importance of such effects are clearly shown by the Columbia and other rivers of the Pacific Northwest. The Colorado River shows effects perhaps less serious in economic terms, but of major importance to biodiversity (*See* Reading 16.1).

The Columbia River is the longest North American river flowing into the Pacific Ocean, and the second largest river, in volume of flow, in the United States. The first dam on the main channel of the Columbia, at Rock Island, was built in 1933. Today 86 dams exist on the Columbia and its major tributaries (Fig. 16.4), and much of the river is simply a chain of reservoirs. These dams have had a profound effect on the river ecosystems. Several anadromous fish migrate into the Columbia River to spawn. Originally, the most important of these were salmonids: the steelhead, chinook, coho, sockeye, and chum salmon. The shad, an anadromous fish from the Atlantic coast of North America, was also introduced to the Columbia.

Damming has adversely affected the migratory fish of the Columbia system. Although all of the major dams on the lower river have fish ladders that enable adults to pass the dams on their upstream migration (Fig. 16.5), not all of the upper dams allow such passage. The Columbia above the Grand Coulee Dam and the Snake River above Brownlee Dam, together with smaller areas of watershed elsewhere in the system, are now unavailable to migratory fish (Fig. 16.6).

THE COLORADO RIVER: A CASE STUDY

The Colorado River has experienced even more severe impacts than the Columbia River from water diversion and damming (Carothers & Johnson 1983; Hickman 1983). Diversion of water for agricultural and urban use consumes the entire flow of the river. The legal allocation of Colorado River water to various states and Mexico equals 14.84 million acre-feet (Miller et al. 1986). The Colorado system, which now loses about 1.89 million acre-feet annually by evaporation, has an annual flow of about 11.0 to 13.8 million acre-feet (Hunt 1988). The only water regularly entering the Gulf of California is now waste irrigation water from southern Arizona and the Mexicali Valley of Mexico. In Colorado, water is diverted to the east side of the continental divide, as well as to irrigated land in the western part of the state. The Central Arizona Project now carries water to Phoenix and Tucson for urban and agricultural use. In California, water is diverted to Los Angeles and San Diego, where it is ultimately discharged into the Pacific Ocean, and to the Imperial Valley, where waste irrigation water flows into the Salton Sea. Mexico takes the remaining water, which is used for irrigation in the Mexicali Valley, and as the municipal supply for the cities of Mexicali and Tijuana.

Damming of the Colorado River has also changed physical and chemical conditions of this river (Carothers & Dolan 1982). From a free-flowing river with enormous spring floods, the lower Colorado has largely become a series of lakes without seasonal change in flow. Once a stream so laden with sediment that it was named, in Spanish, the "Red River," it is now clear for some distance below Glen Canyon Dam, due to deposition of the sediment in Lake Powell. In the free-flowing section of river in the Grand Canyon, the water now remains icy cold in summer because the water in this section comes from below the thermocline of Lake Powell. The release of water through hydroelectric generators is also timed to provide power at peak demand times, and this caused the river below the dam to fluctuate as much as 4 meters in depth during the day (Elfring 1990). This enormous variation in flow can alternately flood and expose the spawning areas of fish. In the hot, arid climate of the Southwest, evaporation from the various reservoirs and the inflow of salts in waste irrigation water have doubled the salinity of the water.

Biotically, the Colorado River has a native fish fauna with a greater degree of endemicity than that of any other North American river (Minckley & Deacon 1968). Loss of the river habitats to which these species were adapted, along with competition from species introduced into the reservoirs, have endangered this fauna almost in its entirety. Nine species of larger native fish, including eight species of river chubs and the squawfish, are now designated as threatened or endangered (Hickman 1983). In the colder portions of the river, these natives have largely been replaced by introduced trout, and in warmer regions by channel catfish.

FIGURE 16.4

Bonneville Dam on the lower Columbia River, completed in 1938, was originally designed without provision for passage of anadromous fish.

(Photo by G. W. Cox)

FIGURE 16.5

Fish ladders were added to the design of Bonneville Dam to allow anadromous salmon and other fish to migrate upstream to spawn in the upper Columbia River system.

(U.S. Bureau of Reclamation photo by J. D. Roderick)

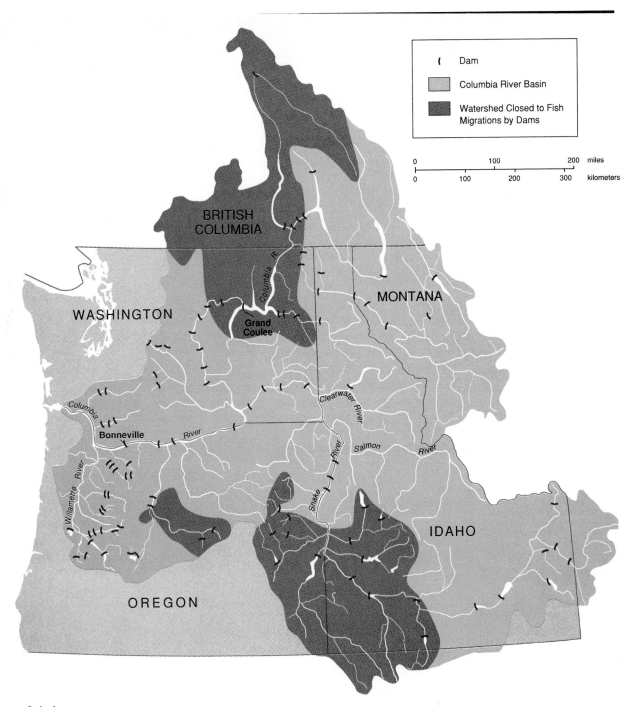

FIGURE 16.6

Portions of the upper Columbia River system are no longer accessible to migratory fish because of the lack of fish ladders on dams.
(Modified from A. Netboy, *Nat. Hist.* 89(7):55)

Several formidable challenges are faced by the young salmon on their downstream migrations. The change from a swiftly flowing river to a series of reservoirs has lengthened the downstream migration time by about one month. At the dams, some young salmon are trapped in the flows entering hydroelectric turbines, where they are killed by the intense turbulence. Ponding of the river has also raised its summer temperatures, which in August now rise above 20° C, well above the range of 5.8° to 12.8° C that is optimum for salmonid fish, and just

below the lethal limit of 22.5° C for juvenile salmon. Where thermal plumes from power plants occur, these limits of tolerance are exceeded. The cooling systems of these power plants draw water from the river, which also kills young salmon.

Dams, deforestation, and inadequately regulated fish harvests have led to decline of sea-run salmonids throughout the Pacific Northwest. From the 1880s through the 1930s, the Columbia River commercial fishery for salmon often yielded over 18,000 metric tons annually (Cummins et al. 1995). In the

FIGURE 16.7

Several stocks of chinook salmon in the Snake River portion of the Columbia River system have recently declined to the brink of extinction.

(U.S. Fish and Wildlife Service photo)

FIGURE 16.8

Many fish in the Amazon River system, such as this freshwater croaker of the genus *Plagioscion*, are adapted to feed on tree fruits that they obtain in the flooded *varzea* forest zone during the high-water stage of the river's annual cycle.

(World Wildlife Fund photo by Michael Goulding)

1920s, a rapid decline began, accelerating with construction of dams on the region's rivers. In the 1960s through early 1990s, harvests have been only a quarter of earlier levels. Several efforts have been made to maintain the fisheries. Loss of access to spawning areas and slowed return of young salmon to the ocean was attacked in several ways. Hatcheries were built to rear salmon fry and reintroduce them in lower river sections, where they could migrate to the ocean. On the Columbia, young steelhead and salmon were collected at John Day Dam and transported by barge to Bonneville Dam, helping them pass this stretch of the river without high mortality.

Overall, these efforts have had little benefit. In spite of expenditures of more than $1 billion, the runs of salmonids in the Columbia River are only about 15% of those in the 1800s, and only a fifth of these are fish hatched in the wild. Hatcheries may even have promoted the decline of wild populations. Today, commercial harvests have mostly been halted. In the Pacific Northwest, an American Fisheries Society committee in 1991 judged 101 of 214 spawning populations of native anadromous salmonid fish of the Pacific Northwest to be at a high risk of extinction, with 58 other populations having a moderate extinction risk (Nehlsen et al. 1991). By 1995, the situation had worsened. Numerous spawning populations are now extinct (Cummins et al. 1995). In particular, chinook salmon in the Snake River and coho and chum salmon along most of the United States coasts have declined toward the brink of extinction (Fig. 16.7).

The salmon crisis has led to great public concern, and to proposals that dams on rivers such as the Elwha, on the Olympic Peninsula of Washington, be removed and restoration of salmon runs encouraged (Wunderlich et al. 1994). Nevertheless, possible future developments, such as diversions of water for irrigation and interbasin transfer, are likely to intensify the challenges facing migratory fish of the Pacific Northwest.

MAJOR TROPICAL AND SUBTROPICAL RIVERS

The largest rivers of the world are in the humid tropics, and like many other tropical ecosystems, they are biotically rich (Davies & Walker 1986). The Amazon River, for example, possesses about 2,000 species of fish (Allen & Flecker 1993), together with endemic and highly distinctive crocodilians, turtles, and a freshwater dolphin. The Mekong River of southeast Asia probably has more than 850 fish species and the Zaire River of Africa about 700 species. The faunas of all three rivers are still not fully described. Most of these species are endemic, so that preserving them is a major conservation priority.

The Amazon, Mekong, and Zaire show large seasonal variations in their flow. During flood seasons, large areas of their floodplains become functional units of the river ecosystem. In the Amazon Basin, seasonally flooded forests of the *varzea* zone bordering the river channel (*See* Chapter 9) form a major feeding habitat for many species of fish (Fig. 16.8), some of which are highly specialized for feeding on tree fruits (Goulding 1980, 1985). In some cases the fish are predators on the seeds of these fruits; in other cases the seeds pass through the digestive tracts of the fish, which act as their dispersal agents. Perhaps 75% of the fisheries productivity of the Amazon River system is believed to depend on this relationship. Because the *varzea* forests are accessible from the river channels and logs can easily be floated out from them during flood periods, logging of these forests has been heavy. Many distinctive reptiles, birds, and mammals are also associated with these tropical rivers. In the Amazon Basin, several of these have been exploited heavily for food, skins, and other purposes. The giant otter, two species of giant river turtles, and two species of caimans have been reduced to endangered status (Johns 1988).

FIGURE 16.9

The Okavango River, originating in the highlands of Angola, ends in an enormous interior delta in the Kalahari Desert of Botswana, creating one of the richest wildlife habitats in southern Africa.
(Photo by G. W. Cox)

Several river systems are linked with major wetland areas of enormous conservation significance. In southern Africa, the Okavango River, arising in Angola, flows south into Botswana, forming an immense interior delta in the Kalahari Desert. The Okavango Delta, consisting of papyrus-bordered channels and vast areas of seasonal marshland (Fig. 16.9), contains a distinctive fish fauna, and a rich mammal and bird fauna. In Brazil, the Paraguay River floods seasonally to create what is perhaps the world's largest wetland complex, the Pantanal do Mato Grosso (*See* Reading 16.2) (Alho et al. 1988). This region is now beginning to experience major impacts of exploitation and development (Mittermeier et al. 1990). The deltas of major rivers such as the Mississippi in the United States, and the Rhone and Danube in Europe are also wetlands rich in wildlife.

Some of the world's most endangered large animals inhabit tropical and subtropical rivers. Dams, boat traffic, and fishing activities are implicated in the decline of the endemic river dolphins of the Ganges, Indus, and Yangtze Rivers of Asia (*See* Chapter 18). Of these, the Chinese river dolphin, *Lipotes vexillifer,* is most endangered, having been reduced to less than 200 individuals (Peixun & Ding 1988).

STREAM ECOSYSTEMS AND GLOBAL CHANGE

Global warming will modify stream ecology, particularly in northern latitudes. The amount, timing, and variability of stream flows will be modified in complex fashion (Grimm 1993). In semiarid regions, for example, the incidence of flash floods may increase. Under a scenario of doubled CO_2 level, northern rivers are predicted to show up to a 28% increase in flow, with the season of maximum flow coming earlier in the year (Van Blaricum et al. 1995).

Continued damming of rivers may create a positive feedback on global warming. In the Amazon Basin, more than 70 new dams are projected (Fearnside 1995). The reservoirs created by these dams will be deforested permanently, thus adding to net enrichment of the atmosphere in CO_2, the principal greenhouse gas (*See* Chapter 23).

CONSERVATION MANAGEMENT OF RIVER ECOSYSTEMS

Central to management of flowing water ecosystems is preserving a flow adequate to maintain the basic features of the aquatic ecosystem. The rights to water in rivers and streams, however, are governed by complex laws that vary from country to country and even state to state (Cooper 1990). Versions of the **riparian doctrine** derived from English common law govern water rights in most of Europe and eastern North America. This doctrine guarantees owners of land along streams the right to an undiminished and undefiled flow, a guarantee that is compatible with preservation of stream wildlife. Under the riparian doctrine, the rights to the use of water cannot be bought or sold separately from the stream-side land, nor can they be lost because water is not used in some active fashion.

In the arid and semiarid western United States, however, a very different type of water law, the **prior appropriation doctrine,** evolved. This doctrine permits diversion of water for beneficial uses, with the rights to the water being governed by the dictum "first in time, first in right." Under this doctrine water rights can be sold, and, if they are not used for beneficial purposes involving diversion, they can be lost (Meyer 1989; Wiley 1990). The appropriation doctrine does not recognize an intrinsic beneficial use of water in the stream itself.

Several strategies are now being used to assure the maintenance of basic flows in important wildlife streams in the western United States. Acquisition of water rights by conservation organizations, coupled with legal recognition of in-stream flow as constituting a beneficial use, is one approach (Wiley 1990). The concept of **reserved water rights,** or implicit rights to the maintenance of basic flows that existed in streams within national forests, wildlife refuges, and recreation areas at the time of their establishment, has been supported by a number of court decisions (Hunt 1988). Application of the **public trust doctrine** is another. This doctrine states that the public's right to natural features such as streams and lakes overrides excessive allocations that could destroy these features (Cooper 1990). These emerging doctrines are aided by new approaches to the economic valuation of ecological resources such as free-flowing streams and their wildlife (*See* Chapter 29).

In addition to maintaining adequate flows, the quality of the water in rivers and streams must be preserved. The United States Geological Survey monitors water quality at a series of 388 stream sampling stations (Smith et al. 1987). The Clean Water Act of 1972, together with several subsequent acts, requires the United States Environmental Protection Agency to establish **national effluent standards** that limit the quantities

T he Paraguay River, a tributary of the Parana River in South America, is associated with the world's largest seasonal wetland: the Pantanal of the Mato Grosso. The Pantanal—a Portuguese/Spanish word meaning marshland—lies in the headwater region of the river, and covers about 136,700 square kilometers in the states of Mato Grosso and Mato Grosso do Sul, west-central Brazil. The southern hemisphere summer (December through February) is the season of rains, when the river rises and spills onto its floodplain, creating a vast marshland dotted by scattered islands and low ridges. These areas remain flooded for 2 to 6 months. The biotic richness of this system is yet to be inventoried fully, but is estimated to include over 400 species of fish and 650 species of birds. The river itself has fisheries of major commercial importance. The typical reptile of the Pantanal is the caiman, and the typical mammal the capybara; hundreds of thousands of both animals live in the marshes. Some 19 species of herons, egrets, ibis, and spoonbills breed in the Pantanal and 24 species of parrots and macaws utilize the forest areas of the region. Many species of migrant birds visit the Pantanal, from North America in the northern hemisphere winter and from Patagonia in the southern hemisphere winter. The richness of life in the Pantanal is associated with its proximity to the diverse biogeographic regions of Amazonian forest, Brazilian savannas, Pampean grasslands, and Paraguayan thorn forest, coupled with the mosaic of upland, seasonal wetland, and permanent wetland habitats of the region itself.

The Pantanal exists principally because the channel of the Paraguay River does not drain the wet season precipitation rapidly. This feature is due largely to the underlying geology of the region, which creates bedrock sills, or natural dams, that the river has not

downcut effectively. Major sills occur at several locations, and influence the slope of the river channel for up to 400 kilometers upstream from their location. Many other minor sills also occur, an average of one every 40 kilometers or so. These natural dams trap water in the upper basin, keeping the flow of water slow enough that it takes 6 months to reach the ocean. Spread out over the vast floodplain, the seasonal flood supports a lush vegetation and creates a low-albedo landscape that absorbs solar radiation. In turn, the heating encourages evaporation, so that discharge to the lower river is a very low fraction of the total rainfall in the upper basin. Heating and evaporation also promote convectional thunderstorm activity, so that the Pantanal recycles its moisture, and is more humid than would otherwise be the case. The landscape acts as a huge sponge, as well, storing water during the flood season, and letting it flow and seep back into the channel during the dry season, thus maintaining a substantial flow even in the driest months.

Proposals have recently been made to develop a Parana-Paraguay Waterway, known regionally as the Hydrovia Project. This plan would create a channel that would allow ocean-going vessels to navigate the river to Caceres, Brazil, a point just upstream from the Pantanal. To do this would require straightening and deepening the channel. A minimum depth of 3 meters would be required, and this would necessitate blasting out the sills that restrict depth in many places. This plan would provide shipping benefits to Brazil, Bolivia, Paraguay, Uruguay, and Argentina, and could be funded by loans from the Inter-American Development Bank. The cost for the project is estimated at $1.3 billion. The United Nations Development Programme has undertaken studies of the economic feasibility and environmental impact of such a project.

At the outset, however, it is clear that such a project would modify the Pantanal irreversibly and extensively (Ponce 1995). Straightening and deepening the channel would speed water flow and increase sediment discharge to downstream areas. Channel erosion would be encouraged, and new rounds of blasting to remove sills that appear would probably be required. The fraction of water outflow from the upper basin would increase, and much of the marshlands would disappear. Nutrient recharge of the floodplain would decrease proportionately with the loss of flood-deposited sediment. Flooding would intensify on lower reaches of the Paraguay and Parana Rivers. Recycling of moisture would decrease, and the climate of the region would likely become more arid. Regional droughts would likely become more frequent and longer in duration. In 1994, the region experienced its worst drought in 20 years, and some scientists speculate that the limited dredging already done may have contributed to drying of the Pantanal region. Large areas of wetland vegetation would be replaced by vegetation types of the drier uplands. Erosion would likely increase, further increasing sediment load in the river and the effects of sediment deposition downstream.

The loss of biodiversity due to such change would be enormous. Many presently endangered species would be affected, including species such as the giant otter, river otter, crab-eating fox, maned wolf, jaguar, and swamp deer. The Inter-Governmental Committee on Hidrovia, representing five countries with economic interests in the waterway, has recognized the ecological importance of the Pantanal. The environmental impact of alternative plans for development of the Hidrovia is now under evaluation.

---

of polluting substances that are discharged into surface waters. In addition, the Environmental Protection Agency is responsible for obtaining compliance with these standards by sewage treatment plants, industries, and other sources of pollutants.

SUMMARY

1. Rivers and streams carry only a small fraction of the world's fresh water, but are important in global hydrology. The dynamics of these ecosystems are intimately tied to their floodplains and watersheds.

2. Rivers have been impacted by varied human activities, including water diversion, watershed modification, fishery overexploitation, pollution, mine drainage, damming, and channelization.

3. In North America, multiple human impacts on rivers of the Pacific Northwest have caused the decline of migratory salmon populations of great economic, as well as ecological, value.

4. Tropical and subtropical rivers contain rich vertebrate faunas, and their seasonal changes create distinctive wetland and flooded forest environments.

5. The recognition of legal rights for maintaining in-stream flows of water is an essential strategy for protecting river and stream ecology.

QUESTIONS FOR DISCUSSION

1. Do you think that restoration of wetlands and removal of levees could be full or partial solutions to the problems of flooding along the Mississippi River? What sorts of societal adjustments would be necessary to make these changes?

2. What ecological and economic values have been lost by the changes that have occurred to the Columbia and other rivers of the Pacific Northwest by the impacts described in the text?

3. Do you think that the biodiversity values of a region such as the Brazilian Pantanal or the Okavango Delta of Botswana can provide an economic base adequate to justify their protection against destructive development?

SUGGESTED READING

Allen, J. D. and A. S. Flecker. 1993. Biodiversity conservation in running waters. *BioScience* **43**:32–43. Patterns of biodiversity in stream ecosystems, the factors threatening stream species, and the prospects for stream restoration.

Hey, D. L. and N. S. Philippi. 1995. Flood reduction through wetland restoration: The upper Mississippi River basin as a case history. *Restoration Ecol.* **3**:4–17. Application of the principle that a river's behavior reflects the dynamics of its entire watershed to the issue of flood prevention.

Maxwell, J. 1996. *The salmon circle*. HarperCollins, New York. The ecology, catastrophic decline, and recovery prospects for the salmon of the Pacific Northwest.

Oceanic Ecosystems

ceans cover almost 71% of the earth's surface, and their behavior thus plays a major role in the dynamics of the biosphere. Scientists once believed that hectare for hectare, the marine environment was equal in productivity to the environment of the continents. This belief supported the view that the oceans were a virtually inexhaustible resource of food for meeting the needs of the growing human population.

Oceanographers have now learned much more about the productivity of various ocean realms. The realities are that productivity varies enormously in different situations and that the average productivity is much below that of the land. The production of the oceans, far from being inexhaustible, is easily overexploited, and must be managed with great care—a task that is proving extremely difficult in the international marine environment.

Dolphins are one of the top carnivores in the long food chains of the open ocean.

Outline

TABLE 17.1

Net Primary Productivity (NPP) of Various Marine Environments, Together with the Potential Production of Harvestable Material and Recent Catches of Traditional Fisheries.*

CHARACTERISTIC	ESTUARY/REEF	UPWELLINGS	COASTAL WATERS	OPEN OCEAN	TOTAL OCEAN
Area (10^6 km^2)	2.0	0.8	27.0	332.0	361.8
Area (% of ocean area)	0.55	0.22	7.46	91.76	100.00
NPP (g C m^{-2} yr^{-1})	1,125	625	210	65	
NPP (g wet wt. m^{-2} yr^{-1})	10,125	5,625	1,890	585	
NPP (10^9 t C yr^{-1})	2.25	0.50	5.67	21.58	30.00
NPP (10^9 t wet wt. yr^{-1})	20.25	4.50	51.03	194.22	270.00
Mean efficiency of transfer between trophic levels (%)	10	10	10	10	
Number of trophic transfers up to harvestable species	2.5	2.8	3.4	4.2	
Harvestable Production (10^6 t wet wt. yr^{-1})	640	71	203	122	1,036
Recent Catch and Bycatch (10^6 t wet wt. yr^{-1})	21.02	20.83	67.12	3.60	112.57
Catch as % of Harvestable	3.3	29.3	33.1	3.0	10.9

*Compiled from various sources.

PRODUCTIVITY OF OCEANIC REGIONS

For purposes of analysis of basic productivity, the oceans can be divided into four zones: the open ocean, upwelling zones, continental shelf waters, and reef-estuary systems (Table 17.1). These regions differ in their extent, basic primary productivity, and food chain characteristics (Ryther 1969; Pauly & Christensen 1995). The total net primary production of the oceans is now believed to equal about 30 billion tons of carbon annually, or about 270 billion tons of organic matter by wet weight (Berger et al. 1989).

The **open ocean,** beyond the edge of the continental shelf, makes up about 91.8% of the total marine environment. The net primary production of this ocean region is low, corresponding to that of desert-edge ecosystems of the continents. The main reason for low productivity in this region is severe nutrient limitation. Dead organisms and particulate wastes sink to the ocean depths, causing a constant loss of nutrient-containing matter from surface waters. These nutrients return only by upwelling at distant locations. Terming the open ocean a "marine desert" is thus not far from the truth. In addition, food chains of this region are unusually long. They begin with microscopic phytoplankton, and involve several food chain links before reaching species of fish that are practical to harvest. These include tunas, bonitos, and billfish, some of which travel in large schools and migrate over great distances. Squid are also important fishery species in the open oceans. Some dolphins also forage in herds in the open oceans. In the eastern Pacific, these dolphins frequently travel and feed in association with schools of yellowfin tuna, an association that has led to high mortality in fishing operations (*See* Chapter 19). Of the great whales, the sperm whale is most at home in the open oceans, where it feeds largely on squid. Among seabirds, the wide-ranging albatrosses are major foragers of the open oceans.

Upwellings occur at scattered locations near the edges of continents and around Antarctica. Along continental coastlines, two special circumstances are required for upwellings to occur: deep waters that lie close to the continental margin and prevailing winds that push surface waters away from the coast. Such conditions commonly are met along the west coasts of continents, and here cold, nutrient-rich waters rise to the surface. The slow, nutrient-rich bottom currents of the Atlantic, Pacific, and Indian Oceans converge in antarctic waters and rise to the surface at the antarctic divergence, a complex and variable zone encircling Antarctica. Upwelling areas are very localized, and amount to perhaps 0.2% of the area of the oceans (Fig. 17.1). Of the offshore regions of the oceans, upwelling zones are by far the most productive. Despite the coldness of the water, the abundant nutrients support a moderately high level of net primary production. In addition, the primary producers are much larger than those of the open ocean. Food chains only two or three links long lead to species such as anchovies, sardines, and invertebrates that can be harvested profitably when they are abundant. Most of the catch of sardines and anchovies is converted to oil and fish meal. **Krill** are large marine zooplankton, primarily the small shrimp, *Euphausia superba,* that abound in cold upwelling waters at high latitudes (Fig. 17.2). Krill are abundant in the antarctic upwelling zone. Marine mammals, such as baleen whales and seals are abundant in upwelling zones. Diving marine birds, such as auks, penguins, cormorants, boobies, and sea ducks are also abundant.

Continental shelf waters, the second most extensive ocean region (Fig. 17.1), have a net primary productivity intermediate between upwellings and the open ocean, largely supported by the inflow of nutrients from continental areas. Food chains here are also intermediate in length, averaging two or three links from the producers to fish and shellfish of harvestable size. Many types of fish are harvested in this region of the

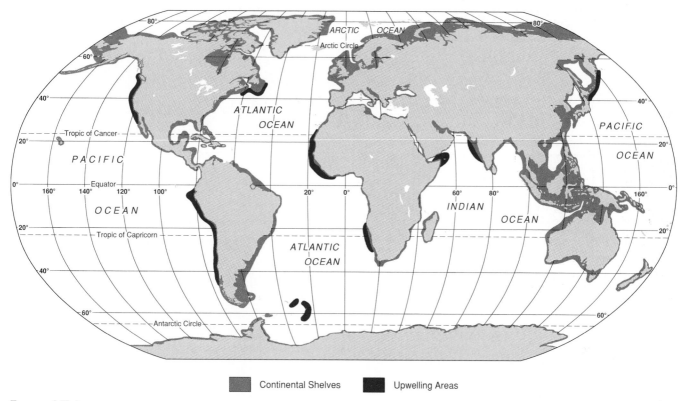

FIGURE 17.1

Continental shelf and upwelling zones of the oceans.

FIGURE 17.2

Euphausid shrimp, or krill, are large zooplankton that are abundant in the antarctic upwelling zone.

(© Photograph by Peter Johnson)

oceans. The most important food fish are pelagic species such as mackerel, salmon, and swordfish, and various bottom-dwelling groundfish, such as cod, flounder, pollock, hake, and croaker. Herring and menhaden, small fish taken primarily for oil and fish meal production, are also harvested in this zone. Many types of shellfish, including clams, shrimp, crabs, and lobsters are also harvested in shallow continental shelf waters. Marine mammals, including seals, sea lions, walruses, porpoises, dolphins, small whales, and baleen whales, concentrate their activities in this zone. Diving marine birds are also abundant.

The net primary productivity of certain other marine environments, such as **estuaries** and **coral reefs,** is even higher than that of upwelling areas, and food chains may average somewhat shorter. Estuaries and reefs are major production areas for finfish and shellfish of many kinds (*See* Chapter 18). Gulls, terns, shorebirds, and wading birds are abundant in estuaries and reefs, but few marine mammals frequent these environments.

These characteristics, coupled with efficiency of transfer of energy from one species in the food chain to the next, determine the potential productivity of the various ocean regions. Food chain efficiency tends to be about 10% for food chains in all regions (Pauly & Christensen 1995). This efficiency means that only 10% of the energy reaching one level in the food chain is passed on to the next level. When net primary production, food chain length, and food chain efficiency are combined, an estimate of the production of harvestable marine fish per unit ocean area is obtained (Table 17.1). This, weighted by the areas of the various regions, gives an estimate of the total production of harvestable fish in the world oceans.

The data in Table 17.1 show several interesting features. First, despite their small areas, estuary and reef systems produce the largest fraction of potential harvestable production. Coastal waters, in contrast, provide the largest portion of recent marine

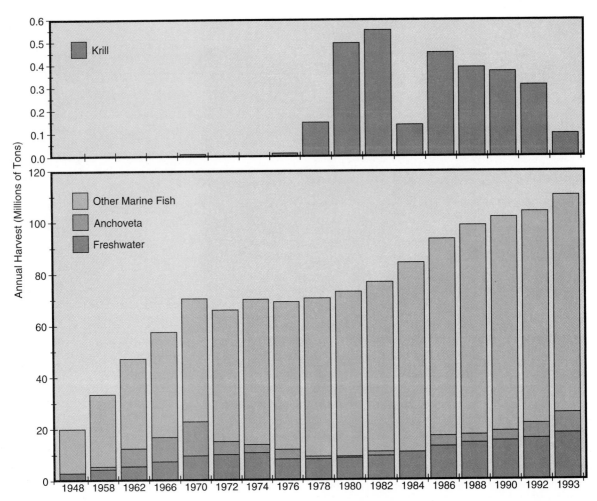

FIGURE 17.3

Global harvests of marine and freshwater fish and shellfish for selected years since World War II.
(Data from annual volumes of the FAO *Yearbook of Fishery Statistics*)

fisheries harvest. Much of the harvest taken in coastal waters, however, depends on production carried out in estuaries (*See* Chapter 18). Second, the total harvestable production is probably about 1 billion tons, far below the billions of tons that some have proposed as the potential harvest from the oceans. Humans are currently taking over 10% of all the production of potentially harvestable species, and in upwelling zones and coastal waters probably more than 25%.

How much of this potential harvest can be taken without overexploiting the target species themselves, and disrupting food web relationships with other members of the ecosystem? The answer to this question is not known with certainty. Ryther (1969) estimated that perhaps 40% of the potential could be taken, if it were harvested in a carefully controlled manner. The major disruptions of marine fisheries stock resulting from current harvest levels, which are quite a bit below this, suggest that Ryther might have been overly optimistic.

TRENDS IN MARINE FISHERY HARVESTS

At the end of World War II, world fishery harvests were just over 20 million tons annually (Fig. 17.3), and some major fisheries, such as that for Peruvian anchoveta, had not even been tapped. Improved marine technologies, however, led to rapid growth of marine fishery harvests, and total fishery yields climbed rapidly. Between 1950 and 1970 nearly a three-fold increase in total harvest was realized. These developments led to highly optimistic predictions of the potential fishery harvest from the world oceans, and to the view that such harvests could solve the protein shortages afflicting much of humanity. Predictions of ultimate annual harvests of up to one billion tons were made.

In the 1970s and 1980s, however, fishery yields began to plateau. In 1987 the United Nations Food and Agriculture Organization (UN FAOa,b) concluded that the era of large, sustained increase in marine fishery harvest was over. Only 25 of 280 fish stocks that the FAO monitored were still only slightly to moderately exploited, whereas 42 were overexploited or depleted. Overexploitation of marine fisheries has been frequent and widespread. In the northwest Atlantic, the cod fishery crashed in the first decade of the 1900s and haddock in the 1930s (Alverson 1978). Overexploitation of Pacific halibut by United States and Canadian fleets in the early 1900s led to the crash of this fishery in 1916, forcing the creation of an international commission to oversee its harvests (Weber 1986). Other fisheries that are now badly overexploited include the Atlantic salmon, the Atlantic striped bass, New England lobster, Gulf of Mexico king mackerel, Pacific albacore, Pacific rockfish, Alaska pollock, Alaska king crab, and others (Weber 1986). Many other stocks are on the verge of overfishing, including sharks (Manire & Gruber 1990), Gulf of Mexico menhaden, Atlantic and Pacific billfishes, and Pacific yellowfin tuna (Weber 1986).

In spite of scientific fishery research programs in most major regions of the oceans, even newly exploited fisheries have proven almost impossible to manage for sustained yield. In Antarctic waters, for example, exploitation of the marbled notothenia and other species near the island of South Georgia began in 1972–1973 with a yield of about 4,000 tons. The yield of this fishery grew to 500,000 tons in 1979–1980, but subsequently crashed to 75,000 tons in 1984–1985, in spite of efforts by the International Commission for the Conservation of Antarctic Marine Living Resources, whose membership includes all 12 of the nations engaged in major fishing activities there. In 1984, the commission was forced to close the fishery for the marbled notothenia because of the rapid decline in catches.

Fisheries of the upwelling zones have proven especially difficult to exploit in a sustained manner. The foremost example of this problem is the Peruvian anchoveta fishery, discussed in detail in Reading 17.1. The Pacific sardine, once abundant from central Baja California to British Columbia, is another example (Browning 1980). Beginning in the 1920s, exploitation of the species by United States and Canadian fishermen increased until, in 1936–1937, over 800,000 tons were harvested. In the years immediately following, the fishery crashed. The factors contributing to this crash are hotly argued. Some biologists believe that overexploitation was the principal culprit, others that changes in currents and water temperature were responsible, and still others that the abundance of the sardine was the high point of a natural, long-term cycle (Browning 1980). In over 50 years, however, the Pacific sardine fishery still has not recovered. The estimated stock of this species in 1989 was 0.5% of its early 1930s level.

As a result, by 1993, commercial marine fishery harvests had reached about 84.2 million tons (Fig. 17.3). Since noncommercial marine catches are about 24 million tons (World Resources Institute 1988) and discarded by-catch (See Reading 17.1) about 27 million tons (Alverson et al. 1994), the total marine harvest for 1993 probably exceeded 135 million tons. In recent years, freshwater fish harvests, much from aquaculture, have grown at a faster rate than those of marine fish, and they now constitute about 17% of all world fishery harvests. Thus, it is likely that many more traditional fish stocks will begin to show signs of decline, and that more efforts will be expended on aquaculture and on harvesting nontraditional resources.

DROPPING DOWN THE FOOD CHAIN

The harvest of marine fisheries could obviously be increased considerably if the harvest could be taken at lower levels in the food chain. In effect, this is the strategy behind efforts to develop a fishery for krill. The effort to develop a krill fishery has been concentrated in the antarctic upwelling zone (Fig. 17.4).

Estimates of the biomass and productivity of antarctic krill are highly uncertain (Knox 1984). Euphausid krill tend to occur in dense swarms, which may extend to a depth of 100 meters over many square kilometers of ocean, and in which animals may reach a density of 10 to 16 kilograms per cubic meter. The true extent of such swarms is difficult to measure, and estimates of standing crop biomass range between 55 million and 7 billion tons (Miller & Hampton 1989), the most probable values lying in the range of several hundred million tons. Surveys in 1981, using sonar techniques, gave a krill biomass estimate of 250 to 600 million tons (El-Sayed 1988). More recently, other workers have suggested that the sonar echoes from krill swarms are much weaker than the values used in this survey, and that krill biomass might be much greater (Everson et al. 1990).

Annual production estimates for antarctic krill are even more uncertain, due to uncertainty about the life history of euphausid shrimp, and range from 135 million to 1.35 billion tons (Nicol & de la Mare 1993). If krill require only two years to reach maturity, the annual productivity would be about half the standing crop biomass, and thus possibly a few hundred million tons (Sahrhage 1989). Most fisheries' biologists now believe that krill have a life span of 6 to 10 years, so that annual production would be a much smaller fraction of standing biomass (Nicol 1990). Annual reproduction is highly variable, however, and in some years is nearly zero (Siegel & Loeb 1995). How much a sustainable, economically sound krill fishery could add to marine fishery harvest is therefore very uncertain.

Krill harvests have been carried out in antarctic waters since 1964 by several countries, including Russia, Ukraine, Poland, Japan, and Chile. The annual harvest in the early 1990s varied from about 90,000 to about 300,000 tons. Krill must be processed immediately after capture to prevent enzymatic deterioration. With limited processing, krill can be converted to protein meal for use in poultry and livestock diets. With more extensive processing, a protein additive suitable for use in human foods can be produced. Altogether, the costs of harvesting and processing krill in remote antarctic waters have not yet become competitive, and krill harvests still remain an experimental enterprise.

HARVESTING INFLUENCES IN COMPLEX MARINE ECOSYSTEMS

The exploitation of marine fisheries involves the harvest of different species, often belonging to different trophic levels, that are interacting members of a single ecosystem (Beddington & May 1982). Harvesting strategies for different fishery species have usually been formulated independently, and usually from theory about the sustained yield harvest of one species (See Chapter 24).

One of the factors contributing to the rapid growth of marine fishery harvest in the 1960s was the emergence of the anchoveta fishery in the Peruvian upwelling zone. Anchoveta are harvested by small purse seiners, boats that enclose schools of fish by a long net that extends downward from the surface, and then purse the deep edge of the net so that it closes, trapping the fish (Reading Fig. 17.1). This fishery began in about 1956 with a harvest of 0.1 million tons and reached 13.1 million tons in 1970. Between 1964 and 1971, anchoveta harvests exceeded 9.5 million tons annually, the level estimated by fishery biologists to be a maximum sustainable harvest (Brown 1985). In 1970, anchoveta constituted almost 20% of the total world fishery harvest, and fish meal manufactured from the anchoveta earned $340 million for Peru, about a third of the country's foreign exchange.

In the Peruvian upwelling zone, occasional incursions of warm tropical water suppress the upwelling phenomenon, killing and dispersing the anchoveta. This event, which is known as El Niño because it tends to occur at Christmas ("El Niño" refers to the Christ child), recurs at intervals of about five to ten years (Philander 1989). In most years, strong trade winds drive equatorial currents westward from the coast of South America, stimulating the upwelling of cold, nutrient-rich water that supports the high productivity of the system. At the peak of this process (now often termed "La Niña") warm surface water becomes concentrated in the western Pacific, where a very strong tropical low-pressure system simultaneously develops. Eventually, this tropical low spreads eastward, reducing the strength of the trade winds and permitting strong ocean currents to carry warm water eastward. When these currents reach South America, they seal off the upwelling, creating an El Niño. A similar but weaker system of this sort occurs in the tropical Atlantic Ocean.

Prior to the development of the anchoveta fishery, the impact of El Niños was mainly on seabirds such as brown pelicans, guanay cormorants, and Peruvian boobies, all of which depend on the anchoveta. Original populations of these birds along the Peruvian coast during normal years are estimated to have been 28 million. Periodic El Niños, however, caused these populations to crash to levels as low as five million.

At the peak of the anchoveta fishery in 1972, an El Niño struck. In an effort to sustain the profitable catch, fishing effort was intensified, the result being that the decimated anchoveta stock was depleted even further. In 1973, instead of showing recovery, the catch fell to less than 2 million tons. The fishery began to recover in 1974–1976, but continued exploitation and subsequent El Niños apparently caused a basic disruption of the breeding stock of the anchoveta. In the 1980s and 1990s, this fishery has remained far below its yields in the early 1970s. In 1993, the anchoveta harvest recovered to 8.3 million tons, the highest level since the early 1970s.

READING FIGURE 17.1

The crew of a Peruvian fishing boat prepares to bring aboard anchoveta captured in a purse seine in the coastal upwelling zone.
(Photo courtesy of the FAO, Rome)

Although such strategies may be suitable for species at the ends of marine food chains, for which humans are the primary predator, they are inadequate for most marine fisheries. Most harvested species are linked with competitors, predators, and prey in a complex system that can respond in unexpected ways to the exploitation of one or more members.

As a result of ecological linkages, heavy exploitation of some fisheries has caused major shifts in the composition of marine fish communities. In the Georges Bank Fishery in the northwestern Atlantic, heavily fished species such as herring, mackerel, and silver hake are being replaced by sand lance, dogfish, and squid (Weber 1986). Decline of the Pacific sardine

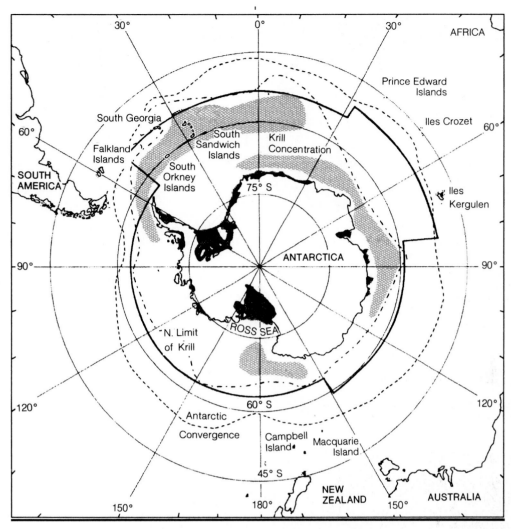

FIGURE 17.4

The antarctic marine ecosystem, showing the prevailing locations of major concentrations of krill and their northern limit of distribution in relation to the antarctic convergence.

(Modified from Siegfried 1985)

fishery off the west coast of North America was coupled with increase in the northern anchovy (MacCall 1986). In the North Sea, even though the overall fisheries' harvest remained relatively constant from the mid-1960s through the late 1970s, the fraction of the harvest made up of herring and mackerel, the most desirable species, declined from about two-thirds to less than one-third (Beddington & May 1982).

The kinds of tradeoffs that involve human marine fisheries' harvests are well illustrated by the antarctic marine ecosystem. The ecosystem is one of the largest, most productive, and most sharply defined ecosystems of the world oceans. Although many uncertainties exist about rates of productivity at the level of primary producers and herbivores, it is clear that euphausid shrimp are the key consumer in the system. They constitute what some ecologists term a **foundation species,** a lower trophic level species whose abundance determines much of the upper trophic level structure of the ecosystem. They form

the primary food of baleen whales, crabeater seals, penguins, and the larger fish and squid. The populations of these species are ultimately determined by krill production.

Populations of krill-eating species have changed markedly in the last century (Laws 1985), due to selective harvest of certain of these animals by humans (Fig. 17.5). For example, in the early 1900s this antarctic region was home to perhaps 1.1 million baleen whales, with a total biomass of about 45 million tons. Whaling has reduced the biomass of whales to about nine million tons. In recent decades, heavy exploitation of several fish stocks has reduced their abundance, as well. These reductions have meant more krill for the remaining species, allowing major increases in their populations. Populations of seals and penguins are probably 2 to 3 times those of the early 1900s, and squid are also believed to have increased in abundance somewhat. The crabeater seal (Fig. 17.6), with a population of 15 to 30 million, is now the most abundant pinniped in the world (Siniff 1991).

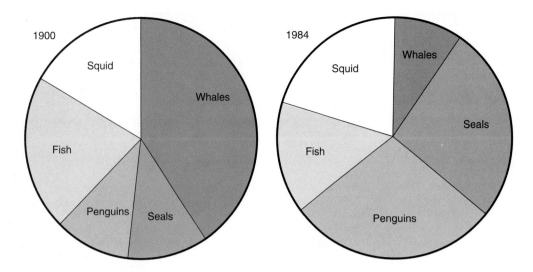

FIGURE 17.5

Major changes in the biomasses of the major krill-feeding animals have occurred in the antarctic marine ecosystem, correlated with human harvests of whales and fish.

(Data from R. M. Laws, 1985, "The Ecology of the Southern Ocean" in *American Scientist*, 73:26–40, Research Triangle Park, NC: Sigma XI Scientific Research Society.)

FIGURE 17.6

Crabeater seals, the most abundant pinnipeds in the world, are one of the major krill predators of antarctic waters.

(Photo by Stephen Leatherwood)

Sei and minke whales, two small species less intensively exploited than the larger whales, may also have increased in numbers.

These changes indicate that a major fishery for krill would create substantial impacts on several or all of the krill-eating species. Depletion of krill stocks would almost certainly lead to declines in populations of whales, seals, birds, and squid. Some scientists fear that reduced numbers of nesting penguins in certain antarctic colonies reflect the effect of krill harvests since 1986 (R. Hewitt, *pers. comm.*). Eventually, krill might even be replaced as herbivores by copepods, salps, or other invertebrates that would channel their food energy along new food chains,

perhaps involving various species of fish and fish predators, so that a major change in community structure might occur.

Similar tradeoffs are evident in the Peruvian upwelling ecosystem and in the Bering Sea ecosystem. In the Peruvian upwelling zone, intense exploitation of anchoveta and occasional El Niños have depleted the anchoveta, the foundation species of this system. In the mid-1950s, before anchoveta harvests began, about 28 million seabirds—Peruvian boobies, guanay cormorants, and brown pelicans—bred on islands along the coast of Peru. These species created the remarkable deposits of guano that were mined on these islands for fertilizer for many years. During El Niño years, shortages of anchoveta would cause their populations to crash, as in 1957, when populations dropped to about five million. Normally, however, recovery of seabirds was rapid. With the growth of the anchoveta fishery, however, bird numbers have never returned to their original levels, reaching only about 16 million in the mid-1960s. With collapse of the anchoveta stock in the 1970s, peak numbers of seabirds declined to only about four million. In 1983, after an El Niño reduced the anchoveta harvest to its lowest level since the fishery began, less than a million seabirds survived along the Peruvian coast. The Bering Sea ecosystem is discussed in Reading 17.2.

DISCARDED CATCHES OF COMMERCIAL FISHING

Another serious problem of commercial fishing is the capture, injury, and discard of nontarget organisms. These include nontarget species of finfish or shellfish, undersize individuals, and animals such as sea turtles, mammals, and birds. This by-catch results from the poor selectivity of the nets, trawls, or traps used in fishing. Most of the by-catch is killed, injured, or stunned in the process, so that even though the animals are returned to the ocean, most die or are quickly taken by predators.

T he Bering Sea lies between the Aleutian Islands of Alaska and the Kamchatka Peninsula of Russia. A wild, cold, and foggy ocean, it is one of the richest marine ecosystems on earth. Whales, pinnipeds, walrus, and a host of seabirds feed on the productivity of its waters. Whales and fur seals have long been harvested in this region. In recent years, however, one of the world's largest single-species marine fisheries has evolved in the Bering Sea.

The walleye pollock, *Theragra chalcogramma,* is a bottom fish that was not harvested heavily until the 1970s. Prior to this, pollock were taken mainly for their roe, and most of the fish was discarded. Development of techniques for processing pollock flesh to produce imitation crab and other products stimulated increased harvest of this species. Pollock are caught by large factory trawlers, using nets with a capacity up to 120 tons. By 1978, the total catch of walleye pollock in the Bering and Okhotsk Seas was 5.3 million tons, exceeding the harvest of Peruvian anchoveta. Establishment of an exclusive economic zone extending 320 kilometers from the American coastline stimulated the growth of a major U.S. fishery in the eastern Bering Sea in the 1980s. This forced Russian and Japanese fleets into the western Bering Sea and an area of international waters known as the "Donut Hole" in the south-central Bering Sea.

Catches in the Donut Hole grew to over 1 million tons in the late 1980s, and the overall catch of pollock increased to 6.7 million tons. But in the early 1990s, catches throughout the Bering Sea plummeted. Many of the fish caught in the Donut Hole appear to come from breeding areas close to the Aleutian Islands. In 1994, Russia, the United States, and four other countries fishing in the Bering Sea agreed to a moratorium on fishing in the Donut Hole.

The enormous harvests of pollock in the 1970s and 1980s, together with a large by-catch of nontarget fish by trawlers, were correlated with declines in populations of a number of marine mammals and birds. The Steller's sea lion, once numbering nearly a quarter of a million animals in the north Pacific, has fallen in numbers to only about 50,000 animals. The decline has been greatest in the Aleutian Islands, where the population fell from about 101,000 in 1960 to 14,000 in 1992 (*See* Chapter 19). The frequency of entanglement of Steller's sea lions in net debris appears to be much lower than for the Alaska fur seal. Thus, depletion of the fish foods of this species seems to be the only likely cause of its decline. Populations of Alaskan fur seals on the Pribilof Islands have fallen to less than half their levels in the 1960s, and their commercial harvest has been suspended. This decline may have multiple causes (*See* Chapter 19).

The Aleutians and Pribilof Islands are a hotspot of seabird biodiversity. This is particularly true for members of the Auk Family, of which 15 species of murres, puffins,

READING FIGURE 17.2

The tufted puffin, a member of the Auk family, is one of the most distinctive seabirds of the Bering Sea.

(Photo by G. W. Cox)

auklets, and murrelets breed in the region (Reading Fig. 17.2). Seabird numbers in the Pribilofs have declined by perhaps as much as 80% since the mid-1970s (Rosen 1991). The bulk of the world population of the red-legged kittiwake, a small gull, nests on St. George Island in the Pribilofs. Its numbers have declined to about 125,000, half the population that nested there in 1976.

Other wasteful fisheries' practices exist. **Roe stripping** has been a common practice of fishing fleets serving Japanese and other Asian markets (Alverson 1990). In this practice, roe (or eggs) are removed from female fish and the rest of the body, together with all male fish, are discarded. **Shark finning,** the removal of fins and the discard of the remaining fish, is a growing practice in shark fisheries (Manire & Gruber 1990). Shark fins are considered a delicacy in certain oriental countries.

Worldwide, discarded by-catch of marine fisheries is estimated at 5.4 to 9.1 million tons (Bricklemyer et al. 1989). Discarded by-catch is probably greatest in shrimp trawling (Fig. 17.7), in which the ratio of by-catch to shrimp is commonly 30:1. The otter trawls used in shrimping are widemouthed, small-mesh nets pulled across the ocean floor, capturing almost all kinds of swimming animals that live close to the bottom. Much of this by-catch consists of valuable species, and often the by-catch of important sport fish is greater than the catch of these species by sport fishermen. In the Gulf of Mexico, shrimp trawling has probably

FIGURE 17.7

By-catch of fish, sea turtles, and other animals is heavy in shrimp trawling.

(Photo by Mike Webber, courtesy of the Center for Marine Conservation)

contributed to the decline of several groundfishes, including croaker and red snapper. Furthermore, by depleting the abundance of small fish, predation on the remaining small animal life, including shrimp, by larger fish may be increased.

By-catch of young halibut was probably the major reason for the catastrophic decline of the halibut fishery in the northeast Pacific in the 1970s (International Pacific Halibut Commission 1987). Following the establishment of the International Pacific Halibut Commission in 1923 to regulate United States and Canadian exploitation of this fishery, yields increased from about 20,000 tons in 1931 to about 32,000 tons in 1962. Harvest then declined, coincident with increased fishing by foreign trawlers that captured a heavy by-catch of small halibut. Harvestable halibut yields fell to about 9,000 tons in 1974 and remained below 12,000 tons through 1979. In 1965 the by-catch of halibut was about 12,700 tons, 50% or more of which was probably killed in the process. Since establishment of exclusive United States and Canadian fishing zones in this region, trawling has been excluded from major halibut nursery grounds. The by-catch of small halibut has been reduced to less than 4,000 tons, and the harvest of legal fish has risen to over 25,000 tons.

Much more information is needed on the types and quantities of by-catch in different fishing operations. Clearly, however, efforts must be made to improve the selectivity of fishing gear. In shrimp trawling, for example, different responses of shrimp and fish to sound, light, or electrical fields might allow fish to be repelled from the trawl mouth (Sternin & Allsopp 1982). Ultimately, much tighter regulation of by-catch must be gained under the United States Fishery Conservation and Management Act and various international conventions.

DRIFT GILL NET FISHERIES

In the 1980s, fishing fleets of several countries began using very large monofilament gill nets for harvesting fish and squid in the open oceans. These nets hang freely from surface floats to a depth of 12–18 meters, and can be up to 50 kilometers long. The nets were deployed in the evening to catch fish or squid that come to the surface to feed at night, and were retrieved the next morning. In the Pacific, about 1,500 fishing boats of Japan, South Korea, and Taiwan were equipped for drift net fishing. The total length of nets for these boats was estimated at 48,000 to 64,000 kilometers.

In the central North Pacific, drift net fishing was ostensibly aimed at squid, but the by-catch of nontarget animals was large (Fig. 17.8). Marine mammals and birds were a significant part of this by-catch (See Chapter 19). Up to 40% of the fish catch was discarded. The United Nations General Assembly imposed a worldwide moratorium in December, 1992.

OCEANIC ECOSYSTEMS AND GLOBAL CHANGE

Global warming will certainly lead to major changes in the world oceans. With warming of surface waters, the extent of arctic and antarctic ice packs may decrease, and the Arctic

FIGURE 17.8

Drift gill-netting has been characterized as "strip-mining the sea" because of the large catch of noncommercial fish, seabirds, and marine mammals, such as this Hector's dolphin calf near New Zealand.

(World Wildlife Fund photo by Steve Dawson)

Ocean may eventually become ice-free (Alexander 1992). The productivity and ecological distinctiveness of the arctic marine ecosystem depend heavily on the influence of the ice pack on ocean circulation and seasonal patterns of productivity. The rich benthic fauna of the Arctic Ocean is essential to the survival of species such as the walrus and gray whale, and the ice itself is an essential habitat for several seals and the polar bear.

Some of the possible impacts of global warming are perhaps presaged by trends in the California Current since the 1950s (Roemmich & McGowan 1995). Although year-to-year fluctuations are great, the average surface temperature of surface waters off southern California has increased about 1.5° C. This appears to have strengthened the pattern of stratification of the coastal ocean, and weakened the upwelling process that brings nutrient-rich water to the surface. This change is correlated with an 80% decline in zooplankton biomass, suggesting that major change has occurred in primary production and lower food chain dynamics.

Changes also may be occurring in the Antarctic Marine Ecosystem. The mean temperature of the Antarctic Peninsula has warmed by 2.5° C over the past 50 years. This warming is correlated with a reduction in the average extent of the winter ice pack, and a reduction in frequency of good krill recruitment years (V. Loeb, *pers. comm.*). Possibly related to these changes is an increase in biomass of other pelagic invertebrates, such as salps, a colonial tunicate with different food chain relations.

MANAGING OFFSHORE MARINE FISHERIES

Ocean fisheries clearly present major management problems because many of them exist in international waters. Since the 1970s, however, most countries have claimed exclusive economic zones extending to a distance of 320 kilometers from shore. This exclusive zone was established by the United States in 1976 under the Fishery Conservation and Management Act. In 1982, the **United Nations Convention on the Law of the Sea,** signed by 119 nations, gave formal recognition to this economic zone (Borgese 1983). Since, in fact, the bulk of the world's most productive fisheries lie within this distance of continental coastlines, this development enables nonmigratory fisheries' stocks to be managed protectively by the country with jurisdiction.

Major fisheries do exist, however, in regions beyond exclusive economic zones, such as the antarctic waters and the central Pacific. Furthermore, many fish are migratory, and thus move through the economic zones of many countries. Thus, many international commissions have been established to promote the sustained yield management of particular fish or the fisheries of particular regions (Gulland 1980). The membership of these commissions includes the various nations exploiting the fisheries. Examples of commissions concerned with particular types of fish are the **Inter-American Tropical Tuna Commission,** which regulates tuna harvesting in the eastern tropical Pacific, and the **International Pacific Halibut Commission,** which coordinates United States and Canadian harvests of halibut in the exclusive economic zones of these two countries. The **Northeast Atlantic Fishery Commission,** which sets quotas for various stocks and established mesh sizes to control the size of fish taken, is an example of a regional commission.

One of the major shortcomings of marine fisheries' management has been the lack of an ecosystem approach. Recently, however, the concept of **Large Marine Ecosystems (LME)** has begun to provide a framework for such management (Sherman 1986). Large Marine Ecosystems are ocean areas with more or less natural boundaries formed by surface currents and submarine topography, and containing species strongly linked into a food web. About 20 such ecosystems have been recognized, primarily in upwelling areas or in partially enclosed ocean basins. The **Antarctic Marine Ecosystem** is one example of an LME (Scully et al. 1986). This ecosystem lies south of the antarctic convergence, where cold antarctic water sinks as it contacts warmer water from farther north. This zone surrounds Antarctica, and corresponds to a sharp biological boundary. The Antarctic Marine Ecosystem is characterized, as described earlier, by the dependence of various fish, squid, birds, and mammals on krill. The **International Commission for the Conservation of Antarctic Living Marine Resources** has undertaken the responsibility for the management of this LME. One of its actions has been to place a preliminary limit of 1.5 million tons on annual krill harvest—a value that has not yet been approached by actual harvests (Nicol & de la Mare 1993).

The Baltic Sea and the North Sea are other recognized LMEs (Sherman & Alexander 1986). These LMEs, however, show the difficulties of achieving integrated management, even where natural ecosystem units can be recognized. In the Baltic Sea, for example, scientists of the **International Baltic Sea Fisheries Commission** in 1981 recommended a cod quota of 197,000 tons, but, for political reasons, the commission itself could establish a quota only as low as 272,000 tons, and could not keep actual harvest below 380,000 tons.

SUMMARY

1. Ocean ecosystems can be grouped into the open ocean, upwelling zones, continental shelf waters, and reef-estuary systems, with the open ocean, the most extensive region, being close to a desert in basic productivity.

2. Human harvest of marine fish stocks has reached or exceeded sustainable levels for most traditional fisheries.

3. The Peruvian upwelling, Bering Sea, and Antarctic upwelling ecosystems illustrate the fact that human harvests greatly influence the abundance of other higher-level members of the marine food chain.

4. Practices such as drift gill-netting and trawling for shrimp and finfish lead to large by-catches of nontarget species, most of which are killed.

5. International organizations have been created to manage some large marine ecosystems, such as the Northeast Atlantic and Antarctic Marine Ecosystems, and some important fish stocks, such as Pacific halibut and tuna.

QUESTIONS FOR DISCUSSION

1. Some economic ecologists have suggested that "resources manage people, rather than people managing resources." Do examples from marine fisheries resources support this view?

2. What ecological, economic, and ethical factors do you think will influence whether or not extensive harvesting of Antarctic krill occurs in the future?

3. Do you think that effective controls on by-catch can be achieved for fishing operations that involve thousands of boats scattered over millions of square kilometers of ocean?

SUGGESTED READING

Nicol, S. and W. de la Mare. 1993. Ecosystem management and the Antarctic krill. *Amer. Scientist* **81**:36–47. The distribution and biology of krill, and a discussion of new approaches to managing the harvest of the krill fishery.

Norse, E. A. (Ed.). 1993. *Global marine biological diversity.* Island Press, Washington, DC. The nature of, threats to, and strategies for the preservation of marine biodiversity.

Weber, M. and J. Gradwohl. 1995. *The wealth of oceans.* W. W. Norton, New York, N.Y. A nontechnical account of the history of human exploitation of the oceans, and of the current crises of human mistreatment and mismanagement of the oceans and their resources.

chapter

18

Coastal Marine Ecosystems

Many distinctive marine ecosystems occur along ocean coastlines. These include estuaries and their associated salt marshes, mangrove swamps, and seagrass beds, as well as kelp beds, coral reefs, and other intertidal and shallow water ecosystems. Of all marine ecosystems, these are the most productive, and, because of their productivity and accessibility, they are among the ecosystems most impacted by human activities.

Coastal ecosystems have long been exploited by humans. Now, however, intensified exploitation of biotic production is coupled with sewage and nutrient discharges, chemical pollution, and siltation from growing human populations increasingly crowded into coastal regions. More and more, the shallow marine environment is being probed for oil, phosphates, building materials, and other nonrenewable resources. Plans for floating airports and other large facilities are being drawn. Yet shallow marine ecosystems, if protectively managed, also have enormous potential for mariculture and sea farming.

Human exploitation and pollution are growing threats to the biodiversity of coral reefs.

ESTUARIES

An **estuary** is a semienclosed body of water occurring where rivers discharge into the ocean, usually with a salinity regime intermediate between fresh waters and the ocean. Estuaries include river deltas where fresh and ocean waters mix, partially enclosed bays and lagoons that are fed by streams and open to the ocean, and the sounds that lie inside barrier island systems. Salinity grades from nearly fresh in the inner parts of an estuary to almost marine where it opens to the ocean. Due to seasonal variation in freshwater inflow, salinities often vary greatly.

Estuaries function as partial traps for nutrients, which are carried into them by stream flows that are confined by barrier islands and barrier beaches, as well as by the tidal prism of high-density salt water, with which the fresh water is often slow to mix. The aquatic portions of an estuary are often in contact with salt marshes that are subject to tidal action (Fig. 18.1). The primary producers of the estuary thus include the salt marsh plants, the algae that live on the surfaces of tidal mud flats, the phytoplankton of the estuary waters, and beds of plants, such as eelgrasses, growing on the estuary bottom. Net primary production of estuaries is higher than that of any other aquatic ecosystem.

Estuaries produce large quantities of fish and shellfish that are harvested by man. In North America, for example, most oyster and blue crab production is carried out in estuaries along the Atlantic and Gulf Coasts. Much of the shrimp production in the Gulf of Mexico depends on estuaries that serve as nurseries for the young shrimp. Along the Texas coast, shrimp utilize the coastal bays and sounds for their early growth. The adult shrimp spawn in the open ocean, but the young larvae migrate into estuaries. Many fish also use estuaries as nursery areas for their young.

The productivity of estuaries thus depends heavily on the normal inflow of streams. Freshwater inflow brings nutrients to the estuaries and maintains the normal salinity gradient. Water of intermediate salinity is essential to many typical estuarine species by excluding their predators, parasites, and diseases. Oysters, for example, are attacked by predatory oyster drills (a carnivorous snail), parasitic boring sponges that erode the oyster's shell, and certain fungal diseases, none of which can tolerate intermediate salinities. Normal river inflow also functions in certain situations to help maintain the passes through barrier island systems—the water connections through which organisms such as shrimp migrate between the oceans and sounds. How essential this inflow is to the ecology of estuarine species is evidenced by studies that show that the shrimp harvest along the Texas coast is directly related to the rainfall in inland Texas during the two preceding years (Copeland 1966).

In spite of the ecological importance of stream inflow, resource policy tends to treat water that enters the ocean as water wasted, particularly in regions such as Texas and California where demands for fresh water for urban use and irrigated farming are great. Increasingly, rivers that feed into coastal bays and sounds are dammed, and the water diverted for other uses. The nutrients in the flow of these streams thus are lost to coastal waters, and the salinity regimes of the estuaries are

FIGURE 18.1

The open waters of estuaries are often bordered by salt marshes, the result being that primary production is carried out by phytoplankton in open-water areas, surface-living algae on mud flats, and rooted salt grasses and other higher plants in the marsh.

(U.S. Fish and Wildlife Service photo)

modified considerably. Sediments that served to maintain beaches and other structural features of coastal ecosystems are trapped by dams. Loss of the inflow of fresh water from the Colorado River to the Gulf of California, for example, is probably one of the major reasons for decline of the totoaba, once an important commercial fish, now listed as endangered (Cisneros-Mata et al. 1995). The head of the Gulf of California is the region of spawning and juvenile growth by totoaba.

Increasingly, the ecology of estuaries is being threatened by introduced exotics (Carlton & Geller 1993). Most of these organisms arrive in bays and estuaries in the ballast water of cargo ships. For example, 367 taxa have been identified in ballast water of ships arriving at Coos Bay, Oregon from Japan. Other species that grow attached to substrates may be transported as fouling organisms on ship hulls.

Temperate zone estuaries are often bordered by areas of salt marsh. The primary producers in these marshes are rooted plants, such as the cordgrasses (*Spartina* spp.) and pickleweeds (*Salicornia* spp.), and algae that form mats on tidal mud surfaces. In some cases, these salt marshes play an important role in the dynamics of open estuarine waters because large portions of marshland are flushed daily by tides. In marshes that open broadly to adjacent areas of water, this action transports detritus produced by decomposition of marsh plants into the open waters of the estuary (Odum 1980). In other cases, marshes may import and trap particulate detritus, augmenting detritus food chains and decomposition within the marsh (Zedler 1982). Salt marshes are among the most threatened coastal habitats, often being used as trash dumps, filled to create building sites or dredged to construct marinas.

FIGURE 18.2

Mangrove ecosystems occupy vast areas of shallow, subtropical, and tropical marine waters. Here, red mangroves cover marine shallows around Big Pine Key, Florida.

(U.S. Fish and Wildlife Service photo by Rex Gary Schmidt)

In tropical regions, the role of salt marshes is largely taken over by mangrove swamps (Fig. 18.2). "Mangrove" refers to more than 50 species of trees and shrubs, belonging to 12 plant families, that are able to live in shallow seawater or soil periodically flooded by seawater. Some species can invade shallow, permanent water; others are adapted for sites only occasionally inundated. Southeastern Asia possesses a rich flora of mangrove species, but most other areas have only one to four species. Mangroves are unable to tolerate prolonged periods of freezing temperatures, and in North America occur only from Florida and central Baja California southward.

Healthy mangrove stands have a primary productivity higher than any other estuarine habitat (Odum et al. 1982). Insect herbivores and tree crabs sometimes harvest a significant fraction of leaf production of mangroves, but most of the primary production of this ecosystem finds its way into detritus food chains. Substantial export of particulate and dissolved organic materials to the open waters of neighboring estuaries also occurs.

Mangrove forests are extremely valuable ecosystems. They stabilize shorelines and protect them against hurricane damage. They provide habitat for valuable fish and shellfish. They are a source of wood, tannins, and other tree products in many areas. Mangrove forests typically have a distinctive fauna of land birds. In Florida, several West Indian birds reach their northernmost breeding localities in mangrove areas. Mangroves are often the sites of major breeding water bird colonies. The only colony of magnificent frigatebirds in Baja California, for example, breeds in a small mangrove forest area in Magdalena Lagoon. In Florida, the American crocodile is also largely confined to this habitat.

Large areas of mangrove forests can be killed by changes in water levels induced by diking, dredging, and impoundment. Oil spills can kill mangroves as well, by coating the surfaces of the aerial roots that carry on gas exchange. The use of herbicides

as forest defoliants in Vietnam revealed that mangroves are unusually sensitive to these chemicals. More than 1,000 square kilometers of mangrove forests were killed by this spraying.

About 1,900 square kilometers of mangrove swamps occur in Florida, with only a small fraction of the original swamps having been lost (Odum et al. 1982). Elsewhere in the world, however, large areas of coastal mangrove swamps have been destroyed. Cutting of mangroves for wood is heavy in areas such as Baja California, where wood for fuel and construction is scarce. Mangroves in many parts of southeast Asia are being cut and converted to wood chips for paper pulp production (Fortes 1988). Clearing of mangrove forests to permit coastal development is also a major cause of destruction of this ecosystem. In Ecuador and the Philippines, for example, many mangrove swamps have been cleared for construction of maricultural lagoons. Elsewhere, mangrove areas have been cleared for urban, industrial, and resort development.

Seagrass beds occupy many sheltered temperate and tropical estuaries and coastal waters with sandy or muddy bottoms (Thayer et al. 1984). Seagrasses are true flowering plants that are rooted in the bottom, sometimes to a depth of 20–30 meters in clear water. In temperate regions the dominant seagrass is eelgrass, *Zostera marina,* and in tropical waters *Thalassia testudinum.* Other seagrasses and large algae often occur in seagrass beds, and the surfaces of the seagrass leaves are covered with an epiphytic community of algae and small invertebrates.

The productivity of seagrass beds is high, and the beds are home to large populations of fish and invertebrates (Thayer et al. 1975, 1984). Scallops, in particular, are bivalve mollusks that are specialized for living in seagrass beds. Some waterfowl, such as brant and tundra swans, feed heavily on seagrasses in temperate estuaries. In tropical waters, seagrass beds are home to endangered sea turtles. Much of the productivity of seagrass beds is converted into detritus that feeds into other estuarine food chains. Dredging and filling often cause high turbidity and sediment deposition that can kill seagrasses. Eutrophication, which reduces light penetration to seagrass beds, can also lead to their demise. Rakes and dredges that are used to harvest shellfish also damage seagrass beds by tearing plants loose, as do boat anchors.

ESTUARINE EUTROPHICATION

Some estuaries, such as Chesapeake Bay in the eastern United States, are suffering from eutrophication. This bay, the largest estuary in the world, contains some of the most valuable fish, shellfish, and wildlife resources in North America. The symptoms in Chesapeake Bay are similar to those observed in freshwater lakes: reduced transparency of the water, blue-green algal blooms, and oxygen depletion in the deeper waters (D'Elia 1987; Cooper 1995). Similar conditions afflict the Baltic Sea in Europe (Cederwall & Elmgraen 1990).

The rivers that flow into the bay drain a heavily populated region of about 164,000 square kilometers, and are the ultimate discharge sites for enormous volumes of urban sewage. Most of the change in transparency, however, is not due to particulate matter, such as silt and phytoplankton, washed into the estuary, but to the growth of phytoplankton within the estuary itself. Large quantities of both phosphorus and nitrogen enter the bay from sewage discharge and runoff from the terrestrial landscape, including fertilized farmland. Studies of seasonal nutrient dynamics show that nitrogen inputs are greatest in winter and early spring, when the freshwater inflow from surface runoff is highest. Because of low temperatures at this season, decomposition is low and large amounts of organic matter accumulate in the sediments. In summer, surface runoff declines, and more of the inflow comes from sewage discharges, which are relatively high in phosphorus. Higher water temperatures favor decomposition, which reduces oxygen levels at the sediment surface. Under these conditions, large quantities of phosphorus are released into the water.

The high concentrations of phosphorus in Chesapeake Bay in summer mean that it is not a limiting nutrient at this season (Fig. 18.3). Thus, summer inputs of nitrogen are critical in increasing algal growth and reducing transparency. Similar observations have been made in Long Island Sound (Ryther & Dunston 1971), and it appears that nitrogen, rather than phosphorus, is usually the key nutrient in eutrophication of coastal estuarine and marine waters.

Reduced transparency of the water is evidently responsible for the decline of eelgrass beds in Chesapeake Bay between 1960 and the early 1980s (Orth & Moore 1983). Between 1971 and 1979, at mapped study locations in the lower bay, the percentage of stations lacking eelgrass increased from less than 10% to more than 60%. Loss of eelgrass is probably a contributing factor to the declines of wintering tundra swans and ducks of several species (Orth & Moore 1983; Terborgh 1989). Deoxygenation of the deeper water of the bay has also become catastrophic at times (Seliger et al. 1985), and has probably contributed to the decline of striped bass, shad, oysters, and other valuable aquatic wildlife.

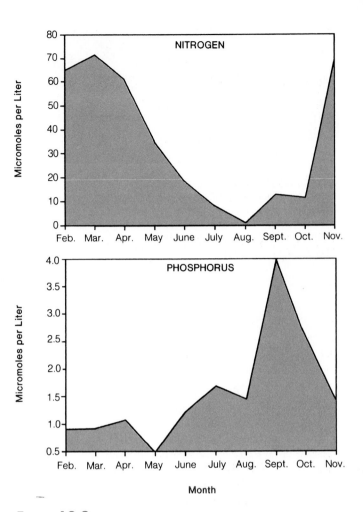

FIGURE 18.3

In Chesapeake Bay, phosphorus is abundant during the summer production period, making nitrogen the critical limiting factor in primary productivity and eutrophication.
(Modified from D'Elia 1987)

SHALLOW MARINE WATERS

Shallow marine waters derive a high productivity from nutrient inflows from the adjacent land, but they also receive the detrimental impacts of human activity in coastal areas. Coastal waters also possess several distinctive ecosystems, such as the complexly zoned systems of the intertidal and the kelp forests of cold, shallow waters.

Intertidal and Subtidal Environments

Intertidal and subtidal habitats range from sandy beaches and exposed rocky coastlines to protected muddy shores. Many kinds of invertebrates and fish occupy these habitats, some of them of considerable commercial value. Various species of clams occur in sandy and muddy substrates, and mussels and abalones on rocky substrates. Lobsters, crabs, and sea urchins frequent subtidal areas. Larval fish of many species occur in these shallow waters as well. Other interesting and ecologically important species also occur

THE SAN FRANCISCO BAY ESTUARY: A CASE STUDY

T he Sacramento and San Joaquin Rivers drain the Central Valley of California and join to discharge into San Francisco Bay, creating the largest, and now the most intensively manipulated, estuary system on the North American Pacific Coast (Nichols et al. 1986; Herbold & Moyle 1989). San Francisco Bay itself is about 100 kilometers from north to south. Farther inland, the Sacramento-San Joaquin delta is a maze of channels and islands extending from Sacramento over 100 kilometers south and 50 kilometers west to San Francisco Bay. Altogether the bay and delta cover an area of nearly 6,000 square kilometers (Reading Fig. 18.1A).

About 80% or more of the freshwater inflow now comes from the Sacramento River system. Large withdrawals of water for irrigation in the San Joaquin Valley and urban use in southern California are taken from the south end of the delta. In winter, flows in all channels normally move down-channel, toward San Francisco Bay. In summer the withdrawals at the south end of the delta exceed the inflow there, reversing flows in certain channels and causing a large cross-delta movement of water from northern and central portions of the delta. These withdrawals substantially reduce the total inflow into the lower delta and San Francisco Bay.

Originally, this estuary system included about 2,200 square kilometers of tidal marsh, of which only about 5% remain today as a result of filling and drainage

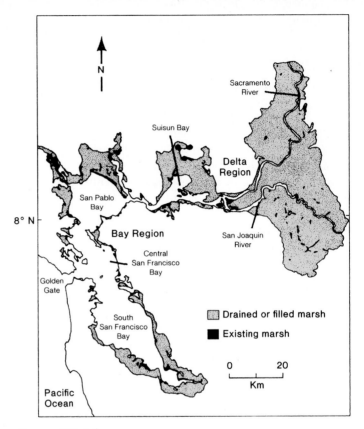

READING FIGURE 18.1A

San Francisco Bay and the Sacramento-San Joaquin Delta form the largest and most heavily impacted estuary on the Pacific Coast of North America.

(Reprinted with permission from F. H. Nichols, et al., 1986, "The Modification of Estuary" in *Science*, 231:567–73. Copyright © 1986 American Association for the Advancement of Science, Washington, DC.)

in these habitats. Rocky intertidal zones, often dotted by tide pools, are one of the temperate environments of greatest biological diversity.

Exploitation of shellfish in intertidal and shallow subtidal areas has severely depleted many populations of clams, abalones, crabs, and lobsters (Weber 1986). Both commercial and non-commercial fishing, along with pollution and physical disturbance of coastal areas, have contributed to these declines. In California, for example, annual commercial harvest of abalones exceeded 1,800 tons as recently as 1968; now less than 400 tons are taken (Tegner 1989). A similar fate has befallen the California spiny lobster, showing the difficulty of managing species attractive for both commercial and sport fishermen in a heavily populated coastal region. Human recreational use of rocky intertidal habitats also can lead to major changes in the biotic community, particularly the under-rock assemblage of invertebrates (Addessi 1994).

Kelp Forests

Kelp forests, dominated by the giant *Macrocystis* species, occur in coastal ocean waters along the Pacific coasts of North and South America and around various islands of the southern oceans (Foster & Schiel 1985). Beds of smaller kelps also occur in cool or cold waters at other locations in the northern and southern oceans. Kelp forests are home to a rich variety of marine invertebrates, as well as many fish, some associated primarily with the kelp canopy, others with the bottom (Fig. 18.4).

Kelp forests are of considerable economic value. In California, up to 170,000 tons of kelp are harvested annually to obtain algin, a substance used in foods and pharmaceuticals (Foster & Scheil 1985). Shellfish such as abalone, lobster, and sea urchins are harvested in kelp forests, and in southern California about 70% of the coastal catch of sport fish comes from kelp forest areas. The red sea urchin, its roe largely exported to Japan, has recently become a major fishery along the North American Pacific coast.

READING FIGURE 18.1B

The Sacramento-San Joaquin Delta in California is a major wintering area for tundra swans.
(U.S. Fish and Wildlife Service photo by Glen Smart)

list of these species numbered 255, and since then other species, such as the European green crab, *Carcinus maenus*, and an Asian clam, *Potamocorbula amurensis*, have arrived. The crab, a voracious predator, is almost certain to have major impacts on populations of a number of molluscs and crustaceans (Grosholz & Ruiz 1995).

The delta region is home to many mammals and is a major wintering area for swans, geese, and ducks. About 10% of the wintering waterfowl in California occur in the delta. These include about 30,000 to 38,000 tundra swans, roughly 85% of the California wintering population (Reading Fig. 18.1B). More than half a million pintail ducks winter in the delta, and over 30% of the statewide hunting harvest of pintails comes from delta counties.

The burgeoning human population of southern California draws much of its water from northern California through the aqueducts of the California Water Project. The growing demand for water in southern California leads to frequent demands for increased diversion from the delta system, and for projects such as a peripheral canal, which would carry water from the Sacramento River around the delta to the head of the aqueduct system leading southward. During the drought years of the late 1980s, plans for protecting the ecological values of the delta ecosystem by retaining freshwater inflows were bitterly criticized by some for ignoring the needs of the southern California population.

(Nichols et al. 1986). About 100 species of invertebrates and 20 species of fish have been introduced to the estuary; one of the latter, the striped bass, is now the most important sport fish in the system. Sedimentation from hydraulic gold mining in the 1800s, and pollution from modern urban and agricultural sources have exerted heavy impacts on the estuary. Due to water diversions for urban and agricultural use, freshwater inflow to San Francisco Bay is less than 40% of its original level.

The native biota of the estuary has been profoundly changed. Six species of salmon originally passed through the delta; only two, the steelhead and chinook, are still common. Four spawning races of the chinook occur in the Sacramento River, but two of them, the winter and spring runs, are on the verge of extinction. American shad and striped bass, introduced to the delta in the late 1800s, are migratory fish that have fostered major sport fisheries; they too have declined with reduction in freshwater flows. The white sturgeon, a native migratory species, also depends on adequate water flows. The delta proper is also home to two endemic fish, the delta smelt and the Sacramento splittail.

A host of exotic invertebrates have become established in San Francisco Bay, brought there mainly in ballast water of ocean ships (Hedgepeth 1993). In 1979, the

In some areas, such as the coast of southern California, kelp forests have experienced severe declines. Some of these declines may be due to coastal marine pollution, and others to incursions of warm ocean water to which kelps are intolerant. In California, declines have been attributed in some cases to intense grazing by sea urchins, which may be sustained at unusually high densities by an ability to assimilate dissolved organic substances introduced into coastal waters by sewage discharges.

Recent studies have revealed a fascinating set of relationships among sea urchins, kelp, and the sea otter, a major sea urchin predator. Sea otters were originally widespread throughout the North Pacific and Bering Sea, south to northern Japan and central Baja California, Mexico (Reidman & Estes 1988). They were harvested intensively in the 18th and 19th centuries by Russian fur hunters. The total population was reduced to perhaps 1,000 or fewer individuals (Estes & Duggins 1995), and the species was completely eliminated from much of its range.

FIGURE 18.4

Kelp forests are a marine habitat important for many coastal fish.
(Photo courtesy of Dan Reed)

Under protection, the sea otter is now increasing in numbers and expanding its range. The total population is now estimated to be more than 150,000 animals, mostly in Alaskan waters.

Sea otters reach a weight of about 20 to 23 kilograms, and in favorable coastal waters their density can reach 20 to 30 animals per square kilometer. They are voracious predators, feeding on sea urchins, mollusks, crustaceans, and fish, more or less in that order of preference (Riedman & Estes 1988). Individual sea otters consume about 20%–23% of their body mass per day, so that a population in favorable habitat consumes about 35 tons of marine animal life per square kilometer annually.

Studies of the effects of sea otter predation suggest that in many places the species is a **keystone predator,** an animal that can cause an almost total restructuring of the ecosystem. In the Aleutian Islands, for example, sea otters are absent from some islands and present around others (Estes & Palmisano 1974). Around the island of Shemya, where otters are absent, kelp beds are absent and sea urchins large and abundant, ranging from 400 animals per square meter in shallow water to 45 per square meter at depths of about 18 meters. Other invertebrates, such as chitons, mussels, and barnacles are also abundant. In contrast, around Amchitka Island, where otters have been present for many years, kelp beds are dense and sea urchins small and scarce. Other invertebrates are also scarce. Along the southern Alaska coast and in the Aleutians, where sea otters have appeared as a result of population growth and range expansion, transformation of the coastal ecosystem has been observed in only a few years (Duggins 1980; Estes & Duggins 1995). When otters appear, sea urchin populations are decimated, and dense, single-species stands of kelps of the genus *Laminaria* appear. Although the keystone predation hypothesis may be too simplistic for conditions throughout the potential sea otter range (Foster & Schiel 1988), sea otters are highly influential predators in the coastal marine environment.

One of the sea otter populations that survived the fur hunters was a group of about 50 to 100 animals off the Monterey Peninsula in central California. This population, too, has been growing and is now estimated at about 1,300 animals. The population has also spread southward, to the dismay of fishermen whose livelihood is lobster, abalone, and sea urchin fishing. Some biologists believe that a recent slowing of the spread of the sea otter has been due to deliberate killing of animals. In spite of the success of the species on its own, under protection from commercial hunting, efforts have been made to introduce the sea otter to San Nicholas Island, one of the California Channel Islands. The purpose of this introduction was to establish a population in an area distant from the present California range, in order to reduce the chance of extinction of the species due to a massive oil spill or other catastrophe. Commercial fishermen operating in the vicinity of San Nicholas Island have opposed this introduction because of the probable severe impact of the sea otter on shellfish populations.

CORAL REEF ECOSYSTEMS

Coral reefs are the tropical rain forests of the sea—marine ecosystems of high productivity and enormous biotic richness. Reefs are also, in a sense, oases of productivity in tropical oceans that are otherwise low in fertility. Their productivity results in large measure because the reef is a flowing-water system. Water from the open ocean is continually flowing across the reef to the lagoon, ultimately to return to the open ocean through deep channels that cut through the reef. As water flows across the reef, however, corals and other reef animals capture and filter living and dead particulate matter from the water, thus harvesting organic food and its contained nutrients from a wide area of ocean.

The nutrients trapped by corals and filter feeders contribute to primary production within the reef ecosystem. The producers themselves are often not obvious. Most reef primary production occurs in **algal turfs,** which are low mats of diverse species of algae, some of them encrusting **coralline algae** that appear more like coral rock than green plants. Other types of algae live in the surface layers of rock, and still others, known as **zooxanthellae,** are single-celled symbiotic organisms that live within the cells of the coral animals. Altogether, these producers contribute to an average net primary production that is higher than that of the oceanic upwelling zones, and only slightly below that of estuaries.

The complex structure of reefs also contributes to the biotic richness of the reef ecosystem. Reefs exhibit a sequence of zones from their ocean face to the backside lagoon (Fig. 18.5). Massive corals form a subtidal buttress zone rising from deep water on the ocean side. These buttresses receive the surge and wave impacts of the open ocean, and by dispersing this energy, protect the inner zones of the reef. Above and behind the buttress zone is often an intertidal coral-alga ridge, which has few living corals and is covered by fragments of coral rock thrown up from the buttress zone by heavy wave action. The coral-alga ridge is home to various encrusting organisms and animals that can hide beneath rock fragments at low tide (Fig. 18.6). Behind the coral-alga ridge are zones of encrusting corals and small coral heads, as the lagoon water deepens. In the deeper waters, massive corals can once again grow, but in this case accompanied by a host of marine organisms that require sheltered waters (Fig. 18.7). Finally, many reefs exhibit sheltered lagoons with a sand-and-shingle bottoms. Each of these zones has a distinctive set of physical conditions and a unique physical structure to which organisms must be adapted. The sequence of diverse microhabitats from exposed outer to sheltered inner sides of a reef is analogous to that from the exposed canopy top to the sheltered ground surface in a tropical rain forest.

Fringing or barrier reefs that lie close to continents or islands face growing problems of pollution and physical damage from the activities of human populations. Worldwide, the most serious threat to reef ecosystems comes from siltation due to soil erosion on nearby land areas. Silt deposition kills most corals and filter-feeding invertebrates by interfering with their systems of food capture and respiration. Locally, sewage discharge has a similar effect. In Kaneohe Bay, Hawaii, for example, sewage discharges led to replacement of pristine, coral-dominated reef by a community of green bubble algae and filter-feeding invertebrates (Hunter & Evans 1995). Marine oil pollution can have serious effects on reefs, especially in heavily polluted areas such as the Red Sea and the Persian Gulf (Loya & Rinkevich 1980). Oil appears to stimulate premature shedding of the planula larvae of

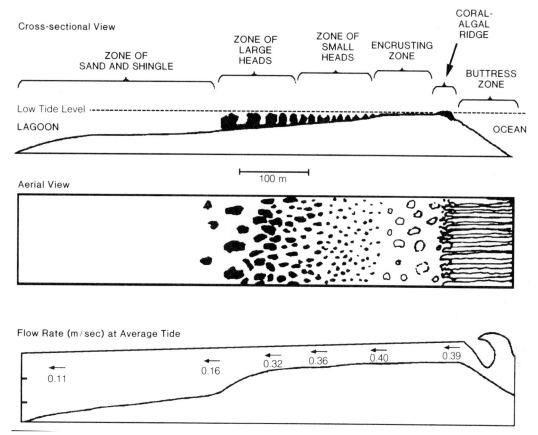

Cross-sectional View

ZONE OF SAND AND SHINGLE

ZONE OF LARGE HEADS

ZONE OF SMALL HEADS

ENCRUSTING ZONE

CORAL-ALGAL RIDGE

BUTTRESS ZONE

Low Tide Level

LAGOON

OCEAN

100 m

Aerial View

Flow Rate (m/sec) at Average Tide

0.11   0.16   0.32   0.36   0.40   0.39

FIGURE 18.5

Zonation and water flow across a coral reef as illustrated by Japtan Reef, part of Eniwetok Atoll in the Marshall Islands of the western Pacific. (Modified from Odum & Odum 1957)

FIGURE 18.6

The coral-alga ridge of a coral reef is typically covered by fragments of coral rock thrown there by storm waves.

(Photo by G. W. Cox)

FIGURE 18.7

Sheltered waters on the inner side of a coral reef are home to many kinds of corals, fish, and other animals.

(U.S. National Park Service photo by M. W. Williams)

corals, retards growth and reproduction, and inhibits colonization of chronically polluted areas. Mining of the coral rock of reefs for construction material and for production of cement, dredging of shipping channels, and the filling of shallow coastal waters to create land for airports and other purposes increasingly take a toll of reef areas in heavily populated coastal areas. On a smaller scale, physical damage to reefs from boat anchors and human trampling also affects many heavily visited reefs.

Fringing reefs, such as the Great Barrier Reef of Australia, are increasingly subject to eutrophication (Bell & Elmetri 1995). Continental runoff from deforested and agricultural lands has enriched the Great Barrier Reef lagoon, increased phytoplankton abundance in the lagoon water, and led to the invasion of coral gardens by dense growths of filamentous algae and macroalgae.

Because of their biotic richness and productivity, reefs are subject to heavy exploitation, even in remote tropical waters. Exploitation has depleted populations of corals, fish, and other reef animals and modified reef ecology over large areas. Reduction of fish that are predators on sea urchins may lead to their explosive increase, which in turn may lead to reduction in algal cover and erosion of the reef framework due to sea urchin grazing activity (Roberts 1995). In the western Pacific, collecting reef fish for aquarium and Asian gourmet dining markets is a major form of reef exploitation. The common collecting technique is shooting a solution of sodium cyanide into masses of large corals to stun the fish inhabitants so they can be captured. The coral animals are more sensitive to cyanide than are fish, and are often killed. The fish themselves frequently are killed or weakened, and many die soon after they are captured or sold. This very wasteful practice has severely damaged many reefs in Philippine waters, and is now spreading over a wide region of the western Pacific.

One of the most widespread threats to reefs in the Pacific and Indian Oceans in recent decades has been the crown-of-thorns starfish, *Acanthaster planci* (Fig. 18.8). This starfish reaches a diameter of half a meter, and possesses 13 to 16 arms with sharp spines up to 5 centimeters in length. The crown-of-thorns is normally scarce, with densities of a few individuals per square kilometer. This voracious coral-eating starfish has exhibited devastating outbreaks that have caused the death of most of the large, reef-building *Acropora* corals over vast areas. During outbreaks, densities can reach 100,000 starfish per square kilometer. Full recovery of reefs from such damage requires about 20 to 40 years. In effect, the crown-of-thorns starfish is a keystone species in the coral reef ecosystem. In this case, the role of man as a causal agent is still uncertain.

Major outbreaks of the crown-of-thorns were first noted in 1962 at Green Island Reef, in the Australian Barrier Reef. From here, over the next 12 years, outbreaks spread throughout much of the Pacific and Indian Oceans. In the late 1970s, when it was thought that the crown-of-thorns had run its course, a new wave of outbreaks began—one that continued into the late 1980s. Many reefs were struck again by this second wave, with Green Island being ravaged again in 1981, for example.

FIGURE 18.8

The crown-of-thorns starfish, a voracious predator on stony corals, has shown long-term outbreaks in the Pacific and Indian Oceans in recent decades.

(Photo by David Zoutendyk)

Several theories of the cause of crown-of-thorns outbreaks have been offered. Some, such as pesticide pollution that kills predators of small starfish or physical damage to reefs that favor settlement of larval starfish, have received no substantial scientific support. One marine ecologist (Endean 1982) has suggested that widespread depletion of a large carnivorous snail, the triton snail, *Charonia tritonis,* and perhaps other such predators, has freed the crown-of-thorns from natural biological control. The triton can kill and consume crown-of-thorns starfish, but many marine ecologists doubt that this predator was originally abundant enough to have been a strong control on the starfish.

Another theory is that unusually high survival of larval starfish during their planktonic life permits the settlement of juvenile starfish on reefs in abnormally large numbers. Birkeland (1982) postulated that heavy rainfall events might wash large quantities of nutrients into coastal waters, fostering a bloom of plankton that provide a rich food source for the crown-of-thorns larvae. He noted that outbreaks in several instances occurred three years after extreme rainfall events in Samoa, Guam, and Palau, and that outbreaks were also more frequent around islands with nutrient-rich volcanic soils than around atolls formed of nutrient-poor coral sand and shingle. The pattern of spread of outbreaks also suggests that some factor has favored unusual survival of crown-of-thorns starfish larvae (Moran 1987). Recent observations of increased phytoplankton abundance in the Great Barrier Reef lagoon are consistent with this hypothesis, as well (Bell & Elmetri 1995). Other workers, however, have noted that outbreaks do occur on isolated reefs that are not near any such source of nutrients.

GLOBAL CHANGE AND COASTAL MARINE ECOSYSTEMS

Inasmuch as global climatic warming will lead to warmer water and rising sea level (*See* Chapter 23), impacts on coastal marine ecosystems are likely to be varied and substantial (Smith & Buddemeier 1992). Coastal erosion, for example, might lead to sediment deposition in subtidal environments. Increased frequency and intensity of hurricanes and typhoons is likely, along with significant shifts in ocean currents.

Coral reefs and kelp forests may be particularly sensitive to global changes in temperature, CO_2, and ultraviolet radiation. On one hand, warmer water may increase the frequency of **coral bleaching,** which involves the loss of their zooxanthellae, and may lead to mortality (Smith & Buddemeier 1992). Warm currents that push farther poleward may eliminate kelp forests from portions of their present range. A generally warmer ocean, however, should expand the latitudinal limits for coral growth. On the other hand, changed ocean currents resulting from warming and sea level rise might bring cold water into some present reef areas, leading to their deaths. Increased ultraviolet radiation might also cause coral death in shallow waters or during low tidal exposure.

ESTUARINE AND MARINE PRESERVES

The marine environment is the last of the major divisions of the biosphere to receive formal protection through the designation of parks, preserves, and sanctuaries. The coastal marine environment has now begun to receive such protection. The United States has initiated systems of national estuarine and marine sanctuaries. Many other countries have also begun to designate marine parks and preserves.

The United States National Marine Sanctuary and National Estuarine Research Reserve Programs were established in 1972. By 1995, eight National Marine Sanctuaries had been designated (Foster & Archer 1988). One of these, the sunken wreck of the U.S.S. *Monitor,* a Civil War ship, is a historical monument. Four of the remaining sites preserve coral reefs, three in the southeastern states and one in American Samoa. The remaining three sites, near the California Channel Islands, surrounding the Farallones Islands off San Francisco, and around several Hawaiian islands, preserve coastal water areas used heavily by marine birds and mammals. The National Estuarine Research Reserves emphasize long-term monitoring and ecological research. Seventeen of these reserves have been designated at representative locations along the Atlantic, Gulf, and Pacific coasts, and in Hawaii and Puerto Rico. Additional sites are being considered for addition to this system.

Many other countries have begun to designate protected coastal and marine areas of various types. By the late 1980s, about 430 marine preserves had been created in 69 countries (World Resources Institute 1988). Australia, for example, has es-

tablished the Great Barrier Reef Marine Park, covering an area of 350,000 square kilometers of ocean and extending along 2,000 kilometers of coastline (Morris 1983). This park is zoned in a comprehensive way, with activities such as fishing being permitted in some areas but not in others. In many cases, however, designated preserves lack effective protection and active management. In other cases, the primary purpose of the preserve is tourism, and heavy use by visitors leads to physical damage and overfishing.

SUMMARY

1. Estuaries occur where rivers discharge into the ocean, creating waters with intermediate salinity and a rich input of nutrients and detritus. Estuaries possess a diverse array of producers, and are the most productive marine environment. Estuaries are being severely impacted in many locations by freshwater diversion, eutrophication, and introduced exotics.

2. Along tropical coastlines, estuarine environments often possess mangrove forests that are essential to the maintenance of estuarine productivity and the protection of the coast against storm damage.

3. Coastal marine ecosystems such as the intertidal zone, shallow subtidal waters, and kelp forests are environments in which keystone predators may play major roles, and in which human exploitation has produced major changes in plant and animal composition.

4. Coral reefs, the most diverse marine ecosystems, possess a variety of inconspicuous primary producers, such as zooxanthellae and coralline algae, but also harvest the productivity of the adjacent ocean by filtering and trapping organic matter from the water flowing across the reef.

5. The need for protection of critical marine ecosystems is leading to the establishment of protected areas such as the United States National Marine Sanctuaries and National Estuarine Research Reserves.

QUESTIONS FOR DISCUSSION

1. Do you think that the ecological and economic benefits of freshwater flow into coastal marine waters often outweigh the economic benefits of diversion of this water for urban and agricultural use?

2. Considering past and present patterns of human exploitation of sea otters, fish, shellfish, kelp, and other marine life along the California coast, what set of ecological conditions do you think should be the goal of conservation management?

3. Considering the fact that very little extinction of marine invertebrates and fish has been recorded, do you think that an extensive system of marine sanctuaries, reserves, and parks is desirable?

SUGGESTED READING

Carlton, J. T. and J. B. Geller. 1993. Ecological roulette: The global transport of nonindigenous marine organisms. *Science* **261**:78–82. A summary of the patterns of introduction of marine organisms to new areas in ballast water of ships.

Estes, J. A. and D. O. Duggins. 1995. Sea otters and kelp forests in Alaska: Generality and variation in a community ecological paradigm. *Ecological Monographs* **65**:75–100. A comprehensive account of the ecosystem reorganization induced by sea otter invasion of coastal marine areas.

Ray, G. C. 1991. Coastal-zone biodiversity patterns. *BioScience* **41**:490–96. A discussion of concepts of landscape and seascape ecology as they apply to the zone where the continent meets the ocean.

five

Special Problems of Aquatic Ecosystems

L ike terrestrial ecosystems, aquatic ecosystems possess special problems. Marine mammals and seabirds, for example, have been among the most carelessly and severely exploited animals on earth. Efforts to protect and manage them have likewise become among the most controversial issues of international conservation.

chapter

19

Marine Mammals and Birds

F ew groups of animals have held greater attraction for humans than marine mammals. Aristotle and Pliny the Elder wrote accounts of the natural history of dolphins and whales in ancient times. The greatest attraction to marine mammals, however, soon became economic: oil, baleen, pelts, and meat. The lure of gain from whaling and sealing drew men in sailing ships to the wildest regions of the Arctic and Antarctic Oceans—many to their deaths—and sustained commercial whaling until near the end of this century. More recently, the seeming intelligence and friendliness of certain marine mammals have gained them increasing prominence in literature, art, and folklore. These characteristics have also helped to give the group as a whole a special prominence in the efforts of animal protectionists and conservation ecologists. Some marine mammals have, in fact, become rallying standards for popular conservation efforts.

Marine mammals belong to four orders (Table 19.1). The cetaceans include 78 species of whales, dolphins, and porpoises (Gaskin 1982; Perrin 1989). The pinnipeds comprise 33 species of seals, sea lions, and the walrus (Riedman 1990; Reeves et al. 1991). The sirenians, so-named because of their fancied relation to the sirens of Greek mythology, include four living species of manatees and dugongs. Two species, the polar bear and sea otter, belong to the Carnivora, an order containing the major families of flesh-eating land mammals. All of these groups contain species that are or were seriously endangered.

Seabirds range widely, visiting the most distant parts of the world oceans, and the impressive nesting colonies that they form have also attracted human interest and exploitation. They likewise belong to several orders, most importantly the Sphenisciformes (penguins), Procellariiformes (albatrosses, shearwaters, and petrels), Pelecaniformes (pelicans, boobies, gannets, cormorants, and frigate birds), and Charadriiformes (auks, puffins, gulls, and terns).

We shall consider first the history of whaling and the current status of efforts to protect and manage populations of the great whales. Then we shall examine conservation issues for other groups of marine mammals and seabirds.

Humans have long hunted marine mammals, but now we are their growing competitors for the productivity of ocean ecosystems.

TABLE 19.1

Major Groups of Living Marine Mammals

GROUP	NUMBER OF SPECIES	NUMBER ENDANGERED	
		U.S. LISTING	IUCN LISTING
Order Cetacea			
Whales, Dolphins, and Porpoises			
Baleen Whales	11	7	4
Sperm Whales	3	1	
Beaked and Bottlenose Whales	19		
Beluga and Narwhal	2		
Pilot and Killer Whales	6		
River Dolphins	5	1	2
Porpoises	6		
Marine Dolphins	26		
Order Pinnipedia			
True Seals, Sea Lions, Fur Seals, and Walrus			
True Seals	18	3	
Sea Lions and Fur Seals	14		
Walrus	1		
Order Sirenia			
Sirenians			
Manatees	3	2	
Dugong	1	1	
Order Carnivora			
Carnivores			
Sea Otter	1	1 (partial)	
Polar Bear	1		

TABLE 19.2

Distribution and Estimated Original and Present Populations of Great Whales

SPECIES	RANGE	MAXIMUM LENGTH (M)	ORIGINAL POPULATION	PRESENT POPULATION
Baleen Whales				
Gray	N. Pacific	14	(24,000)	22,000
Minke	Cosmopolitan	10	(350,000)	(900,000)
Sei	Cosmopolitan	18	(105,000)	(25,000)
Bryde's	Cosmopolitan	15	(92,000)	(90,000)
Blue	Cosmopolitan	30	(196,000)	(9,000)
Fin	Cosmopolitan	26	(464,000)	(119,000)
Humpback	Cosmopolitan	15	(120,000)	(10,000)
Bowhead	Arctic Ocean	18	(55,000)	8,000
N. Right	N. temperate and polar oceans	18	(50,000)	350
S. Right	S. temperate and polar oceans	18	(100,000)	1,500
Sperm Whales				
Sperm	Temperate and tropical oceans	18	(2,770,000)	(1,810,000)

Source: Compiled from various sources. Population figures in parentheses represent rough estimates for present populations, and no more than educated guesses for original populations.

WHALES AND WHALING

The great whales, the principal objects of commercial whaling, are derived from an early evolutionary line allied with ungulates. Whales first appeared about 70 million years ago, following the disappearance of giant marine reptiles. By 30 million years ago, the two modern groups of whales—toothed and baleen whales—had appeared. Of the modern great whales, one, the sperm whale, is a toothed whale and the remaining ten are baleen whales (Table 19.2).

The sperm whale, a medium-sized species, is a toothed whale that feeds on larger fish and squid (Fig. 19.1). The largest males rarely exceed 16 meters in length, and few females exceed 11 meters. Sperm whales largely confine their activities to the open temperate and tropical oceans. They are deep divers, and can remain submerged for up to two hours. Adult females travel in breeding herds of 10 to 15 animals, which are visited for short periods by individual bulls for mating. Nonbreeding males travel in bachelor herds. The present population is thought to be of the order of 1.8 million animals, or about two-thirds its

FIGURE 19.1

A sperm whale is butchered at a whaling station in Taiji, Japan. The sperm whale, largest of the toothed whales, is an open-ocean, deep-diving animal that feeds primarily on squid.

(Photo by Howard Hall, courtesy of Stephen Leatherwood)

FIGURE 19.2

Iceland continued to take a scientific harvest of small numbers of fin whales and other species until 1989.

(U.S. Fish and Wildlife Service photo by G. Buterbaugh)

prewhaling estimated number. Two related species, the dwarf and pygmy sperm whales, are poorly known and probably rare forms.

The baleen whales lack true teeth, and feed mostly on krill, small fish, and other small marine animals, which they capture by use of the specialized **baleen apparatus** suspended from their upper jaws. This structure, made of fibrous plates, acts as a filter. Several of the baleen whales have "pleated" throats that are enormously expandable, enabling them to take a large volume of water, containing food organisms, into their mouth cavity. The water is then expelled through the baleen plates, so that the food organisms are retained and swallowed. Others, such as the gray and right whales, take food-rich materials into the baleen cavity by sucking or skimming. Baleen whales seem to have a promiscuous breeding system and live either alone or in herds containing varying numbers of individuals of one or both sexes.

Baleen whales are varied in size and distribution. They frequent coastal waters and upwelling zones where their food is abundant. The largest is the blue whale, which reaches 30 meters in length and 160 tons in weight—the largest animal ever to live on earth. Whaling has reduced this species to less than a tenth of its original number. It is also one of the most endangered of the great whales, with its total number probably being 5,000 to 10,000. Like all of the baleen whales, it is migratory, moving to higher latitudes in the summer. The fin, northern and southern right, bowhead, sei, humpback, Bryde's, and gray whales are progressively smaller species, all of which were hunted heavily and are greatly reduced in abundance compared to their original numbers. The smallest of the commercial baleen whales is the minke whale, which is just over 9 meters in maximum length and about 10 tons in maximum weight. Minke whales, the least persecuted of the baleen whales, are believed to number about 900,000, and are probably more abundant now than when commercial whaling began. The smallest baleen whale, the pygmy right whale, is a rare, poorly known species that reaches a length of about 6 meters.

The objectives of whaling have been to obtain meat, oil, baleen, and other materials (Allen 1980). Aboriginal peoples have long sought whales for subsistence harvests of meat, blubber, and oil, and this type of whaling is still practiced by a few groups, such as the Inuit of Alaska, Canada, and Greenland. Early commercial whaling concentrated on oil for use in lamps and candles and for lubrication. The sperm whale yielded oil that was prized for use in oil lamps and as a high-quality industrial lubricant. **Spermaceti,** a waxy material from the oil deposit in the whale's head, was used in making the finest candles. A ton or more of spermaceti could be obtained from a large male sperm whale. Another substance, **ambergris,** formed in the intestine of the sperm whale, was used as a chemical stabilizer in perfumes. In modern time, sperm whale oil was used as an industrial lubricant, but the meat was used mostly as animal feed. Baleen whales were sought originally for their oil, used in lamps, and for baleen. A large right whale could yield over 300 barrels of oil. The cartilaginous baleen apparatus was also used to manufacture corset stays, buggy whips, and other products. In recent years, the oil of baleen whales was used in margarine and cosmetics. The meat of some baleen and toothed whales has become a popular food in Japan.

Commercial whaling began in the 12th century, when Basque fishermen began hunting the right whales in the Bay of Biscay, off the northern coast of Spain (Gambell 1976). Whaling spread to several other North Atlantic locations in the 17th and 18th centuries, and to the rest of the world oceans in the 19th century. The greatest destruction of whales, however, has occurred during the modern era. Modern whaling dates from the development of steam-powered ships and the harpoon gun in the 19th century. These advances enabled whalers to hunt whales effectively in the most remote areas of the oceans. Modern harpoon guns can fire either a "cold" harpoon, lacking an explosive charge, or a harpoon with an explosive charge that kills or stuns the animal, facilitating its capture. The explosive harpoon is regarded as the more humane version, since it kills the animal quickly. In recent years, use of the explosive harpoon was required for all species except the small minke whale, this exception reflecting the fact that an explosive charge damaged such a high percentage of the flesh of this species.

Whaling harvests increased until the mid-1900s. By tonnage, the harvest peaked in 1930–1931, when 3.6 million tons were taken. The numbers of whales taken continued to increase until the early 1960s, when harvests reached about 90,000 animals per year. This pattern reflected the progressive depletion of whales from large to small. Beginning in the 1930s, the average size of the animals taken began to decline, as the largest species and individuals were overexploited. In 1903, the average whale taken weighed 66 tons, by 1970 only 23 tons.

Concern over the rapid decline of whale stocks led to the formation of the **International Whaling Commission (IWC)** in 1946 (Gambell 1990). The members of this commission are nations with territorial waters containing whales, and include most whaling nations, as well as many nonwhaling countries. In 1979, the commission designated a sanctuary zone in the Indian Ocean, giving partial protection to several species, and began to set quotas on whale harvests. Unfortunately, these quotas were not always respected by member nations of the commission. In the 1960s, for example, the former Soviet Union took numbers of right, humpback, and blue whales far in excess of their quotas, and reported false data to the commission. In the 1970s and early 1980s, as whale numbers continued to decline, the commission reduced quotas rapidly, and shifted them more and more onto the minke whale, which had not been heavily exploited because of its small size. This trend culminated in adoption of zero quotas beginning in the 1984–1985 whaling season. In 1994, a Southern Ocean Whale Sanctuary, covering oceans south of 40°S, was designated.

After the IWC set zero quotas, Japan, Norway, and Iceland continued to take small numbers of several species, many in so-called "scientific" harvests (Fig. 19.2). Iceland withdrew from the IWC, but a boycott of Icelandic fish products forced it to abandon scientific harvests in 1989. Norway has continued to take minke whales—301 in the 1994–1995 season—in defiance of the IWC moratorium. Other illegal whaling evidently still occurs. DNA tests of whale meat in Japanese markets in 1994, for example, revealed that some came from humpback and fin whales. As of 1995, the IWC has continued to maintain zero quotas.

Due to the great difficulties of estimating the numbers of diving marine animals, the present populations of the great whales are only roughly known, and their original numbers can only be guessed at (*See* Table 19.2). Even so, the present numbers of all species except the northern right whale appear to be at least a few thousand, and their recovery appears to be possible if the moratorium on harvests is indeed respected. Some economists believe that once terminated, whaling will be uneconomical to resume due to the high costs of outfitting vessels for the enterprise and the fact that alternate sources of oil and meat will have become entrenched in the marketplace. Continued interest in whaling by several countries, however, indicates that this may not be true.

Figure 19.3

Pantropical spotted dolphins are one of the species that swim in association with schools of yellowfin tuna in the eastern tropical Pacific.

(U.S. National Marine Fisheries Service photo)

Small Cetaceans

Several species of small cetaceans have now become more endangered than most species of great whales, and once-important populations of many species have been extirpated or threatened with extirpation (Brownell et al. 1989; Wursig 1989). Annually, more than 150,000—perhaps as many as 500,000—small cetaceans are killed, deliberately or incidental to other fishery activities. Others are captured alive for research or display (Meith 1984). Populations of several species have suffered from depletion of their foods or disturbance of their habitat. The International Whaling Commission has generally ignored the conservation issues of small cetaceans, except for certain populations of bottlenose and killer whales that have been taken by commercial whalers. Hunting some species of small cetaceans for meat has been carried out for centuries or millennia. North American and Greenland Inuit hunt the beluga (white) whale and the narwhal (Hertz & Kapel 1986). In recent decades, commercial whaling vessels equipped with harpoon guns also shifted their attention to some of the smaller whales, including the killer whale and the Baird's beaked whale. Large harvests of several species of dolphins and porpoises are taken by Japan in the western Pacific. In the late 1980s, after the IWC placed zero quotas on whales, Japanese hunters severely overexploited Dall's porpoises over a three-year period. Along the coast of Chile, killing of dolphins and seals to obtain meat for baiting crab traps has nearly extirpated the Commerson's dolphin, one of the least common delphinids, in this region.

Hunting harvests, however, are second in total numbers to the mortality of several other species incidental to tuna, squid, and salmon fishing. In the eastern Pacific Ocean, schools of spinner, striped, and spotted dolphins often swim above schools of tuna (Fig. 19.3). For years, tuna boats equipped with purse seines used these schools of dolphins to locate tuna, and set their nets around both tuna and dolphins. Many dolphins were

ABORIGINAL WHALING

Aboriginal whaling is still practiced for several species of the great whales. In Alaska, hunting of bowhead whales is carried out in eight villages near and above the Bering Strait (Braham 1989). Whaling by Inuit people predates western contact, and was carried out perhaps as long as 4,000 years ago (Stoker & Krupnik 1993). The original technique was to pursue whales in sealskin boats and strike them with stone-bladed spears to which skin ropes and sealskin floats were attached.

Whaling by the Alaskan Inuit uses a mix of modern, 19th century, and traditional techniques (Reading Fig. 19.1). The bowhead whales are captured with a small explosive harpoon that is thrown by hand or shot from a shoulder gun. Whaling teams operate during the spring, when the offshore ice begins to shift and retreat. At this time, movements of the ice open up long gaps, or "leads," through which the bowheads migrate. Whaling teams haul their boats, usually made of sealskins sewn over a wooden frame, out to the edge of these leads, which are up to several miles from shore. When a whale appears, the boats are launched and driven by paddle or sail to a point from which the harpoon can be thrown or shot. If the whale is taken, it is hauled onto the ice and butchered. Some whaling is also pursued in fall, during the return migration period.

During the late 1970s, concern over the decline in whale populations coincided with a sharp increase in whaling activity by the Alaskan Inuit. This increase reflected the prestige of whaling and the fact that more money was flowing into the villages from governmental programs and activities

READING FIGURE 19.1

In spring, Alaskan Eskimos capture small numbers of bowhead whales in open-water "leads," using sealskin boats and explosive harpoons thrown by hand or shot from shoulder guns.
(Photo by David Withrow, U.S. National Marine Mammal Laboratory)

such as petroleum development. Increased affluence enabled many individuals to acquire equipment and funds to support a whaling crew, in spite of a lack of whaling experience. One consequence of increase in whaling activity was an increase of the total number of whales taken from 10–15 per year to 25–50 per year. A second was an even greater increase in the number of animals struck with explosive harpoons, but not recovered. In 1977, for example, 79 animals were struck and lost. As a result, in the late 1980s Inuit whaling was restricted by quotas both on the number of animals taken and on the number struck. The most recent quota (1992–1994) allows a maximum annual harvest of 41 animals or a maximum of 54 animals struck.

The broader concern over Alaskan Inuit whaling centers on the fact that the bowhead may still be one of the most depleted of great whales. Populations in four other northern ocean areas probably amount to only a few hundred individuals each (Braham 1989). That of Alaskan waters, estimated at 7,800 individuals, is the only healthy population.

Aboriginal whaling is also conducted for minke and humpback whales along the western coast of Greenland and for sperm and some baleen whales in Indonesia. A similar form of traditional whaling was practiced for humpback whales near the Lesser Antillean island of Bequia, with an annual quota of three animals, until 1990.

trapped and drowned when the net was closed. From 1959 through 1972, for example, purse seining killed an average of 347,000 dolphins annually (Wade 1995).

Techniques were developed to permit the release of dolphins over the edge of the purse seine before it is closed and hauled aboard (Fig. 19.4). The Marine Mammal Protection Act of 1972 also placed a quota on incidental kill by United States tuna boats. Many animals were killed in purse seines by tuna boats of other countries, however, and in 1990 the total dolphin mortality incidental to tuna fishing was still over 50,000. In 1990, several tuna packing companies announced that they would not market tuna captured by setting purse seines on herds of dolphins, and that their products would be labelled "dolphin

safe." By 1995, this action had effectively closed United States and many European markets to tuna captured by setting on dolphins. In 1994, incidental mortality had declined to only 4,080 animals, mostly by Mexican and Venezuelan boats that were catching tuna for domestic consumption.

The policy of not setting on dolphins has unfortunately led to fishing strategies that tend to capture smaller, immature tuna. Heavy exploitation of juvenile tuna might cause a decrease in the total biomass of the tuna population (Punsly et al. 1994). This has led some marine ecologists to suggest that the dolphin-safe policy be abandoned, and efforts be concentrated on perfecting techniques for releasing dolphins from purse seines (Joseph 1994).

FIGURE 19.4

Mortality of dolphins in purse seining for tuna can be reduced by a "backing down" procedure in which the surface edge of the net is pushed down to allow them to escape before the net is brought on board.

(U.S. National Marine Fisheries Service photo)

Gill nets also have been a major cause of accidental mortality of dolphins and porpoises. In this case, the animals simply become entangled in the nets and drown. In the 1980s, drift gill-netting for tuna, squid, and salmon in the Pacific killed tens of thousands of small cetaceans annually, mainly right whale dolphins and Dall's porpoises. A worldwide moratorium on drift gill-netting (*See* Chapter 17) has reduced these kills.

Gill nets set in coastal waters continue to be a cause of mortality for some porpoises (Jefferson & Curry 1994). In the northwestern Atlantic, harbor porpoises have fallen victim to gill nets in the Bay of Fundy (Read & Gaskin 1988). The estimated annual mortality in these nets is about 7.5% of the minimum population, a loss close to the maximum rate of potential population growth for animals of this type. Thus, net mortality could be a major contributor to population decline. In the Gulf of California, the endemic vaquita, a close relative of the harbor porpoise, was driven to the brink of extinction by mortality in gill nets used to catch totoaba, a tuna-like fish (Brownell et al. 1989).

Several other cetaceans have become endangered as a result of fishing mortality or disturbance of their river, estuary, or mangrove swamp habitats (Meith 1984; Brownell et al. 1989). In Asia, the endemic river dolphins in the Indus and Yangtze Rivers have been reduced to a few hundred individuals or fewer, and those of the Ganges and Brahamaputra to a few thousand. Both species of humpbacked dolphins, which inhabit estuaries and mangrove swamps, have been severely affected by habitat destruction and disturbance. In the lower St. Lawrence River, Canada, the beluga whale may be succumbing to the effects of chemical pollution from urban-industrial sources (*See* Chapter 22).

PINNIPEDS

Almost all species of pinnipeds have been hunted for meat, oil, fur, or ivory, although only three species of monk seals (one, the Caribbean monk seal, is probably extinct) are classified as endangered by the U. S. Fish and Wildlife Service. The Mediterranean monk seal, with a declining population of about 300 individuals, is severely threatened (Reeves et al. 1991).

Two species of pinnipeds have been exploited on a commercial basis in the Northern Hemisphere in recent decades: the northern fur seal and the harp seal. Northern fur seals, which breed almost entirely in the Pribilof Islands of the Bering Sea, were first exploited in the 18th century by Russian fur hunters. During the 18th and 19th centuries most fur seals were hunted at sea, where both males and females were taken. The result of heavy exploitation was serious depletion of fur seal numbers. In 1911, under a treaty between the United States, the former USSR, Japan, and Canada, pelagic sealing was halted, and an annual harvest of young male fur seals was scheduled to be taken from the Pribilof Islands. The proceeds from sale of pelts was divided among the treaty nations, with the meat being utilized by the Aleut residents of the islands, who also provided the work force for the hunts. The annual harvest of fur seals under this arrangement was 20,000 to 60,000 animals annually, from a population of 2.0 to 2.5 million animals.

For many years, the harvest of fur seals was considered to be one of the most successful programs of sustained yield management of a wild animal population. However, in the early 1970s the numbers of fur seals unexpectedly began to decline, falling to about 900,000 animals in the late 1980s. Between 1973 and the early 1980s the decline in young fur seals born in the Pribilof colonies averaged 6% annually. This decline, together with decreasing profitability of the harvest and pressure from environmentalists, led to a cessation of the harvest in 1983. The exact cause of the population decline is still uncertain. The population appears to be below carrying capacity, but the birthrate of young is high, parental feeding seems to be normal, and the animals are healthy (Loughlin et al. 1987). One likely contributing cause of the population decline is high mortality of fur seals in drift gill nets and by entanglement in discarded fragments of trawl nets used for other fish in the Bering Sea and North Pacific (Fig. 19.5). Estimates of possible net mortality are as high as 30,000 animals per year (Alexander 1986). Disease, chemical

FIGURE 19.5

Entanglement in active fishing nets and discarded net fragments may be a major cause of recent declines in northern fur seal numbers in the Pribilof Islands.

(Photo by Brent S. Stewart)

FIGURE 19.6

A walrus breeding colony.

(U.S. Fish and Wildlife Service photo)

pollutants, predation, depletion of seal foods by commercial fishing, and the lingering effects of an experimental harvest of female seals from 1956 to 1968 have not been ruled out as partial causes, however (Scheffer & Kenyon 1989).

In the Bering Sea, populations of Steller's sea lions (Stirrup 1990) and harbor seals (Pitcher 1990) have also declined. Net entanglement is believed to be less common for these species than for fur seals. Greatly increased harvests of finfish, particularly pollock, in the feeding grounds of these species may be the prime contributor to their decline.

Harp seals, which occur in the North Atlantic, have also been harvested in large numbers. This species is quite abundant, with an estimated total population of 2.2–3.0 million (Ronald & Dougan 1982). In this case, commercial harvest was largely centered on the newborn pups, because of their dense, white fur. Young seals are taken in early spring on the offshore ice pack, where they are born. Until 1983, about 50,000 animals were harvested in Norwegian and Russian waters and about 170,000 taken in Canadian waters. An additional 10,000 or so harp seals were also taken by Canadian Inuit, so that the overall harvest of harp seals was about 230,000. Strong pressure from environmental and animal rights groups led the European Economic Council to suspend the import of white harp seal pelts in 1983, and the Canadian government to ban large-scale commercial hunting of seal pups in 1987. Small numbers of both pups and older seals, which have lost the white coat, continued to be taken on a subsistence basis. The harp seal has also become the object of ecotourism in Canada's Magdalen Islands, with about 2,000 tourists annually taking helicopter trips to see the newborn seals on the ice pack. In 1992, however, the Canadian government tried to encourage increased harvest of harp seals, arguing that they were contributing to the decline of the cod fishery.

The walrus, limited in distribution to the Arctic Ocean, has been harvested by Inuit and other northern peoples for meat, skin, and ivory (Fig. 19.6). The world population of the species

FIGURE 19.7

Only about 1,000 West Indian manatees remain in their Florida range.

(U.S. Fish and Wildlife Service photo by Gaylen Rathburn)

is estimated at 235,000 to 245,000, and the annual harvest in Asia and North America at about 12,000 (MacKay 1987). The worldwide increase in value of ivory has also affected walrus ivory, which, in 1989, brought about $90 per kilogram. Reports in 1989 indicated that an average pair of walrus tusks was being sold for $500 to $700 in Alaska.

OTHER MARINE MAMMALS

Manatees and dugongs, which occupy fresh and saltwater areas along tropical and subtropical coastlines (Fig. 19.7), have been greatly reduced in numbers, largely by habitat disturbance and hunting for their meat. In the 1980s, the Florida population of the West Indian manatee had been reduced to

FIGURE 19.8

Injuries from boat propellers are a major cause of mortality of manatees in Florida.

(U.S. Fish and Wildlife Service photo)

FIGURE 19.9

Populations of northern elephant seals, such as these animals on San Miguel Island, and several other pinnipeds have increased greatly off the California coast since passage of the Marine Mammal Protection Act in 1972.

(Photo by Brent S. Stewart)

about 1,000 animals, compared to an estimated 10,000 animals in the 1940s (St. Aubin & Lounsbury 1990). In addition to drainage and conversion of their habitat, Florida manatees now suffer heavy mortality from collisions with motorboats (Fig. 19.8). Another member of this group, the Steller's sea cow, occurred in the North Pacific but, in the late 1700s, was hunted to extinction for its oil.

The polar bear is a true marine mammal, spending most of its life on the arctic ice pack. The species is circumpolar in distribution, but its overall population size is still uncertain. About 3,000–5,000 are believed to occur in the Alaskan region. Under a 1973 international agreement, the world population is protected, except for subsistence harvests by northern native peoples. In Alaska, this type of harvest has involved fewer than 50 animals annually.

Another marine mammal, the sea otter, was discussed in detail in Chapter 18.

MARINE MAMMALS AND FISHERY CONFLICTS

The Marine Mammal Protection Act of 1972 established a moratorium on the harvest or intentional killing of pinnipeds, sea otters, and other marine mammals in United States waters, along with the importation of such animals or their products (Hofman 1989). Exceptions were made for limited "unintentional" killing of porpoises in tuna fishing, the Inuit harvest of certain species, harp seal harvests in Canada, and captures for research and public display. The act has since been amended to cover other unintentional harvests. The act further specifies that the populations of these animals are to be managed to maintain "optimal sustainable populations," a level defined as lying between the maximum supportable population and that at which annual growth is greatest.

Not surprisingly, under this protection, marine mammal populations have increased substantially in many locations. In California, for example, populations of California sea lions, harbor seals, northern fur seals, and elephant seals have grown rapidly (Fig. 19.9). California sea lions have increased from about 3,000–4,000 animals in 1930 to over 100,000 animals in the 1990s. Elephant seals, reduced to about 100 animals on Guadalupe Island, Mexico, in the early 1900s, have increased in numbers and spread north into California (Cooper & Stewart 1983). The California population of elephant seals had grown to about 120,000 animals in the late 1980s. The Guadalupe fur seal is now increasing in numbers, and could move into California in the near future.

The growing numbers of these species, together with the activities of sea otters and several small cetaceans, are leading to conflicts between marine mammals and fishermen. In effect, the Marine Mammal Protection Act gave priority for protection to the top carnivores of the marine ecosystem (Manning 1989). In 1984, pinnipeds along the California coast consumed 953,000 tons of sea life, nearly five times the 202,000 tons harvested by commercial fishermen in the same area. A similar situation prevailed in the Bering Sea in the 1970s, where pinnipeds alone consumed nearly 2.8 million tons of finfish (Lowry & Frost 1985). In addition to the consumption of finfish and shellfish, the larger mammals often rob or damage nets and other fishing gear. Along the California coast, such losses amounted to about $600,000 in 1980 (DeMaster et al. 1985). Some of the mammals that breed in California also migrate north into Oregon, Washington, and British Columbia, where they prey on salmon and other fish.

Similar problems of competition between marine mammals and human fishing interests, both commercial and sport, have appeared in Maine, Scotland, Norway, and Chile (Manning 1989). Ultimately, it appears that active control of some marine mammal populations will be needed to balance the interests of commercial and sport fishing with those of wildlife conservation.

Marine Birds

Little attention has been paid to conservation problems of marine birds. We have considered some of the problems faced by seabirds in their coastal breeding colonies (*See* Chapter 11). Oil pollution, one of the most serious threats to certain species, will be examined in detail in Chapter 22. Other problems faced by seabirds include competition with commercial fisheries and mortality in fishing nets.

Estimates of the fish harvest by seabirds in productive marine waters suggest that they often take about 20% of fish production (Furness & Ainley 1984). Considering that commercial fishing often takes 50%–70% of fish production, these harvests must reduce bird populations in heavily fished areas. The changes in the seabird populations of the Peruvian upwelling are one example of such an impact (*See* Chapter 17). Many other cases of actual or potential impact of commercial fishing on seabird numbers are recognized, however. Human harvests of sand eels, herring, and capelin in the North Atlantic are likely to impact auks, terns, and kittiwakes that depend on these fish. Harvests of sardines and anchovies in various upwelling areas already impact several kinds of diving seabirds. The growth of squid fisheries near New Zealand is likely to affect shearwater and petrel numbers in this region. If krill harvests become major in the Antarctic Ecosystem, penguin numbers will surely be reduced, as well.

Drift gill-netting for squid, salmon, and other fish was a major cause of mortality of many diving seabirds, particularly shearwaters and members of the Auk Family, until the practice was banned by the United Nations in 1992 (Atkins & Heneman 1987). In United States waters in the North Pacific and Bering Sea, Japanese drift gill-netting for salmon killed 96,000 to 251,400 seabirds annually. Most of these were short-tailed shearwaters and tufted puffins, but some 19 other species were also recorded in nets. In addition, Japanese fisheries based on land in the western north Pacific were estimated to kill 134,000 to 171,000 additional seabirds. The mortality of seabirds in drift nets set for squid by Japanese, South Korean, and Taiwanese boats farther south in the Pacific probably lay between 30,000 and 400,000 birds. Damaged and discarded nets, left floating in the ocean, took a possible toll—so-called "ghost fishing"—of 68,000 birds annually. Smaller losses occurred, and still occur, in gill nets along the coasts of Canada and the United States. Altogether, gill net losses added up to between half a million and a million birds annually. Although these numbers are large, the North Pacific breeding seabird population is about 100 million birds (Bartonek & Nettleship 1979), and the total population of short-tailed shearwaters, which nest in the southern hemisphere, alone is about 15 to 16 million.

Along the California coast, however, gill-netting has contributed to the decline of breeding seabird populations. Here, gill nets are used to catch halibut, white croaker, and other bottom fish. The Farallon Islands, off San Francisco, originally had breeding colonies of about 400,000 common murres (Carter 1986). Uncontrolled commercial harvests of eggs in the late 1800s reduced the colonies to fewer than 20,000 birds in 1910. Marine oil pollution and disturbance by visitors drove the colonies to a low of a few thousand birds in 1937. With greater protection of the colonies in the late 1930s, the murre population rebounded, reaching 88,000 in 1982. At about this time, however, gill-netting began along the California coast. In 1983, 25,000 to 30,000 murres were estimated to have died in gill nets in the area (Atkins & Heneman 1987). Restrictions on the use of gill nets in shallow coastal waters, and in the Gulf of the Farallones National Marine Sanctuary, have reduced murre mortality in nets. However, from 1984 to 1986, annual mortality was still 6,000 to 8,000 birds, and breeding murre populations were again declining.

Conservation of Marine Mammals and Birds

The fact that heavy mortality of many marine mammals and birds occurs in spite of their having legal protection emphasizes the need for global programs of conservation. In 1984 the United Nations Environmental Program and U.N. Food and Agriculture Organization initiated the Global Plan of Action for the Conservation, Management, and Utilization of Marine Mammals (Nielsen 1984). This effort is being developed in coordination with other international commissions and agencies that deal with marine fisheries and wildlife.

Summary

1. Commercial whaling, which reached its peak in the middle of the 20th century, has reduced several of the great whales to the brink of extinction. Although a general moratorium on whale harvests was adopted by the International Whaling Commission, small numbers are still taken in aboriginal and scientific operations, and by ships of non-IWC countries.

2. The small cetaceans are subject to some commercial hunting, particularly in Japan and Russia, and have suffered heavy accidental mortality in fishing operations such as purse-seining for tuna and open-ocean drift gill netting.

3. Pinnipeds have been subject to both aboriginal and commercial harvests, the latter for their fur. Populations of North Pacific fur seals and Stellar's sea lions have declined in recent decades, possibly due to net entanglement and losses of their prey to commercial fishing.

4. In some regions, the protection of pinnipeds has led to rapid growth of their populations, leading to increasing conflicts with human fishing interests.

5. Seabirds, members of four major orders of birds, are wide-ranging forms that feed on small fish and invertebrates, such as krill, and nest on islands or sea cliffs. Principal threats to these species are colony disturbance, oil spills, gill nets, and harvest of their food by humans.

QUESTIONS FOR DISCUSSION

1. Should aboriginal whaling, practiced by people with an increasing reliance on modern technology, continue to be permitted in locations such as Alaska, when commercial whaling for the same species is prohibited?

2. Assuming that minke whales are twice as numerous as they were originally, do you think that a sustained yield harvest is possible, and might even be beneficial to the recovery of other, more endangered whales?

3. Do you think it is reasonable to cull populations of marine mammals, such as seals and sea lions, in order to increase the availability of their prey for sport and commercial fishermen?

SUGGESTED READING

Commission on Life Sciences. 1992. *Dolphins and the tuna industry*. National Academy Press, Washington, DC. An analysis of the ecology, history, and economics of tuna fishing in the eastern tropical Pacific, and of prospects for reducing the impact of fishing on dolphin populations.

Payne, R. 1995. *Among whales*. Charles Scribner's Sons, New York. A popularly written, personal account of the author's remarkable experiences in studying whales, and of his philosophy of living in harmony with them.

Reeves, R. R., B. S. Stewart, and S. Leatherwood. 1992. *The Sierra Club handbook of seals and sirenians*. Sierra Club Books, San Francisco. A concise, abundantly illustrated, species-by-species account of pinnipeds, sirenians, marine otters, and the polar bear, with an informative introduction and extensive bibliography.

six

Special Problems at the Biosphere Level

M any human activities have impacts that affect both terrestrial and aquatic ecosystems. The invasion of ecosystems by exotic species has now become one of the most serious threats to biodiversity. Migratory animals, both terrestrial and aquatic, are being threatened by intensified use of land, freshwater, and coastal marine ecosystems. Environmental pollution by toxic chemicals, thought by many to have been conquered in the 1960s and 1970s, has grown in extent and complexity as a global problem. The exponentially increasing magnitude of energy use, industrial activity, and destruction of forests and other vegetation is leading to change in the earth's climate.

chapter

20

Management of Exotic Species

T
he deliberate or inadvertent introduction of plant and animal species to new regions has been both beneficial and detrimental. The transport of crop plants to new regions, where their diseases and pests are absent, is one of the most successful strategies of agriculture and horticulture. Some of these deliberate introductions, however, have become serious pests. Many recognized pests also have found ways to hitchhike to new regions aboard the vehicles of human commerce and transportation. Some of these exotics have invaded natural ecosystems, replacing native species and contributing to the loss of biotic diversity. The potential for introduction of exotics increases as human populations and economic activity grow. Counteracting the detrimental impacts of problem exotics is a growing responsibility of conservation biology.

We shall focus our attention on several major questions: What determines whether or not an introduction is successful? What processes lead to the invasion of exotics? What determines the impact of a successful introduction? How can the detrimental effects of exotics be countered? Finally, we shall consider the criteria that should be met before exotics are deliberately introduced to new areas.

Outline

Introduced animals and plants threaten the integrity of many terrestrial, freshwater, and coastal marine ecosystems.
(U.S. Bureau of Land Management Photo.)

DETERMINANTS OF SUCCESS

Four major factors largely determine if a group of individuals introduced to a new location will become established. These are 1) suitability of habitat and ecological niche, 2) adequacy of the introduction unit, 3) degree of escape from diseases, parasites, and predators, and 4) competitive regime of the new location.

Suitability of Habitat and Ecological Niche

To become established, the individuals that reach a new area must find both suitable **habitat** and an available **ecological niche.** Habitat suitability relates partly to abiotic conditions such as temperature, humidity, and substrate conditions in terrestrial environments; or salinity, current, and sediment chemistry in aquatic environments. Suitability of the habitat also involves biotic conditions such as vegetation structure and the kinds of competitors and enemies present. To succeed, plants must find climates, substrates, and a biotic community favorable for germination, growth, reproduction, and dispersal. Animals must be able to grow and reproduce under the abiotic conditions of their new environment, and to respond adaptively to the other living organisms.

Niche conditions—the resources that a species must exploit to survive and reproduce—must also be suitable. For plants, these include soil nutrients, carbon sources for photosynthesis, and moisture. For animals, particular types of food materials, water sources, and nest sites must be available.

Introduced species that have found suitable habitats and niches in North and South America include many plants and animals adapted to urban and agricultural environments that were created in the New World after European settlement. Terrestrial plants from agricultural ecosystems in the Old World have become dominant members of New World farmlands, particularly in regions of grassland and Mediterranean climate (Heywood 1989; Mack 1989). The house sparrow and European starling, deliberately introduced to the New World, also found habitats very similar to those they inhabited in Europe. Among deliberately introduced game animals, the ring-necked pheasant has shown similar success. Introduced from China to the Willamette Valley of Oregon, and from there to many other localities in North America, this species has become an important game bird. The adaptation of all of these species to the intensively farmed landscapes of Europe and Asia preadapted them to the farmlands of North America.

The cattle egret (Fig. 20.1) is native to Africa and Asia, where it forages for insects and other small animal life in grassland and savanna habitats, usually in company with large grazing animals. Livestock ranching greatly expanded such habitats in the Americas. Eventually, in the 1930s, a handful of cattle egrets crossed the Atlantic Ocean from West Africa to Brazil and found ideal habitat. The species has now spread over much of tropical and warm temperate North and South America, showing how rapidly a new colonist can occupy a vast area of favorable habitat (Bock & Lepthien 1976).

FIGURE 20.1

The cattle egret, which reached South America from Africa by natural dispersal, has rapidly colonized regions of North and South America where suitable grasslands and pastures with livestock exist.
(Photo by G. W. Cox)

Adequacy of Introduction Unit

For establishment to be successful, the introduced individuals also must be adequate in health, behavior, and simple numbers. Many attempts have failed because individuals were in poor health or lacked the behaviors necessary for life in the wild. Animals reared in captivity often fail to survive in the wild because of inability to find or capture food, avoid predators, or combat diseases. Recent attempts to reintroduce the red wolf to the wild in the Alligator River National Wildlife Refuge, North Carolina, for example, have been hampered by a lack of resistance of captive-reared animals to bacterial infections (Rees 1989). Similarly, efforts to establish thick-billed parrots in mountain areas of southern Arizona have been hampered because captive-reared birds lack effective raptor-avoidance behavior.

The introduced individuals must be numerous enough to form a reproductively effective unit. Many attempted introductions have failed because so few individuals were released that they became lost in the new region, and failed to create an interacting population. For plants, successful reproduction requires pollination, seed production, seed dispersal, and germination. Weeds are ideal colonists, and excel in these abilities. Baker (1974) listed 12 characteristics of an "ideal" weed, ten of which pertain to reproduction and dispersal. Some of these include rapid growth to flowering, continued flowering and seed production for as long as conditions are favorable, self-compatibility or reliance on unspecialized cross-pollination mechanisms, high seed output and the ability to produce seed under varied conditions. For perennials, a vigorous vegetative reproduction system is also advantageous. Adaptations for dispersal, seed longevity, and ability to germinate under varied conditions are also beneficial. Pest animals, such as many agricultural insects, possess many similar characteristics, and are thus effective colonists.

FIGURE 20.2

Introduced to Australia deliberately, the European rabbit has become
a serious pest of agricultural ecosystems.

(Photo courtesy of CSIRO, Canberra, Australia)

For vertebrates, the ability of introduced individuals to establish a normal social system is often important to success. Failure of the effort to establish a new whooping crane breeding
population at Gray's Lake National Wildlife Refuge in Idaho
(See Chapter 26) was perhaps due to the lack of an established
social system of adults for the young cranes to join.

Escape from Diseases, Parasites, and Predators

When species are transported to a new region, the degree to
which they escape natural enemies and find prey that are easy to
capture is also important to their success. Every species in a
community stimulates a suite of evolutionary adjustments by the
other species with which it interacts—a set of adjustments
termed **counteradaptation.** Natural selection favors, for example, the ability of predators and parasites to exploit the species
available in a community as prey or hosts. At the same time, it
favors the improvement of defenses by the potential prey and
host species. The result of this interaction is often an evolutionary "standoff," the pressures of counteradaptation restricting
each species to a narrower habitat and niche than it might otherwise occupy. When such a species is transported to a new location, however, these counteradaptive constraints may be removed, and it may become successful and abundant in many
habitats.

Many examples of release from counteradaptive constraints exist. European rabbits (Fig. 20.2), introduced to Australia, were released from the constraints of various predators and
parasites. In Australia, introduced rabbits carried lighter parasite
loads than they did in Spain, where the introduced animals originated (Myers 1986). In addition, native predators had been reduced or eliminated over vast areas of Australian farmland. Thus,
the rabbit rapidly increased to plague abundance, becoming a serious agricultural pest.

On oceanic islands, where plants are not adapted to grazing and browsing mammals and the enemies of such mammals
are absent, introduced mammalian herbivores often become disastrous successes. On Pinta Island in the Galápagos, for example,
three goats introduced in 1959 grew to a population of
5,000–10,000 animals by 1970. Browsing by these goats severely
damaged the native vegetation (Coblentz 1978). Horses, burros,
cattle, and sheep have become feral in other locations in the
Galápagos, also causing severe damage to the native vegetation.
In New Zealand, where natural enemies of large animals are also
absent, many introduced mammals have become established in
the wild. These include the Australian brush-tailed opossum, the
European hare, and some 14 species of wild and domestic ungulates. The weak defenses of native New Zealand plants against
excessive browsing or grazing has contributed to the success of
these herbivores and to the severity of the damage they cause.
Brush-tailed opossums have been linked specifically to the decline of endemic trees of the genus *Metrosideros* and to many
species of mistletoes, which grow as semiparasites in the canopies
of trees (Norton 1991).

In South Africa, trees such as acacias and hakeas from
Australia and pines from the Mediterranean region have proved
to be destructive invaders of the fynbos shrublands of the Cape
Region (Shaughnessy 1986), which contain many endemic
species. Many of these invaders seem to flourish, however, only
where populations of ungulates are low (Macdonald 1985), suggesting that their success depends partially on the fact that they
do not invest heavily in herbivore defenses, as do many of their
African competitors. The success of these plants also depends in
part on their adaptations to fire and low-nutrient soils, which
characterize both their native regions and the South African
Cape Region.

Escape from Competitors

The competitive regime that a species encounters in a new area
also affects its probability of establishment. In undisturbed, continental ecosystems, native species normally exploit resources almost completely, leaving few "empty niches," or unexploited
types of resources, that an exotic species can preempt with ease.
In isolated regions, such as oceanic islands, however, species capable of exploiting certain resources may neither have colonized the area nor evolved from earlier colonists. Disturbance
of an indigenous ecosystem, in addition, often changes patterns
of resource availability, either by altering the overall abundance
of resources, or by changing the relative abundance of various
types of resources (Fox & Fox 1986). Such changes often create
underutilized resources that can be exploited by exotic species.

Analysis of successes and failures of attempted introductions shows that success is greatest when the invaded system has
few species (Moulton & Pimm 1986; Brown 1989). In large
measure, this probably reflects competitive pressure: the fewer
the competitors, the more likely it is that an exotic will succeed.

FIGURE 20.3

The nutria, brought to the United States from South America for fur farming, has escaped and colonized marshes along the Atlantic and Gulf coasts, often severely disturbing marsh vegetation by its grazing.

(Photo by G. W. Cox)

FIGURE 20.4

The red fox, introduced to Australia to control the introduced European rabbit, has also become a serious predator on native small mammals.

(Photo courtesy of CSIRO, Canberra, Australia)

This principle contributes to the high success of exotics on islands and in regions such as Florida, which has a much less diverse native biota than other, less isolated subtropical areas.

The success of so many urban and agricultural exotics in the New World is certainly due in part to the inability of native North American species to exploit the resources of these habitats fully. Cities have existed in the Old World for thousands of years, and several birds and mammals evolved to exploit this habitat and its resources. In most of North America, cities did not exist in pre-Columbian times, and urban species thus did not evolve. European settlement created cities with unoccupied habitats and unexploited resources, assuring the success of urban exotics from the Old World. The same is true of cultivated croplands in North America. On the other hand, repeated attempts to introduce the Eurasian *Coturnix* quail to the New World have probably failed in part because native quail exist in all areas otherwise suitable for *Coturnix*.

MECHANISMS OF INTRODUCTION

Detrimental introductions have been both deliberate and inadvertent (Brown 1989; Heywood 1989). Domestic animals, both livestock and pets, have been carried to all parts of the world. Often they establish feral populations, or populations that live and reproduce in the wild. In many places, feral animals have wreaked havoc on the native vegetation, as ungulates have in the Galápagos Islands, or on the fauna, as cats have in parts of Australia. Undomesticated species, such as the South American nutria (*Myocastor coypus*) that was imported to the United States for fur farming, have also escaped and established wild popula-

tions (Fig. 20.3). Predators have often been imported to control previously introduced exotics, such as the red fox to control rabbits in Australia (Fig. 20.4) and mongooses to control rats on various oceanic islands. Often, these predators have become even more serious pests. Settlers to new continents and islands have carried crop and horticultural plants with them, as well. Some of these species have invaded the native vegetation aggressively, such as the guava and quinine tree have in the Galápagos Islands (Macdonald et al. 1988).

Inadvertent introduction of weeds to new continents or islands has occurred commonly by seed mixed with that of crop plants or carried in the food, bedding material, or pelage of livestock. Some weeds also may have arrived as seed in soil used as ballast by early sailing ships. Animals have inadvertently been carried along with their plant hosts, or, in the case of some of the most destructive rodents, as unwelcome guests on ships and other forms of transport. The brown tree snake probably reached Guam (*See* Chapter 5), for example, in shipments of fruit from Southeast Asia.

DETERMINANTS OF IMPACT

Urban and agricultural ecosystems are particularly prone to invasion of exotics, and these invaders often directly damage the interests of humans: their lawns, farms or gardens, pets or livestock, and homes. The impacts of exotics in natural ecosystems range from direct and obvious to indirect and subtle. The factors that determine their impact include their ecological distinctiveness and their potentials for competitive displacement, disease or disease transmission, and genetic swamping.

Ecological Distinctiveness

Species that are highly successful in invading natural communities are usually different in structure, physiology, or behavior from native forms. As a result, counteradaptation to them by native species is often absent. Consequently, exotics with distinctive features can become **keystone exotics**—invaders that can cause almost complete biotic reorganization of the ecosystem. Keystone exotics, being so different ecologically from native forms, are often able to invade undisturbed natural systems, effectively pursuing patterns of resource exploitation that native species have not evolved. Keystone exotics can exert their effects at the population, community, or ecosystem level (Macdonald et al. 1989; Vitousek 1990). Some disrupt major patterns of coevolved mutualism among native species. Others exert trophic effects that restructure the food web. Still others modify basic processes of nutrient cycling or hydrology, or encourage unusual patterns of disturbance by fire or other factors.

Among plants, keystone exotics are exemplified by invasive species that become the dominants of the vegetation. In northern Australia, for example, a prickly leguminous shrub, *Mimosa pigra,* has invaded wet ecosystems ranging from open sedge marshes to monsoon woodlands (Braithwaite et al. 1989). Open sedge marshes tend to be invaded first, and the mimosa soon forms a dense, monospecific stand of shrubs up to five meters high. Invasion in some areas has been promoted by soil disturbance by feral water buffalos, another exotic species. Later, woodlands are invaded, the mimosas forming a dense understory that prevents the reproduction of native trees. The almost complete restructuring of wetlands vegetation by this species threatens to cause a massive change in animal communities of the region.

In the southeastern United States, vines introduced from the Far East, such as Japanese honeysuckle (*Lonicera japonica*) and kudzu (*Pueraria lobata*), have smothered native forest areas in a manner similar to the banana poka (*Passiflora mollissima*) in Hawaii (*See* Chapter 12). In southern Florida, an Australian paperbark tree (*Melaleuca quinquenervis*) (Fig. 20.5), an aggressive invader of the borders of cypress swamps, is replacing the native pond cypress (*Taxodium ascendens*) and other species over wide areas (Ewel 1986). The paperbark is not only tolerant of prolonged flooding, but is highly adapted to recovering from fire. It thus poses a major threat to the vegetation of the Everglades area. The Brazilian peppertree (*Schinus terebinthefolius*) is also an aggressive invader of disturbed upland sites in southern Florida. With bird-dispersed seeds and the ability to resprout after fire, it is a potential invader of almost all upland habitats.

Among animals, introduced livestock have played a keystone role in the transformation of extensive areas of perennial grasslands into annual grasslands in North America, South America, and Australia (Mack 1989). On the Columbia Plateau, annual Eurasian grasses such as cheat (*Bromus tectorum*), which is adapted to both heavy grazing and soil disturbance, have become the dominants of once-perennial grasslands. In this region, where bison were originally absent, the original bunch grasses were adapted to a light grazing regime. These grasses, which do not possess rhizomes, unlike many of the grasses of the Great

FIGURE 20.5

The paperbark or cajeput tree, *Melaleuca quinquenervis,* has become a destructive invader of Florida wetlands, forming dense, single-species stands that replace open wetlands and displace native swampland trees.

(Photo by G. W. Cox)

Plains, are unusually sensitive to the effects of grazing and trampling by large ungulates (Mack & Thompson 1982). A similar transformation followed livestock introduction to California (*See* Chapter 7). Dominance of the grasslands by annuals has led in turn to significant changes in animal life. In Manitoba, for example, the numbers of several grassland birds were increased in areas dominated by introduced Eurasian plants, whereas the abundance of others was reduced (Wilson & Belcher 1989).

In the Andean foothills of northern Patagonia, introduced red, fallow, and axis deer have severely disturbed evergreen beech and cedar forests of two of Argentina's most important national parks (Veblen et al. 1989). Browsing by deer has nearly eliminated one understory tree species and reduced the abundance of many shrubs and herbaceous plants. Regeneration of the over-story beeches and cedars in tree-fall gaps has also been inhibited. These changes pose a serious threat to efforts to encourage the recovery of populations of two species of small native deer, the pudu (*Pudu pudu*) and the huemul (*Hippocamelus bisulcus*). In New Zealand, damage by red deer to high-elevation tussock grasslands has contributed to the decline of the takahe (*Notornis mantelli*), a flightless rail that feeds on the bulblike bases of these grasses (Mills et al. 1989). Once thought to be extinct, this rail was rediscovered in the mountains of southern South Island, an area now included in Fiordland National Park.

In aquatic ecosystems, introduced plants such as water hyacinth (*Eichornia crassipes*) and hydrilla (*Hydrilla verticillata*) have transformed open waters to weed-choked swamps in many locations in the southern United States. Introduced fish such as the sea lamprey in the St. Lawrence Great Lakes, the peacock bass in Gatun Lake, Panama, and the Nile perch in Lake Victoria, East Africa also illustrate the effects of keystone exotics (*See* Chapter

MOUNTAIN GOATS IN OLYMPIC NATIONAL PARK

The mountain goat, *Oreamnos americanus,* is the single member species of a genus of ungulates endemic to North America (Reading Fig. 20.1). Not a true goat, it belongs to a group of goat-antelopes, which includes the chamois of mountains of Europe and the Middle East. The natural range of this species extends from southern Alaska and the Yukon Territory south through the mountains of Canada, to the Cascades of Washington State, and the Rocky Mountains of Idaho and Montana. Mountain goats have been introduced to mountain areas in several states, including, for example, the Black Hills in South Dakota.

Mountain goats were introduced to the Olympic Mountains of Washington State in the late 1920s (Scheffer 1993; Houston et al. 1994). Several introductions, totalling 11–12 individuals, apparently were made between 1925 and 1929. By 1937, the species had become established, but not increased greatly in abundance, numbering only an estimated 25 individuals. By the 1970s, though, the population had grown and spread throughout the higher elevations of the mountains. In 1983, the population was estimated to be about 1,175 individuals (Houston et al. 1994).

The rapid growth of the goat population in the 1970s and 1980s triggered concern by the U.S. National Park Service that mountain goats might threaten some of the endemic plants of the Olympics. A goat removal program was begun, and about 509 animals were removed between 1981 and 1989, most captured and translocated to other locations, but some killed. In 1990, the population remaining in the Olympics was estimated to be 389 animals (Houston et al. 1994).

Mountain goats are foraging generalists, and their impact on alpine meadow vegetation is potentially extensive. Studies of the ecology of mountain goats in the Olympics have revealed significant effects on alpine meadow vegetation due to grazing, trampling, and wallowing. Trampling has reduced the cover of mosses and lichens in some meadow areas. Grazing has also reduced the abundance of some native

READING FIGURE 20.1

The recommendation by the U.S. National Park Service that the mountain goat be removed from Olympic National Park as an exotic species that threatens several endemic plants has been challenged by the Fund for Animals. These animals at Oyster Lake, near Atherton Pass, were being studied by park biologists in 1994.

(Photo by G. W. Cox)

bunch grasses, such as Idaho fescue, and increased that of disturbance-oriented forbs, such as yarrow. Mountain goats create bare-soil wallows that favor the establishment of weedy, disturbance-oriented species. Given the impacts of feral domestic goats in the Hawaiian and Galápagos Islands, national park ecologists are concerned about the ultimate impact the mountain goat might have on endemic plants. The Olympic Mountains have eight endemic vascular plants, seven of which occur in mountain goat habitat. A number of other species are endemic to the Olympics and Vancouver Island or the Queen Charlotte Islands to the north. Four of these endemics are known to be eaten by goats.

In 1995, the National Park Service released a draft environmental impact statement on goat management. This report concludes that the goat population is exotic and should be removed completely. Because of the rugged terrain, the report

identified the most practical method of removal to be shooting.

The Fund for Animals, an animal rights organization, has contested the proposed removal of mountain goats. Representatives of the Fund for Animals contend that the goat has not been proven to be exotic, and that vegetational damage has been overstated (Anunsen & Anunsen 1993). Their argument for the species being native rests on several early accounts that mention goats or mountain goats. Schultz (1994), however, has carefully reviewed the historical evidence relating to mountain goat occurrences throughout the Puget Sound region, and concluded that these early reports are erroneous or refer to other animals.

This situation raises a basic policy issue of the role of national parks in the preservation of species of a general region. Should Olympic National Park, for example, be responsible for preservation of species that occur in the general region, but are not native to the park area itself?

15). In the United States, more than 40 species of fish have become established by intentional or accidental releases (Taylor et al. 1984). Of these, 37 are of tropical or subtropical origin, their establishment being primarily in the southern and southwestern states. In Florida, some 20 species have become established, including the pike killifish (*Belonesox belizanus*), a voracious predator that has virtually eliminated native fish from many waters.

Potential for Competitive Exclusion

Invaders are sometimes ecologically similar to native species, and able to utilize the same habitats and resources with greater efficiency, particularly if the natural system has been disturbed. Introduced riparian shrubs and trees, such as salt cedars (*Tamarix* spp.) have proved to be formidable competitors to native willows, cottonwoods, and other species, replacing them along streams in much of the southwestern United States. Sika deer (*Cervus nippon*) have been introduced to several locations in North America from Japan. On Assateague Island, Maryland, the Sika deer (Fig. 20.6) has increased in numbers at the apparent expense of the native white-tailed deer (*Odocoileus virginianus*) (Keiper 1985). The impact of feral burros on desert bighorn sheep in Death Valley is also an example of the detrimental competitive influence of an exotic species (*See* Chapter 8). Even the ring-necked pheasant, usually regarded as a beneficial game species, may have a detrimental competitive effect on native prairie chickens. Vance and Westemeier (1979) found that pheasants tended to dominate prairie chickens in behavioral encounters and to displace them from booming grounds. Pheasants often laid eggs in prairie chicken nests, reducing the reproductive success of the latter.

Potential for Disease or Disease Transmission

Introduced species may be reservoirs or vectors of diseases that affect native forms. Exotic birds introduced to Hawaii, for example, carried avian malaria, and introduced mosquitos provided a system of vectors to transmit the disease to native land birds (*See* Chapter 5). Fungal diseases introduced to North America have substantially altered the composition of the eastern deciduous forests (von Broembsen 1989). The chestnut blight (*Endothia parasitica*), accidentally introduced to North America in the 1890s on nursery stock of Asian chestnuts, has eliminated the American chestnut as a component of mature forests. The Dutch elm fungus, brought to North America from Europe in the early 1900s on logs for veneer production, has caused extensive mortality of native elm species.

Potential for Genetic Swamping

Some exotics are also genetically similar to native species, and thus able to interbreed with them. The result, genetic swamping, can be loss of the genetic identity of the native species, a form of extinction (*See* Chapter 5). In New Zealand, for example, the introduced mallard duck (*Anas platyrhynchos*) has interbred so extensively with the native grey duck (*Anas superciliosus*)

FIGURE 20.6

The sika deer has become established in several locations in North America, and is sometimes a strong competitor of the native white-tailed deer.

(U.S. Fish and Wildlife Service photo)

that less than 5% of the combined population of these two forms is representative of the pure grey duck (Gillespie 1985). It appears unlikely that the grey duck will survive in the wild, given the extent of interbreeding.

Genetic swamping is a major threat to many fish species. Miller et al. (1989) concluded that hybridization with other native or introduced fish was a contributing factor in the extinction of 15 species or subspecies in North America. Several forms of the cutthroat trout (*Oncorhynchus clarki*), for example, have disappeared because of interbreeding with introduced rainbow trout (*O. mykiss*). Eight other forms of the cutthroat are endangered, threatened, or reduced to the point of concern, largely as a result of hybridization (Williams et al. 1989).

DEALING WITH DETRIMENTAL EXOTICS

Several strategies can be used to avoid and reduce the impacts of exotics on natural ecosystems. Nowhere is the need for such measures greater than on oceanic islands, where exotics rank as the most serious threats to survival of endemic species.

1. Prevention of Entry.
   This strategy includes prohibition of entry, quarantine of living plants or animals before or at entry, inspection of biological materials to detect unwelcome associates, and fumigation of imported materials to kill hitchhiking invaders. Human inspection at ports of entry is being augmented by use of more effective canine or electronic "sniffers" to detect biological materials that might contain unwelcome exotics (Stone & Loope 1987). Strict rules against the importation of new species of plants and animals have been adopted in some locations. In the Galápagos Islands, most of which constitute an Ecuadorian national park, for example, the importation of exotics is completely prohibited.

2. Control of Spread.

Once an exotic has gained a foothold, its spread can sometimes be slowed by direct control or by control of vectors that assist its dispersal. This approach is being used against the Argentine ant in several locations. In Hawaii, the Argentine ant has been introduced to most islands, and has invaded Haleakala National Park on Maui and Hawaii Volcanos National Park on Hawaii. The ant preys on native insects and other small invertebrates, many of them endemic. Some of these forms have disappeared from areas invaded by the ant. In South Africa, Argentine ants have invaded fynbos shrublands, which contain many species of plants with ant-dispersed seeds (Bond & Slingsby 1984). Native ants harvest the seeds of these species and store them underground, effectively "planting" them. Argentine ants displace this native ant fauna, but do not bury these seeds, leaving them vulnerable to small mammals and other seed predators. In both regions, efforts are being made to control this ant, which has flightless queens and spreads by the division of colonies, with the use of toxic baits.

3. Protection of Pristine Areas.

Location and protection of areas that have not yet been invaded by exotics is also an effective strategy. Small islets lying offshore from inhabited islands are sometimes free of exotic species. In parts of Hawaii, islands of native vegetation, or **kipukas**, exist where extensive lava flows have isolated them from the main vegetated landscape. Fencing, monitoring, and weeding such sites may enable small areas of pristine nature to be maintained with relatively small effort.

4. Local Eradication.

Exotics can often be removed and then excluded from small areas, such as small islets or fenced exclosures. Rabbits and goats were eliminated from Round Island, an islet of 151 hectares near Mauritius in the Indian Ocean (Atkinson 1988), providing an excellent opportunity for preservation of many endemic plants and animals of the Mauritius Islands. Eradication of goats, dogs, cats, black and Norway rats, and five exotic plants from Nonesuch Island, a 6-hectare islet off Bermuda, has provided a site for the attempted restoration of native Bermudian ecosystems, as well as for nesting of the endangered cahow petrel (Wingate 1985). In Hawaii, exclosures for goats and pigs have been created in some 51 sites to protect specific examples of native ecosystems or to promote recovery of native vegetation (Stone & Loope 1987). In some of these exclosures plants thought to have been locally extinct have reappeared, apparently from seeds lying dormant in the soil.

5. Protection of Individuals.

Exclosures or chemical repellents may be necessary to protect individuals of critically reduced species. Some exclosures in Hawaii function to protect the last individual plants from destruction by goats or pigs (Stone & Loope 1987). Chemical repellents have also been used to deter rats from climbing into canopies of certain native trees to feed on flowers and fruits, in an effort to increase reproduction of critically threatened species.

6. General Population Reduction.

Reduction of keystone exotics over large areas has been attempted for only a few species. Eradication is usually difficult, if not impossible, with highly successful invaders of habitats that possess wilderness characteristics. Direct control, by killing or capturing of animals and weeding of undesirable plants, has been successful in a few cases. Efforts have been made to reduce feral goat and pig populations in the Hawaiian Islands by hunting, both by citizen hunters and by hired hunters (Stone & Loope 1987).

Biological Control

Biological control is the establishment of a strong, negative biological interaction against a designated species. Although biological control is sometimes portrayed as "reestablishing the balance of nature," it really involves the creation of a strong "imbalance of nature." The goal is to reduce an undesirable species to such a low abundance that its impact is minor. Biological control is often more desirable than eradication, since, as long as the detrimental species survives at a low abundance, its control agent also survives. If a pest is eradicated locally, its biological control agent may also disappear, favoring reinvasion of the pest from a source area.

Biological control employs several different strategies. Protection and promotion of native predators, parasites, and disease species is one strategy. Under this strategy, pesticides are used in ways that do not kill such species, and in ways sought to improve the microhabitats and resources on which they depend. A second strategy is the release of native biological control agents at critical times, either to reestablish populations in habitats where they do not persist well, or to overwhelm the target pest at a critical stage of its life cycle. The third strategy, sometimes termed "classical" biological control, is introduction of exotic biological control species, either from the native area of the pest species or from some other location.

Reduction of the impact of the introduced perennial Klamath weed (*Hypericum perforatum*) is one of the major successes of biological control in North America (Dahlsten 1986). Klamath weed was introduced to western North America from Europe about 1900, and by the 1940s had invaded hundreds of thousands of hectares of rangeland, crowding out desirable range species and creating health problems in livestock that ate the plant. Major infestations of the weed have also occurred in Australia, South America, and South Africa. Eight species of herbivorous insects that fed on Klamath weed were identified in its European homeland, the most effective being a beetle of the Family Chrysomelidae, *Chrysolina quadrigemina*. Introduction of this beetle, and often one or more other species, has led to varying degrees of biological control of Klamath weed. In California, biological control has reduced this weed to less than 1% of its former abundance.

CRITERIA FOR DELIBERATE INTRODUCTIONS

Introduction of exotic species should obviously be done only under special circumstances, and then only after careful study. A clear need or benefit should be evident, such as the benefit of achieving biological control of a pest species. Suitable habitat and resources for the proposed introduction must exist. Strong evidence that the introduced form will not have a detrimental impact on other species, or on basic ecosystem properties, must be provided. Finally, it must be shown that the introduced form can be confined to the region for which its suitability has been evaluated.

SUMMARY

1. Whether or not a species introduced to a new area becomes established depends largely on suitability of habitat and ecological niche, adequacy of the introduction unit, and the degree of escape from diseases, parasites, predators, and competitors.

2. Introductions of detrimental exotics have been both deliberate and inadvertent, but most are the consequence of human transport and commerce.

3. The impact of exotic species often depends on how different they are in structure, physiology, or behavior from native forms. Exotics with distinctive features can become keystone species that can cause almost complete biotic reorganization of a native ecosystem.

4. The eradication of species, such as the North American mountain goat, from areas outside their original range, or species, such as *Eucalyptus,* that have become well integrated into native ecosystems, presents issues that must be addressed individually, by careful consideration of impacts of the species and of their removal on the ecosystem as a whole.

5. Strategies that can be used to prevent and reduce impacts of exotics include prevention of entry, control of spread, protection of pristine areas and surviving individuals, local eradication, and general population reduction using techniques such as biological control.

QUESTIONS FOR DISCUSSION

1. What criteria do you think should be used to determine whether a particular exotic species should be controlled, eradicated, or simply left alone?

2. What are the implications of increased freedom of international trade to the problem of introduced exotic species and their impacts on native ecosystems?

3. Consider a policy that considers native North American species, such as the mountain goat, that have been introduced to areas outside their historical range as undesirable exotics. What are the implications of such a policy to conservation strategy in the face of global climatic change?

SUGGESTED READING

Houston, D. B., E. G. Schreiner, and B. B. Moorhead. 1994. *Mountain goats in Olympic National Park: biology and management of an introduced species.* Scientific Monograph NPS/NROLYM/NRSM-94/25, U.S. Dept. of Interior, National Park Service. An account of the ecology of the Olympic Mountains, and of the history and impacts of mountain goats on high-mountain ecosystems of the Olympics.

McKnight, B. N. (Ed.). 1993. *Biological pollution: the control and impact of invasive exotic species.* Indiana Academy of Science, Indianapolis, IN. Case histories of the ecological effects of introduced species and of efforts to control them.

Ruesink, J. L., I. M. Parker, M. J. Groom, and P. M. Karieva. 1995. Reducing the risk of nonindigenous species introductions. *BioScience* **45**:465–77. The application of risk assessment techniques to introduction of exotic species: A policy of guilty until proven innocent.

Disruption of Migrations

Seasonal migration is the regular movement of individuals between a breeding and one or more nonbreeding ranges separated by distances many times greater than the scale of their normal home ranges. Seasonal migration is one of the major phenomena of animal biology. Although best developed in birds, other migratory land animals, such as bats, wildebeest, tundra caribou, and the monarch butterfly, and aquatic animals, such as whales, pinnipeds, sea turtles, and salmon, carry out extensive migrations. Billions of birds of more than a thousand species seasonally move distances varying from a few kilometers to more than 10,000 kilometers between breeding and nonbreeding ranges.

Migratory animals are believed by many to be at special risk from effects of environmental change because they require different habitats in different geographical areas during their annual cycle (Brower & Malcolm 1991). The high mobility of migrants may give some species flexibility to adjust to changes in their environment, but it may also make them vulnerable to impacts on any of the habitats on which they depend.

In this chapter, we shall examine migratory species and their conservation challenges. First, we shall consider migratory birds, and then other animal groups that face similar challenges.

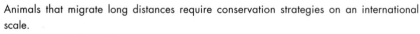

Animals that migrate long distances require conservation strategies on an international scale.
(Photo by M. Philip Kahl.)

MIGRATORY BIRDS

Birds show all degrees of development of seasonal migration, but for convenience we can group these into three categories: weather migrants, short-distance migrants, and long-distance migrants. **Weather migrants** are species that move varying distances in direct response to severe weather conditions. In the temperate zone, many species move southward, or to lower elevations in mountainous areas, in response to severe winter weather. The timing, distance of movement, and length of residence in nonbreeding areas are all highly variable from year to year. **Short-distance migrants,** in contrast, show more regularly timed movements between breeding and nonbreeding ranges separated by distances of a few hundred kilometers. Many such species breed in the higher temperate latitudes or in the arctic, and winter at lower latitudes within the temperate zone. Often the breeding and nonbreeding ranges of these species overlap latitudinally. **Long-distance migrants** are those that annually move between breeding and nonbreeding ranges thousands of kilometers apart. Many long-distance migrants breed at high temperate or arctic latitudes and winter in the tropics or in the opposite hemisphere. The timing and pattern of their movements are even more regular than those of short-distance migrants. Altitudinal and short-distance migrations by birds are also common in the tropics.

For many, if not most, long-distance migrants, the timing, route, and distance of migration are genetically programmed. Furthermore, strong fidelity is shown by many species to natal and nonbreeding sites that were utilized during their first year, including areas visited during migration (Rappole & Warner 1978). The demography of long-distance migrants is also specialized (Cox 1985). Many of these species have extended their breeding ranges into high latitudes where total annual fecundity is severely limited by the short summer period. At higher latitudes, migrants can usually rear only a single brood. Often, poor summer weather conditions cause many nesting efforts to fail. Thus, species migrating to high latitudes must compensate by spending the nonbreeding season in areas where survival is high. The migratory flights themselves are also risky, and long-distance migrants are adapted to carry these out as quickly as possible. Thus, environmental changes that reduce breeding productivity, increase time along the migration route, or reduce nonbreeding survival can all contribute to the decline of migrant populations.

Three major systems of long-distance migration exist (Rappole 1995). In the Old World, the **Palearctic-African Migration System** involves about 185 species of land and freshwater birds that breed in Europe and western Asia. These species move southward to winter in Africa south of the Sahara Desert. An estimated five billion birds undertake this southward migration, with about half of these surviving to return to their breeding grounds the following summer (Moreau 1972). The **Palearctic-Asian Migration System** includes about 338 species that breed in eastern Asia and parts of western Alaska, and winter in southeast Asia, Australia, and Pacific island regions. Finally, the **Nearctic-Neotropical Migration System** involves 338 species that breed in North America and winter in the New World tropics or in temperate South America. Of these, 98 winter exclusively in the New World tropics or in temperate South America. In all probability, this migration system involves as many individuals as the Palearctic-African system.

Staging and Stopover Areas

Species that make flights of hundreds or thousands of kilometers must store large amounts of body fat just prior to these flights to provide the needed energy. In some cases, as much as 40% of the weight of a bird departing on a migratory flight consists of fat. Species that carry out their migrations in one or a few long flights are particularly dependent on habitats in which abundant foods suitable for fat deposition are available. Areas that provide these needs are known as staging and stopover areas.

Staging areas are locations at which birds gather prior to the first major leg of migration, whereas **stopover areas** are those at which birds stop to replenish their fat stores partway through migration. Shorebirds, in particular, are highly dependent on staging and stopover areas (Myers 1983; Myers et al. 1987). The Hudsonian godwit, for example, breeds in arctic Canada and then gathers at a staging area along the western shores of the Hudson and James Bays. Most of the individuals then fly nonstop about 4,500 kilometers to the northern coast of South America (Morrison & Harrington 1979). Other species of shorebirds move shorter distances at a time, and use several stopover areas, where they remain for a few days and store fat (Fig. 21.1). Staging and stopover areas are used both during autumn and spring migration periods. In the spring, the last stopover before arrival on the breeding ground must provide for the last leg of the migration and often for the period of territory establishment and mating, when weather may be poor and food supplies low.

In eastern North America, Delaware Bay (Fig. 21.2) is an especially important stopover area for shorebirds on northward migration (Clark et al. 1993). Some 300,000–600,000 shorebirds visit Delaware Bay on spring migration, many of them having completed flights of 2,800–3,200 kilometers from South America. There, to rebuild their fat stores, they feed primarily on eggs of horseshoe crabs, which spawn in enormous numbers at the same time (Botton et al. 1994). In southern Alaska, the Copper River delta is the stopover site for about 20 million shorebirds prior to the last leg of migration that takes them to arctic tundra breeding areas (Senner 1979). The fat stored at this last stopover site also serves to support the birds during early phases of the nesting cycle, when weather conditions on the tundra can be unfavorable to feeding. The Cheyenne Bottoms in Kansas are a critically important stopover site for shorebirds in the interior of North America (Zimmerman 1990).

Although staging and stopover areas are clearly important for water birds, less is known about their importance for migrant land birds. Some coastal areas are known to be of major importance, however. Coastal chenier woodlands of Louisiana are important stopover sites for birds that have crossed the Gulf of Mexico in spring migration (Moore & Kerlinger 1987). Many

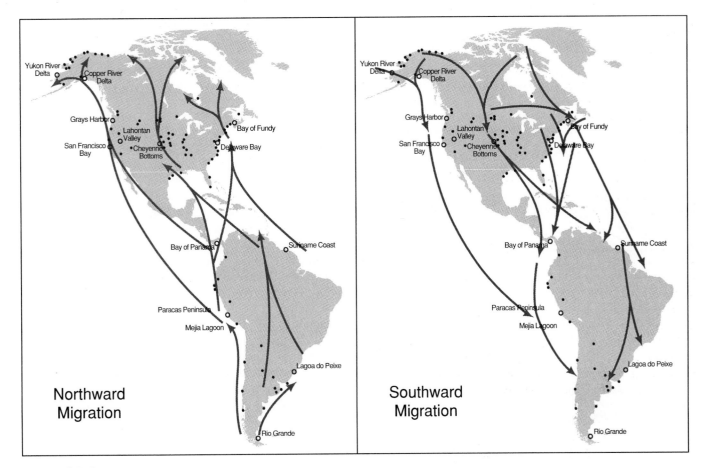

FIGURE 21.1

Fall and spring staging and stopover areas for migrating shorebirds in North and South America. The localities designated are part of the developing Western Hemisphere Shorebird Reserve Network. Circles indicate areas visited by 250,000 or more birds annually (or over 30% of a particular species' population), and dots, areas visited by at least 20,000 birds (or at least 5% of a particular species' population).

Modified from Myers et al. (1987)

FIGURE 21.2

Delaware Bay is a major spring stopover area for migrating shorebirds, such as these ruddy turnstones and sanderlings, which feed heavily on the eggs of horseshoe crabs before moving on north to their arctic breeding grounds.

(VIREO photo by Douglas Wechsler)

individuals spend several days in these woodlands and rebuild their fat stores (Fig. 21.3). In the autumn, Atlantic coastal areas from Nova Scotia south to North Carolina are staging areas for a number of songbirds that fly long distances over water to wintering areas in the Bahamas, West Indies, or South America (Terborgh 1989). The blackpoll warbler, a small songbird, is believed to make a fall flight from the coast of Nova Scotia and New England directly to South America (Nisbet 1970). The birds fly southeast, toward the open Atlantic, until they reach the belt of northeast trade winds, which they ride southwest to their destination. Inland, where forest migrants must rest and feed in whatever wooded areas remain, small woodlots may not serve as adequate stopover areas. In eastern Illinois, for example, Graber and Graber (1983) found that migrating warblers were unable to gain weight when foraging in small woodlots, whereas they were able to store fat when foraging in the more extensive forest areas of southern Illinois.

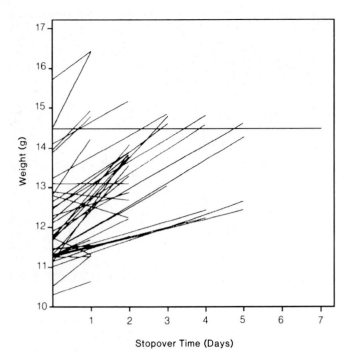

FIGURE 21.3

Changes in weights of migrant Kentucky warblers after arrival at a stopover site on the coast of Louisiana, following a spring migration crossing of the Gulf of Mexico.

(Modified from Moore & Kerlinger 1987)

Long-Term Trends of Migratory Birds

Many migratory songbirds, particularly long-distance migrants breeding in forest habitats, appear to be declining in parts of eastern North America (Aldrich & Robbins 1970; Hall 1984; Leck et al. 1988; Terborgh 1989) and probably also Europe (Grimmett 1987). In the Hutcheson Memorial Forest, a 24-hectare woodlot in New Jersey, for example, a substantial decline occurred in long-distance forest migrants between 1960 and 1984 (Fig. 21.4). Even in virgin mountain forests in West Virginia, Hall (1984) found that 6 of 14 neotropical migrants had disappeared between 1947 and 1983, with the total density of the remaining species being only 63% of that in 1947.

In North America, one of the strongest sources of evidence about trends in populations of migrants comes from the North American Breeding Bird Survey (BBS). The BBS is a system of roadside counts taken annually at about 2,000 locations in Canada and the United States. Each count consists of observations at 50 points located at 0.8-kilometer intervals along a 40-kilometer route. BBS counts have been carried out since 1966. Analysis of these data, collected by many volunteer observers, is complicated by many factors (Thomas 1996). Some analyses suggest that most long-distance, forest migrants declined between 1978 and 1987, after having shown stable or increasing abundances from 1966 through 1977 (Robbins et al. 1989). Other analyses of BBS data show that great variation exists in population trends of migrants over time and space

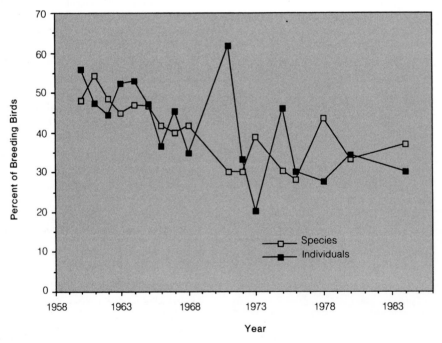

FIGURE 21.4

Changes in the percentage of long-distance migrants in the avifauna of the Hutcheson Memorial Forest in New Jersey between 1960 and 1984.

(Data from C. F. Leck, et al., 1988, "Long-Term Changes in the Breeding Bird Populations of a New Jersey Forest" in *Biological Conservation*, 46:145–57. Essex, England: Elsevier Applied Science Publishers Ltd.)

(Villard & Maurer 1996), and even raise doubt about a general decline of neotropical migrants at large (James et al. 1996).

Data from other sources suggest that real declines have occurred. Hill and Hagan (1991), for example, noted significant declines in 18 of 26 species, based on counts of migrating birds over 50 years in New England. Other evidence for a substantial decline of migrants wintering in the neotropics comes from radar observations of spring migrants arriving on the northern coast of the Gulf of Mexico (Gauthreaux 1992). Flocks of migrants that depart from the Yucatan Peninsula early in the night on one day reach the northern coastline late in the afternoon of the next day. Arriving flocks are evident as masses of "blips" on weather surveillance radar screens. In the early 1960s, major flights were noted on 95% of days between early April and mid-May. From 1987 to 1989, however, major flights were noted on only 44% of all days, suggesting an overall trans-gulf migration only half as extensive.

Declines may also be occurring in some short-distance migrants in eastern North America. Using BBS data, Whitham and Hunter (1992) found that many species breeding in early successional habitats in New England and wintering in similar habitats in the southeastern states had declined in numbers from 1966 to 1988. The decline of these species may relate to the disappearance of early successional habitats in the Southeast as biotic succession and commercial forestry convert more of the landscape to forests. In the Midwest, BBS data show the most extensive declines for grassland birds that winter north of the tropics, probably as a result of habitat reduction (Herkert 1995).

Migrant Birds on Their Breeding Grounds

Several studies have shown that forest fragmentation in eastern North America leads to a major change of the kinds of breeding birds (*See* Chapter 13). Long-distance migrants are most abundant in landscapes that retain large areas of forest and wetlands, and possess large units of forest with small amounts of edge habitat (Flather & Sauer 1996). They make up 80%–90% of the breeding birds in extensive tracts of eastern deciduous forest, but usually less than half of the breeders in small woodlots (Whitcomb et al. 1981). In eastern Illinois, for example, Blake and Karr (1984) found that forest interior, long-distance migrants were completely absent from woodlots less than 24 hectares in area. In Maryland, Whitcomb et al. (1981) found that several long-distance migrants were absent from portions of the state 25 square kilometers in area where forests had been fragmented, even though these areas had some woodlots 50 hectares or more in size, indicating that fragmentation was leading to general regional extinction.

Much of the loss of forest interior species in eastern North America appears to result from increased predation and nest parasitism (*See* Reading 13.1). Most neotropical migrants have open, cuplike nests (Fig. 21.5), and several of the species most sensitive to forest fragmentation nest on the ground or low in the vegetation. In addition, neotropical migrants tend to have small clutches and only one brood per season. Wilcove (1985)

FIGURE 21.5

The hooded warbler is a neotropical migrant that breeds in forest areas of eastern North America, builds an open-cup nest, and winters in mature tropical forest in southern Mexico and Central America. (U.S. Fish and Wildlife Service photo)

showed that predation on open-cup bird nests was much heavier in small woodlots than in large areas of forest, apparently because many kinds of avian and mammalian predators associated with open country or forest edges do extend their hunting some distance into the forest. As woodlots become smaller, this penetration zone constitutes a larger and larger fraction of the total area. Nest parasites, such as the brown-headed cowbird in North America, also contribute to the decline of forest interior species in small woodlots. Competition from generalist, forest-edge birds may also tend to depress forest interior species in small woodlots.

The maintenance of populations of migrants in individual woodlots in fragmented landscapes depends both on survival of adults that have previously bred in that unit and by settlement of first-time breeders, many of which are birds reared in other locations. Thus, most migratory species exhibit regional metapopulations, within which dispersal plays a major role in maintaining local populations.

Migrants on Their Nonbreeding Grounds

Wintering migrants in the tropics often constitute a major component of the bird communities of their nonbreeding areas. In the New World, for example, about half the birds that come from an area of 16.2 million square kilometers in North America crowd into an area of about 2.2 million square kilometers in Mexico, the Bahamas, and the West Indies. In these areas, therefore, their densities must often be greater than in their breeding ranges. Wintering migrants also form a major portion of the total bird population in such areas (Lynch 1989; Terborgh 1989; Rappole 1995).

During the nonbreeding season, migrants use many different kinds of habitats, which may or may not be similar to those used during the breeding season. In some parts of the tropics, such as western Mexico, wintering land bird migrants seem to concentrate in disturbed habitats, in successional communities, or near superabundant food sources such as fruiting trees (Hutto 1989). Resident tropical birds are probably unable to exploit any of these situations fully. In other places, such as the Yucatan Peninsula, wintering migrants appear to use the full spectrum of habitats available, including mature forest (Lynch 1989; Rappole 1995). For the Neotropics, Terborgh (1980) estimated that at least 55 species of North American migrants winter partly or entirely in mature tropical forests. These include many flycatchers, thrushes, vireos, warblers, and tanagers (Fig. 21.5). Many of these birds, particularly those that are primarily insectivorous in winter, establish individual territories. Other insectivorous species may become members of mixed foraging flocks of resident and migrant species, the flock often acting as a territorial unit. On the other hand, species that are highly frugivorous in winter may form flocks that move from place to place and exploit a changing spectrum of fruit resources. In any case, wintering migrants become intimately integrated into the tropical bird community (Terborgh 1989; Rappole 1995).

Tropical deforestation has major implications for many migratory land birds, particularly in southeast Asia and the Neotropics. In analyses of BBS data, Robbins et al. (1989) noted that declines of neotropical migrants between 1978 and 1987 were significantly greater for species wintering in forests than for those wintering in more open vegetation types. The extent to which tropical deforestation has contributed to the decline of neotropical migrants remains controversial (Holmes & Sherry 1988; Hutto 1988). Rappole and McDonald (1994) presented 14 hypotheses about relationships that should exist if declines of neotropical migrants were related to breeding-ground changes. Although the evidence required for confident tests of these hypotheses was mostly lacking, that available conflicted with 10 of these 14 hypotheses. Sherry and Holmes (1966) also found that in Jamaica intraspecific competition among wintering warblers was strong, indicating that habitat was a significant limiting factor in winter.

In any case, continued destruction of tropical forests will reduce the habitat of certain species (Lynch 1989), and sooner or later, tropical deforestation will become a substantial contributor to the decline of long distance migrants. In the southeastern United States, the Bachman's warbler probably became extinct because its tropical forest winter habitat in Cuba was converted to sugar cane fields (Powell & Rappole 1986).

Other changes in the landscapes of wintering areas may also pose problems for northern hemisphere migrants. Fragmentation is also occurring in tropical forest habitats, and some wintering migrants are absent from small woodlots that otherwise seem suitable (Rappole 1995). In Africa, the major wintering areas of many land birds lie in the sub-Saharan savanna zone. Here, desertification appears to have caused declines of several long-distance migrants (Grimmett 1987). Throughout the world, waterfowl, shorebirds, and other water birds utilize coastal or freshwater shores and marshes as wintering sites. Some shorebirds, storks, and other water birds also winter in open grasslands and savannas. The destruction of coastal marshlands, especially in North America and Europe, and the conversion of native grasslands to cultivated cropland, particularly in South America, also threaten wintering populations of many species. Extensive changes are also occurring in African wetlands, where large numbers of water birds from western Eurasia winter (Grimmett 1987).

Hunting of Migratory Birds

Several migratory species are legal gamebirds in North America. These include most ducks and geese, sandhill cranes, shorebirds such as Wilson's snipe and woodcock, and various doves and pigeons. The largest harvests are of waterfowl and doves. The annual harvest of mourning doves in the United States is about 50 million birds. Before the ecological roles of raptors were appreciated, migrating hawks and eagles were often killed for sport (See Reading 21.1).

North American waterfowl migrations follow four major pathways, or **flyways** (Fig. 21.6): **Atlantic, Mississippi, Central,** and **Pacific** (Hawkins et al. 1984). These flyways are not as sharply distinct as one might imagine, but birds breeding in the northern portions of a particular flyway do tend to stay within the overall flyway system in their movements during the nonbreeding season. The flyways are perhaps as much administrative units as they are biological units. Surveys of reproductive success in the nesting regions of the different flyways have traditionally been used to determine bag limits and the length of the hunting season throughout the respective flyways.

During the 1970s, the total North American breeding population of ducks, including sea ducks and mergansers, averaged about 62 million birds, and yielded a fall flight of more than 100 million birds (U.S. Fish and Wildlife Service 1986). The population generally declined during the 1980s and early 1990s, but have recovered rapidly in the mid-1990s. Continental goose populations range between 4 and 6 million, and those of tundra swans about 129,000. In addition, about 10,000 trumpeter swans occur in western North America. These populations, particularly those of ducks, are subject to great fluctuations in reproductive success. Many ducks lay clutches of eight to ten eggs, so that in good years reproduction can increase the total population threefold or more. Because many ducks nest in marshlands or on small prairie ponds, however, drought years can be equally disastrous.

Hunters take large numbers of waterfowl and many other migratory birds, both legally and illegally. Recent hunting harvests of waterfowl in Canada and the United States have ranged from 10.8 million ducks in 1968 to 20.2 million in 1970, and from 1.9 million geese in 1974 to 2.5 million in 1980. About 80% of the total North American harvest occurs in the United States. The greatest harvests of ducks are in the Mississippi and Pacific flyways, and of geese in the Pacific flyway. Subsistence harvests by native Americans in Alaska and Canada constitute about 5% of the total duck harvest and 7% of the total goose

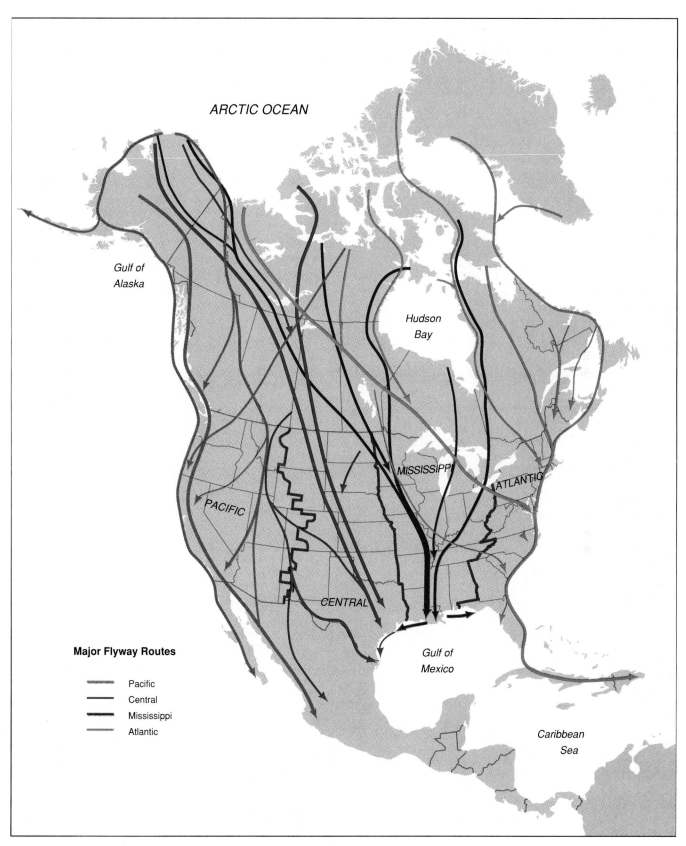

ARCTIC OCEAN

Gulf of
Alaska

Hudson
Bay

MISSISSIPPI

ATLANTIC

PACIFIC

CENTRAL

Gulf of
Mexico

Caribbean
Sea

Major Flyway Routes

Pacific
Central
Mississippi
Atlantic

FIGURE 21.6

Waterfowl flyways in North America.

(Sources: U. S. Fish and Wildlife Service, *North American Waterfowl Management Plan,* 1986, U. S. Department of the Interior, Washington DC, and *Environment Canada,* Ottawa, Canada.)

F ormerly, when most raptors were considered to be enemies of both wildlife and domestic fowl, hawk shooting was a popular autumn practice in some parts of North America. Shooting was concentrated in areas where southward migrations of these birds were concentrated. Major southward flyways of raptors follow the Atlantic coastline and the shorelines of other large water bodies such as the Great Lakes (Reading Fig. 21.1A). Large numbers thus become concentrated in locations such as Cape May, at the southern tip of New Jersey, and Point Pelee, a peninsula on the northern shore of Lake Erie. Other major flyways follow the ridges of the Appalachian Mountains. Here, the birds glide southward in the updrafts created as winds strike the western face of the major ridges. Since the major Appalachian ridges extend unbroken for hundreds or thousands of kilometers, they provide a "free ride" for raptors from New England to Georgia. In the early 1900s, hawks were shot at many ridge peaks along the Appalachians.

In 1934, one of the most notorious of these shooting locations (Reading Fig. 21.1B) was converted into **Hawk Mountain Sanctuary** (Brett 1991). Hawk Mountain has become both a center for study of raptor migrations, and a center of conservation effort for protection of raptors. Now that these birds are protected, many of the former shooting points have become observation sites for ornithologists and bird watchers interested in raptors (Reading Fig. 21.1C).

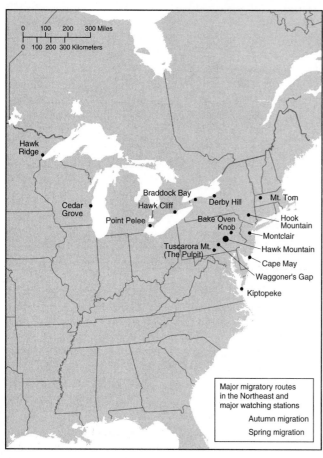

READING FIGURE 21.1A

Major spring and fall migration routes of raptors in eastern North America, with important observation stations.

(From J. J. Brett, *The Mountain and the Migration: Revised/Expanded Edition.* Copyright © Cornell University Press. Used by permission of the publisher.)

READING FIGURE 21.1B

Hawk Mountain, Pennsylvania, was once a major shooting gallery for hawks during autumn migration. These are part of more than 230 hawks killed near the main lookout on one day in October, 1932.

(Photo by Richard Pough and Henry Collins, Jr., courtesy of Hawk Mountain Sanctuary Association)

READING FIGURE 21.1C

Today, Hawk Mountain is a sanctuary dedicated to observation and study of raptor migration.

(Photo courtesy of Hawk Mountain Sanctuary Association)

FIGURE 21.7

The Sacramento National Wildlife Refuge is a major wintering area for snow geese that have moved south along the Pacific Flyway.

(U.S. Bureau of Sport Fisheries and Wildlife photo by P. J. Van Huizen)

harvest. Several thousand tundra swans also are harvested for subsistence, and a very small recreational harvest occurs in the United States. For ducks, hunting mortality is estimated to be roughly 50% of total over-winter mortality, and that for geese somewhat higher.

In addition to birds killed and recovered by hunters, mortality includes birds that are crippled, but not recovered. Most of these ultimately die. For ducks, crippling losses are estimated to be 20%–40% of birds hit by gunfire, or 1.6 to 4.4 million ducks in 1992 (Norton & Thomas 1994). Waterfowl also ingest shotgun pellets that fall into the shallow waters of marshes and ponds, taking them in as they would grit to promote the action of the gizzard. The intake of lead pellets, formerly used in shotgun shells designed for hunting waterfowl, caused lead poisoning, which killed about 1.6–2.4 million ducks annually. Waterfowl hunters have been required to use steel shot since the 1991–1992 season, so that lead poisoning is declining.

The **National Wildlife Refuge System** of the United States, and the comparable system in Canada, function largely for protection of migratory birds, both game and nongame species. National Wildlife Refuges exist in all 50 states, and include 500 sites with a combined area exceeding 13.7 million hectares. Most of these refuges have the primary purpose of protecting breeding, migration, or wintering areas of waterfowl and other migratory birds (Fig. 21.7). Many others combine preservation of migratory birds and protection of other forms of wildlife.

In other parts of the world, hunting of migratory birds is practiced in a manner unfamiliar to most North Americans. In Europe, North Africa, and the Middle East, enormous numbers of migratory songbirds are hunted for food. In Italy, for example, about 240 million songbirds are harvested annually. Birds are shot, captured with nooses, taken in nets made of fine nylon mesh, and trapped with birdlime. More than 15 million are also taken in Spain, France, and Lebanon. Large numbers are killed on the islands of Mallorca, Cyprus, and

Malta, and some harvest occurs in all countries bordering the Mediterranean. All told, about 500 million small birds are harvested annually in this region.

Other Migration Hazards

Tall lighted structures create a hazard to songbirds migrating at night, especially during their fall migration. The structures involved include television and radio towers (Crawford 1981), lighthouses, refinery flare stacks (Bjorge 1987), and tall, lighted structures such as the Washington Monument. Migrating birds evidently become disoriented in the vicinity of these structures, and are killed by colliding with them. Formerly, when most airports used fixed-beam ceilometers, birds often became disoriented and were killed by collision with the ground. Most such mortality occurs on overcast nights when concentrated southward flights have been triggered by the passage of a cold front. The precise cause of disorientation is not understood, but may involve temporary blinding or orientation of birds to the light as if it were the moon.

At the WCTV tower near Tallahassee, Florida, scientists from the Tall Timbers Research Station and Florida State University have monitored mortality since 1956. Over this period, the tower, 308 meters tall, has claimed over 42,000 birds of 189 species, or about 1,600 birds annually. Occasional massive kills have been recorded at other locations. On the nights of September 18 and 19, 1963, for example, about 30,000 birds were killed at one tower near Eau Claire, Wisconsin. Overall, perhaps half a million birds may be killed annually in this manner in North America.

Conservation of Migratory Birds

In North America, migratory birds have been the subject of agreements among several countries, particularly Canada, the United States, and Mexico. The earliest agreement was the **Migratory Bird Treaty Act** (1918), in which Canada and the United States protected nongame species and specified coordinated management policies for game species. In 1980, the United States passed the **Fish and Wildlife Conservation Act,** which requires the Fish and Wildlife Service to monitor populations of all nongame birds, and to identify the conservation needs of any that are approaching endangerment. Specific funding was not provided for this effort, although the United States Congress appropriated $1.75 million for nongame activities in 1988 and 1989. The Office of Migratory Bird Management in the Fish and Wildlife Service devoted about 10% of its $8 million basic budget to nongame species in 1989 (Gradwohl & Greenberg 1989). Specific objectives for coordination of research on nongame birds by the Fish and Wildlife Service and other private and state groups have been developed, however (Office of Migratory Bird Management 1990).

To promote the preservation of migratory waterfowl, the United States and Canada have developed the **North American Waterfowl Management Plan** (U.S. Fish and Wildlife Service 1986). This plan emphasizes habitat protection and restoration, and aims to restore duck populations at the average levels prevailing in the 1970s, and to maintain goose populations at close to the high levels of the 1980s. The plan identifies specific regions of major importance to waterfowl, in which habitat conservation activities would be concentrated. It covers the period from 1986 through 2000, and is subject to review at five-year intervals. To implement this plan, Canadian and United States agencies must spend $1.5 billion, two-thirds of it in Canada. By 1993, slightly over $1 billion had been invested in this effort (Patterson 1994).

For migratory shorebirds, an international consortium of private and governmental organizations is working to develop a **Western Hemisphere Shorebird Reserve Network** (Myers et al. 1987). The goal of this effort is to promote a network of sites throughout North and South America that will guarantee adequate breeding, staging, stopover, and wintering habitats for shorebirds (Fig. 21.1). By 1987, 23 state, provincial, or national agencies in Canada, the United States, and Peru had become involved in this effort, together with several private conservation groups. As of 1994, 21 reserves in seven countries had been incorporated into this network (Patterson 1994).

Other Migratory Animals

A number of other migratory animals face risks associated with their concentration in special sites during nonbreeding or breeding stages of their seasonal cycle. A number of bats, for example, migrate to cave locations where they spend the winter months. In some cases, as for the gray bat of the southeastern United States, a large fraction of the total species' population may be concentrated in a few caves (Tuttle 1979). Similarly, 90 million or so monarch butterflies congregate for the winter in enormous aggregations (*See* Chapter 5) in forest areas in the mountains of central Mexico and in coastal California (Brower & Malcolm 1991). Other species, such as sea turtles (*See* Chapter 11), some ungulates, several whales, and many pinnipeds depend on special sites for reproduction. The Porcupine Herd of tundra caribou, for example, migrates to the coastal plain portion of the Arctic National Wildlife Refuge to calve (*See* Chapter 7). The entire population of the California gray whale, for example, winters, calves, and breeds in a few large lagoons along the Pacific coast of Baja California.

Other migratory species face risks because they have very confined migrationways. Anadromous fish such as Pacific salmon must migrate through long stretches of rivers such as the Columbia to reach their headwater spawning areas (*See* Chapter 16). For much of their southward and northward migration, gray whales closely follow the Pacific coastline of North America. In these cases, it seems that the species are at risk not so much because they are migratory, but because of the specialized or restricted environments that they have come to use at different seasons.

Conservation efforts for these species are focused largely on protecting and restoring the special sites that are required. Mexico has protected five monarch butterfly wintering sites, and in 1995 Canada established three reserves in areas along the Great Lakes where monarchs congregate in migration.

Migratory Species and Global Change

Migratory species will certainly be affected by global warming and sea level rise, but their mobility may give them greater opportunity to respond. The breeding areas of long-distance migrant shorebirds lie at high latitudes, where climatic change is expected to be several times as great as in the tropics. The consequences of such change for breeding habitats of these species are difficult to predict. Climatic change might also upset the timing relationships that now exist between the use of stopover areas and arrival on the breeding grounds. If the arctic breeding grounds become favorable earlier in the season, but horseshoe crab spawning in Delaware Bay remains at about its present date, for example, shorebirds may tend to reach the breeding grounds too late.

Global warming, and the associated sea level rise, will severely reduce the area of many coastal wetlands, particularly in temperate and tropical regions. This change may threaten many coastal staging and stopover areas for migratory birds (Smit et al. 1987). Coastal wetlands, already pressured by development and pollution, will be constricted between the rising water level and developed portions of the landscape that humans will try to protect against flooding (Lester & Myers 1989).

Summary

1. Birds show all degrees of development of seasonal migration, but long distance migrants are highly specialized, genetically programmed species that are intimately integrated into the communities occupied both during the breeding and nonbreeding seasons.

2. Many long-distance migrants, especially in the Nearctic-Neotropical System, appear to have declined in abundance, probably due to a combination of factors such as habitat conversion and fragmentation on their breeding ranges and deforestation in their tropical wintering areas.

3. For migratory waterfowl, hunting is a major component of annual mortality, but regulated harvests, together with protection of wetlands habitats, may allow populations to be restored to levels that permit continued large hunting harvests.

4. Many migratory mammals, reptiles, and fish appear to be at high risk because migration has required the use of confined migrationways, or been coupled with the use of specialized breeding or nonbreeding habitats.

5. Migratory species most affected by global change will be those breeding at high latitudes, since global warming will be greatest in these regions, and those that depend on coastal staging and stopover areas, which will be reduced in area by sea level rise.

QUESTIONS FOR DISCUSSION

1. What characteristics of migratory animals might enable them to survive and even flourish in the face of global change? What characteristics might put them at severe risk?

2. If declines in abundance of migratory species are partially offset by increases in abundance of nonmigratory species using similar resources and habitats, do you think that this might impede the recovery of the migrant forms?

3. Do you think that human transformation of natural habitats or direct killing by humans poses the greater risk of extinction for most migratory animals?

SUGGESTED READING

Brett, J. J. 1991. *The mountain and the migration. Revised edition.* Cornell Univ. Press, Ithaca, NY. A history of the Hawk Mountain Sanctuary in Pennsylvania, and its transformation into a center for the study of the ecology of migratory raptors.

Line, L. 1993. Silence of the songbirds. *National Geographic* **186(6):**68–91. A popularly written, well-illustrated account of declines in migratory land birds of eastern North America, their varied causes, and some of the efforts being made to slow these declines.

Rappole, J. H. 1995. *The ecology of migrant birds: A neotropical perspective.* Smithsonian Institution Press, Washington, DC. A concise, readable summary of what we know about the ecology of migratory birds, and the implications of this knowledge for conservation.

chapter

22

Chemical Pollution

Human activities have introduced many toxic and biologically active chemicals into the global ecosystem and have greatly increased quantities of actively circulating, natural substances with similar properties. These include pesticides, synthetic industrial chemicals, petroleum compounds, heavy metals and other toxic trace elements, radioisotopes, and plastics. The effects of these toxins are both direct and indirect, and they have the potential for interactions that may be complex and synergistic.

In this chapter we shall consider the extent and ecological consequences of pollution by toxic chemicals. Finally, we shall examine some of the efforts to control chemical pollution.

Chemical pollutants are still one of the most serious threats to wildlife throughout the biosphere.

Outline

PESTICIDES

Synthetic organic pesticides have created enormous problems in ecosystems, both aquatic and terrestrial. Hundreds of basic pesticide chemicals exist, and these are utilized in thousands of formulations and brand-name products. Several pesticides have become general contaminants of the biosphere, having been transported worldwide by air and water circulation.

Although pesticides are usually aimed at a single species or group of ecologically similar species, many nontarget organisms can be affected. Furthermore, some long-life pesticide chemicals establish biogeochemical cycles that carry them to distant localities and concentrate them in unexpected ways. Repeated use of these very powerful chemicals has also made them major agents of evolutionary change in many animal and plant species. Pesticide impacts are a continuing concern of conservation ecology.

We shall first consider the chlorinated hydrocarbon pesticides as an example of the far-ranging impacts that pesticides can have. Then we shall consider some of the dangers posed by pesticides of other types.

Chlorinated Hydrocarbon Pesticides

The pesticidal properties of DDT (dichloro, diphenyl, trichloroethane), were discovered in 1939 by a Swiss biochemist (Dunlap 1981). After World War II, it rapidly came into use for the control of agricultural and forest pests and insect vectors of human diseases. DDT belongs to a group of chemicals known as **chlorinated hydrocarbons (CHs).** CHs include dieldrin, aldrin, endrin, chlordane, heptachlor, toxaphene, mirex, kepone, and many others. The members of this chemical group share four characteristics that, in combination, make them serious environmental pollutants (Wurster 1969). These characteristics are:

1. Chemical stability or persistence. CHs degrade very slowly in nature. DDT and its major metabolites, for example, have an environmental half-life of 10–15 years, meaning that after this time, half the quantity applied in a particular location still survives somewhere. Much of the CHs ever produced still exist somewhere in nature.

2. High mobility. CHs in vapor, dissolved, or particulate form are transported by movements of air and water. Some are very volatile, and are dispersed widely by atmospheric circulation (Simonich & Hites 1995). Several of the CHs have become biosphere-wide in distribution, and their residues are detectable in mid-ocean waters and in antarctic snow.

3. High solubility in lipids. CHs have a very low solubility in water, but a high solubility in lipids (fats and oils). This means that wherever water comes in contact with lipids, CHs tend to move into the lipids. Since lipids are a basic component of living tissues, organisms tend to take up CHs from the water that surrounds them, or from the food or water they ingest.

4. Toxicity and biological activity. Pesticides are usually toxic to organisms other than the targeted pests. Even when they are not lethal, these chemicals can be biologically active, disrupting physiological processes such as those involved in reproduction and development.

Soon after DDT came into general use, **direct side effects** were observed on nontarget organisms in areas of heavy applications (Carson 1962; Rudd 1964). Heavy spraying of DDT against the elm bark beetle (vector of the Dutch elm disease fungus) and the Japanese beetle resulted in the mortality of songbirds in areas of the eastern and midwestern United States. Spraying to control spruce budworm outbreaks in forests of Canada and the northern United States led to heavy mortality of salmonid fish in streams exposed to the applications or receiving runoff carrying DDT. Rachel Carson documented these events dramatically in her 1962 book, *Silent Spring.*

Rachel Carson also recognized some of the first **indirect side effects** of pesticide use—effects distant in time or far removed in space from the actual applications of pesticides. Indirect effects have proved to be the greatest threat to wildlife. They involve the uptake and transfer of CHs along food chains, often with the concentration increasing at each step. This process involves both **bioaccumulation,** the concentration of CHs from an organism's physical environment, and **biomagnification,** the concentration from one link to the next in a food chain. Phytoplankton, for example, bioaccumulate CHs from water because of the solubility relationships of CHs in water and fats. Zooplankton that feed on phytoplankton retain most of the CHs from their food, as do fish and the fish-eating birds and mammals on up the food chain. In a salt marsh on Long Island, for example, Woodwell et al. (1967) found that DDT had contaminated all members of the ecosystem, but that concentrations varied through three orders of magnitude. In zooplankton, the concentration was about 40 parts per billion, while in various water birds values ranged from 3 to 75 parts per million (Fig. 22.1).

Bioaccumulation and biomagnification have been recorded in many other situations, and for several other CHs. Bioaccumulation alone can increase the concentrations of CHs in aquatic organisms by a factor of 10^3 to 10^6. Biomagnification at one food chain transfer tends to be less, but may be as much as thirtyfold.

Contamination of ecosystems by CHs has many effects other than direct mortality of adult animals. The most widespread and insidious effect of CHs is eggshell thinning in birds at the ends of long food chains. Shell thinning was first recognized for the peregrine falcon and other raptors in Great Britain (Ratcliffe 1967). The decline of populations of these species was correlated with an unusually high frequency of egg breakage in nests. Comparison of the thicknesses of eggshells that had been collected and placed in museum collections prior to 1945 with those of eggs taken from the wild revealed a substantial thinning.

Subsequent studies showed that shell thinning was extensive in four groups of birds: 1) bird-eating raptors such as the peregrine falcon (Fig. 22.2), 2) fish-eating raptors such as the bald eagle (Fig. 22.3), 3) fish-eating water birds such as pelicans,

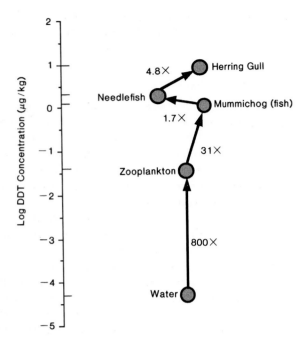

FIGURE 22.1

Bioaccumulation and biomagnification of DDT in the food chain of an estuary on Long Island, NY, based on studies in the 1960s.

(Data from G. M. Woodwell et al. 1967, "DDT Residues in an East Coast Estuary: A Case of Biological Concentration of a Persistent Pesticide" in *Science*, 156:821–823. Washington DC: American Association for the Advancement of Science.)

FIGURE 22.2

The peregrine falcon disappeared from much of eastern North America due to eggshell thinning induced by DDT contamination.

(World Wildlife Fund photo by E. Hosking)

FIGURE 22.3

The bald eagle was one of many fish-eating raptors and water birds to suffer population decline due to eggshell thinning caused by DDT.

(U.S. Fish and Wildlife Service photo by W. A. Troyer)

and 4) carrion feeders such as the European carrion crow. For raptors, most species that experience an average shell thinning of more than 18% are unable to maintain stable populations (Lincer 1975). Many species were affected to this extent in the 1960s and 1970s, so that their numbers declined. Counts of migrating hawks at Hawk Mountain Sanctuary, Pennsylvania (*See* Reading 21.1), for example, showed that bald eagles, golden eagles, sharp-shinned hawks, Cooper's hawks, and peregrine falcons all declined significantly from 1946 to 1972 (Bednarz et al. 1990).

With restriction of the use of CHs in North America and Europe in the 1970s and 1980s, many bird species affected by eggshell thinning are recovering (Blus 1982; Anderson & Gress 1983; Chapdelaine et al. 1987). Several of the most seriously affected raptors have recently shown increasing numbers in annual counts at Hawk Mountain (Bednarz et al. 1990). The osprey populations of Long Island Sound, one of the first populations to show DDT-induced decline, have recovered rapidly (Spitzer & Poole 1980). In Ontario, the reproductive success of bald eagles, which had fallen to less than one young fledged for every two nests in 1966, climbed to over one bird per nest by 1981 (Grier 1982).

CH problems still persist in many places, however. In North America, snowy egrets breeding in Idaho (Findholt 1984) and white-faced ibis breeding in the Great Basin (Henny & Herron 1989) still show thin-shelled eggs and reduced breeding success due to DDT residues picked up on their winter ranges in Mexico. In Florida, fulvous whistling ducks show substantial body burdens of several CHs, including chemicals such as aldrin that are rapidly converted to distinct metabolites (Turnbull et al. 1989). In both North America and Europe, lingering contamination of river delta regions still leads to reduced breeding success in some bird species (Anthony et al. 1993; Dirksen et al. 1995).

In many developing countries, however, use of CHs remains high, or is increasing, and environmental pollution is serious (Lincer et al. 1981). One cause has been the resurgence of malaria throughout the tropics, which has led to the revival of DDT use against *Anopheles* mosquitos in countries such as India

FIGURE 22.4

Brown pelicans off the coast of southern California suffered almost complete reproductive failure due to eggshell thinning caused by industrial pollution of coastal waters by DDT.

(Photo by G. W. Cox)

and Brazil. Reports of declining reproduction by raptors in Zimbabwe, due to effects of DDT and other CHs (Tannock et al. 1983), indicate that pesticide pollution is far from eliminated on a worldwide basis.

The southern California coastal marine environment experienced massive contamination by DDT in the 1950s and 1960s. During this period, the peregrine falcon and bald eagle disappeared as nesting birds on the Channel Islands, and populations of brown pelicans (Fig. 22.4) experienced almost complete nesting failure due to shell thinning. The source of the DDT turned out to be a chemical plant in Los Angeles that had been discharging chemical wastes into the Los Angeles sewer system (MacGregor 1974). From there they passed directly into the Pacific Ocean. Once this source was discovered, the discharge was terminated. Since 1970, the populations of brown pelicans have slowly recovered (Anderson & Gress 1983), and are once again common along the southern California coast.

Even in the 1990s, however, DDT effects persist in southern California waters. In 1990 the California State Department of Health recommended closure of the white croaker fishery because of lingering DDT contamination. The National Oceanic and Atmospheric Administration has filed suit against several companies for damages to fisheries and the costs of cleaning up the DDT-contaminated marine sediments.

In 1970, a similar episode of industrial pollution occurred at Wheeler National Wildlife Refuge on the Tennessee River near Huntsville, Alabama (O'Shea et al. 1980). In this case, an estimated 4,000 tons of waste DDT from a plant at the Redstone Arsenal were discharged into the refuge. The effects of this pollution episode will probably also continue for many decades.

Other Pesticides

Carbamates and **organophosphates** are groups of pesticides that offer alternatives to many of the CHs. These compounds are very toxic, but their environmental halflife is short, and they do not show biological concentration or transport to distant locations. Thus, most of their side effects are direct (Smith 1987), although some sublethal effects on behavior and reproduction have been noted (Grue et al. 1983).

Carbamates such as carbofuran, and organophosphates such as parathion, methyl parathion, and diazinon, sometimes kill herbivorous birds such as waterfowl or sage grouse feeding in recently treated areas, and sometimes of other birds that have fed on insects killed by these pesticides (Balcomb et al. 1984; Smith 1987; Blus et al. 1989; Fox et al. 1989). Carbofuran and a number of other pesticides are also commonly applied as granules, which birds sometimes ingest directly as grit (Mineau 1988). The problems with carbofuran have led the United States Environmental Protection Agency to recommend that it be banned (Thomas 1989). The use of diazinon to control insects on golf courses and sod farms was banned in 1990 in the United States because of frequent mortality of Canada geese and other birds.

Herbicides, like insecticides, belong to several different chemical groups. Most herbicides have a short environmental life. Although not without toxicity to nontarget organisms, few herbicides have caused major direct or indirect side effects (Freemark & Boutin 1995). Some indirect effects have been noted on survival of young game birds that feed on arthropods, apparently because of the reduction of arthropods that depend on weeds in cropland. Some of the **phenoxy herbicides,** which are similar in structure to plant growth hormones, contain trace quantities of dioxin, an intermediate-life chemical that may be concentrated along food chains. Dioxin is both cancer-inducing and a cause of birth defects in laboratory animals, so that herbicides containing dioxin are presumably dangerous to all vertebrates. Reduced reproduction and developmental abnormalities have been documented in ducks nesting in areas contaminated with wastes from herbicide plants (White & Seginak 1994).

MARINE OIL POLLUTION

Petroleum enters marine ecosystems in various ways, some conspicuous, some not so conspicuous. The best estimate of total input is 3.2 million tons per year, but the limits of certainty for this estimate are 1.7 to 8.8 million tons (National Academy of Sciences 1985). Almost half of all input is from land-based sources. Much of this is oily wastes that are discharged from municipal and industrial sources into the ocean, or into rivers that carry them into the ocean. The rest is from evaporation and combustion that introduce petroleum compounds into the atmosphere, from where they are precipitated into the oceans. These sources of pollution are less concentrated than sea-based sources, so their effects are usually not apparent.

Sea-based inputs of petroleum are usually concentrated and conspicuous. In some places, natural petroleum seeps exist, but human activities lead to far larger inputs. What was probably the largest single marine oil pollution event occurred in January, 1991, when over 1,700,000 tons of oil spilled into the Persian Gulf during the war to liberate Kuwait. Accidents at offshore production platforms also cause severe pollution. A blowout at the Mexican platform, Ixtoc I, in 1979 released between 500,000 and 1,400,000 tons of crude oil into coastal waters. Tanker accidents also cause severe, localized oil pollution. The grounding of the *Amoco Cadiz* off the coast of France in 1978 spilled 220,000 tons of oil into coastal waters, and the 1989 grounding of the *Exxon Valdez* in Prince William Sound, Alaska, spilled 41,000 tons into waters rich in wildlife (*See* Reading 22.1). Other major spills occurred in the British Isles in 1994 and 1995. The potential for tanker accidents is now greater than ever, with over 2 billion tons of petroleum being moved by ocean transport annually, and with many tankers capable of carrying loads of over 400,000 tons.

Nevertheless, most oil from sea-based sources comes from normal operations of tankers and other ships (National Academy of Sciences 1985). These operations include the discharge of water and oily wastes from the bilges of ships of all types and the discharge of ballast water by tankers. When a tanker is carrying a load of petroleum, the weight of the cargo causes it to ride low in the water, making the ship stable. When the ship has discharged its load, the cargo holds must be partially filled with water to make the ship stable. This ballast water becomes mixed with the oil residues coating the walls of the cargo chambers. When the ballast water is discharged, substantial amounts of oil are released into the ocean.

A procedure known as the **Load-On-Top (LOT)** system has been developed to minimize oil discharge in ballast water. This involves pumping the ballast water into certain tanks, where the oily residues separate and concentrate at the surface. The ballast water is then released from the bottom of the tanks. The residues are thus retained and the new load pumped in "on top" of them. Most modern tankers are equipped to follow this procedure. The LOT system, however, takes time and requires calm seas to be fully effective.

Recent estimates of marine petroleum pollution suggest that it is declining somewhat. Part of this can be attributed to improved technology, such as the LOT system, and part to enforcement of regulations against discharges of petroleum wastes at sea. Samples of oil from a pollution event can be analyzed to obtain a chemical "fingerprint" that can identify the geographical source and state of refinement of the oil. Combined with information on ship movements, the source of an oil slick can usually be determined. A 10,000-ton spill off the California coast in 1984 was traced to its source, and a $1,700,000 judgment was obtained against the ship owners.

Several areas of the world's oceans show very heavy oil pollution. These include enclosed marine basins such as the Mediterranean Sea, Gulf of Mexico, Caribbean Sea, Red Sea, and, of course, the Persian Gulf. Spills in these enclosed basins are less easily dispersed and are more likely to affect coastal ecosystems than elsewhere. Major tanker routes also experience heavy oil pollution. Thus, both the east and west coasts of Africa receive heavy pollution from tanker traffic between the Middle East and Europe. The route from the Middle East across the Indian Ocean and through the South China Sea to far eastern ports is another area of heavy pollution.

Oil discharged into the ocean forms slicks, which range from a fraction of a millimeter to 20 millimeters in thickness. The lightest components evaporate within a few days, reducing the total amount of petroleum by 60% or more. The light, non-tar components rapidly dissolve or disperse, as well, and their concentrations usually fall to background levels in two to six months. Heavier components may be churned into a viscous emulsion, known as chocolate mousse, that is about 75% water. Mousse forms readily in cold, icy waters, and may persist for a long time. Over time, photochemical and microbial processes gradually break down petroleum compounds. Oil is also precipitated gradually from surface waters by mixing with mineral sediment, adsorption on particulate matter, incorporation in zooplankton fecal pellets, or uptake by phytoplankton, all of which create particles that eventually sink. The heaviest components eventually form tar balls, which float indefinitely but are inert chemically.

The effects of oil pollution vary greatly in different marine environments (Teal & Howarth 1984; Neff 1990). In the open ocean, oil slicks may kill plankton and fish, but severe effects on these organisms usually last only for days to a few weeks. Sea birds and mammals, however, may be fouled badly by these slicks.

When oil reaches coastal areas, longer-term impacts occur. On sandy beaches, the effects of oil pollution may last for weeks to months. Oil may be buried for periods by sand deposition, and exposed later by normal beach erosion. Ultimately, the reworking of beach sands leads to chemical and mechanical dispersal of the petroleum residues.

Rocky coastlines experience still longer impacts; here the effects may last for months to years. Deposition of oil on rocky intertidal surfaces kills almost all organisms, and coats the surfaces with tars that weather very slowly, inhibiting recolonization. The inhabitants of rocky intertidal areas, however, are adapted to colonization of disturbed sites, and in time recovery occurs without serious damage.

The most prolonged impacts of coastal oil pollution occur in estuary and salt marsh areas, where they may last from years to decades. This long-term effect occurs because large quantities of oil may become buried in anoxic sediments where little or no chemical degradation occurs. After months or years, this oil may be reexposed and exert toxic effects anew.

A striking example of the long-term impact of oil pollution on an estuary ecosystem followed the breakup of a barge near Buzzard's Bay, Massachusetts in September, 1969 (Sanders, et al. 1980). This small spill, involving only 630 tons of fuel oil, caused severe immediate effects. Much of the oil became buried in the sediments, however, to be reexposed and reburied time after time by tidal action. In 1989, some 20 years later, undegraded oil could still be found in places (Teal 1993).

On 24 March 1989, the tanker *Exxon Valdez* struck a reef in Prince William Sound, Alaska, spilling 41,000 tons of crude oil, only a part of its total load (Reading Fig. 22.1A). Currents carried the oil southwest, fouling about 16% of the sound's shoreline and about 13% of shoreline on spill-affected coasts of Kodiak Island and the Kenai Peninsula. About 40%–45% of the spilled oil was deposited on shores; the remaining oil evaporated, dissolved, or was carried to the open ocean.

Cleanup activities were carried on for four summers. These efforts included washing rocky shores with hot water, mechanical collection of oil and tar, and bioremediation. Most of the cleanup, however, resulted from natural processes. Cleanup efforts in some cases proved destructive. The hot-water washing of rocky coasts killed or buried many invertebrates and carried the oil to lower beach levels that often had not been affected severely (Reading Fig. 22.1B). As a result, treated coastlines were more heavily damaged than untreated areas. All told, about $2.1 billion were spent in cleanup efforts.

In the days after the spill, many birds and mammals were killed by the oil slicks. About 350,000 to 500,000 seabirds, mostly common and thick-billed murres, 200 or more bald eagles, 3,500 to 5,500 sea otters, and 350 harbor seals were estimated to have been killed. Of the 35 killer whales living in the sound, 14 may also have succumbed. Breeding by seabirds in colonies affected by the spill was also reduced.

Longer-term effects included major impacts on salmon fisheries. Fry of the pink salmon, which spawn in the mouths of streams along the coastline of the sound, were largely sterilized in 1989 and 1990. Sockeye salmon, which could not be harvested for human consumption because of oil contamination, returned to their natal streams and spawned in unusual volume. A large fraction of the young sockeye fry starved, however, because food supplies in the streams was inadequate for their enormous numbers.

In several proceedings from 1989 through 1994, Exxon was assessed or agreed to pay about $1.7 billion in civil and criminal penalties. In 1994, Exxon Corporation was ordered to pay $5 billion in punitive damages to 34,000 Alaskans whose interests suffered damage from the *Exxon Valdez* spill. In 1996 this punitive judgment was still under appeal.

READING FIGURE 22.1A

The oil tanker *Exxon Valdez* went aground on a reef in Prince William Sound, Alaska, in March, 1989, spilling 41,000 tons of crude oil and polluting waters and shorelines over 25,000 square kilometers.

(Photo courtesy of A. W. Maki and Exxon Company, U.S.A.)

READING FIGURE 22.1B

Efforts were made to clean rocky shores contaminated with oil in Prince William Sound after the *Exxon Valdez* spill, but these may have been more harmful than beneficial to marine invertebrates.

(Photo courtesy of A. W. Maki and Exxon Company, U.S.A.)

FIGURE 22.5

Diving birds such as western grebes can become completely oiled by repeated passage through the ocean surface in their feeding activities.

(U.S. Fish and Wildlife Service photo)

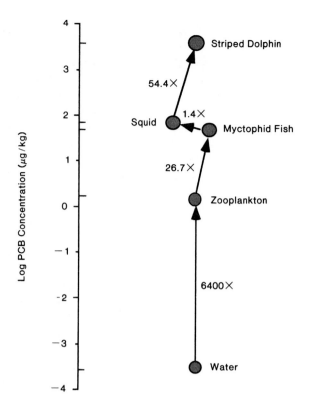

FIGURE 22.6

Data from a food chain in the western North Pacific Ocean show the strong food-chain concentration and magnification pattern of PCBs, similar to that of chlorinated hydrocarbon pesticides.

(Data from S. Tanabe, et al., 1984, "Polychlorophenyls, DDT, and Hexachloro-cyclohexane Isomers in the Western North Pacific Ecosystem" in *Archives of Environmental Contamination and Toxicology*, 13:731–738. New York: Springer-Verlag.)

The most serious impacts of marine oil pollution are on four groups of seabirds: alcids, cormorants, sea ducks and eiders, and penguins (Ohlendorf et al. 1978). Alcids, or members of the auk family, include puffins, murres, auklets, murrelets, and the razorbill. Members of these four groups are diving birds that feed on fish or mollusks. In areas with oil slicks, their diving carries them back and forth through the oil-covered surface, causing heavy oiling of the plumage. Other diving birds, such as western grebes, are similarly affected (Fig. 22.5).

Oiling of the plumage leads to loss of flight ability and buoyancy. Birds unable to remain afloat drown or are forced ashore, where they are unable to feed. The oiled plumage is also less effective as insulation, leading to rapid loss of body heat and increased metabolic consumption of body fat reserves. Preening of the oiled plumage results in ingestion of oil, which may cause toxic effects on the pancreas, liver, and kidneys, as well as reducing intestinal water absorption, leading to dehydration. Sea otters fouled by oil also lose the insulative value of their pelage, and the ingestion of oil in efforts to clean it leads to toxic effects on internal organs (Geraci & Williams 1990). Most heavily oiled birds and sea otters die, even when efforts are made to clean and rehabilitate them.

SYNTHETIC ORGANIC POLLUTANTS

Several other chemical pollutants are now global in their distribution. **Polychlorinated biphenyls (PCBs),** a group of industrial chemicals used as insulation fluids in transformers and as plasticizers and fire retardants in various materials, have become widespread in aquatic environments through leakage and trash disposal (Waid 1986; Tanabe 1988). PCBs are close relatives of CHs, and share their persistence and high solubility in lipids. Thus, they show food chain magnification, sometimes reaching

concentrations of hundreds to thousands of parts per million in birds and mammals at the ends of aquatic food chains (Fig. 22.6). PCBs are known to impair reproduction in a number of birds and seals (Kubiak et al. 1989; Reijnders 1986). Because PCBs are widespread in the marine environment and show food chain magnification, marine mammals, especially the polar bear, are likely to be most affected (Tanabe 1988; Addison 1989).

Still other petroleum derivatives, the **polynuclear aromatic hydrocarbons (PAHs),** have recently been recognized as major environmental pollutants. PAHs are stable molecules consisting of two to several fused benzene rings. Some are natural components of petroleum, others are formed by combustion of fossil fuels. They enter aquatic environments through oil pollution, urban runoff, and atmospheric precipitation. Many of these compounds are highly carcinogenic, and are thought to be responsible for liver cancers and fin deterioration of fish in polluted areas.

HEAVY METALS AND CYANIDE

Mercury and other heavy metals such as copper, cadmium, chromium, lead, nickel, and zinc all occur naturally in fresh and marine waters. Their use by humans, however, has greatly increased their rates of entry into natural ecosystems. Human activities have roughly doubled the rate of entry of mercury into active circulation in the biosphere. Mercury shows food chain magnification, and on several occasions has reached concentrations in marine fish, such as tuna, that exceed levels considered safe for human consumption. Failure of egg hatching in white-tailed sea eagles, a fish-eating species, has been attributed in Sweden to mercury contamination. Cadmium is also a heavy metal of increasing concern. Ohlendorf and Fleming (1988) identified it as a pollutant of particular concern for birds in both San Francisco and Chesapeake Bays.

Mercury and cyanide also have become serious pollutants in gold mining areas. In the Amazon Basin, mercury is being used in large quantities to separate gold from lighter minerals. About 130 tons of mercury are discharged annually into the Amazonian environment (Reuther 1994), and food chains in several tributaries of the Amazon and Paraguay Rivers are heavily contaminated. In the western United States, cyanide is being used to leach gold from old mine tailings and low-grade ore. Large numbers of birds and mammals have died after feeding or drinking at ponds of cyanide-containing water used in leaching.

Tributyltin (TBT), an organic tin compound, is a highly effective ingredient of antifouling paints for marine vessels (Champ & Lowenstein 1987). TBT paints protect hulls against growth of fouling organisms much longer than copper-based antifouling paints. TBT is highly toxic to marine organisms in concentrations of less than 1 part per billion. In France, TBT caused mortality and deformation of oysters in bays where it was used in antifouling paints on fishing boats. Concentrations of TBT in the range of potentially serious impact have been detected in bays and harbors in several locations in North America (Goldberg 1986; Stewart & Thompson 1994). TBT has been banned or its use restricted in North American and many European countries. In 1988 the United States Environmental Protection Agency proposed restrictions that included a ban on use of TBT paints on most boats under 20 meters in length, and restrictions on the rate of release of TBT from the treated surfaces of larger vessels (Weiss 1988).

SELENIUM

Selenium is an element required by most vertebrates in trace amounts, but toxic to them in high concentrations (Ohlendorf 1989). Selenium is low in concentration in most soils, because it is easily removed by leaching, but is abundant in some desert soils. Farming in desert regions and the leaching of selenium in irrigation drainage waters has created severe pollution problems in several wildlife areas in western North America. Selenium shows bioconcentration and possibly some degree of biomagnification.

Kesterson National Wildlife Refuge, California, received irrigation runoff water from irrigated land in the San Joaquin Valley. In 1983, the water entering the refuge contained selenium in concentrations about 200 times normal (Ohlendorf 1989). Plants and animals accumulated high concentrations of selenium. American avocets and black-necked stilts experienced

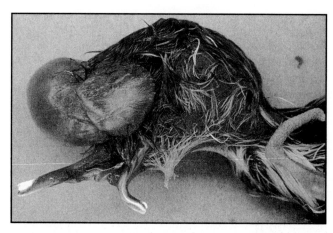

FIGURE 22.7

A deformed black-necked stilt chick from Kesterson Reservoir in California is an example of the effects of selenium pollution in several locations in western North America.

(U.S. Fish and Wildlife Service photo by Samuel Woo)

nearly complete reproductive failure at Kesterson Reservoir in 1984 and 1985 (Williams et al. 1989). Eared grebes and American coots nesting at Kesterson in 1983 also experienced low nesting success, and these species, together with black-necked stilts (Fig. 22.7) showed frequent embryonic abnormalities (Ohlendorf et al. 1989). Many adult water birds that died at the reservoir showed toxic selenium levels. As a result of these observations, the reservoir has been closed until the selenium contamination problem is overcome.

RADIOISOTOPES

The worldwide proliferation of nuclear technology has created the threat of environmental pollution by illegal disposal of radioactive wastes and release of radioisotopes by plant accidents. Dangerous radioisotopes include those of cesium, strontium, and phosphorus, which have moderate to long half-lives and which also show biological concentration and food chain magnification. Unfortunately, some accidental releases of these substances have been kept from public knowledge.

In the United States, radioactive wastes from Department of Energy plants in South Carolina, Tennessee, and Washington have contaminated wildlife, both terrestrial and aquatic (Holloway 1990). At the Hanford, Washington plant, radioisotopes entered the Columbia River by waste discharge and the discharge of contaminated reactor cooling water to the river from 1945 through the early 1960s. The entire food chain of the river was contaminated to the extent that human health impacts became a serious concern.

The 1986 disaster at Chernobyl in the former Soviet Union led to contamination of marine life in the northern part of the Black Sea. In this case, surface runoff carried radioisotopes into the Black Sea from regions receiving fallout from the plant explosion.

FIGURE 22.8

Plastic articles and their fragments have become a global form of marine pollution.

(Photo by Anthony F. Amos, University of Texas Marine Science Institute)

FIGURE 22.9

Sea turtles, such as this hawksbill, often die after they have become entangled in discarded vegetable bags, or after ingesting plastic bags, sheets, or strips that they apparently mistake for food organisms.

(Photo by Anthony F. Amos, University of Texas Marine Science Institute)

PLASTICS

Plastics are still another group of pollutants. Plastics make up more than half the man-made debris littering the ocean (O'Hara 1988). The plastics found at sea and on beaches include plastic products, their fragments, and industrial pellets (Fig. 22.8). Over time, plastic products entering the ocean by trash disposal become brittle and break into small, irregular pieces. These fragments remain buoyant and float until they are washed up on beaches or become trapped in other materials that eventually sink. Industrial pellets are of two types. **Polyethylene pellets** are beads 2 to 5 millimeters in diameter that are the raw material for the production of molded plastic products. They are transported by ships designed to carry bulk products of various sorts. Quantities of these pellets enter sea waters when the holds are cleaned or by ballasting practices like those used by oil tankers. Polyethylene pellets remain buoyant indefinitely, and are now found worldwide in oceans and on ocean beaches. **Polystyrene pellets,** 0.1 to 3 millimeters in size, are the raw material from which styrofoam products are made. They are also shipped by bulk carriers, and enter ocean waters in ballast water discharge. Polystyrene pellets ultimately become waterlogged and sink, so that they are not as widespread as polyethylene pellets.

The consequences of plastic pollution for marine wildlife are uncertain, although these items are ingested by many kinds of invertebrates, fish, and birds. Sea turtles, in particular, frequently ingest plastic bags, sheets, or other bulky items, apparently reacting to them as if they were jellyfish or other food items (Fig. 22.9). In the North Pacific, over 75% of the plastics ingested by seabirds are industrial pellets, and ingestion of plastics is a growing problem (Robards et al. 1995). Laysan albatrosses, nesting on one of the most remote islands in the Pacific Ocean, feed large numbers of plastic items to their chicks, and most dead chicks of this species have such items in their stomachs (O'Hara 1988). In the tropical Pacific, Spear et al. (1995) found a generally low level of ingestion of plastics by seabirds. For birds that had ingested such materials, however, they found a trend for body mass to decline with the quantity of plastics ingested.

INTERACTIONS OF POLLUTANTS

In most fresh and marine waters, wildlife is exposed to several pollutants, each potentially serious by itself, but having potential additive or synergistic effects. In the lower St. Lawrence River, for example, beluga whales have been dying in large numbers in recent years. The tissues of these animals contain high levels of CHs, PCBs, heavy metals, and an industrial chemical, benzo-a-pyrene, that might have come from an aluminum plant on a tributary river. Many of the animals appeared to have died of infectious diseases, suggesting that their immune systems had failed, possibly as a result of pollutant effects.

Large die-offs of aquatic mammals have occurred in several other locations in recent years. These have involved bottlenose dolphins along the eastern seaboard of the United States, gray and ringed seals in the Baltic Sea, harbour seals in the Dutch Waddensee, and the Lake Baikal seal in the former Soviet Union. These die-offs are the apparent result of disease epidemics or, in the case of bottlenose dolphins, toxins produced by red tide organisms. In all cases, these animals were exposed to substantial levels of several chemical pollutants. Pollutant-induced failure of the immune system is likely in some of these episodes (Crowe 1995).

PESTICIDE REGULATION AND ALTERNATIVES

In the United States, pesticides are regulated under the **Federal Insecticide, Fungicide, and Rodenticide Act (FIFRA)** of 1947, which requires registration of pesticide chemicals, and the **Federal Environmental Pesticide Act** of 1972. Registration of pesticides, as well as cancellation of those found to be harmful, is carried out by the Environmental Protection Agency, which publishes a list of pesticides that have been suspended, restricted in use, or banned (Office of Pesticides and Toxic Substances 1985).

At the international level, however, regulation of pesticide use is still very lax (Prabhu 1988). Most pesticides are produced in the United States, western Europe, and Japan—countries which, in general, have agencies that regulate pesticide use. Companies in the United States, for example, produce 45% of the pesticides used throughout the world. Much of this production is exported, including chemicals that are banned or severely restricted for use in the United States. The use of pesticides in many developing countries often has been subsidized heavily by agencies such as the World Bank and the United States Agency for International Development (Repetto 1985). Most developing countries, however, do not have agencies that regulate pesticide use. Several international organizations are attempting to develop ways to regulate trade and use of pesticides. Nevertheless, many pesticides, such as several of the CHs that have adverse environmental impacts, are still in widespread use.

Exclusive reliance on chemical pesticides for control of agricultural pests and disease vectors is a dangerous strategy (Cox & Atkins 1979). Pesticides disrupt natural biological control relationships, allowing rapid recovery of the pest itself. Damage to natural biological control species often permits **secondary outbreaks** of other species that were formerly not pests. Frequent pesticide applications strongly select for mechanisms of **pesticide resistance,** the ability to avoid, tolerate, or inactivate the pesticide chemical. Hundreds of insects and mites have evolved resistance to insecticides of almost all major chemical types, and many species of weeds are now resistant to herbicides (LeBaron & Gressel 1982; National Research Council 1986). Resistance is also shown by vertebrate pests such as rats, as well as by many fungal pathogens and nematodes that attack crop plants. These problems with chemical control have stimulated the development of **integrated pest management,** a strategy that combines cultural, biological, and chemical approaches to pest control.

CONTROLLING CHEMICAL POLLUTION

Reducing marine oil pollution is one of the major challenges for the last decades of the petroleum age. Most of the oil and gas basins remaining to be tapped lie offshore, and many of these are in high latitudes. In Alaska, for example, petroleum basins occur in the Gulf of Alaska, along the Aleutian Peninsula, in the Bering Sea, in Norton Sound, and along the Arctic Ocean (Bolze & Lee 1989). Other potential basins lie along the coasts of California, North Carolina, Florida, and other states. Developing the technology for safe exploitation of these reserves is essential.

Since the *Exxon Valdez* accident, efforts to reduce the risk of tanker accidents have centered on a requirement that oil tankers be constructed with double hulls to reduce the chance of oil spillage in groundings. In 1990, the United States passed legislation requiring double hulls on new tankers and the retrofitting of double hulls for existing tankers.

Several national and international efforts have been made to control the discharge of hazardous materials into lakes and oceans. The broadest international agreement is the **London Dumping Convention,** drafted in 1972, and ratified by 64 countries, including the United States, as of 1989 (Duedall 1990). This agreement bans the deliberate discharge of CHs, PCBs, certain heavy metals, plastics, petroleum substances, high-level radioactive wastes, and some other materials. Many other materials can be dumped only by permit, which essentially serves to control the location of dumping.

The **International Convention for the Prevention of Pollution from Ships,** commonly known as the **MARPOL Convention,** prohibits the dumping of plastics at sea. The United States signed this agreement, effective in December, 1988. The MARPOL Convention has been signed by 39 other nations.

SUMMARY

1. Chemical pollutants exhibit both direct side effects, seen near the time and place of pollution, and indirect side effects, seen later and at distant locations due to movement of the chemical by air and water currents and its accumulation in living tissues and magnification along food chains.

2. Chlorinated hydrocarbon pesticides and PCBs have become serious environmental pollutants because of their chemical stability, ability to move through the environment, tendency to concentrate in lipids, and capacity for toxic action and disruption of reproductive and developmental physiology.

3. Oil pollution can have severe immediate impacts on many organisms in all marine environments, but in protected estuarine situations oil may be locked up in anaerobic sediments from which it is released anew at intervals for years or decades.

4. Many animals, especially those at higher levels in food webs, are now challenged by many biologically active pollutants that may exhibit synergistic effects.

5. Efforts are being made to develop international controls on the release of toxic pollutants into the environment, but few agreements exist other than those involving disposal of materials in the ocean.

Questions for Discussion

1. Based on the demonstrated evolutionary ability of pests, both animal and plant, to evolve resistance to pesticides, do you think the pesticide pollution problem is likely to get worse because new and more toxic chemicals will come into widespread use?

2. What are the implications of additive or synergistic effects of multiple pollutants on organisms or ecosystems for the regulation of emissions, and for the establishment of liability of polluters?

3. In cases involving environmental pollution from private industrial operations, such as large oil tanker accidents, or from major government programs, such as nuclear weapons development, do you think that a danger exists for massive funding of research biased toward conclusions of negative impact?

Suggested Reading

Burger, J. (Ed.). 1994. *Before and after an oil spill: The Arthur Kill.* Rutgers Univ. Press, New Brunswick, N.J. A comprehensive evaluation of the effects of an oil spill on a well-studied estuary near New York City, and of the cleanup efforts undertaken.

National Research Council. 1995. *Clean ships, clean ports, clean oceans: Controlling garbage and plastic wastes at sea.* National Academy Press, Washington, DC. A summary of the sources and effects of vessel trash, and of strategies to implement the MARPOL treaty on reduction of trash pollution.

Pimentel, D. and H. Lehman (Eds.). 1993. *The pesticide question: Environment, economics, and ethics.* Chapman and Hall, New York. Eighteen chapters by various authors examining environmental and societal aspects of pesticide use.

Global Climate Change

H uman activities have modified the chemistry and physics of the earth's atmosphere, as well as the physical characteristics of the land surface. These modifications are affecting global climate, ocean circulation patterns, and ecosystem dynamics, and have profound implications for both biodiversity and human well-being. Unless their causes are identified and addressed vigorously, humankind will face serious challenges from global change within the lifetime of most of its living members. In this chapter we shall examine the causes and patterns of global change, with particular reference to biodiversity and ecosystem function.

The scale of human activities is modifying the composition of the earth's atmosphere and altering its radiation regime and overall climate.
(Photo by Alan Torreneuva.)

Outline

CAUSES OF GLOBAL CHANGE

Global change has many causes (Fig. 23.1). Some involve basic changes in the atmosphere and oceans, and are thus truly global, whereas others involve activities concentrated in particular re-gions but having impacts that extend over broad portions of the biosphere (Vitousek 1992). Some of these changes are chemical, involving release of gaseous or water-soluble pollutants into the atmosphere or into aquatic ecosystems. In the atmosphere, pol-lutants influence processes at several levels and in a very complex

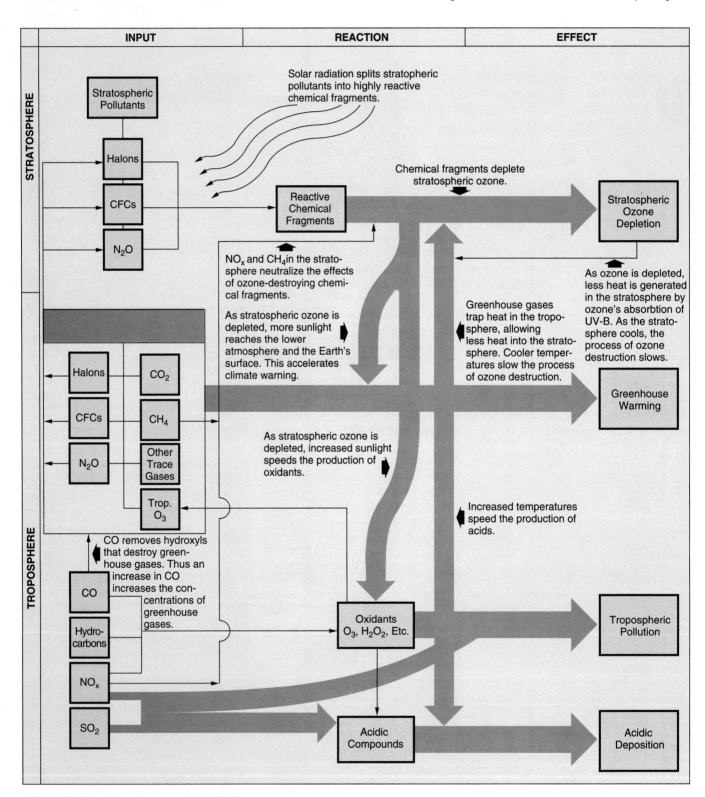

FIGURE 23.1

The sources, reactions, and effects of gaseous pollutants at various elevations in the earth's atmosphere.

TABLE 23.1

Greenhouse Gases, their Rate of Annual Increase in the Atmosphere, and their Estimated Contribution to Greenhouse Warming					
GAS	CONCENTRATION	ANNUAL INCREASE (%)	CONTRIBUTION TO WARMING PER KG**	OVERALL	DISAPPEARANCE TIME (YR)
CO_2	362 ppm	0.5	1	60%	120
CH_4	1.7 ppm	1.0	70	15%	10
N_2O	310 ppb	0.2	200	5%	150
O_3*	30 ppb	0.5	1,800	8%	0.1
CFCs	0.76 ppb	4.0	5,300	12%	120

*Troposphere
**Relative to CO_2
Source: Estimates based on data from Rodhe (1990).

manner. Other changes are physical, involving the introduction of particulates into air or water: aerosols, ash from combustion, and wind- or water-eroded soil. Physical changes may also occur in the land surface, leading to changed **albedo,** or reflectivity of the surface to incoming solar radiation.

THE GREENHOUSE EFFECT AND GLOBAL WARMING

The **greenhouse effect** results from increase in concentration of gases that are transparent to incoming shortwave solar radiation, but absorb outgoing long-wave heat radiation. This absorbed heat radiation is reradiated, partly outward and partly back toward the earth's surface, thus slowing the rate of net heat loss from the surface. The surface and the lower levels of the atmosphere are thus warmed. **Greenhouse gases** thus act in a manner similar to the glass in a greenhouse roof, letting sunlight pass through but absorbing and reradiating back some of the heat radiation from within the greenhouse. In nature, water vapor acts as a greenhouse gas, as one can easily verify by the warmer temperatures under a cloudy night sky compared to a clear night sky.

In addition to water vapor, greenhouse gases include carbon dioxide (CO_2), methane (CH_4), nitrous oxide (N_2O), ozone (O_3), chlorofluorocarbons (CFCs), and other halogen compounds. These gases differ greatly in their concentrations and in their abilities to absorb heat radiation (Table 23.1). Per unit mass, O_3 and CFCs have the greatest capacity for absorbing heat radiation, and even the low concentrations of these gases account for perhaps 20% of the total greenhouse effect. The most abundant gases, CO_2 and CH_4, however, account for about 75% of greenhouse warming.

The various greenhouse gases have been increasing in concentration in the atmosphere due to human activities. Since about A.D. 1750, CO_2 has increased from about 280 parts per million (ppm) to over 360 ppm, and continues to increase exponentially (Fig. 23.2). The main cause of increase in CO_2 is the burning of fossil fuels such as coal and petroleum, but deforestation and other reductions in plant biomass in global ecosystems have also contributed. Methane has more than doubled over this same period, due to venting of gas from petroleum wells, release

of intestinal gas by cattle and other ruminant livestock, and production of gas in anaerobic mud in flooded rice paddies. Nitrous oxide has increased only by about 10%, mainly due to the action of denitrifying bacteria on nitrogen fertilizers and natural nitrogen compounds in anaerobic agricultural soils. CFCs, used as industrial solvents, refrigerants, and propellants, are entirely of human production. Other halogens include methyl bromide, derived from industrial production for use as a pesticide, and from biomass burning (Mano & Andrae 1994). Tropospheric ozone is largely a photochemical product of pollutants released by internal combustion engines. Several of these gases have atmospheric lifetimes of over a century, so that their increases in concentration cannot be reversed quickly (Vitousek 1992).

The influences of global changes in atmospheric composition and land surface characteristics are leading to significant changes in climate and oceanic circulation patterns. These changes are being studied with the aid of **general circulation models (GCMs)** that simulate the global system of atmospheric and oceanic circulation. Several different GCMs now exist, and their predictions differ somewhat. These models describe global relationships for grid units approximately 500 kilometers square, so that their ability to predict changes on a scale useful to regional planners is very limited. All models, however, predict that substantial climatic warming will occur.

Climatic warming is typically projected under conditions in equilibrium with a doubled level of CO_2 and corresponding increases in other greenhouse gases. An evaluation of the predictions of various GCMs (Houghton et al. 1990) has concluded that an average global warming of 2.5° C is likely to result from doubled CO_2, and that this warming might occur in 50–100 years, or between A.D. 2040 and 2090.

The basic predictions of GCMs are not uncontested. Lindzen (1994), for example, argues that these models incorporate assumptions about the behavior of atmospheric moisture that amplify greatly the influence of doubled CO_2. He believes that moisture relations in the atmosphere, such as increased development of very high altitude cirrus clouds that reflect incoming light radiation, may act as a negative feedback that opposes warming at lower levels in the atmosphere. These criticisms are discounted by most other atmospheric scientists (Schneider 1992).

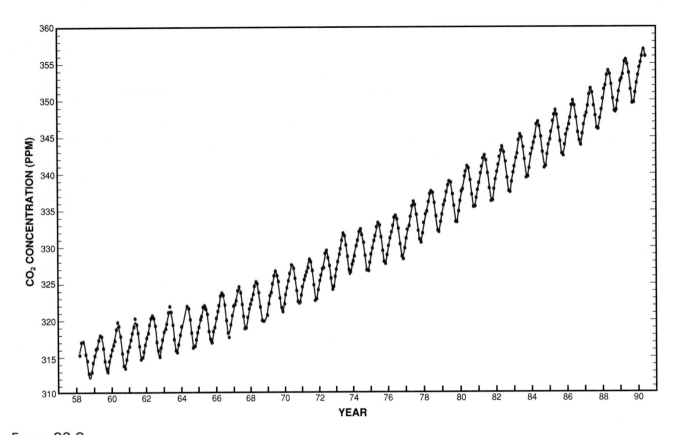

FIGURE 23.2

Burning of fossil fuels and deforestation of vast areas of the tropics are contributing to a rising concentration of carbon dioxide in the atmosphere, as measured at Mauna Loa Observatory in Hawaii.

(Data from C. D. Keeling, et al., "A Three Dimensional Model of Atmospheric CO_2 Transport Based on Observed Winds: Observational Data and Preliminary Analysis" Appendix A, in *Aspects of Climate Variability in the Pacific and the Western Americas*, Geophysical Monograph, American Geophysical Union, Vol 55, 1989 (Nov).)

Climatic warming is projected to be much greater at high latitudes than in temperate and tropical regions. In polar regions, summer temperatures are expected to average about 2.0° C warmer and winter temperatures about 10° C warmer than at present (Hom 1995). Warmer temperatures are expected in the interior regions of continental areas in the northern hemisphere. In some interior continental areas in the Arctic, such as central Alaska, summer temperatures may be 4.0° C warmer and winter temperatures 8°–12° C warmer than at present (Hom 1995). In other areas, such as the areas of northwestern Europe now warmed by the Gulf Stream, decrease in mean temperature may occur.

Global climate change will also involve precipitation and moisture relations (Schneider 1992). Warmer temperatures will increase evaporation, and, consequently, mean global precipitation. More moisture will likely be carried to high latitudes, so that the increase in precipitation may be greatest there. At midlatitudes, interior regions of continents, on the other hand, may experience drier soil conditions in summer because snowmelt will be completed earlier in the spring, and summer temperatures will be higher.

Climatic warming will also affect the oceans. Warming of the earth's surface will also include the surface water of the oceans. Among other things, this will lead to reduced areas of polar sea ice (Schneider 1992). Under a climate in equilibrium with a doubled CO_2 level, sea level is expected also to rise by 20–140 centimeters. Most of the rise will be due to thermal expansion of warmer ocean water, the rest to increased melting of continental glaciers and ice caps. Changes in ocean currents are likely, with the Gulf Stream perhaps flowing in a more western path.

Evidence is accumulating that global environmental change has already caused climatic change. Bradley et al. (1987) note that mid-latitude precipitation has increased and low-latitude precipitation has decreased over the past 30–40 years in the general manner predicted by most GCMs. A report issued in 1995 by the U.S. Intergovernmental Panel on Climate Change concluded that the apparent global warming trend observed since 1900, and especially since 1980, is probably due to human activities (Stevens 1995). Based on data from weather stations around the world, 1995 was calculated to be the warmest year ever recorded.

STRATOSPHERIC OZONE DEPLETION

In the stratospheric ozone layer, at an elevation of 15–35 kilometers, O_3 is formed by the splitting of oxygen, O_2, molecules by ultraviolet (UV) radiation, and the subsequent reaction of individual O atoms and O_2 to form O_3. The O_3 atoms absorb UV, as well, so that this layer of the stratosphere filters out most

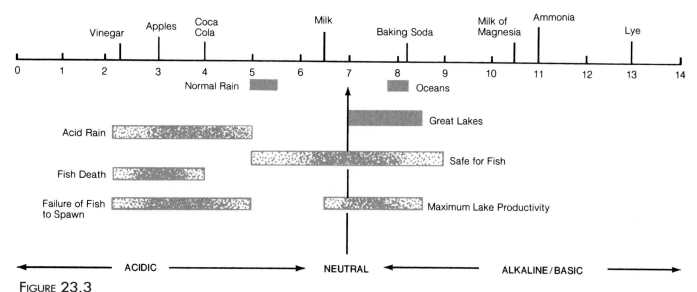

FIGURE 23.3

The pH scale.

(Modified from *Fish and Wildlife Leaflet No.1*, U.S. Fish and Wildlife Service)

of the UV radiation, which, if it reached the earth's surface, would kill all terrestrial life. This layer has consequently gained the popular name of the "ozone shield." Some UV radiation, in the wavelength range designated as UV–B, reaches the surface, and is the cause of sunburn in fair-skinned people.

In the stratosphere, CFCs, N_2O, and other halogen compounds such as methyl bromide contribute to the breakdown of O_3. This process of O_3 breakdown was only proposed in the mid-1970s, but it quickly became the focus of both scientific and political controversy. Doubt that human activity is leading to decrease in stratospheric O_3 concentration no longer exists.

One major symptom of the decrease in stratospheric O_3 is the development of an **"ozone hole"** over the Antarctic region. Since 1979, extensive depletion of ozone has been recorded over Antarctica during the southern hemisphere spring (Makhijani & Gurney 1995). From 1992 to 1994, the ozone hole exceeded 23 million square kilometers in extent, an area roughly equal to that of North America. At the fringes of this hole, substantially reduced levels of O_3 existed above southern Australia, New Zealand, and South America. Measurements of O_3 taken from satellites from 1979 through 1993 also reveal increased UV-B levels reaching the earth's surface at high and mid-latitudes in both hemispheres (Madronich et al. 1995).

TROPOSPHERIC POLLUTION AND ACID DEPOSITION

The release of sulfur (S) and nitrogen (N) oxides into the atmosphere leads to reactions with water to form sulfuric and nitric acids, which enter terrestrial or aquatic ecosystems by **acid deposition.** Deposition may occur as **acid precipitation—** acidic rain and snow. It may also occur as **dry deposition,** the fallout of ash or dust particles containing S and N oxides, or the direct absorption of gaseous oxides by the soil, the water of streams and lakes, or the tissues of organisms. These oxides are derived mostly from combustion. Oxides of S result largely from the combustion of fossil fuels and the smelting of ores, with electrical power plants being the greatest contributor in the eastern United States (MacDonald 1985). Oxides of N result largely from the burning of organic matter, the union of N and oxygen (O) under the high pressures and temperatures of vehicle engines, and the conversion of N from fertilizers to N oxides by bacteria. In the tropics, nitrogen and sulfur oxides enter the atmosphere from burning of savannas or newly cleared forestlands (Rodhe & Herrera 1988). In the industrial belt of eastern North America, perhaps two-thirds of the oxides are those of S; where vehicle emissions predominate, the N oxides are more important. Dry deposition tends to be great in localities close to major combustion sources, such as smelters or power plants.

Unpolluted rain and snow show low acidity. On the **pH scale** (Fig. 23.3) from 0 to 14, acidity increases as the scale value decreases. Each unit change on this scale represents a tenfold change in the concentration of acid ions (H^+). A pH value of 7.0 is neutral, meaning that equal numbers of acid and basic (OH^-) ions are present. Pure water has a slightly acid pH of about 5.6 due to dissolved carbon dioxide (CO_2) from the air. CO_2 reacts with water to form carbonic acid (H_2CO_3), which ionizes to release H^+ ions. In nature, furthermore, the pH of precipitation in areas unaffected by human activities and free of

FIGURE 23.4
Regions of eastern North America affected by acid precipitation.
(Modified from National Research Council 1986)

calcareous dust is usually about 5.0, due to acids of natural origin (Schindler 1988). Thus, acid precipitation as an unnatural phenomenon is rain or snow with a pH less than about 5.0.

Acid deposition is greatest in areas downwind from regions where large S and N emissions occur. In North America, the northeastern United States and eastern Canada are most seriously affected, but most of the area east of the Mississippi River experiences heavy acid deposition (Fig. 23.4). In Europe, the Scandinavian countries and the portions of eastern Europe downwind from the heavily industrialized parts of the United Kingdom and central Europe receive the heaviest levels of acid deposition. A smaller area of acid deposition is centered on Korea and Japan.

Acid rain, snow, or fog may have pH values down to 3.0 or lower. The pH of rain and snow now averages 4.0 to 4.4 in many locations in eastern North America and Europe (Park 1987), and mean values as low as 3.8 have been recorded in parts of the Netherlands. In the White Mountains of New Hampshire, rainstorms have been recorded with pH values as low as 2.85; in Pennsylvania a storm value of 2.70 was noted. The moisture in clouds bathing the spruce-fir forests on Mount Mitchell, North Carolina, has registered a pH as low as 2.1.

Why has acid deposition only recently become serious? Manufacturing, heating, transportation, and farming activities that release S and N oxides have grown enormously since World War II. The release of new pollutants that catalyze the oxygenation of S and N compounds has also contributed (Schindler 1988). In addition, industries such as smelters, power plants, and steel mills have sought to relieve intense local air pollution problems by building taller stacks. Discharge

FIGURE 23.5

Tall stacks like these in Sudbury, Ontario, have helped transform severe local acid pollution problems into regional acid deposition syndromes. The stack of the Inco smelter (background) is over 400 m tall, and in 1989 released about 637,000 tons of sulfur oxides.

(Photo by Alen Torreneuva)

of combustion wastes at higher levels in the atmosphere leads in their dispersal by winds over a much broader area. Many industries and power plants now have stacks over 300 meters high, and in a few cases over 400 meters high (Fig. 23.5). These tall stacks have helped convert intense local air pollution problems into the general regional problem of acid deposition.

ECOLOGICAL IMPACTS OF GLOBAL CHANGE

The effects of global change will result from the combined influences of the varied global and regional impacts of human activity (Wyman 1991). Many uncertainties exist in the predictions of these effects, several relating to the extent to which positive feedbacks could act to increase global warming. For example, arctic soils contain immense amounts of carbon as peat, so that warming at high latitudes could cause large additional releases of CO_2 due to increased decomposition. Oechel et al. (1993) give evidence that the North American arctic tundra has changed from being a net storage system for carbon to being a net source. Northern wetlands may similarly become major sources of added CO_2 and CH_4 (Bridgeham et al. 1995).

Global Change and Biodiversity

Global change will modify the geographical distribution of climate to which species are presently adapted. For the magnitude of change projected by the middle of the 21st century, shifts in climatic belts of several hundred kilometers are likely (Peters 1991; Davis & Zabinski 1992). The ability of many terrestrial and freshwater species to move as climate changes is in many cases quite limited. Many trees, for example, require 20–30 years to gain reproductive maturity, so that their rate of geographical

spread is slow. In other situations, such as complex montane environments, dispersal of species may be easy, and shifts in dominance, as from herb-dominated meadows to montane shrublands, may be rapid (Harte & Shaw 1995).

Climatic shifts have particular danger for species in preserves surrounded by human development (Peters 1991). Without corridors or direct human assistance in following their optimum climatic zone, they may be trapped and decline to extinction as the climate changes. In the case of alpine and arctic ecosystems, climatic warming may eliminate the high mountain or high latitude climates on which species depend, pushing them upward and poleward until the land habitat ends (Peters 1991). Data on altitudinal ranges of plants on alpine peaks in Europe suggest that some upward retreat has already occurred (Grabherr et al. 1994).

The shift of climates appropriate to various plants and animals is likely to change the combination of species that occur together in communities (Peters 1991). For example, the responses of North American breeding birds to yearly variations in conditions and to environmental trends over a few decades are highly individualistic (Bohning-Gaese et al. 1994). Thus, it is not likely that all members of present communities will respond similarly to future changes in continental climate patterns.

Increases in UV-B radiation may affect the composition of terrestrial and aquatic plant communities. UV radiation affects plants in several ways, causing damage to DNA and influencing plant growth, morphology, and biomass accumulation (Caldwell et al. 1995). Tolerance of UV differs for terrestrial plant species, and even though overall production at the community level may not be depressed, the competitive balance among species might be shifted. Thus, shifts in relative abundance and composition of species in communities are likely to occur. Marine organisms also show differential sensitivities to UV radiation, with serious effects being observed for reproductive stages of many invertebrates and fish (Hader et al. 1995).

UV-B radiation is also known to damage amphibian DNA, an effect that is mitigated by the enzyme photolyase, which acts to repair damaged DNA. In the western United States, increased UV-B radiation has been suggested as a causal factor in the decline of certain montane species of frogs and toads that lack high levels of photolyase (Blaustein et al. 1994).

Acidification of aquatic ecosystems reduces the abundance and diversity of both zooplankton and benthic invertebrates (Hendry et al. 1976). In Ontario, lakes with pH values above 5 had 9–16 species of zooplankton, 3–4 of which were abundant. Below pH 5, only 1–7 species were present, of which only 1–2 were abundant. Among larger invertebrates, snails are highly sensitive to acid conditions because of their calcium carbonate shells. Below a pH of 6.6 the diversity of snails declines, and below pH 5.2, snails are absent. Insect larvae decline in diversity as pH drops, but acid-resistant species tend to replace acid-sensitive species (Hall & Ide 1987). Loss of fish also permits the dominance of certain groups of insect predators, such as dragonfly larvae, backswimmers, and midge larvae (Evans 1989; Stenson & Eriksson 1989). These invertebrates, in turn, modify the zooplankton community, reducing the abundance of some cladocerans and favoring dominance of certain copepods.

Loss of fish from acidified lakes and streams has attracted the greatest concern. In several areas of eastern North America, such as the Adirondack Mountains and Nova Scotia, acidification has led to loss of fish from many lakes and streams (Haines 1981, 1986; Lacroix 1987). Fish species differ in sensitivity to acid conditions, but at pHs of 4.5 or lower, most species are unable to reproduce, and lakes and streams with such pH values usually lack fish. Amphibians are also sensitive to acid conditions, especially during their embryonic or larval stages (Freda 1986).

Impacts of acidification on aquatic invertebrates and fish have been implicated in declines of populations of several waterbirds (Mitchell 1989). Fish-eating birds such as loons and mergansers have declined in eastern North America in recent decades, and show reduced reproductive success in lakes without fish. Many waterfowl feed primarily on aquatic invertebrates during the breeding season, and declines of some of these species, such as the black duck, may be related to decreased food availability in acidified aquatic systems. Songbirds, such as swallows and flycatchers, that nest near water and feed heavily on insects with aquatic larvae, have been reported to show thin-shelled eggs and reduced reproductive success in areas with acidified waters. In the Netherlands, eggshell thinning in woodland insectivorous birds may be due to low calcium availability caused by acidification of the forest ecosystem (Huckabee et al. 1989).

The potential impact of acid deposition on aquatic ecosystems in sensitive regions downwind from major urban and industrial regions is enormous. Hundreds of thousands of lakes and streams in eastern Canada and the northeastern United States are at risk. In the United States, about 1,000 lakes have been severely acidified and 3,000 or so are somewhat affected (French 1990). The most severe impact has been in the Adirondack Mountains of New York State, where about 50% of the lakes above 600 meters in elevation have pHs less than 5.0 and lack fish. Acidification exists on a similar scale in Scandinavia. In Norway, for example, 200,000 lakes are sensitive to acidification, and nearly 78% of the lakes in the southernmost counties are now barren of fish (Henricksen et al. 1989).

Forest communities are also being affected by acid deposition. Increased mortality and reduced growth, collectively termed **forest decline,** have been observed over large areas of western Europe and eastern North America (*See* Reading 23.1).

Global Change and Ecosystem Function

Global change will influence terrestrial ecosystems through increased CO_2 availability and the direct and indirect effects of climatic change. Many studies have examined the effect of increased CO_2 on photosynthesis and water use by plants. Laboratory studies show that photosynthesis is stimulated and water loss by transpiration reduced (Field et al. 1992). Warmer temperatures, however, may tend to increase water use (Baker & Allen 1994).

Together, CO_2 and temperature change may increase growth rates and shorten the lifetime of some plants. Changes of this sort might already be evidenced by increased turnover rates in tropical forests (Phillips & Gentry 1994). In nature, however, plant production is often limited by nutrient availability, so that the response of many ecosystems will depend on how climate change affects decomposition and other soil processes controlling nutrient availability (Shaver et al. 1992). The adjustments in structure of plant communities will also be complex, involving, for example, changes in the allocation of production to roots versus shoots by various species, and shifts in the abundance of species with differing photosynthetic systems.

Major effects will occur in aquatic ecosystems, as well. In the Temperate Zone, warmer temperatures and greater precipitation will tend to increase flood flows in streams, both from storm rains and spring snowmelt. In turn, these may lead to decreased streamflow in summer. In general, greater precipitation and runoff are likely to carry more nutrients and sediment into aquatic ecosystems, increasing sediment pollution and eutrophication.

Depletion of the stratospheric ozone layer may also affect ecosystem properties. Both laboratory and field studies have shown that UV-B radiation reduces photosynthesis by marine phytoplankton. Some believe that this may lead to a reduction in marine phytoplankton production by as much as 5% (Hader et al. 1995). Others, who have considered the effects of increased UV-B radiation on antarctic marine phytoplankton production over the full depth range of production, suggest that little reduction in total productivity is likely (Arrigo 1994).

Acid deposition has modified the structure and function of many aquatic ecosystems. Lakes and streams are sensitive to acidification in regions where watersheds are weakly buffered against pH change due to increased rates of input of acids. In fact, many aquatic ecosystems in such regions are naturally quite acid (Patrick et al. 1981). Acid bogs and bog lakes are common in the coniferous forest region of the Northern Hemisphere. Many lakes and streams that were originally nearly neutral in pH, however, are very sensitive to the effects of increased acid deposition. Sensitive aquatic ecosystems tend to be those of regions underlain by granitic or other parent materials low in carbonate minerals. Resistant ecosystems occur in regions with limestones, shales, or other formations rich in carbonates. Carbonate minerals give rise to bicarbonate ions in water, and these can react with acids to neutralize them. Watersheds of larger size, and those with deep soil or glacial till layers, are also more resistant to acidification.

In North America, the Cascades, Sierra Nevadas, Rocky Mountains, Appalachians, and Adirondacks—all areas underlain largely by granitic rocks—tend to be sensitive to acid deposition. Most of eastern Canada, underlain by the granites of the Canadian Shield, is also sensitive. The eastern and southeastern lowlands of the United States are also sensitive to acid deposition due to shortage of carbonate minerals in their waters. In eastern North America, the buffering capacity of watersheds now averages about 40% less than normal, and for many lakes this capacity is completely exhausted (Schindler 1988). Depletion of buffering capacity is the result of decades of acid deposition, another reason why acid deposition has not become a serious problem until recently. In Europe, large areas of Scandinavia are sensitive to acid deposition.

I n North America, forest decline has affected red spruce and Fraser fir at high elevations in the Appalachian Mountains (Reading Fig. 23.1A) and sugar maple in New England and eastern Canada (Vogelmann et al. 1985; Pitelka & Raynal 1989). In Europe, forest decline was first noted for spruce and fir in the Black Forest of West Germany in the 1960s. In the 1970s and 1980s, decline spread to other countries and began to affect pines, beech, and other hardwoods (Park 1987). By 1988, forest decline in Europe had affected about 35% of all forest area (French 1990). In Czechoslovakia, over 71% of forests were affected, and in West Germany 52% were damaged. The estimated loss in timber production for 1990 amounted to over $30 billion for Europe as a whole. In Sweden, where soil pH declined about 0.8 pH units over a period of 35 years, substantial change in composition of the herbaceous stratum of forests has also occurred (Falkengren-Grerup 1986).

Forest decline is apparently the result of complex ecosystem-level influences. Four major hypotheses about how acid deposition might contribute to forest decline have been offered. These involve 1) direct effects on the foliage, 2) leaching of nutrient cations from the soil, 3) aluminum toxicity in the soil, and 4) increased frost vulnerability due to excess N availability (Foster 1989). Acidic cloud moisture increases the sensitivity of red spruce foliage to winter cold, and acid deposition contributes to soil acidification and nutrient leaching. In addition, many of the stands showing decline are old, overstocked, or stressed by insects or drought. It is likely that forest decline is not a single acid-induced phenomenon, but the result of multiple stresses involving the effects of acid deposition, heavy metals, and other air pollutants, weather conditions, insects, diseases, stand age, and repeated harvesting without fertilization (Chevone & Linzon 1988; Klein & Perkins 1988; Garner et al. 1989). North American forest areas are being monitored to determine the input of pollutants (Reading Fig. 23.1B), and greenhouse studies are being conducted to determine the role of specific pollutants in forest decline.

READING FIGURE 23.1A

Acid precipitation may be a major contributor to mortality of red spruce on Camel's Hump Mountain, Vermont.

(Photo courtesy of H. W. Vogelmann)

READING FIGURE 23.1B

Wet (left) and dry (right) precipitation collectors on Whiteface Mountain, New York, are one of many sets of collectors in operation to measure the input of acids and other pollutants to forests in eastern North America.

(U.S. Forest Service photo)

The effects of acidification result not only from the acid ion itself, but also by increased leaching or chemical activation of aluminum and perhaps other heavy metals (Haines 1981). Heavy metals, in general, are enzyme poisons, and even slight changes in concentration of their biologically active forms can cause major effects. Aluminum is almost always present in soils in large quantities, and in its ionic form it is toxic to many plants and animals (Schindler 1988).

The response of aquatic ecosystems to acidification involves all major groups of organisms: producers, decomposers, and consumers (Stenson & Eriksson 1989). Among the producers, filamentous algae tend to increase in abundance, often forming mats on the bottom and on submerged surfaces. Increase in filamentous algae is largely due to the disappearance of snails, which graze on them. Increase in filamentous algae is often at the expense of phytoplankton, reflecting competition for nutrients between these two groups (Hendrey et al. 1976). Rooted higher plants, which utilize bicarbonates as their source of carbon for photosynthesis, also tend to decline in acid waters, and sphagnum mosses, which are more efficient at using dissolved carbon dioxide, increase in abundance. Moderate acidification seems not to reduce overall primary productivity (Schindler et al. 1985). But when acidification becomes severe enough to slow nutrient recycling, primary production does decline.

In some cases, decomposition in aquatic ecosystems is not affected by moderate acidification (Schindler et al. 1985). In other cases, it appears to be reduced because of reductions in detritus-feeding invertebrates, termed "shredders." Shredders fragment leaves and other bulky units of dead tissue into smaller bits that are more easily attacked by bacteria. When the rate of decomposition declines, undecomposed material accumulates, and nutrient regeneration is slowed. Accumulation of undecomposed organic matter may seal the surface of the bottom sediments, also reducing the regeneration of nutrients. Ultimately, this results in the creation of a bog system, where peat deposits accumulate.

COMBATTING GLOBAL CHANGE

Slowing or stopping global warming is likely to be the greatest challenge of environmental management in the coming century. The agreement reached at the Earth Summit in Rio de Janeiro in 1992 committed developed nations to limit CO_2 emissions at 1990 levels by A.D. 2000. This goal seems increasingly unlikely to be achieved. Stabilizing CO_2 at any level short of 750 ppm will evidently require reducing emissions well below those of 1990.

Some scientists have proposed that massive reforestation programs could create a storage system for CO_2, reducing its accumulation in the atmosphere. Even if such a program were undertaken, the maximum storage capacity could not be realized for about 60 years, and its maximum capacity to lock up CO_2 would equal less than a third of the current net annual addition of CO_2 to the atmosphere (Nilsson & Schopfhauser 1995).

Discovery of the Antarctic ozone hole raised international concern about depletion of stratospheric ozone by CFCs and other halons. The Montreal Protocol, which took effect in 1989, specified a timetable for reducing and eliminating the production of most ozone-depleting chemicals. By 1995, 148 countries had ratified this protocol (Makhijani & Gurney 1995). Since it took effect in 1989, however, evidence has accumulated that ozone depletion is more extensive and serious than believed at the time the protocol was written. Amendments to the protocol have been added to speed the phase-out of CFCs and to extend the protocol to cover other halogens, such as methyl bromide, a fumigant pesticide. Even these changes, however, will be inadequate to halt the increase in stratospheric halogens; to do this would require an immediate 85% reduction in their use (Hively 1989).

Reduction of acid deposition can permit the recovery of acidified lakes and streams. In Ontario, for example, reduced sulfate and heavy metal emissions have led to increases in pH and decreases in toxic metal levels of many lakes (Schindler 1988). Whitepine Lake, 90 kilometers north of Sudbury, has shown both chemical and biological recovery (Gunn & Keller 1990). In 1980, it had a pH of 5.4, and acid-sensitive fish were not reproducing. By 1988, lake pH had increased to 5.9, lake trout had resumed breeding, and recovery of invertebrates typical of nonacidified lakes had begun. Ontario has required major smelters, steel mills, and power plants to reduce sulfur oxide emissions 43% between 1989 and 1994 (APIOS 1991).

Estimates of the reductions that must be made in acid deposition to permit recovery of lakes in sensitive areas suggest that about 9–14 kilograms of sulfate per hectare is the maximum annual deposition allowable in areas with sensitive lakes (Schindler 1988). This compares to levels of 20–50 kilograms per hectare that are being received annually in most of eastern North America and Europe. Given reduced acid deposition, however, recovery may be slow where the buffering capacity of lake and stream watersheds has been exhausted by decades of acid leaching. Also, biological recovery may require a long process of recolonization and ecological interaction before food webs typical of nonacidified ecosystems are restored.

On the short term, local treatments to mitigate the effects of acid precipitation on a small scale are possible for lake ecosystems. Where fish reproduction is declining due to moderate acidification, stocking of young fish that are past the most acid-sensitive age can be carried out. Species with greater acid tolerance can also be introduced as the pH of the water declines, although at the risk of a permanent change in the biotic structure of the lake ecosystem (Porcella 1989).

Liming of lakes or their watersheds to restore some measure of buffering capacity can also be carried out (Fig. 23.6). In Sweden, many lakes and streams have been treated with ground limestone. Studies of the effects of liming lakes have been carried out in Ontario and New York (Porcella 1989; Keller et al. 1990). In small lakes in which the water volume is replaced within a short time by inflow and outflow, the effects of liming are transitory. In larger lakes, liming mitigates acidification for 3 to 6 years. Even in these lakes, liming is probably too

FIGURE 23.6

Experimental liming of lakes in the Adirondack Mountains as part of the Lake Acidification Mitigation Project suggests that the costs of maintaining normal acidity in this way are about $1,000–$3,000 per hectare over a 10-year period.

(Reprinted from the *EPRI Journal,* photo by D. B. Porcella)

expensive to be used except in special circumstances (Huckabee et al. 1989). Treating the watersheds of 1,419 acidified lakes in the northeastern United States would total $128 million if treatments were effective for five years.

Over the long term, however, acid deposition can only be resolved by reducing regional air pollution. The international aspects of acid deposition and its relation to basic industrial and urban economics have made resolving this problem very difficult. Effective international controls have yet to be devised, either in Europe or North America. In Europe, efforts to reduce acid deposition are coordinated by the United Nations Economic Commission for Europe (ECE). The ECE has called for a 30% reduction of S oxide releases over 1980 levels and for N oxide emissions to be held at 1987 levels. Most western European countries have agreed to this plan. In North America, Canada and the United States have argued over acid deposition and the need for reduction in oxide emissions. Emissions of S oxides in the United States are over five times those in Canada, and about 50% of the S deposition in Canada is of United States origin (Schindler 1988). In New England, however, some of the S deposition is of Canadian origin. The Canadian government has pressured the United States to reduce emissions, but the United States has generally responded that further study must be conducted before regulatory actions are taken. In the United States, similar controversy has flared between the New England states, where release of oxides is low but acid deposition heavy, and the industrial midwestern states where most oxides origi-

nate. In 1990, reauthorization of the **Clean Air Act** included amendments to reduce S oxide emissions from power plants to 8.9 million tons by A.D. 2000, a reduction of 10 million tons over the 1980 level, and to reduce N oxide emissions to 2.0 million tons below the 1980 level by 1997. These reductions may go far to protect acid-sensitive ecosystems.

SUMMARY

1. Global change results from human modification of the chemical and physical composition of the atmosphere and physical changes in the land surface.

2. Greenhouse gases, which include H_2O, CO_2, CH_4, N_2O, O_3, and CFCs, act to trap heat in the lower atmosphere, creating climatic warming. Additionally, CFCs catalyze the breakdown of O_3, in the stratosphere, reducing the atmosphere's absorption of ultraviolet radiation.

3. Acid deposition results from the increased input of sulfur and nitrogen oxides into the atmosphere, from where they enter terrestrial and aquatic ecosystems by precipitation or dry deposition.

4. Global warming will lead to shifts in terrestrial climate, modification of ocean currents, and significant sea level rise, all of which will change the distribution of habitats for which species are now adapted.

5. Acid deposition has lowered the pH of many lakes and streams to the point that extirpation of many invertebrates and fish has occurred. Acid deposition is also implicated in forest decline over wide areas.

QUESTIONS FOR DISCUSSION

1. Since soils are the product of thousands of years of interaction of parent materials, climate, and living organisms, do you think that global change might shift climatic conditions optimal for important plants, such as beech and sugar maple, to regions where soils may remain unsuitable for hundreds or thousands of years?

2. Based on world patterns of industrialization, urban growth, and economic development in different countries, what patterns of change in acid deposition do you think are likely to occur over the next half century?

3. Do you think that the basic structure of federal and state endangered species laws that are designed to protect species in their historical ranges are adequate to preserve biodiversity under changing global climate? If not, what changes do you think will be needed?

Suggested Reading

Kareiva, P. M., J. G. Kingsolver, and R. B. Huey (Eds.). 1993. *Biotic interactions and global change*. Sinauer Associates, Sunderland, MA. The results of a workshop sponsored by the U. S. National Science Foundation, covering almost all ecological aspects of global change and their implications for policy-makers.

Makhijani, A. and K. R. Gurney. 1995. *Mending the ozone hole: Science, technology, policy*. MIT Press, Cambridge, MA. The state of our knowledge about the nature, origin, and effects of chemicals that are degrading stratospheric ozone, and discussion of efforts to regulate their emissions.

Peters, R. L. and T. E. Lovejoy (Eds.). 1992. *Global warming and biological diversity*. Yale Univ. Press, New Haven, CT. A set of 26 separately authored chapters exploring the possible consequences of global warming on terrestrial and aquatic ecosystems and their members, together with approaches to mitigation of these consequences.

seven

Conservation Theory and Practice

T o conserve the earth's biological diversity, we must go beyond the diagnosis of problems, difficult as this can often be. We must define strategies of managing nature that are both ecologically sound and acceptable to society. In this section we shall consider some of the major areas in which the development of conservation strategy is centered. These range from very applied topics such as sustained yield management to theoretical topics such as how to maintain the genetic health of populations and how to protect and aid the recovery of endangered species. At an even more general level, strategies must be developed for building comprehensive systems of preserves to protect as much of the earth's biodiversity as possible.

chapter

24

Harvesting Natural Populations

H arvesting populations of wild plants and animals is still a major human activity. Some human populations, such as the North American Inuit, still rely heavily on wild plants and animals for food. Aside from a few intensive aquacultural operations, most fisheries' harvests are taken from wild populations of fish, shellfish, and plants such as kelps. Likewise, although modern forestry increasingly emphasizes the production of timber on intensive "tree farms," timber harvests worldwide still mostly involve the cutting of natural forest stands. Can such harvests be made on a sound, continuing basis? And if so, how?

In this chapter we shall consider some of the scientific theory on which the harvesting of natural populations is based. Application of such theory, even if it is sound, involves social, economic, and political considerations that extend beyond the realm of science. Indeed, some resource ecologists believe that it is more realistic to view resources as managing people rather than people as managing resources (Ludwig et al. 1993). The benefits or profits that can be gained from rapid, overexploitation of many resources often fuel political actions opposed to long-term, sustainable management plans. In other cases, societal desires may oppose harvesting resources even when sustainable harvests appear possible.

Major sectors of the human economy depend on harvesting resources that are renewed primarily by natural ecological processes.

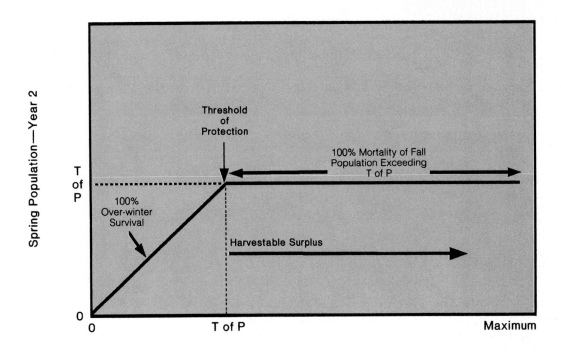

Figure 24.1

The threshold-of-protection (T of P) concept assumes that the habitat can support a certain density of animals over an unfavorable season, such as winter in the midwestern United States, and that the excess population entering that season is a harvestable surplus.

HARVESTING TERMINOLOGY

Applied ecologists have defined several basic concepts of sustainable yield for wild populations. **Sustained yield** is any level of harvest that can be taken from a population indefinitely. **Maximum sustained yield (MSY)** is the greatest harvest that can be taken indefinitely. MSY does not take into account factors such as the cost of taking such a harvest relative to the various benefits that may result. Nor does it consider the benefits that may result from leaving some or all of the MSY unharvested, such as recreational benefits for people who come to view and photograph the unharvested plants or animals. The yield that provides the best compromise of all of the economic and societal considerations is termed the **optimum sustained yield.**

In its most general terms, harvesting theory suggests that up to some threshold, animals or plants that are removed are partially or completely replaced by the increased reproduction or survival of those remaining (Rosenberg et al. 1993). If the threshold is exceeded, the population will decline toward extinction.

EARLY HARVEST MODELS

One of the earliest sustained yield formulations was suggested by Paul Errington (1946), based largely on his studies of bobwhite quail populations in the midwestern United States, and popularized by Durward Allen (1954). Errington visualized midwestern farmland as having a certain **threshold of protection** for quail during the winter: a certain average capacity, based on availability of food and cover, to carry birds through to the next breeding

season (Fig. 24.1). This threshold was regarded as being essentially constant from year to year. If the fall population was below this threshold, over-winter survival would be very high, whereas if the fall population was above this level, the excess would certainly die from factors such as starvation or predation. Thus, the portion of the fall population in excess of the threshold of protection constituted a harvestable surplus—birds that would otherwise die anyway due to lack of winter food and cover.

The threshold-of-protection concept assumes that hunting mortality and natural winter mortality are almost perfectly compensatory. When this is true, and if a population follows strong seasonal changes in resources such as food and cover, the concept may provide a good guide to harvesting, especially for a species like the bobwhite quail with a high reproductive potential. For less seasonal environments and for species whose populations fluctuate less from season to season, the threshold-of-protection idea fails to recognize the great adjustment that most populations can make to mortality, and thus greatly underestimates the harvest that can be taken (McCullough 1979).

CARRYING CAPACITY AND POPULATION REGULATION

More recent harvesting models have been based on the concepts of carrying capacity of the habitat and of how populations grow relative to this carrying capacity. The **carrying capacity** is the maximum population that can exist in equilibrium with conditions of resource availability—the supplies of food, nest sites,

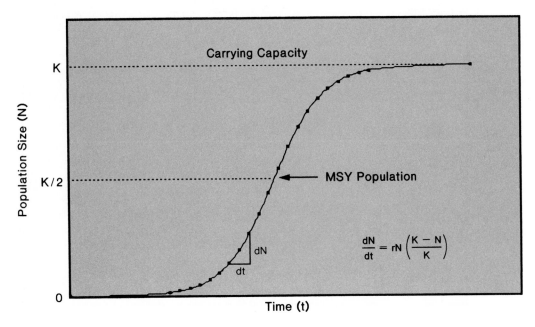

FIGURE 24.2

The logistic growth curve assumes a population whose growth toward a certain carrying capacity (K) is continuously and gradually slowed by various restraints. According to this equation, the point of maximum sustained yield (MSY) is halfway between zero and the carrying capacity.

cover, and similar environmental features that are used by organisms. For most plants, resources comprise water, nutrients, and sunlight, with occupation of an area of substrate often providing the guarantee of these resources in the case of rooted or attached plants. For animals, the supply of appropriate foods is often the primary determinant of carrying capacity, but the availability of nest sites or sites that provide protection from predators or bad weather may force carrying capacity to be less than that possible on the basis of food.

Although carrying capacity can be defined as an average level of resource availability, it is, in fact, a variable. Actual resource conditions change seasonally and from year to year. Often, as well, long-term trends in resource conditions occur due to biotic succession, climatic shifts, or human impacts on the environment. In addition, populations that exceed a habitat's carrying capacity for one reason or another can themselves damage resource relationships so that the carrying capacity becomes reduced.

Many factors affect the growth of populations. The population in an area of habitat grows when reproduction and immigration of individuals exceeds mortality and emigration of individuals to other areas. When populations are low, relative to carrying capacity, competition for resources is likely to be low, and individuals find it easy to protect themselves against predators or normal fluctuations of weather. Reproduction is active, mortality infrequent, individuals from neighboring areas may be attracted to the site, and the population growth rate is high. When populations are close to the carrying capacity, however, competition for food and other resources is strong; individuals often may be unable to find cover against predators or bad weather. Reproduction is likely to be low, mortality high, and emigration to less densely populated areas common. Thus, some of the factors that affect population growth increase the intensity of their action as the population approaches the carrying capacity. These are termed **density dependent factors,** because the intensity of their effects is determined by the density of the population relative to carrying capacity.

Other factors may act quite differently. Diseases, catastrophic weather events, episodes of pollution, and other factors may reduce reproduction, kill individuals, or trigger the movement of individuals in ways quite unrelated to population density. These are termed **density independent factors** because the intensity of their effects is not related to the size of the population relative to its carrying capacity.

Obviously, density dependent and independent factors are the ends of a continuum; one factor might be strongly dependent on population density, another weakly related to density. Generally, biotic factors such as competition and predation tend to show strong density dependence and abiotic factors such as severe weather and chemical pollution show density independence. However, the action of a disease agent may be determined by the genetic composition of a population, rather than its density. The action of a pollutant might also be more severe in a dense rather than in a sparse population if individuals are weakened because of competition for food. Populations may thus be subject to influences that are primarily density dependent over certain periods, and largely density independent over other periods (McCullough 1990).

In any case, the size of any population at a given instant is the result of the action of both density dependent and density independent factors. For large, long-lived organisms such as most vertebrates, density dependent factors are usually assumed to be the strongest determinants. Small, short-lived organisms such as insects and other invertebrates are usually much more strongly influenced by density independent factors.

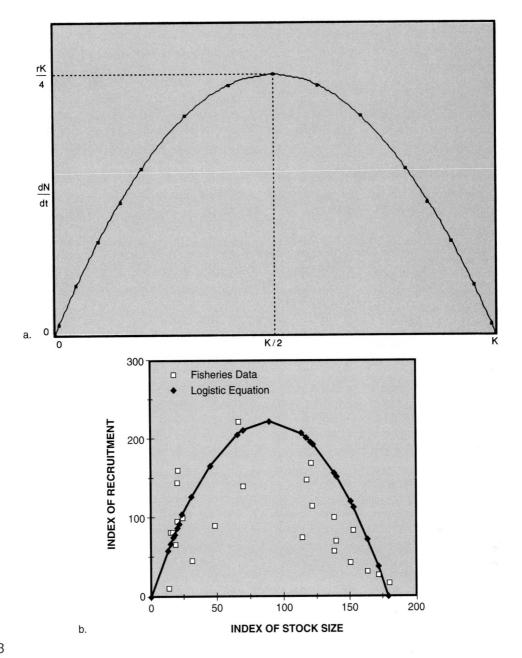

FIGURE 24.3

a. Population growth per unit time (dN/dt) in relation to population size (N), according to the logistic equation, indicating the point of maximum sustained yield (K/2). b. The logistic growth curve fitted to an index of recruitment of young fish in relation to an index of stock size for cod in northern European waters, with an estimate of maximum sustained yield.

(a. Data from D. J. Garrod. 1967. *J. Fish. Res. Bd.* Canada **24:**145–90)

LOGISTIC HARVESTING MODELS

Much sustained yield theory has been based on the logistic family of equations, which describe an S-shaped curve of growth of a population in a habitat with a certain carrying capacity, and a pattern of growth determined primarily by density dependent factors (Fig. 24.2). The simple logistic equation is given below:

$$\frac{dN}{dt} = rN\left(\frac{K-N}{K}\right)$$

In this equation, N is the number of individuals in the population at a given time, K is the average carrying capacity, r is the per capita potential population growth rate (under nonlimiting

conditions), and t is time. The symbol d means "change in," so that the left side of the equation reads, "change in number of individuals divided by change in time," or growth per unit time. The right side of the equation states that this growth is equal to rN multiplied by $(K-N)/K$. The term rN is the potential growth that would occur if resources were not limiting, based on the number, N, of individuals present and the growth potential, r, associated with each. The remaining term applies a "degree of realization" factor to potential growth. This term, in effect, expresses the increasing intensity of density dependent factors as the population approaches carrying capacity. When N is very small relative to K, this expression yields a numerical value near 1.0; most of the potential growth is thus realized. When *N* is near *K*, the value of the expression drops to near zero; most of the potential growth is not realized. Thus, the logistic equation can be stated:

Realized growth = Potential growth × Degree of realization.

If we graph realized growth against *N*, as it increases from 0 to *K*, we see a dome-shaped curve with a peak halfway between 0 and *K* (Fig. 24.3a). In other words, population growth is greatest when *N* equals *K/2*. If exactly this number is harvested, the population will not increase, and during the next time interval it will once again produce this maximum growth (which could once again be harvested). This suggests that MSY is possible when a population is about half its carrying capacity. How many individuals does this represent? When *N* is half of *K*, only half the potential growth is realized; this equals a quarter of the potential growth at *K*, or *rK/4*. What harvesting effort (fraction of the *K/2* population) does this involve? If realized growth is only 50% of potential growth at *K/2*, this equals *r/2* times *N*; harvest effort thus equals *r/2*. In summary:

MSY should be possible at a population of   *K/2*
The MSY, in numbers of individuals, equals   *rK/4*
For MSY, the fraction of the population to harvest is   *r/2*

A curve of this sort can be constructed for index of recruitment of young fish in relation to index of stock size for cod in northern European waters, giving an estimate of maximum sustained yield (Fig. 24.3b). This graph emphasizes the fact that data from real populations show great variation about any theoretical relation, such as the logistic curve.

Having determined the maximum sustained yield from a population, how should this yield be taken to maintain a population at its maximum production level? Two possibilities exist, the removal of a fixed quota per unit time, or the application of a fixed harvesting effort per unit time. In a hunted population, a fixed quota is illustrated by the removal of a certain number of animals, regardless of the effort involved, and fixed effort by control over the total number of hunter days, regardless of the number of animals removed.

The fixed quota option, illustrated in Fig. 24.4a, assumes the removal of the same number of individuals regardless of the population size. If the population is somewhat above the *K/2* level, this means that the harvest will exceed the actual growth of the population, and the population will thus decline toward the *K/2* level of highest growth rate. However, if the population has fallen somewhat below the *K/2* level, harvest will also exceed growth, and the population will therefore decline further. As long as the same quota is applied in this situation, the population will be forced downward, ultimately to extinction.

Application of a fixed harvest effort (Fig. 24.4b), however, acts in a different fashion. If the population is above the *K/2* level, harvest exceeds growth, and the population declines toward the level of greatest growth rate. If the population falls below the *K/2* level, however, the harvest resulting from the same effort becomes less than the actual population growth, and the population increases toward the level of maximum growth.

Thus, a fixed effort strategy of harvest should have the effect of regulating the population at the desired level of *K/2*. A fixed-effort harvesting technique is therefore, in principle, a sounder approach than a harvest by quota.

In some cases the curve of population growth per unit time may have more than one intersection point with a line representing fixed-effort harvest (Fig. 24.5). In this case, if the population falls below the lower point of intersection, the tendency of continued harvesting is to drive the population toward zero. The continued depressed state of the Peruvian anchoveta harvest following its decimation by heavy exploitation, in combination with a severe El Niño, may be an example of this phenomenon.

How can wildlife managers determine where a population stands relative to the MSY level, given the difficulties of determining the carrying capacities of wild populations? At any sustained yield level, recruitment into the population must equal mortality. For animals in which yearlings can be distinguished from older adults, the maximum ratio of yearlings (the component just recruited into the adult population) to total adults is thus an estimate of the fraction of the population that can be harvested. This practice assumes that, at the MSY population density, most of the unrealized growth of the population involves poor reproduction and preyearling survival, and that hunting harvest and natural mortality of older adults are nearly fully compensatory.

Sustained yield theory based on the simple logistic equation is usually too oversimplified to be applied in most cases, however. The attractive simplicity of the logistic equation hides several unrealistic assumptions about the members of a population. For example, it assumes that the rate of supply of resources that define the carrying capacity is unaffected by the population (Caughley & Sinclair 1994). Also, representation of a population by *N* assumes that all individuals are the same; each counts "one." Real populations of 100 individuals might in one instance consist of young, highly reproductive animals, and in another of old, largely postreproductive individuals, however. The potential population growth rates, *r*, for these two groups would be quite different. The model also assumes that there is an exact, 1:1 trade-off among different causes of mortality in harvested populations; e.g., complete compensation of hunting mortality by reduced natural mortality.

The populations of most higher plants and animals violate several assumptions of the simple logistic equation. If reproduction is concentrated in certain age classes, the MSY population tends to lie between half and three-quarters of the carrying capacity, *K* (Goodman 1980). Populations in which reproductive potential differs for individuals of different age are common. If the carrying capacity is set by consumable resources whose renewal rate can be decreased by their exploitation, the MSY population also is pushed above *N/2* (Caughley & Sinclair 1994). Many animals have a carrying capacity set by food or other resources that they can deplete, causing a substantial drop in mean fecundity of individuals in populations close to the carrying capacity. Thus, the MSY population is frequently greater than *K/2* (Fig. 24.6).

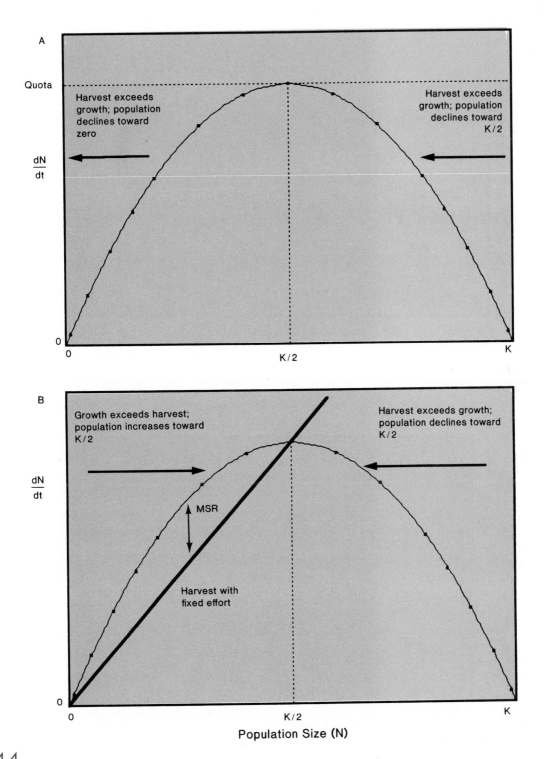

Figure 24.4

a. The effect of fixed quota harvests on population size when populations stray from the MSY level. The MSY level is unstable since deviations below this level initiate a sequence leading to extinction. *b.* The effect of fixed-effort harvest on population size when populations stray from the MSY level. The MSY level is stable since deviations both upward and downward return the population toward this level. If the harvest effort line is equivalent to the cost of harvesting, maximum sustained revenue (MSR) occurs at the point where the difference between effort and harvest curves is greatest.

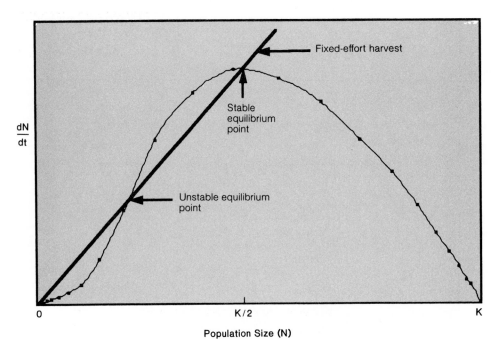

FIGURE 24.5

Curves of population growth per unit time may have more than one equilibrium point, the lower of which is unstable, as indicated in Figure 24.4A.

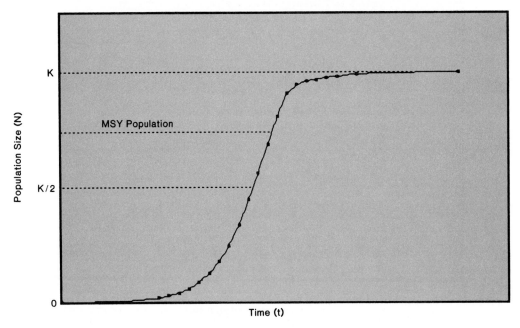

FIGURE 24.6

The shape of population growth curves of many species causes the MSY population to lie closer to K than predicted by the simple logistic relation.

DYNAMIC POOL MODELS

Simple logistic models do not explicitly take the age structure of the population into account, nor do they explicitly consider such relationships as recruitment rate, body growth rate, natural mortality rate, and harvest rate of individuals of various ages or sizes. Dynamic pool models (Pitcher & Hart 1982; Getz & Haight 1989) describe the numbers or biomass of different age (or size) classes as a function of these different processes, each of which is described by one or more equations.

Dynamic pool models can become very complicated, but are structured in the general fashion indicated in Figure 24.7.

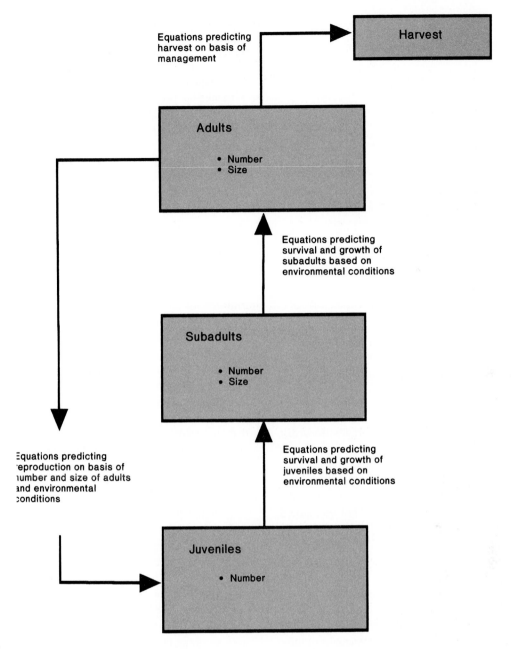

Harvest

Adults

- Number
- Size

Equations predicting
survival and growth of
subadults based on
environmental conditions

Subadults

- Number
- Size

Equations predicting
reproduction on basis of
number and size of adults
and environmental
conditions

Equations predicting
survival and growth of
juveniles based on
environmental conditions

Juveniles

- Number

FIGURE 24.7

A simple example of a dynamic pool model for a population with three age classes.

The population is divided into several age or size classes. Recruitment into the youngest or smallest class is calculated as a function of the population of adult organisms (older or larger size classes) and other important environmental variables. The growth of individuals is described mathematically; together with mortality factors, this determines the number and biomass of individuals in older or larger classes. Natural mortality is related mathematically to factors such as predation, physical conditions, and degree of compensation with harvesting mortality. Mortality due to harvesting is determined by management practices, such as regulation of net size, bag limit, length of harvest season, and so forth. Thus, a dynamic pool model consists of a set of many equations.

HARVEST MANAGEMENT IN PRACTICE

How effective is regulation of harvest in promoting high productivity in exploited populations? Does reduction of populations to some level below the carrying capacity increase reproduction, improve health, and reduce natural mortality? From a conservation standpoint, does controlled harvesting promote healthier populations and greater ecosystem stability? How effective are short-term adjustments in harvest as management tools? Unfortunately, although many examples can be cited of how populations have behaved under particular management schemes, almost no carefully designed experiments address these questions (Caughley 1984).

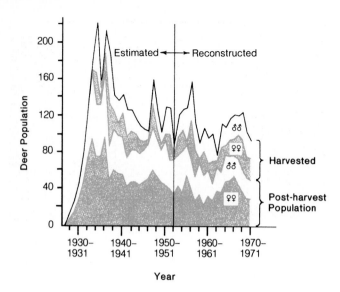

FIGURE 24.8

The white-tailed deer population on the University of Michigan's George Reserve has been subject to carefully controlled harvesting since the early 1930s in an effort to determine maximum sustained yield and to determine the relationship between hunting harvest and recruitment of deer into the adult population.

(Modified from McCullough 1979)

For large mammalian herbivores, many studies show that populations can reach, or even exceed, the maximum level that the habitat can support, leading to poor reproduction and high mortality. Often, such overpopulations develop when natural predators are absent. These populations may severely disturb the ecosystem. On St. Matthew Island, Alaska, for example, a herd of 29 introduced reindeer grew to about 6,000 animals in only 19 years (Klein 1968). By this time, the tundra vegetation had become severely overgrazed, and, in three years, starvation had reduced the herd to only 42 individuals. The vegetation following the crash showed severe damage, with the lichen component being reduced to about 10% of normal.

White-tailed deer in the George Reserve of the University of Michigan (Fig. 24.8) provide one of the best examples of these facts, and of how harvesting affects population dynamics and ecosystem conditions (McCullough 1979). The reserve, completely fenced, is a 4.64 square kilometer area of forest, grassland, and wetlands. In 1928, two male and four female deer were introduced to the area, at that time a private estate. In 1933, the first effort to estimate the number of deer was made by counting the animals passing a line of 30 people walking across the reserve. A count of 162 animals was obtained, a value later corrected to an estimate of 181 deer. This, itself, is one of the most remarkable examples of exponential growth by a large animal population. During the late 1930s, the population was estimated to have exceeded 220, and severe browsing effects became evident. Substantial culling of the herd did not begin until the late 1930s, by which time considerable natural mortality must have begun to occur.

Systematic study of the George Reserve deer herd began in 1952, after the reserve had become a research area of the University of Michigan, and has continued until present. These studies have led to an estimate of carrying capacity of about 176 deer, and of 99 deer in the postharvest population at which maximum sustained yield, about 49 animals per year, is obtained. Under this management regime, virtually the entire mortality for the population is from hunting. Based on a model of the dynamics of the population, a harvest level that permitted the population to remain at 176 animals would result in about 18 adult and 26 preadult deaths from natural causes (e.g., malnutrition, disease).

Thus, strong compensatory responses in mortality and recruitment—close to those predicted by basic harvesting theory—indeed are shown by the George Reserve deer. Members of the George Reserve population are also healthier than deer in unharvested populations (McCullough 1984). Storch (1989) also found that this held true for chamois in a heavily harvested population in Bavaria, in which animals of all age classes weighed more than those in a lightly harvested population in similar habitat. For mule deer in Colorado, Bartmann et al. (1992) obtained evidence for compensation in mortality from starvation, coyote predation, and hunting kill. Thus, for some ungulates, the basic premises of harvesting theory appear to be sound.

How effective are short-term adjustments in harvest or bag limits to maintaining populations at MSY levels? Because hunting harvest of waterfowl is such a large fraction—about half—of total winter mortality, adjustments in bag limits have been considered to be essential means of management to assure that an adequate breeding population survives to produce the next season's crop of waterfowl. This practice assumes that lower hunting harvests in years of low populations lead to increased over-winter survival, and thus to larger breeding populations in the following summers.

Some studies of the relation of bag limits for mallard ducks (Fig. 24.9) to hunting harvest have raised questions about the efficacy of this approach. Anderson and Burnham (1978) found that in years of low numbers, lower bag limits reduced the percentage of the population taken by hunters, but that this reduction did not result in increased over-winter survival. Nichols et al. (1984) reached similar conclusions after a review of studies of various species of ducks. Studies of population fluctuations of the pintail also suggest that habitat conditions rather than changes in bag limit are the major determinant of population size (Raveling & Heitmeyer 1989). These observations suggest that hunting harvest and other causes of mortality are strongly compensatory; an increase in one is balanced by a decrease in the other.

More recently, however, Reynolds and Sauer (1991) found that low harvest rates favored increased mallard populations in the following breeding season, although this relationship was highly variable. Smith and Reynolds (1992) compared periods of years when bag limits were liberal versus periods when they were restrictive in order to determine if hunting and natural mortality were additive or compensatory. They found that the relationship was not purely additive or compensatory. Their analysis most strongly rejected the model of compensatory mortality, however, leading them to conclude that restrictive bag limits may be a useful tool to increase mallard over-winter survival.

FIGURE 24.9

Analyses of the influence of liberal and restrictive bag limits for mallard ducks suggest that natural mortality and hunting harvest are neither perfectly additive nor completely compensatory.

(U.S. Fish and Wildlife Service photo by Glen Smart)

FIGURE 24.10

Clear-cutting has been the primary technique of forest harvest in the Pacific Northwest.

(U.S. Forest Service photo)

Thus, for waterfowl, it appears that simple concepts of purely additive or compensatory mortality from hunting and natural causes are inadequate. The exact nature of this relationship is uncertain, and is evidently influenced greatly by many factors, such as breeding and wintering habitat quality. As a consequence, a new approach, termed **adaptive management,** is being advocated for waterfowl management (Nichols et al. 1995; Williams & Johnson 1995). This approach is one of management under uncertainty, with the goal of reducing uncertainty. Major alternative management options are defined, one is applied, and the effect on population dynamics is monitored and used to refine options for subsequent management.

SHORTCOMINGS OF POPULATION MODELS

A major shortcoming of most models is that they usually focus on the population dynamics of a single exploited species, and do not consider the ecosystem as a whole. Harvests that reduce the population of one herbivore to 50% of its carrying capacity, for example, may free food resources for use by other herbivores. Increase in the populations of these species may ultimately utilize the initial excess of resources, in effect establishing a new, lower carrying capacity for the exploited species. Or, as in the case of the Antarctic krill fishery, exploitation of a species at one trophic level may reduce populations of other species of higher trophic levels, some of which may be of equal or even greater value in the long run (Beddington & May 1980).

Few models have attempted, however, to consider the role of an exploited species in relation to other species in its food chain, or to species that are competitors or have close symbiotic or commensal ties. Simple predator-prey models, like the logistic single-species model, contain many hidden assumptions that may violate ecological reason (Yodzis 1994).

On top of this, populations of some species are subject to exploitation in different locations. The Pacific whiting, *Merluccius*

productus, a migratory fish of the Pacific coast of North America, illustrates the complexity of managing migratory fish (Getz & Haight 1989). Whiting spawn in waters off southern California and Mexico. As young fish, they are prey to various other fish, but as they grow they become predators on young of some of these same species. As they migrate north in summer, they are harvested by fisheries based at several ports in the United States and Canada. Thus, harvesting strategy must consider a complicated set of ecological, economic, and political relationships.

HOW SHOULD FORESTS BE MANAGED?

Ecosystems dominated by trees range from pristine old-growth forests to intensively managed tree plantations. Enormous controversy has arisen over the management of forests that retain high diversity of structure and composition, both on public and private lands. Whether lightly or intensively managed, however, their ecosystem character must be recognized if sustainability is to be achieved (Maser 1994).

For forests from which timber harvests are taken, two major systems of management have been used: even-aged and uneven-aged stand management. Even-aged management emphasizes plantations of single tree species of the same age, which are managed to maximize growth to some predetermined size and harvested at about the same time. The most common pattern of harvesting is clear-cutting, in which all trees are cut and all harvestable timber removed at one time (Fig. 24.10). Regeneration may occur from the natural seed bank, from seed broadcast onto the site after cutting, or from seedlings planted after cutting. About two-thirds of United States timber is harvested in this fashion.

Modified forms of clear-cutting include **shelterwood cutting** and **seed-tree cutting.** In shelterwood cutting the mature trees are removed in several loggings over several years. This retains a partial canopy that favors germination and early growth of a new generation of trees, and is used for species that require shade during their early life. All mature trees are nevertheless

Recognition that survival of many wildlife ecosystems depends on the creation of economic benefits for local people has led conservationists to view harvest of wildlife with renewed interest. In Zimbabwe, for example, efforts are being made to promote sustained use of wildlife resources on communal lands that were created from the tribal reserves of colonial days (Metcalfe 1944). Communal lands amount to almost 42% of Zimbabwe, and have a population of about 5 million people. Much of this land is farmed, but about 20% has significant populations of wildlife.

The new program, the Communal Areas Management Programme for Indigenous Resources (CAMPFIRE), began in 1989 and now involves 12 of the 55 districts of the country. CAMPFIRE's objective is to enable people in the communal areas to benefit from use of wildlife and other nat-

ural resources by giving them ownership and management responsibility for these resources. Government agencies, such as the Department of National Parks and Wildlife Management, provide technical help, such as determining hunting quotas and setting fees for trophy hunting. Communal area administrations are responsible for protection of wildlife and the management of activities such as wildlife-centered tourism, trophy hunting, commercial game and fish harvests, and subsistence harvests. Under this arrangement, profits and benefits from wildlife resources accrue to the local population, encouraging the sustained management of wildlife and giving strong incentives for reduction of poaching. The income from these activities is distributed to local villages, which are then free to decide whether it will be used for special projects, such as school or clinic developments, or distributed to households as cash payments.

In 1993, the participating districts earned over $1.6 million from wildlife-related resource use, mainly trophy hunting fees (Butler 1995). This is estimated to represent an increase of 15%–25% in household income for these communal areas. Poaching also seems to have declined in many areas. Although these results are encouraging, it is still uncertain if a genuinely sustainable pattern of resource use can be achieved. Many local communities are not actively participating in planning and management of wildlife use (Metcalfe 1994). Human population growth and migration of settlers into the communal lands are increasing pressure on resources of these areas. Unless these pressures are managed, farming and livestock grazing are likely to increase at the expense of wildlife. Nevertheless, CAMPFIRE represents an important test of whether managed use of wildlife resources can fill an important niche in the conservation scheme.

---

eventually cut, and a young, even-aged stand replaces them. In seed-tree cutting, all trees are cut except for a scattering of mature trees that are left to provide the seed for forest regeneration.

Uneven-aged management involves harvesting of individual trees or small groups of trees at frequent intervals (Fig. 24.11). **Selective harvesting** is most suitable for tree species whose seedlings are tolerant of shaded conditions. For species whose seedlings require strong sunlight, groups of trees—in effect, small clear-cuts—must be done. Uneven-aged management has long been the basic technique of forest management in Europe.

It is often argued that clear-cutting represents the most efficient and profitable technique of forest management for timber production. Haight (1987), however, showed that this does not necessarily apply to management initiated in forests of mixed composition and age structure. In a specific example involving management of California white and red fir forests, Getz and Haight (1989) found that the profitability of uneven-aged management was greater than that resulting from conversion of the stands to even-aged management by initial clear-cutting.

GAME CROPPING AND WILDLIFE CONSERVATION

It is often suggested that sustained yield harvesting of wildlife, or **game cropping,** in ecosystems such as the East African savannas, can produce greater yields of meat and other animal products than can livestock ranching in the same areas. The logic of this suggestion is that a diverse

fauna of wildlife species specialized for feeding on different plants and at different heights can exploit plant production more fully than can a single domestic species such as cattle. Furthermore, native wildlife are often resistant to endemic diseases, such as *trypanosomiasis* in Africa, to which domestic animals are susceptible. The advantage of game cropping for conservation, according to this argument, is that it gives strong economic justification for retention of native ecosystems rather than replacing them by ranching operations (*See* Reading 24.1).

Several efforts have been made to test the economic viability of game cropping. However, few long-term, economically viable game cropping operations have yet been established. **Game farming,** a ranching operation that mixes livestock with selected species of wildlife, has proved successful in several areas (Pollock 1969). The game animals are usually species for which trophy hunters are willing to pay large fees, or animals that provide meat and skins for tourist restaurants or markets. In South Africa, however, several species of medium to large ungulates are commonly raised for meat production on private farms (Skinner 1985), and some of these operations appear to be equally profitable to beef ranching.

Despite the logic about harvest of primary production by a diverse fauna of native species, a larger sustained yield by game cropping than by ranching has never been demonstrated. This is largely due to the fact that a ranching operation is able to control the age and sex composition of a herd of domestic animals very closely, maintaining the herd in a highly productive state with

FIGURE 24.11

In the Panhandle National Forest, Idaho, selective logging is carried out with the aid of a helicopter that lifts logs from steep slopes to waiting trucks.

(U.S. Forest Service photo by DelMar Jaquish)

regard to the desired product (Caughley 1976). Thus, the MSY of a herd of domestic animals seems to be higher than that possible from a wild fauna harvested in any practical fashion. According to this view, the economic arguments for wildlife conservation relate to tourism and sport hunting, rather than to game cropping.

On the other hand, productive livestock ranching depends on the maintenance of vegetation that domestic animals can utilize fully (McCullough 1979). In areas that cannot be maintained as open grasslands, cattle are unable to harvest plant production as effectively as can a diverse community of wild species. In many places livestock ranching may also be difficult because of shortage of water or presence of poisonous plants (Skinner 1985). Efforts to extend ranching activities into such regions are likely to destroy wildlife communities, but create inefficient ranching operations.

SUMMARY

1. Sustained yield is any rate of harvest that can be continued indefinitely, maximum sustained yield the greatest such rate, and optimum sustained yield the harvest rate that considers both harvest and nonharvest values associated with the resource.

2. The theory of maximum sustained yield has been based on the logistic family of equations, which predict that harvest rate should be possible when the population is about one-half to three-quarters the carrying capacity.

3. More complex dynamic pool models take into account age structure, and age-specific recruitment rate, body growth rate, natural mortality rate, and harvest rate, each described by one or more equations.

4. In forests, even-aged management by clear-cutting or similar harvest techniques is used by most commercial operators, but uneven-aged management, involving selective harvest techniques, is less disruptive to the total landscape ecosystem and may be economically advantageous.

5. Game cropping, the harvest of animals from wild populations for their meat, may help justify retention of native ecosystems, especially if these ecosystems are managed so that they retain tourism and sport hunting values.

QUESTIONS FOR DISCUSSION

1. Do you think that failures to achieve sustained yield harvest of fisheries and wildlife are due primarily to inadequacy of ecological models of the population and ecosystem relationships or to inadequacy of the procedures to regulate harvests according to these models?

2. Do you think that it is fair to say that intensively managed tree plantations differ ecologically from natural forests in manner and degree similar to the ways an alfalfa field differs from a native prairie?

3. If wildlife harvest becomes important for meeting food needs of human populations in areas such as rural Africa, do you think the survival of wildlife will be encouraged, or will this approach make it easier to displace wildlife with domestic animals that are more easily managed for food production?

SUGGESTED READING

Costanza, R. (Ed.). 1991. *Ecological economics: The science and management of sustainability*. Columbia University Press, New York, NY. An examination of broad, interdisciplinary approaches to the management of ecological resources under conditions of high uncertainty.

Levin, S. A. (Ed.). 1993. Forum: Science and sustainability. *Ecol. Applications* **3**:545–89. Eighteen short essays on sustainability in the harvesting of ecological resources and in the overall relationship of humans and the biosphere.

Maser, C. 1994. *Sustainable forestry: Philosophy, science, economics*. St. Lucie Press, Delray Beach, FL. An analysis of the ecological characteristics of forests, the shortcomings of present forestry practice, and the prospects for practicing sustainable forestry into the next century.

chapter

25

Conserving Genetic Diversity

T he genetic structure of a population of any species is, in a real sense, its evolutionary life insurance policy. The variability that exists in the gene pool of the population is the raw material for natural selection, and is essential to permitting the species to continue adapting to its changing environment (Selander 1983).

When populations decline to a few individuals, however, much of this genetic variability can be lost. This means that the evolutionary adaptability of the species is also being reduced. In addition, in remnant populations of a few individuals, breeding with close relatives becomes more frequent. Such inbreeding often leads to detrimental genetic effects. **Conservation genetics** is concerned with evaluating the dangers of genetic impoverishment and inbreeding, and with devising ways to maintain as high a degree of genetic variability as possible in populations of species that have been pushed to the edge of extinction. The principles of conservation genetics apply both to populations in nature and to those in captivity.

Many species, including the cheetah, have lost much of their genetic variability, and are at risk in the race to adapt to changing environmental conditions.

Outline

Genes in Populations

Genes are units of the genetic molecule, DNA (deoxyribose nucleic acid), that contain the coded instructions for assembly of particular proteins, such as enzymes, that in turn determine the structure and function of organisms. An organism has many genes, which are located on chromosomes. Humans, for example, have 23 different chromosomes, which contain about 100,000 genes.

The gene for a particular protein may exist in slightly different forms, known as **alleles,** that consequently tend to produce different versions of the protein, or sometimes a protein so different that it is nonfunctional. One human gene, for example, produces an enzyme that mediates the conversion of tyrosine to melanin, a dark pigment present in epidermal cells of the body. The form of the gene that produces the functional enzyme is one allele, but another allele exists that produces a nonfunctional enzyme. Humans who have only this allele are unable to produce melanin, and are albinos. Alleles are produced by **mutation,** which is nothing more than an accident in the replication of the gene during the process of cell division. Mutations may be spontaneous, or produced through the action of agents, such as ionizing radiation or highly reactive chemicals, that can interfere with molecular replication.

One component of genetic variability is thus whether few or many of the genes in a population are **polymorphic,** having two or more alleles, rather than being **monomorphic,** or having only one allele. Polymorphic genes also vary in the number of alleles they possess. In principle, no limit exists, so that in a large population, many different alleles might exist for a particular gene. **Polymorphism,** the percent of genes that have more than one allele, is one common measure of the genetic makeup of a population, or what is termed the **gene pool** of the population. Normally, about 30%–50% of the genes of an organism are polymorphic, although this figure varies somewhat among different groups of plants and animals. An allele that arises by mutation in one location within the range of a species, provided it is not so deleterious that selection eliminates it quickly, can spread to other individuals through dispersal and interbreeding of individuals. The spread of alleles in this way is termed **gene flow,** and is one process that maintains a characteristic level of polymorphism in the overall species population. The movement of only a few individuals from one subpopulation to another each generation, carrying their alleles, can be quite effective in maintaining a high polymorphism level.

The body tissues of higher plants and animals normally contain chromosomes in pairs, one member of which came from each parent. Humans, for example, possess 23 pairs of chromosomes that are distinct in size, shape, and the genes they carry. Each member of a pair holds one version of each of the many genes located on that chromosome (Fig. 25.1). These might be the same allele, in which case the individual is **homozygous** for that allele, or they might be different alleles, in which case the individual is **heterozygous.** At most, an individual can have two alleles, but in a population, of course, many different alleles can occur.

Mean heterozygosity, the average percent of the genes that are heterozygous in individuals, is thus a second component of genetic variability. Normally, 5%–7% of genes are heterozygous, but again, this varies somewhat from group to group. Although changes in mean heterozygosity in natural populations are not easy to relate to specific selective advantages, many studies show increases in mean heterozygosity with increasing age or following periods of severe weather. This suggests that individuals with high mean heterozygosity survive better than those with low mean heterozygosity (Allendorf & Leary 1986). Watt (1983), for example, found that heterozygosity of the gene for a particular enzyme important in energy mobilization enabled the sulfur butterfly, *Colias philodice,* to be active over a wider range of weather conditions than could homozygous individuals. This resulted in increased survival of heterozygous individuals.

Thus, the components of genetic variability of a population include 1) the fraction of genes that are polymorphic, 2) the average number of alleles for polymorphic genes, and 3) the average percentage of genes that are heterozygous in individuals.

Genetic Changes in Small Populations

As the overall abundance of a species declines, its population typically changes from one that is relatively continuous and has only small gaps separating subpopulations to one consisting of widely separated, small groups of individuals. With increased isolation of such subpopulations, the rate of gene flow decreases, due to reduced interchange of individuals, and mutation becomes the only process working to increase genetic variability within individual subpopulations.

When the number of individuals in an isolated population declines to less than 100 or so, some of the alleles of polymorphic genes tend to be lost from the population by a process termed **genetic drift,** the random fluctuation in the frequency of an allele due to accidents that affect the survival and reproduction of individuals (Fig. 25.2). At first, genetic drift affects the very rare alleles that may be carried only by a small percentage of the individuals in the population. Death or failure of the few individuals to transmit the rare allele, simply by chance, may cause these alleles to disappear. But as the population becomes reduced to a handful of individuals, such accidents may cause an erratic fluctuation from generation to generation in the frequencies of even the more common alleles. In time, such fluctuations may cause the frequency of one allele to reach 100%, for much the same reason that flipping a coin a few times will occasionally yield a run of all heads.

An example of the relation of genetic variability to population size is provided by the coniferous tree, *Halocarpus bidwillii,* in New Zealand (Billington 1991). Isolated populations of this tree in the mountains of South Island range from about 400,000 individuals to only 20 to 25. In populations of 10,000 or more individuals, polymorphism is 25%–35%, whereas in the smallest populations it is only 0%–5% (Fig. 25.3). Many examples are available for animals. For example, two small populations of

A. GENES AND ALLELES

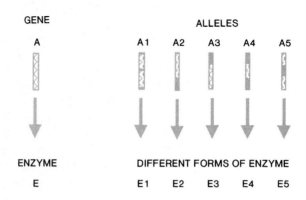

GENE

A

ENZYME

E

ALLELES

A1   A2   A3   A4   A5

DIFFERENT FORMS OF ENZYME

E1   E2   E3   E4   E5

B. ALLELES ON PAIRED CHROMOSOMES

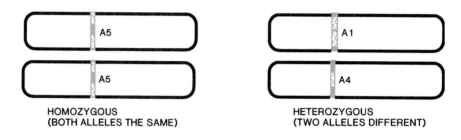

A5

A5

HOMOZYGOUS
(BOTH ALLELES THE SAME)

A1

A4

HETEROZYGOUS
(TWO ALLELES DIFFERENT)

C. GENE POOL OF POPULATION

POLYMORPHISM

GENES

A	5 alleles	
B	2 alleles	
C	1 allele	Polymorphism
D	1 allele	= 40%
E	1 allele	

HETEROZYGOSITY

INDIVIDUALS

A1/A1	
A2/A3	
A1/A3	Heterozygosity
A2/A4	= 60%
A2/A2	

FIGURE 25.1

Alleles are different forms of the gene for a particular enzyme or other protein. Each individual has two copies of each gene, which might be the same (homozygous condition) or different (heterozygous condition). The gene pool of the population is characterized by the number and relative abundance of the alleles present, and by the mean fraction of heterozygous individuals.

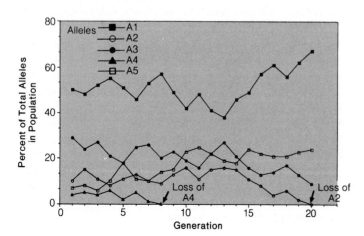

FIGURE 25.2

Genetic drift is the random fluctuation of relative abundances of alleles due to accidents of reproduction and survival. Drift can often lead to the loss of alleles in small populations.

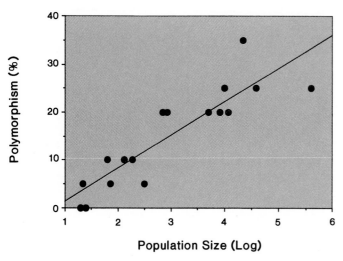

FIGURE 25.3

Percentage of polymorphic genes in isolated populations of the tree, *Halocarpus bidwillii*, in mountains of South Island, New Zealand, in relation to population size.

(Data from H. L. Billington, 1991, "Effects of Population Size on a Genetic Variation in a Dioecious Conifer" in *Cons. Biol.*, 5:115–119. Cambridge, MA:Blackwell Scientific Publications, Inc.)

FIGURE 25.4

The northern elephant seal is an example of a species that experienced a recent population bottleneck that has apparently caused the loss of most of the genetic variability in its population.
(Photo by G. W. Cox)

brown bears survive in the Italian Alps (5–10 animals) and the Italian Apennines (80–100 animals). Mitochondrial DNA analysis showed that both populations had lost all genetic variation. The genetic sequence patterns of the two populations were different, however, indicating that drift had eliminated different sets of alleles in the two populations (Randi 1993).

When the influence of drift is considered in relation to time measured in years or decades, rather than generations, one can see that the rate of loss of variability by genetic drift is faster in species with short generation times than in a population of an equal number of individuals of a long-lived, slow-maturing species (Franklin 1980). Over a century, for example, a population of a small, short-lived desert pupfish may be replaced perhaps 50 times, each replacement carrying a certain risk of loss of alleles by genetic drift. Over the same period, a population of elephants would replace itself perhaps four times, running a much lower overall risk of loss of genetic variability.

Over the entire gene pool, the loss of genetic variability by genetic drift is quite predictable. The rate of loss of heterozygosity, *H*, for example, is predicted by a simple exponential relationship (Lande & Barrowclough 1987). The mean loss per generation depends only on the size of the breeding population, *N*. Heterozygosity at a time *t* generations from t_0 is given by the equation:

$$H_t = H_0 e^{-t/2N}$$

In this equation, generation time, *t,* corresponds to the mean age of all individuals that are engaged in reproduction.

Using a modeling approach, Lacy (1987) examined the roles of genetic drift, immigration, and mutation, together with that of natural selection, in determining genetic variability in small populations. For populations less than about 100 individuals, drift proved to be a very strong force, much more

important, for example, than common intensities of selection. Mutation, because of its low incidence, was unable to maintain polymorphism by overriding drift, and was essentially unimportant. Immigration, from a large source population, on the other hand, was a strong force for maintaining genetic variability in the face of genetic drift, even when only a few immigrants arrived per generation.

Occasionally, populations of organisms drop to a low level, and then recover. At their low point, however, they may have experienced severe effects of genetic drift, passing, in a sense, through a **genetic bottleneck.** How much variability is lost depends on the size of the population and the number of generations that the bottleneck condition lasts (Allendorf 1986). During a bottleneck of short duration, say a single generation, some alleles may be lost, but the general level of heterozygosity may not be reduced much. Populations that are reduced to only a few individuals for several generations may lose all variability, however. The northern elephant seal (Fig. 25.4), a species that once occurred along much of the Pacific coastline of North America, was reduced by Russian seal hunters to fewer than 100 individuals—perhaps as few as 20 individuals—in the 1890s on Guadalupe Island, Mexico (Bonnell & Selander 1974). Among these individuals, reproduction involved perhaps only a few harem males, so that the effective breeding population was very small. Although the population has now recovered to 120,000 or more animals, and the breeding range of the species has spread north to central California waters, the genetic variability within the population is very small. Of the 21 genes that can be analyzed, none shows polymorphism. Southern hemisphere populations of elephant seals were not reduced to such low levels, and, in contrast, show about 30% polymorphism and 3% heterozygosity, values that are essentially normal for mammals.

FIGURE **25.5**

The cheetah may have experienced two genetic bottlenecks, one prior to the separation of eastern and southern African populations, and a second in the southern African population in very recent time.

(Photo by Darla G. Cox)

FIGURE **25.6**

The bison in Badlands National Park, South Dakota, were derived from only about nine founder animals, and show a very low level of genetic polymorphism.

(U.S. National Park Service photo by Jack Boucher)

The African cheetah (Fig. 25.5), particularly the population in southern Africa, shows a similar absence of genetic variability (O'Brien et al. 1985, 1987). One evidence of the genetic uniformity of these animals is the fact that skin grafts can be made successfully between unrelated individuals. This indicates that there are essentially no differences in the skin proteins, which might cause antibodies to be produced against the grafts. The low genetic variability suggests that cheetahs in southern Africa passed through one, or possibly two, genetic bottlenecks. The most recent bottleneck might have occurred during the early European settlement period, prior to the establishment of wildlife preserves, when wildlife of all sorts was decimated by settlers.

Many other examples of bottlenecked populations have been reported in both plants and animals. The 75 to 125 lions of Ngorongoro Crater, Tanzania, are derived from an estimated 15 founder individuals; members of this population show lower heterozygosity and higher frequencies of sperm abnormality than animals in the nearby Serengeti Plains (Packer et al. 1991). The Arabian oryx and Pere David's deer, species eliminated in the wild and reduced to small captive populations, show little genetic variation (Woodruff & Ryder 1986). The 700 or so bison in Badlands National Park, South Dakota (Fig. 25.6), which were derived from about nine founder animals, show polymorphism for only one of 24 genes (McClenaghan et al. 1990). Populations of the Brazilian lion tamarin, nearly extinct in the wild, exhibit reduced genetic variability (Forman et al. 1986). Local populations of desert pupfishes in the Death Valley area of California and Nevada (Turner 1983), along with topminnows in small desert springs of Arizona and northern Mexico (Vrijenhoek et al. 1985), also show loss of genetic variability. Among plants, the Torrey pine, a tree restricted to a small range on the southern California coast and on one of the California Channel Islands, shows no genetic variability (Ledig 1986). The Washington fan

FIGURE **25.7**

Washington fan palms, limited to small populations in desert oases in the American Southwest, lack any genetic polymorphism.

(U.S. Bureau of Land Management photo)

palm (Fig. 25.7), limited to small populations in the deserts of California and nearby areas, has greatly reduced polymorphism (McClenaghan & Beauchamp 1986).

Loss of variability by genetic drift in small populations is often coupled with an increased degree of **inbreeding,** or mating of closely related individuals. Closely related individuals are much more likely to carry certain rare, recessive alleles for deleterious expressions of a gene, which are usually masked by the presence of a dominant, normal allele. Inbreeding greatly increases the chances of pairing of such recessive alleles.

The effects of paired deleterious alleles are varied, but many lead to poor survival or reproduction, a syndrome known as **inbreeding depression.** Ralls et al. (1988), for example, compared the survival of the offspring of inbred and noninbred parents for 44 species of animals at the National Zoological Garden in Washington, DC. They found that survival was higher in noninbred offspring in 41 of the 44 species, and that survival of offspring of inbred crosses averaged 33% lower. The low genetic variability in the cheetah is coupled with some possible inbreeding depression. Cheetahs show very high percentages of abnormal sperm, as well as a relatively high infant mortality rate among captive animals. In addition, captive cheetahs show low resistance to disease (O'Brien et al. 1985, 1987). In the wild, inbreeding problems may occur in the small wolf population on Isle Royale, Michigan (*See* Chapter 12), which now consists of the offspring of a single female (Wayne et al. 1991), and in the relict Florida panther population, in which males show very high levels of abnormal sperm (U.S. Fish & Wildlife Service 1987).

GENETIC PROBLEMS IN CAPTIVE POPULATIONS

Captive populations are subject to the patterns of genetic drift and inbreeding that operate in small populations, and to a number of additional genetic influences (Hedrick 1992). When a captive population is established, for example, another source of reduced genetic variation operates: the **founder effect.** A small group of individuals used to found a captive breeding population will possess only part of the total variability of the wild population. If the individuals are closely related or come from a small part of the total species range, they may lack much of the allelic variation in the population. Thus, if possible, a captive population should be founded with unrelated individuals drawn from various parts of the wild population.

The okapi, an African forest antelope, illustrates how difficult it is to obtain a good representation of the variability in a wild population (Fig. 25.8). Okapis in captivity around the world are derived from only 75 individuals captured in the wild (Lacy 1989). Of these, only 30 have bred, and only 23 have living offspring. Those that have bred vary considerably in the number of offspring they have left. The result is that the present genetic structure of the captive population is equivalent to that which would be produced by only about 12 wild individuals contributing equally in their breeding efforts.

In small captive populations, these problems are also coupled with a peculiar selective environment. Since individuals are artificially protected from predators and are provided food and shelter, selection to maintain natural survival behaviors is nonexistent (Lacy 1987). On the other hand, selection may be strong for characteristics relating to tolerance of confined quarters and absence of normal social contacts.

Still another problem that may arise in artificially managed populations is **outbreeding depression,** or reduced fitness of offspring of distantly related individuals (Templeton 1986). The clearest cases of outbreeding depression are hybrid offspring of individuals from populations adapted to different environments. In such situations, local populations may have become adjusted genetically to local conditions. Hybrid offspring of individuals from two such situations may be adapted to neither. Introduction of individuals of two foreign subspecies to the ibex popula-

FIGURE 25.8

Although some 75 okapi, an African forest antelope, have been brought into captivity in zoos around the world, the gene pool of the present zoo population shows variability equivalent to that of only 12 equally productive parent individuals.

(Photo by Craig W. Racicot, courtesy of the Zoological Society of San Diego)

tion in the Tatra Mountains, Czechoslovakia, for example, reportedly resulted in the birth of hybrid young in midwinter (Greig 1979). The result was high mortality, and eventual extinction of the ibex population.

In zoo populations of many animals, the interbreeding of individuals belonging to different subspecies, or occasionally species, has been frequent. Inaccurate understanding of the taxonomy of the species, poor record keeping, and shortage of possible breeding individuals have all contributed to this sort of hybridization. The result, however, is that some zoo populations are unsuitable for use in captive breeding programs aimed at survival of species and their distinct subspecies. Analysis of the genetics of animals in captive populations, to enable breeding programs to avoid such hybridization but maintain genetic diversity within natural taxa, is a major priority.

CONSERVING GENETIC VARIABILITY

The first strategy in conservation of genetic variability, of course, is to prevent populations from reaching "bottleneck" levels. Efforts have been made to determine what constitutes a minimum population size to prevent loss of evolutionary adaptability (Shaffer 1981; Gilpin & Soulé 1986). Obviously, this minimum population size must apply to the breeding individuals. Often, however, the **genetically effective population size** is much smaller than the actual population (Vrijenhoek 1989). As a formal genetic concept, the genetically effective population size is the ideal breeding population (a 1:1 sex ratio, random mating pattern, and random variation in number of offspring per mating) to which an observed population is equivalent (Lande & Barrowclough 1987). Consider a hypothetical population of nine male and nine female elephant seals, in which only one male mates with the nine females each generation. In addition to the fact that eight males are excluded from breeding, the genetically effective breeding population may be reduced still more because of the close relation of individuals. The various females, in this example, would all be full or half siblings. In this example, the chance of losing alleles by genetic drift would be about equal to that in a population of only two males and two females (Frankel & Soulé 1981).

A rough estimate of the genetically effective population size, N_e, of a population in which generations do not overlap (such as insects with a single summer breeding effort) can be obtained from the numbers of breeding males, N_m, and breeding females, N_f, by the equation:

$$N_e = (4N_mN_f)/(N_m + N_f).$$

More precise estimates require information on the variability of reproduction among individuals, and can be calculated for populations with nonoverlapping or overlapping generations by more complicated equations (Lande & Barrowclough 1987). In most cases, genetically effective population sizes are between a quarter and a half the number of adults of breeding age (L. Nunney, *pers. comm.*).

Estimates of the minimum genetically effective population size necessary to maintain the evolutionary vitality of a species vary considerably. Franklin (1980) suggested that a minimum effective breeding population of 50 would be necessary to prevent deleterious short-term effects due to inbreeding—the so-called "rule of 50." Applied to the preservation of genetic variability, however, one must specify the percentage of variability and the length of time in order to determine the minimum population required. For example, one might specify that 95% of the existing heterozygosity must be retained for 100 years. In such a case, the minimum genetically effective population is related to the generation length of the organism, and would be much greater for small, short-lived organisms than for large species with long generation times. For an endangered population of a species of kangaroo rat, a century might be equivalent to 50 generations, whereas for the African elephant it might equate to only four generations. Based on the equation for the loss of heterozygosity through genetic drift, discussed earlier, a population of 487 kangaroo rats would be necessary to retain 95% of the original level of genetic variability for 100 years, but a population of only 39 elephants would suffice. Frankel and Soulé (1981) suggested that a minimum of 500 breeding individuals—the total population, including nonbreeding individuals, would be larger—is a rough

first estimate of a genetically healthy population. Considering both genetic and ecological factors, however, a single minimum population required for long-term survival in nature cannot be specified, and must be determined for each species (Soulé 1987). We shall examine this problem in detail in Chapter 26.

Many cases exist, however, of populations that have fallen well below the 50-individual level below which serious inbreeding problems may occur. In North America, populations of the whooping crane, California condor, Florida panther (*See* Reading 25.1), and black-footed ferret have all dropped below 50 individuals. Some of these populations have recovered to higher levels, but many still remain below the size at which losses of variability by genetic drift are likely to occur.

Captive breeding programs for endangered animal species run the risk of all of the genetic problems noted above (Ralls & Ballou 1986). In establishing such populations, efforts should be made to maximize genetic diversity by selecting unrelated individuals. Careful management of the breeding population can also maximize the genetically effective population size. With detailed pedigrees of the members of captive populations, various pairing strategies can be used to promote genetic variability (Ballou 1991). Pairings of the most genetically distinct individuals, and regulation of the number of offspring entering the next generation from different pairings, in fact, can create an effective population size greater than the actual number of individuals. Breeding exchanges between institutions having separate colonies of species can reduce the chance of inbreeding effects and expand the effective breeding population. Improved technologies of sperm and embryo banking will eventually allow an effective breeding population to include individuals that cannot be exchanged, or that are no longer alive. Obviously, a well-integrated, international system of breeding records is essential for efforts to create minimum genetically effective populations of species in captivity.

MANAGING THE GENETICS OF SMALL WILD POPULATIONS

Growing human populations and shrinking wildlife habitat mean that, even in nature, increasing numbers of species will need active human management not only to maintain minimum numbers but also to maintain a healthy genetic structure. Fragmentation of the range of species in nature, and the accompanying reduction of gene flow, can be alleviated by translocation of individuals (Griffith et al. 1989). Even though small, relict subpopulations tend individually to lose variability due to genetic drift, altogether they often retain much of the normal variability of the original, large population (Lacy 1987). Regular transplants of a few individuals from one population to another can offset much of the loss of variability by genetic drift. The tiger (Fig. 25.9), for example, now survives in small, widely scattered populations in eastern and southeastern Asia. These populations are subject to inbreeding, and it is unlikely that they will survive indefinitely unless deliberate efforts are made to translocate breeding animals among them (Wemmer et al. 1988).

THE FLORIDA PANTHER

T he Florida panther, *Felis concolor coryi,* is a federally endangered subspecies of the widespread mountain lion, cougar, or puma of North and South America (Reading Fig. 25.1). The panther has become reduced to 70–80, or perhaps even fewer, individuals in southern Florida, most of the animals living in Big Cypress National Preserve, the Florida Panther National Wildlife Refuge, Fakahatchee Strand State Preserve, and nearby private lands west of the Everglades (Maehr & Cox 1995). The rapid development of southern Florida has reduced the habitat of the panther, and the construction of high-speed highways has led to frequent roadkills.

The Florida population has probably been isolated from other populations of the species for 15–25 generations, long enough that genetic bottlenecking and inbreeding are likely to have had significant effects (Hedrick 1995). The remaining panthers have begun to show abnormalities of several sorts, which are believed to be the general result of close inbreeding among the animals in this relict population. These abnormalities include heart deformities, sterility, and, in males, undescended testicles and a very high percentage of malformed sperm. Analyses of genetic diversity in the remaining animals also suggest that the population has lost about 90% of its variability.

The dire condition of the panther population has led to the first genetic restoration effort for endangered animals in North America. Under this program 6–8 female Texas cougars were released

READING FIGURE 25.1

Inbreeding problems and loss of genetic variability have led to a genetic restoration program for the small population of Florida panthers that survives in southwest Florida.

(U.S. Fish and Wildlife Service photo)

into southern Florida in 1995. The Texas form is considered to be very similar genetically to the Florida animals, so that problems of outbreeding depression were considered unlikely. Female animals were chosen for releases on the assumption that they would not create strong territorial challenges to resident male panthers. Essentially, this introduction is intended to replace the natural gene flow that once existed in panther and cougar populations

that were continuous from Florida to Texas through the Gulf Coast region.

The released females, and many of the resident males, are equipped with radio collars, so that movements of the animals can be followed and contacts between males and females noted. Early results indicate that at least one of the released females had mated and appeared to be preparing a den for the birth of young.

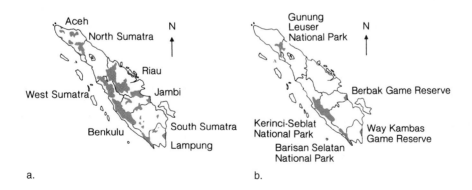

FIGURE 25.9

Tiger populations are fragmented throughout much of the species' range, and the small populations in many locations may lead to inbreeding and consequent genetic problems leading to their extinction. (*a*) Less than 1,000 animals occur in the highly fragmented range on Sumatra, for example. (*b*) Formal protection exists only in five parks and game preserves, themselves small and isolated from each other.

(Modified from Santiapillai & Ramono 1987)

Translocations or breeding exchanges should only be carried out, however, when evidence exists that the isolated populations were once part of a continuous, freely interbreeding ancestral population. Determining the actual degree of genetic divergence among isolated populations and different subspecies is thus essential to conservation management (Lacy 1988). Long-isolated populations that are classified as the same species may actually have evolved many genetic differences that adapt them to their respective areas. Transplants between such populations may thus do more damage, through outbreeding depression, than benefit.

Just how much genetic diversity must be maintained to assure the evolutionary health of populations is uncertain. Soulé et al. (1986) have recommended the goal of preserving heterozygosity of managed populations at 90% of normal over a course of 200 years. A goal of this nature is a general guideline, however, and no species should be abandoned because it has fallen below the criterion. The successful recovery of bottle-necked populations such as the northern elephant seal, in fact, suggests that genetic diversity *per se* may be less important than elimination of deleterious recessive alleles that may appear in homozygous condition through inbreeding. Nevertheless, until the genetics of the species in question are thoroughly inventoried, the strategy of preserving as much diversity as possible is the only reasonable alternative.

ZOOS, BOTANIC GARDENS, AND GENE BANKS

Because of the rapid disappearance of natural ecosystems and their species, zoos, aquaria, botanic gardens, arboretums, and other facilities will bear a heavier responsibility for preserving genetic resources and diversity. Zoos, which focus their efforts on terrestrial vertebrates, will probably be faced by the need to maintain about 2,000 species of larger vertebrates by some time within the next 200 years (Soulé et al. 1986). At best, existing zoo facilities have a capacity of housing about 1,000 such species with minimum populations of 250 individuals (Office of Technology Assessment 1988). Considering the space requirements and costs of zoo maintenance, this is a major challenge.

New technologies may assist the conservation of animal germ plasm. **Cryogenic storage,** long-term preservation of semen, ova, or embryos at subfreezing temperatures, is an established technique for domestic ungulates, and has been proved applicable for many other mammals. The temperatures at which these materials are stored range from −160° C to −196° C. How long animal materials will remain viable under such storage is uncertain, but domestic animal semen and embryos have been stored successfully for 10–30 years. Little has been done to extend this technique to other vertebrates and invertebrates, however.

Botanic gardens and arboreta face a similar challenge. In the United States alone, for example, about 5,000 rare to endangered taxa of plants exist in the wild (Falk 1990). Typically, botanic gardens focus attention on particular families or genera of plants for which their environment is best suited. The plant groups best represented are often those of horticultural interest, such as orchids and palms. Agricultural research institutions often maintain collections of plants of agronomic importance. In such collections, space is a severe constraint on the number of living individuals that can be preserved, particularly for shrubs and trees. In the United States, the **Center for Plant Conservation,** located in Jamaica Plain, Massachusetts, coordinates the activities of 20 botanical gardens and arboreta in the propagation of endangered plants.

Seed storage facilities have also been developed for higher plants, both those of recognized economic value and other species. Conventional facilities, such as the **National Seed Storage Laboratory (NSSL)** in Fort Collins, Colorado, store seeds at about 4.4° C and low relative humidity. The stored samples must be tested for viability at about five-year intervals, and propagated in the field to produce new seed when viability begins to decline. Seeds of most crop plants must be propagated about every 15 years. Conventional facilities like the NSSL exist in many countries, primarily to protect genetic lines of plants of agronomic value (Cox & Atkins 1979).

Cryogenic storage of many seeds is also possible. For this, seeds must be dried to reduce moisture to about 5% and stored at temperatures below −160° C. The viability of seeds stored at such temperatures needs to be checked only at about 50-year intervals, and the seeds probably need to be propagated anew only at intervals of hundreds of years. Unfortunately, the seeds of many tropical and aquatic plants cannot tolerate the drying necessary for cryogenic storage. The **Royal Botanic Gardens** at Kew, England, has pioneered cryogenic storage of the seeds of wild species (Koopowitz & Kaye 1983). Several regional centers for cryogenic storage of seed of wild species have now been established, such as the **Iberian Gene Bank** in Madrid, Spain, which has become the center for storage of seed of endangered plants for the entire Mediterranean region. In 1979, the **Botanic Gardens Conservation Coordinating Body** was created to help integrate the efforts of such centers on a worldwide basis.

SUMMARY

1. Genetic variability in a population includes 1) the fraction of genes that are polymorphic, 2) the number of alleles for polymorphic genes, and 3) the average percentage of genes that are heterozygous in individuals.

2. In reduced and fragmented populations, gene flow becomes restricted and alleles tend to be lost by genetic drift, leading ultimately to the loss of all genetic variability, a situation known as a genetic bottleneck.

3. Inbreeding, the mating of closely related individuals, and outbreeding, the mating of individuals from genetically distinct populations, can also lead to reduced reproduction and survival of individuals in nature or in captivity.

4. Management of the genetics of endangered or captive populations increasingly emphasizes the interchange of individuals and the pairing of mates to minimize inbreeding and foster high levels of polymorphism and heterozygosity.

5. Traditional techniques of propagating species in zoos, aquaria, and botanic gardens are being supplemented by procedures such as cryogenic storage, the preservation of reproductive tissues at temperatures below −160° C.

QUESTIONS FOR DISCUSSION

1. Do you think that global change poses a special threat of extinction to species that have lost much of their genetic variability?

2. Do you think that translocating individuals from one wild population to another to increase genetic diversity is a safe strategy, considering that people can only assess variability for a small fraction of all genes of any species?

3. Do you think that *ex situ* strategies can be counted on to carry a major responsibility for preserving biodiversity over the next century, given the increased difficulty that most governments are having in financing basic services to humans?

SUGGESTED READING

Avise, J. C. and J. L. Hamrick (Eds.). 1995. *Conservation genetics: Case histories from nature.* Chapman and Hall, New York. A survey of the conservation genetics of various plant and animal groups, and discussions of how genetics is relevant to the preservation of biodiversity in nature.

Frankel, O. H., A. D. H. Brown, and J. J. Burdon. 1995. *The conservation of plant biodiversity.* Cambridge Univ. Press, Cambridge, England. A description of genetic variability in wild and domesticated plants and how to protect it in nature and in botanical gardens and gene banks.

Vogel, J. 1994. *Genes for sale: Privatization as a conservation policy.* Oxford Univ. Press, New York. A discussion of issues relating to ownership of genetic resources in developing nations, and of the potential for income as royalties on such resources to aid the preservation of biodiversity.

chapter

26

Protecting Endangered Species

H uman activities have pushed many species to the brink of extinction. These species need special protection, but for some, recovery to a healthy status is possible. International, national, and state endangered species programs are thus designed to identify species that are threatened with extinction, provide them with immediate protection, and develop plans for their recovery.

The management of endangered species management has become one of the most interdisciplinary areas of conservation biology. Since socioeconomic and political forces are largely responsible for the endangerment of species, conservation biologists have been forced into political and legal arenas (Kellert 1985). In addition, efforts to achieve the recovery of endangered species have forced them to draw on the expertise of fields such as veterinary medicine, zoo biology, agronomy, and horticulture.

In this chapter we shall first examine the history and current status of federal and state endangered species programs in the United States. Next, we shall consider some of the recovery programs that are under way, emphasizing some of the special techniques that are being employed. Finally, we shall examine other private and international programs relating to endangered species.

Legal protection and special recovery efforts are essential for survival of many species that are declining rapidly or have reached the brink of extinction.

United States Governmental Programs

Although conservationists have long been concerned about endangered species, the first comprehensive federal policy of protection of such species in the United States was not established until 1966, with the passage of the **Endangered Species Preservation Act** (Bean 1987). In 1969, this act was replaced by the **Endangered Species Conservation Act,** which strengthened several aspects of the program. In 1973, Congress passed the much more comprehensive **Endangered Species Act;** this act, with various amendments, has been reauthorized on several occasions, most recently in 1988, and currently governs United States federal endangered species policy. Federal endangered species law recently has become the focus of some of the most bitter political controversy relating to conservation (Tobin 1990; Mann & Plummer 1995).

Under the 1973 act, responsibility for management of endangered species is assigned to the Fish and Wildlife Service (Department of Interior), in the case of terrestrial and freshwater species, and the National Marine Fisheries Service (Department of Commerce), for marine species. The major duties of endangered species offices in these agencies are 1) designation of endangered species, 2) designation of critical habitat for endangered species, and 3) development and implementation of recovery plans for these species.

The way in which species are designated and protected makes the endangered species act a very powerful conservation instrument. Species are designated as **endangered,** if they are determined to be in danger of extinction throughout all or a significant portion of their range, or **threatened,** if they are likely to become endangered within the foreseeable future. In its legal usage, the term *species* is not restricted to the entire biological species, but can refer to any subspecies of animal or plant, or to any distinct population of a vertebrate animal. This broad definition of the term is one of the features that makes the 1973 act so powerful: a portion of the biological species population can be designated as endangered even though other portions are still abundant.

The 1973 Endangered Species Act covers not only vertebrate animals, but also invertebrates (exclusive of pest insects) and plants. Both species occurring within the United States and foreign species are covered (Fig. 26.1). The listing of foreign species as endangered or threatened strengthens the ability of the United States to participate in international agreements regulating commerce in such species. The federal list of endangered and threatened species for 1995 contained about 1,525 forms representing 962 biological species (Table 26.1).

The procedure for placing species on the list of endangered and threatened species can be initiated by the offices of endangered species themselves, or by petitions from private groups. Petitions for placing species on the list are reviewed for merit by scientific panels. If addition of a species to the list is recommended, an announcement of the proposed action is published in the *Federal Register* and other appropriate places to solicit additional information and public comment. Following a

Figure 26.1

The U.S. listing of endangered and threatened species includes many foreign species, such as the jaguar, which is classified as endangered. (Photo by Y. J. Rey-Millet, courtesy of the World Wide Fund for Nature)

year-long review period, a final decision on listing is made by the agency panel. Congressional action, however, has at times placed a moratorium on new listings.

Other important activities of the offices of endangered species include designation of **critical habitat** and the development of recovery plans. Critical habitat consists of areas, either within or outside the range of the species, that are essential to its conservation. Essential areas within the range consist of occupied habitat; those outside the range might, for example, consist of the watershed on which a small area of aquatic habitat occupied by an endangered species depends. As of 1995, critical habitat had been designated for less than a quarter of federally listed species (National Research Council 1995). **Recovery plans,** which we shall examine in detail shortly, are designed to increase the abundance and distribution of species so that they can be removed from the list. As of 1995, such plans had been developed for 521 species. Information on these activities, as well as on the listing activities, is published monthly or bimonthly in the *Endangered Species Technical Bulletin.*

The federal endangered species program of the United States has become a model for many other governmental programs. In the United States, almost all state and territorial governments have developed some type of endangered species program, and most of these have established cooperative agreements with the federal endangered species offices. California, for example, has a system for designating species as *endangered, threatened,* and in the case of plants, also as *rare* (Table 26.1).

International Endangered Species Listings

The **Convention on International Trade in Endangered Species of Wild Fauna and Flora (CITES)** is an international agreement that regulates or prohibits commercial trade in globally endangered species or their products. This convention

TABLE 26.1

Numbers of Species Listed as Endangered and Threatened by the U.S. Office of Endangered Species (1995) and by the California Department of Fish and Game (1996)

	UNITED STATES				CALIFORNIA	
	Endangered		Threatened		Endangered	Threatened or Rare
	USA	Foreign	USA	Foreign		
Mammals	55	252	9	19	8	9
Birds	75	177	16	6	18	5
Reptiles	14	65	19	14	3	5
Amphibians	7	8	5	1	2	6
Fish	68	11	37		14	3
Snails	15	1	7			1
Clams	51	2	6			
Crustaceans	14		3		2	
Insects	20	4	9			
Arachnids	5					
Plants	435	1	92	2	124	86
TOTAL	**759**	**521**	**203**	**42**	**171**	**115**

was developed in 1973, and by 1994 some 122 countries were party to its provisions (Hemley 1994). Appendix I of CITES lists about 675 species for which trade is prohibited, and Appendix II lists about 3,700 species of animals and over 21,000 species of plants for which trade is regulated. In the United States, responsibilities for general management, scientific decisions, and enforcement activities relating to CITES are assigned to separate divisions within the Fish and Wildlife Service.

The **World Conservation Union,** located in Gland, Switzerland, is a private organization that coordinates many global efforts in wildlife conservation. The World Conservation Union sponsors the publication of the **Red Data Books,** which list species according to their degree of endangerment and give summaries of information on distribution, populations, habitat ecology, cause of endangerment, and conservation measures taken and proposed. Red Data Books are now available for many taxonomic groups worldwide, as well as for certain major taxa in specific world regions.

STRENGTHENING ENDANGERED SPECIES LAW

A major weakness of the U.S. Endangered Species Act has been the development and implementation of recovery plans. Several years usually pass between listing of a species and the development of a recovery plan, and some species remain without recovery plans after many years. Many plans lack detailed biological information about the species, and many set recovery goals that are no better, and sometimes worse, than the status of the species when it was listed (Tear et al. 1995). Few are able to draw together information to conduct a formal analysis of the viability of existing populations and those projected as the goal of recovery.

In 1991, several members of the U.S. Congress requested the National Research Council to examine scientific issues relating to the Endangered Species Act. In the resulting report (National Research Council 1995), the committee appointed for this purpose concluded that the act was based on sound scientific principles, but also that it could be improved in several ways. They recommended that any taxon proposed for listing be an **evolutionary unit,** a population genetically isolated from other such units and having a shared evolutionary history and unique biological characteristics. Because of the essential nature of habitat for any species, they recommended that at the time of listing, **survival habitat** be designated. Survival habitat is simply the habitat necessary to support the endangered population. The emergency "survival habitat" would be replaced by critical habitat when a recovery plan is prepared. Additional recommendations included strengthening and speeding the development of recovery plans and making more effective use of models for assessing risk of extinction.

The National Research Council's committee also noted that the Endangered Species Act was basically a "safety net," and that additional approaches are needed to preserve biodiversity. Examples of broader approaches include total ecosystem management, ecosystem restoration, new approaches to wildland management, and improved assessment of economic values of biodiversity.

The Ecological Society of America also appointed a committee to consider the same issue. Their report (ESA Committee on Endangered Species 1996) argued for giving priority for listing to species whose protection would also benefit many other species (so-called **umbrella species**), to species most immediately threatened, and to species that are taxonomically unique. It gave strong support to strengthening the framework for developing ecosystem-based plans that would protect critical habitat for many species.

TABLE 26.2

	EXTINCT POPULATIONS		SURVIVING POPULATIONS	
Group	Median	95% Confidence Limit	Median	95% Confidence Limit
Lagomorphs	3,276	702–56,952	70,889	34,720–173,150
Artiodactyls	241	3–1,273	792	429–1,504
Small Carnivores	256	122–880	1,203	908–1,704
Large Carnivores	24	14–68	108	70–146

Estimated Initial Populations of Species Belonging to Various Mammal Groups in National Parks in Western North America, Summarized Separately for Populations That Have Subsequently Become Extinct and Those That Have Survived

Source: Data from W. D. Newmark, "Mammalian Richness, Colonization, and Extinction in Western North American National Parks" Ph.D. Dissertation, University of Michigan, Ann Arbor, MI.

POPULATION VIABILITY ANALYSIS

Central to long-term management of endangered species is estimating the probability of survival of existing populations and determining what would constitute a "safe" population size (Boyce 1992). This effort is known as **population viability analysis (PVA).** A frequent objective of PVA is to specify a **minimum viable population (MVP),** or the smallest number of breeding individuals that has a specified probability of surviving for a certain time, without losing its evolutionary adaptability (Gilpin & Soulé 1986). Specific probabilities and specific time periods are the essential features of PVA. Indeed, PVA is an effort to force endangered species management to become objective in its procedures. Using this approach, for example, one would specify the population of Kirtland's warblers that must be maintained to, say, ensure the survival of the species with 95% probability for 100 years.

PVA assumes that the population in question is protected against factors that might drive it progressively toward extinction (*See* Chapter 5). Thus, given favorable habitat conditions, an MVP is the population size necessary to prevent extinction from various stochastic processes (*See* Chapter 5). These include accidental variation in natality and mortality, the impacts of random variation in environmental conditions, and chance events affecting genetic makeup of the population (*See* Chapter 25).

Several approaches have been taken to PVA. One approach has been the empirical determination of the size of populations that have persisted for a known period following isolation. Newmark (1986), for example, estimated the population sizes of various mammals in national parks in western North America at the time of their establishment (Table 26.2). He then compared populations of those that have become extinct with those that have survived, the length of intervening time being roughly 75 years. The median sizes of surviving populations varied considerably with the size and trophic level of the mammal. Populations of hundreds, thousands, or tens of thousands were not always enough to guarantee survival of lagomorphs (hares and rabbits), which tend to show high levels of natural fluctuation (including major cycles of abundance and scarcity). On the other hand, the median size of surviving

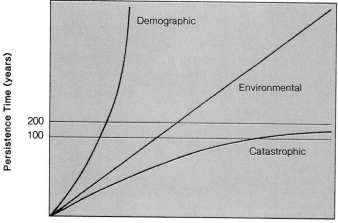

FIGURE 26.2

The relationship of survival probability to population size for populations subject to demographic stochasticity, moderate environmental stochasticity, and catastrophic environmental effects. (Modified from Shaffer 1987)

populations of large carnivores was only 108 individuals. Perhaps systems of territoriality of these animals stabilize their populations well within the limits of prey availability, making them less prone to extinction. More detailed analysis of data sets such as this may allow more precise estimation of probability of survival of populations of different size.

A second approach to determining MVPs has been through modeling population demography (Goodman 1987; Belovsky 1987). The basic strategy of this approach has been to work with *r*, the per capita population growth rate, and an estimate of the variability of *r* due to demographic or environmental influences. Results of such modeling (Fig. 26.2) indicate that different types of variability affect the MVP estimate to very different degrees. The risk of extinction from random accidents affecting births and deaths can be minimized at a relatively small population size. Moderate levels of environmental variability, in contrast, lead to a more or less linear reduction in extinction risk with increase in population size. Catastrophic patterns of

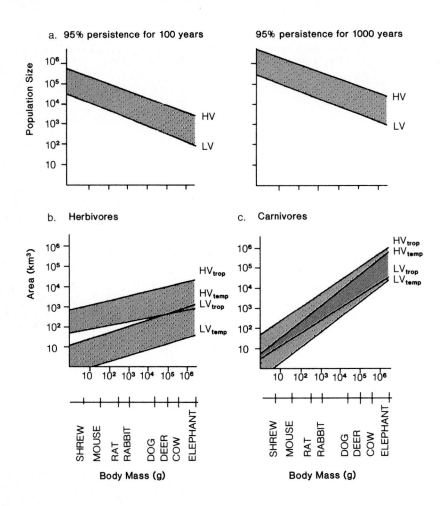

a. 95% persistence for 100 years

95% persistence for 1000 years

b. Herbivores

c. Carnivores

SHREW MOUSE RAT RABBIT DOG DEER COW ELEPHANT

Body Mass (g)

SHREW MOUSE RAT RABBIT DOG DEER COW ELEPHANT

Body Mass (g)

FIGURE 26.3

A. Populations of mammals of different body size required to provide 95% probability of survival for 100 and 1,000 years. A range of values is given for species with low (LV) to high (HV) population variability. *B.* and *C.* Habitat areas needed to sustain herbivorous and carnivorous mammals for 100 years with 95% probability. Different ranges are shown for temperate (temp) and tropical (trop) species.

(Modified from Belovsky 1987)

environmental variation, however, create an extinction risk that declines very little with increase in population size above a certain level (Shaffer 1987).

Belovsky (1987) attempted to estimate rough MVPs for mammals of various body sizes, using allometric relationships of body mass with *r* and its measure of variability (Fig. 26.3). These estimates show that the MVPs of small animals can sometimes be very large. For a 25-gram mammal, for instance, the MVP would range from about 8,000 individuals, if the variability in *r* were low, to about 150,000, if variation in *r* were high. For an animal with a body mass of 100 kilograms, the comparable range would be 300 individuals (low variability in *r*) to 8,000 individuals (high variability in *r*).

A third approach to estimating MVPs has been to determine the minimum population required to maintain most of the genetic variability of the species. By this approach, an MVP corresponds to the genetically effective population size (*See* Chapter 24) necessary to retain, say, 95% of the original genetic variability (e.g., heterozygosity) for 100 years. This approach is not re-

ally an alternative to the first two methods, but rather a second basic criterion: an MVP must satisfy criteria both of demographic adequacy and of genetic and evolutionary health.

Although it is obvious from the above that MVPs must be determined for individual species, the very rough rule of thumb that a genetically effective population of at least 50 breeding individuals must exist to avoid serious inbreeding problems and at least 500 to prevent serious loss of variability by genetic drift has become incorporated in many species recovery plans. The recovery plan for the red-cockaded woodpecker, for example, incorporates such a requirement (Reed et al. 1988). Analysis of the populations of woodpeckers on major areas of protected habitat, however, showed, first, that an adult population of about 1,323 total birds is needed to give a genetically effective population of 500, because of the breeding system and demography of the species. Further, the analysis revealed that genetically effective populations in the nine major protected habitats of the species ranged only from 95 to 378 individuals.

RECOVERY OF ENDANGERED SPECIES

The long-range objective of endangered species programs is to develop programs for the recovery of endangered species, so that they can eventually be removed from the list. Recovery programs for species that have declined to the brink of extinction are often forced to utilize unusual or extreme techniques in efforts to encourage the reestablishment of viable populations. These techniques include captive propagation, reintroduction to the wild, and translocation of individuals to new habitat areas. These activities, in turn, employ several special techniques to increase reproduction and to encourage establishment of individuals in the wild.

Ex Situ Care of Endangered Species

Ex situ care refers to the maintenance of populations outside their native habitat, as in botanical gardens, zoological parks, and aquariums. **Captive propagation** is the breeding of species in confinement for their preservation, display in zoos, and reintroduction to the wild. Increasingly, such institutions are being called on to maintain populations of species that are being driven to extinction in nature. Relative to the growing need, however, the space in these facilities is extremely limited, and coordination of the efforts of institutions is needed to increase their effectiveness.

An **International Species Inventory System** has been established to maintain records of animal species and subspecies that exist in captivity. Beginning with this inventory system, the American Association of Zoological Parks and Aquariums has developed the **Species Survival Plan,** an outline for the coordination of the captive management of endangered animals. This plan provides for identification of priority species for captive management, and for cooperation in captive breeding activities by participating institutions so that the genetic variability in the total captive population is retained. By animal exchanges, breeding loans, sperm and embryo banks, and the maintenance of comprehensive stud books, genetically effective populations that can reduce the loss of genetic variability to a very low level can be achieved with many fewer individuals than in wild populations.

Captive Propagation for Reintroduction to the Wild

The goal of ex situ care for an increasing number of species is to permit their eventual reintroduction to the wild. This goal faces many challenges, the first often being establishing a captive population. In several instances, captive populations of birds have been established by taking the first clutches of eggs from nests in the wild, stimulating **recycling** by the wild parents (Simons et al. 1988). Recycling is simply initiation of a new breeding cycle and the production of a replacement clutch. This procedure has been used extensively with the bald eagle, and was employed with California condors before the last wild birds were brought into captivity (*See* Reading 26.1).

To produce individuals suitable for reintroduction to the wild, the demands of captive propagation are special. Founder stocks and captive breeding populations may lack the genetic variability appropriate to producing viable reintroduction stock (Philippart 1995). Selection acts in the environment of captivity, just as it does in nature, and may, for example, favor individuals adjusted to confinement and less able to live in the unbounded natural habitat. Rarely can the social environment of captive animals be made as rich as that of a wild population. Nor is it easy to provide learning experiences, such as how to find food, how to recognize and avoid predators, and even how to locomote in natural habitats such as that of a forest canopy. For reintroductions to be successful, all of these challenges must be overcome.

Several species of mammals and birds have been reduced to populations that have survived only in captivity. In North America these include the black-footed ferret and California condor (*See* Reading 26.1). The black-footed ferret, native to central and western North America, was reduced to low population levels by control of prairie-dog populations (*See* Chapter 7). Ultimately, only a single population of ferrets remained, and when a plague epidemic devastated the prairie-dog population, an emergency captive breeding program was begun. Unfortunately, many of the ferrets captured died of distemper, reducing the total population of the species to 18 individuals in 1987, all in captivity. By 1995, however, the captive population had grown to 1,400 animals in breeding colonies in six locations. As of 1995, efforts were being made to reintroduce ferrets to prairie-dog towns in several locations in Wyoming, South Dakota, and Montana. Survival of released animals has been poor, but reintroduced animals had reproduced successfully in one of the South Dakota sites. The long-range goal is to establish a wild population of 1,500 animals in at least 10 sites.

A few species have been propagated in captivity for many years, and then reintroduction attempted to their original ranges. The red wolf, extinct in the wild by the 1970s, has been bred in captivity since 1975 at the Point Defiance Zoo in Tacoma, Washington. Animals from this population were reintroduced to Alligator River National Wildlife Refuge, located on a peninsula on the North Carolina coast (Rees 1989). Several desert ungulates have been reintroduced to their original ranges in Africa and the Middle East (*See* Chapter 8). The Socorro Island dove, a species closely related to the mourning dove of continental North America, was reintroduced to Socorro Island, off the west coast of Mexico, in 1988.

Captive propagation has also contributed to the recovery of other species, such as the peregrine falcon in North America. Captive breeding facilities for this species are located in New York, Idaho, and California, and since the mid-1970s over 1,000 birds have been released to the wild (Peakall 1990). These releases are credited with establishing 30–40 breeding pairs in regions from which the species had disappeared due to DDT contamination (*See* Chapter 22).

CAN THE CALIFORNIA CONDOR BE REINTRODUCED TO THE WILD?

T he California condor recovery plan illustrates the technical difficulties and political problems that confront a recovery effort for a species that has reached the brink of extinction.

The California condor (Reading Fig. 26.1A) is the largest raptor in North America, reaching a weight of over nine kilograms, and a wingspan of 2.75 meters. California condors lay a single egg at a time, and care for the young for over a year, so that successful nestings can occur only at two-year intervals. The young birds require 6–8 years to mature, but adults are believed to live for 20–40 years.

Like the whooping crane, the California condor has been at a critically low level for decades. The condor appears to be a relic of the Pleistocene megafauna that survived in North America until about 11,000 years ago. Fossil remains show that the species occurred in Utah and Arizona, and the species was probably widespread in mountainous areas of western North America. In this century, however, the condor population declined rapidly, largely because of poisoning, shooting, collision of birds with power lines, and lead poisoning (Snyder & Snyder 1989). Annual census efforts beginning in 1965, however, showed that the population was restricted to a limited U-shaped range in the southern coast ranges, Tehachapi Mountains, and southern Sierra Nevadas in California, and that total numbers were about 50–60. Two condor sanctuaries were established in areas of the Los Padres National Forest where the birds were known to nest.

READING FIGURE 26.1A

The California condor, reduced to a small population in the Los Padres National Forest in California, is the object of a major captive breeding and reintroduction program.

(U.S. Fish and Wildlife Service photo by Carl B. Koford)

A condor recovery plan was prepared in 1975. This plan had the objective of creating at least two wild populations, one in the area of the existing population and one somewhere else within the historical range. The plan also stated an objective of a total population of at least 100 birds in the wild. These objectives were to be achieved by reducing mortality, protecting nesting habitat, and improving feeding habitat. Captive propagation and the release of birds to the wild was one of the techniques proposed.

In the late 1970s, it became apparent that the wild population was continuing to decline, and that shell thinning from DDT, lead poisoning from shot in carcasses eaten by condors, and illegal shooting were major causes of mortality and poor reproductive

A research and propagation center for raptors, known as *The World Center for Birds of Prey,* has been established in Boise, Idaho. Captive propagation of several species of hawks and owls is being carried out at this location.

Reintroduction of individuals produced in captivity to the wild is usually done gradually, by a process known as **soft release** or **hacking** (a falconry term) in the case of birds. Individuals are first placed in large pens or flight cages in a seminatural setting to enable them to acclimate to outdoor conditions. After this, they are released, but food is usually provided for awhile at the acclimation site.

A second technique, **cross-fostering,** is the rearing of young of one species by adults of another. For birds, cross-fostering involves the exchange of eggs between species, so that foster parents incubate the eggs and rear the young. This proce-

dure exposes the moved eggs to a higher mortality risk than is likely in captive rearing conditions, and may also result in the young becoming imprinted on adults not of their species (Powell & Cuthbert 1993). Cross-fostering was used in an unsuccessful effort to create a second wild breeding population of whooping cranes (*See* Reading 26.2). For captive mammals, cross-fostering may involve gestation by foster parents, accomplished by means of embryo implants.

Population Translocation

Translocation is the movement of individuals from one location to another to create or enlarge a wild population. Translocations have been attempted for many vertebrate species, with varying degrees of success. Many biological and ecological

READING FIGURE 26.1B

At the San Diego Wild Animal Park, young California condors are fed by hand puppets designed to simulate the appearance of an adult condor. This chick, named Molloko, was the first offspring of birds paired in captivity.

(Photo by Ron Garrison, courtesy of the Zoological Society of San Diego)

performance. The recovery plan was revised in 1980 to intensify efforts at captive propagation. Propagation centers were established at the Los Angeles and San Diego Zoos. Initially, young birds and eggs from nests in the wild were brought into these centers, but eventually the decision was made to bring all remaining wild birds into captivity. This decision was bitterly contested by some conservation groups, in part because it removed all condors from areas considered important for management as condor range.

Reason for optimism existed in regard to the captive breeding program. In the early 1900s, two female condors at the National Zoo in Washington, DC, had regularly laid eggs (infertile), indicating that conditions of captivity did not inhibit reproductive behavior of females. Beginning in 1965, the United States Fish and Wildlife Service had conducted a captive breeding program for Andean condors, a different but related species, with the objective of learning techniques of propagation and release of birds of this type. This program has successfully reared many birds, and has successfully returned several birds to the wild in South America (Wallace & Temple 1987).

Captive breeding efforts have had good success. By 1995, 104 birds existed in captivity at the Los Angeles Zoo, the San Diego Wild Animal Park, and the World Center for Birds of Prey in Boise, Idaho. Reintroduction of California condors to the wild began in 1992, and by early 1996 27 captive-reared birds had been released in California. The first releases were carried out in the Sespe Condor Sanctuary in Ventura County, California, where captive-reared Andean condors had been introduced to the wild, using the hacking technique (Reading Fig. 26.1B). Several of these birds died from collisions with power lines, and one after drinking from a puddle of antifreeze. The release site was eventually changed to a location in Santa Barbara County, and birds were given training designed to encourage them to avoid both humans and power line poles.

The poor success of reintroduction efforts in California has led to proposals to release birds in Arizona. The Vermillion Cliffs area, north of the Grand Canyon, is one of the proposed areas. This area lies within the original range of the species.

factors affect the likelihood of success of translocations (Griffith et al. 1989). Most animal translocations have involved game species, which tend to be more successful than nongame species, and most have involved individuals recently captured and fully adapted to life in the wild. Translocations of mammals have usually been more successful than those of birds. Species that breed early in the season and have large broods or clutches are usually most successful. Quality of habitat, location well within the historical range of the species, and releases spanning several years and involving large numbers of individuals also favor success. Many procedural factors also affect the likelihood of success, and conservation biologists working with reptiles, amphibians, and fish, in particular, recommend that translocation be undertaken only after carefully planning and with assured support (Dodd & Siegel 1991; Minckley 1995).

Successful translocations of large ungulates, such as bison, elk, and mountain sheep, have been made to parts of their former ranges in western North America. Muskoxen, for example, have been translocated successfully to areas in Siberia, Alaska, eastern Canada, and western Greenland (Klein 1988). The sea otter was successfully reintroduced to areas of its original range along the coast of Oregon, and recent efforts have been made to reestablish a population on one of the California Channel Islands by translocation. An endangered, nearly flightless bird, the saddleback, was probably saved from extinction in New Zealand by translocation of populations to small islands that were free of predators (Merton 1975).

Translocation has also been used successfully with endangered plant species. In California, where temporary pond habitats support a rich, highly endemic flora, artificial pond basins

RECOVERY OF THE WHOOPING CRANE

he recovery program for the whooping crane (Reading Fig. 26.2A) illustrates the use of these techniques, as well as the overall nature of recovery plans for endangered species. The whooping crane is a large bird with a low reproductive potential. Adults typically lay two eggs annually, but normally only one chick can be raised successfully. This species was originally widespread in North America (Reading Fig. 26.2B), breeding from Illinois northwestward through the Great Plains to the Northwest Territory of Canada, and wintering in several locations along the Atlantic and Gulf coasts and in north-central Mexico (McMillen 1988). The decline of whooping crane populations was due to hunting in wintering areas and on migration, conversion of southern wintering habitats to rice fields, the drainage of marshland breeding areas in the Great Plains, and egg collecting. The decline was rapid from 1890 through 1910, and by 1918 the total estimated population was only 47 individuals. This population migrated between a breeding range in the southern Mackenzie District, Northwest Territory, Canada (now included in Wood Buffalo National Park) and a wintering range in coastal Texas (now centered on Aransas National Wildlife Refuge).

By 1938–1941, the population of whooping cranes had dropped to only 14–15 individuals. At this point, protection and publicity gained the upper hand, and a slow recovery began. A captive nesting colony was also established at the United States Fish and Wildlife Service research center at Patuxent, Maryland. With the creation of the federal endangered species program, a recovery plan was developed for the species. This plan had two specific objectives. First, the Wood Buffalo/Aransas Refuge population was to be increased to at

READING FIGURE 26.2A

The whooping crane declined to 14–15 individuals between 1938 and 1941, before beginning a gradual recovery in numbers.

(Data from J. L. Macmillan, 1983, "Conservation of North American Birds" in *American Birds*, 42:1212–1221 and U.S. Fish and Wildlife Service, 1986.)

least 40 breeding pairs. Second, two additional breeding populations with at least 20 breeding pairs were to be established.

To increase the population breeding in Wood Buffalo National Park and wintering at Aransas Refuge, efforts have concentrated on reducing mortality from predators in nesting areas, and expansion and improvement of wintering habitat on the Texas coast. In wintering areas, pairs of cranes defend large feeding territories, limiting the number of birds that can be accommodated on a given refuge area. The use of controlled burning to improve winter habitat, as well as the acquisition of additional coastal marshlands to be managed as crane winter habitat, are two efforts aimed at in-

creasing the carrying capacity of the wintering area. By 1992, this population had increased to about 145 birds.

An effort was also made to establish a second wild breeding population, using the cross-fostering technique. Whooping crane eggs were taken from nests at Wood Buffalo National Park (one egg from each set of two) and from the captive breeding colony at Patuxent, Maryland, and placed in nests of sandhill cranes at a breeding colony at Gray's Lake National Wildlife Refuge, Idaho. The young whooping cranes were reared by sandhill cranes, and followed them in migration south to a wintering area centered on Bosque del Apache National Wildlife Refuge, New Mexico. From 1975 through 1989, 288 whooping crane eggs were taken to the Gray's Lake site. Of these, 210 hatched and 85 fledged. This population built up to 33 individuals in 1984–1985, but the young whooping cranes suffered heavy mortality from accidents, disease, and predation. No breeding efforts were attempted and this program was terminated in 1989.

In addition, following the mortality of four whooping cranes in the Patuxent colony from disease in 1987, a plan to divide the captive breeding population was developed. In 1989, 22 of the 54 birds in captivity were moved to facilities of the International Crane Foundation in Wisconsin, where breeding populations of several other species of cranes exist. By 1992, the total captive population had risen to 92 birds at the two locations.

Beginning in 1992, a new effort was begun to establish a resident breeding population at the Three Lakes Wildlife Conservation Area near Kissimmee, Florida. The birds initially released suffered heavy predation by bobcats, but by 1995 some 23 birds were surviving and one pair had constructed a nest.

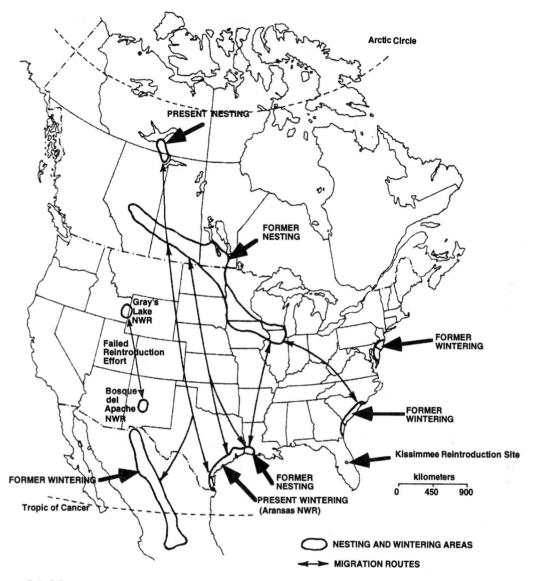

Arctic Circle

PRESENT NESTING

FORMER
NESTING

Gray's
Lake
NWR

Failed
Reintroduction
Effort

Bosque
del
Apache
NWR

FORMER
WINTERING

FORMER
WINTERING

Kissimmee Reintroduction Site

kilometers

0   450   900

FORMER WINTERING

FORMER
NESTING

PRESENT WINTERING
(Aransas NWR)

Tropic of Cancer

⬭ NESTING AND WINTERING AREAS

◄──► MIGRATION ROUTES

READING FIGURE 26.2B

Original and present ranges of the whooping crane in North America.
(Modified from McMillan 1988)

have been created and successfully inoculated with members of this flora, including a federally endangered annual plant, the San Diego mesa mint (P. Zedler, *pers. comm.*)

Careful evaluation of possible consequences of translocation must be conducted, especially if the area lies outside the range of the species (Conant 1988). A proposal was made to translocate the Nihoa millerbird, an insectivorous bird endemic to Nihoa Island in the western Hawaiian Islands, to nearby Necker Island. Necker Island, however, possesses a fauna of highly endemic arthropods that has never been exposed to an avian insectivore. This proposal, fortunately rejected, might have led to the loss of some of these arthropods.

SUMMARY

1. Under the U.S. Endangered Species Act a species is designated as *endangered* if it is in danger of extinction throughout all, or a significant portion of, their range, or *threatened* if it is likely to become endangered within the foreseeable future.

2. Protection of endangered species includes designation of *critical habitat,* consisting of areas either within or outside the range of the species that are essential to its conservation, and the development of *recovery plans,* which are designed to return the species to unendangered status.

3. Population viability analysis is designed to assess the probability of survival of particular populations and to determine the minimum viable population, or the smallest number of breeding individuals with a specified probability of surviving for a certain time.

4. Increasingly, endangered species are being propagated in captivity in botanic gardens, zoological parks, and aquariums with the intention of reintroduction to the wild. Computerized systems are being used to record information on captive populations and plan breeding activities to maintain genetic variability.

5. Translocation, the release of individuals into the wild to create or enlarge a wild population, has often been successful with game species, but efforts to reestablish

many endangered species show that adaption to life in the wild is critical and that success is usually best when individuals are translocated from the wild rather than being released from captivity.

QUESTIONS FOR DISCUSSION

1. Do you think the terms "endangered" and "threatened" are adequately defined in the 1973 Endangered Species Act? Are all species that are rare actually endangered? How would you strengthen the definition of "endangered"?

2. Do you think "last ditch efforts" to save species such as the California condor are worth the cost and effort, or would this money and effort be better directed at preventing many other species from falling to such a critical level?

3. If a species survives only in captivity, separated from its normal social and ecological interactions in nature, does it still exist in a meaningful sense, or is it essentially the same as a preserved specimen in a museum case?

SUGGESTED READING

Ecological Society of America Committee on Endangered Species. 1996. Strengthening the use of science in achieving the goals of the Endangered Species Act: An assessment by the Ecological Society of America. *Ecological Applications* **6**:1–11.

Gibbons, E. F., B. S. Durrant, and J. Demarest (Eds.). 1995. *Conservation of endangered species in captivity*. State Univ. of New York Press, Albany, NY. Discussions of the needs, problems, procedures, and outlook for captive propagation of animals from invertebrates to primates.

National Research Council. 1995. *Science and the Endangered Species Act*. National Academy Press, Washington, DC. History of endangered species protection in the United States, and an evaluation of the strengths and weaknesses of the current Endangered Species Act.

Designing Biodiversity Preserves

A s human populations have grown and their exploitation of natural resources has intensified, natural ecosystems have become reduced in extent and fragmented. Once a continuous matrix that surrounded islands of landscape where humans were active, natural ecosystems have mostly been reduced to islandlike preserves and remnants in a sea of disturbance and development (*See* Chapter 13). Yet we expect these small, isolated areas to bear the responsibility for preserving the bulk of natural diversity. In addition, most preserves were established in an *ad hoc* manner, with little consideration of how they contributed to conservation on a regional basis. We now realize that systems of preserves must be designed carefully, in order to prevent large numbers of species from slipping into endangerment, where their survival depends on extraordinary and expensive efforts. Many

features of preserves, including size, shape, structure, and location, influence their effectiveness in protecting biodiversity.

The design of systems of natural preserves falls in the emerging field of **landscape ecology,** which focuses on the structure and function of portions of the biosphere containing a mosaic of different ecosystems. Like theory dealing with the effects of fragmentation, much of the early theory about design of natural preserves came from the field of island biogeography (*See* Chapter 13). We shall first examine the application of principles of island biogeography to the design of systems of preserves. Finally, we shall consider how modern systems of data acquisition, storage, and analysis are contributing to the design of effective systems of preserves.

New strategies are needed to design systems of preserves for species pressured by expanding human populations and land use.
(Photo courtesy of Abby N. Powell.)

Outline

SIZE, SHAPE, AND SPACING OF PRESERVES

Island biogeographic theory was first used (Diamond 1975) to try to predict desirable features in systems of natural preserves (Fig. 27.1). First, large preserves were predicted to contain more species and to show lower extinction rates than small preserves, and thus be more desirable. Second, assuming a homogeneous distribution of species throughout a region, a single large preserve was predicted to protect more species than several small preserves with the same total area. Sets of preserves were predicted to be more effective if they were near each other and spaced in an equidistant manner than if they were widely spaced or strung out in a linear fashion. Close, equidistant spacing was presumed to facilitate the recolonization of species, should extinction occur in one of the preserves. Similarly, habitat corridors, or stepping stone arrangements of small preserves between larger preserves, were predicted to facilitate biotic interchanges and recolonizations within a system of preserves. Finally, a circular shape that minimizes edge influences was predicted to be better than an elongate shape for survival of obligate interior species of individual preserves. These simplistic predictions, many of them assuming biotic homogeneity of the region in which the preserves are located, stimulated several controversies.

THE SLOSS CONTROVERSY

The suggestion that large preserves are most desirable for conservation purposes spawned a major controversy: the SLOSS argument, or whether a **S**ingle **L**arge **O**r **S**everal **S**mall preserves guarantees the survival of most species (Simberloff & Abele 1976; Diamond et al. 1976). Simple island biogeographic theory assumes that the biota is uniform throughout the region in question. Under this assumption, a single large preserve might contain more species than several small preserves. Even in this case, however, Simberloff and Abele found that a set of small preserves usually held more species unless the single large preserve contained 96% or more of the total species in the region in which the preserves were located. When a region contains major habitat gradients or several centers of biotic diversity resulting from historical influences, a single large preserve obviously will not hold more species than several smaller preserves located in different parts of the region (Simberloff & Abele 1976, 1982). A greater total number of species in small islands or preserves than in a single area of equal total size is usually the case, in fact. Blake and Karr (1984), in their study of breeding bird communities of eastern Illinois woodlots, found that two woodlots of a certain size usually supported more species than a single woodlot of equal total size. Quinn and Harrison (1988) found that this was true both for several groups of organisms of the biotas of oceanic islands and for the faunas of United States National Parks.

Spreading the population of a species among several preserves, rather than concentrating it in a single large preserve, also may reduce the risk of complete extinction of the species. Several workers have studied this question by mathematical model-

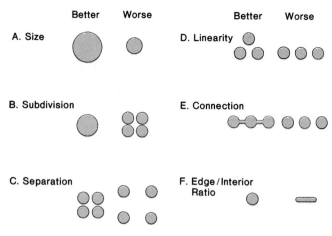

FIGURE 27.1

Some "better-worse" comparisons of design of preserve systems as derived from island biogeographic theory.
(Modified from Diamond 1975)

ing. Kobayashi (1985) concluded that smaller reserves protected more species if extinction rates are low and if different groups of species tend to occur in the different small preserves. On the other hand, if rates of extinction are high, a single large preserve will protect more species unless the restriction of different groups of species to different preserves was very strong. Quinn and Hastings (1987) found that when only random fluctuation in population sizes due to accidents of reproduction and mortality was considered, species survived best in a single large preserve. When the populations were affected by chronic or catastrophic environmental variation, however, a certain degree of subdivision of preserves provided the best guarantee that species would survive somewhere in the set of preserves.

In addition to the number of species, the kinds of species that are protected are an important consideration. Using total species diversity as an index of preserve value is certainly not always appropriate (Diamond 1976; Jarvinen 1982). Preserves are not needed for many species adapted to disturbed or developed landscapes, yet in diversity analyses these species are often counted equal to species with special requirements for undisturbed natural habitats. In other words, preserves are not needed for weedy species such as the house mouse or English sparrow. Native species that are endemic to a region, particularly those with restricted distributions, specialized habitat requirements, or low population densities, are the real focus of conservation concern. Thus, considerations of the size and spacing of preserves must really focus on those native species that are adversely affected by disturbance and development.

The particular scale of the ranges of organisms and the extent of ecosystem types also influence the appropriate size of preserves. Large mammals, such as elk and mountain lions, cannot be maintained on preserves a few square kilometers in area—viable populations of such species require larger areas. Small preserves, on the other hand, may be appropriate for ecosystems of limited extent, such as bogs, caves, or desert springs. In some regions the remaining examples of some ecosystems, such as those of tall grass prairie in Illinois, are small, and the choice is one of a small preserve or nothing (Schafer 1995). Also, rare species do not always occur in locations of high diversity that are good candidates for major reserves. Small reserves thus have important roles to play in biodiversity conservation.

SPACING AND INTERCONNECTION OF PRESERVE UNITS

Increasingly, study is being directed at the potential value of designs that maintain or improve **landscape connectivity,** the degree to which absolute isolation is prevented by features that allow organisms to move among preserve units. Most attention has centered on **corridors,** which can be defined as linear habitat features that differ from their surroundings and connect units of similar habitat type (Hobbs 1992). These features may be thought of as having a certain **corridor capability,** an effectiveness of delivering species from source to target preserves (Soulé & Gilpin 1991). This capability is influenced by characteristics such as the width, length, shape, and intensity of edge effects, as well as by the movement capabilities of the species (Newmark 1993).

Much of the information on corridor capability has been anecdotal and many of the studies lacking in experimental design (Nicholls & Margules 1991). In Maryland, for example, MacClintock et al. (1977) found that certain forest birds were able to occupy a single small woodlot that was connected by a narrow forest strip to a large area of nearby forest, whereas other small woodlots often lacked these species. More recently, however, good evidence of the use of corridors has been obtained for many species. Haas (1995) showed that several migratory bird species tended to use North Dakota shelterbelts that were linked by wooded drainages more frequently than they used shelterbelts that were fully isolated by farmland. In southern California, Beier (1993) radiotracked mountain lions in fragmented wildland areas and documented their use of quite narrow habitat corridors to move between wildland areas in different mountain ranges.

Nevertheless, the idea that corridors are effective and desirable has been accepted rather uncritically (Hobbs 1992; Simberloff et al. 1992). Habitat corridors can potentially act as routes for dispersal of exotic species and diseases (Hess 1994), as well as for endangered species that are the focus of conservation concern. Wildfires might also be carried through corridors. Corridor habitat might also act as an "attractive sink" by attracting animals out of larger habitat units and into locations where breeding success or survival is impaired by strong edge influences. Furthermore, the costs of acquiring land for maintenance or creation of corridors in landscapes with extensive human developments can often be very great. In the absence of good evidence that species of critical conservation concern will use such corridors, available funds might better be spent to enlarge existing preserves or establish new ones where land cost is less.

SHORTCOMINGS OF ISLAND BIOGEOGRAPHIC THEORY

Dealing, as it does, with numbers of species and rates of colonizations and extinctions, island biogeographic theory tends to lose sight of the specific ecological requirements of individual species (Boecklen & Simberloff 1986). The seemingly simple colonization and extinction curves of MacArthur and Wilson (1967) have also spawned a mass of theory that can yield widely divergent predictions when applied to a particular case. Boecklen and Gotelli (1984), for example, show that species-area curves often do not account for a high percentage of variability in data, and are strongly affected by unusual data values. These authors question the use of such weak mathematical relations in predicting the features that preserves should possess. Nevertheless, island biogeographic theory does address questions that are of basic importance to long-term preservation of biotic diversity. Like all mathematical models, as well, those of survival of species in insular habitats are subject to improvement, and are making a significant contribution to conservation strategy.

BUILDING PRESERVE SYSTEMS

The objective of a system of preserves is to protect as much of the biodiversity of a region as possible. To this end, several approaches have been taken to determine objectively the sites that should be protected to achieve such a goal. Margules et al. (1988), for example, suggested procedures to identify the minimum number of sites that would protect all species in a region, and also to protect all species and all habitat types. To protect all species, the suggested procedure was to select all sites with species that occur only in single locations, and then add sites with the rarest remaining species until all species are represented. To protect all species and all habitat types, the procedure was to select the site of each habitat type that contained the most species, then add subsequent sites of each type that add the most unrepresented species, until all species are represented. Lomolino (1994), however, found that procedures that built preserve systems by a rigid procedure of accumulating preserves in this general manner did not give the smallest area that contains a specified number of species. This could be found easily by a computer program that examined all combinations of possible sites and tallied the species present.

On a larger scale, general strategies have been proposed for linking preserves of various types into **conservation networks** that allow viable populations of large or scarce animals to be maintained more effectively (Salwasser et al. 1987). In the United States, 10 potential networks involving national parks, monuments, forests, recreation areas, and wildlife refuges can be outlined (Fig. 27.2). These units would range in size from 5,650 to 75,350 square kilometers. Networking state and private natural areas into such systems could allow further enlargement of these networks and the development of still others. The larger effective units created by such interconnection might support populations of species that cannot now survive in the separate units. Harris (1988) has outlined how the acquisition of critical

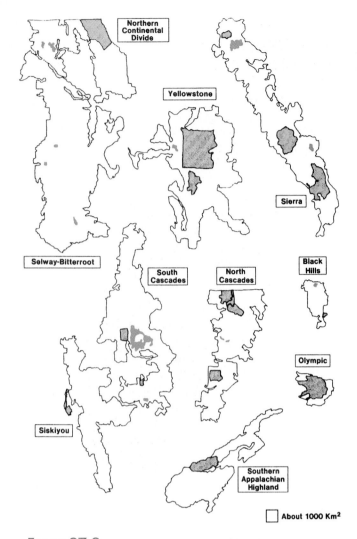

About 1000 Km²

FIGURE 27.2

Proposed conservation networks of natural areas now under federal management in the United States.

(Modified from Salwasser et al. 1987)

lands or easements could create such networks in the southeastern United States. Contiguous tracts of at least 2,000 square kilometers are required in this region, for example, to maintain populations of black bears, and a still larger minimum area to maintain the Florida panther. By linking several federal and state areas, four areas of 2,000 square kilometers could be created where only one now exists.

Noss and Harris (1986) have suggested a more formal strategy for structuring networks of preserves. This proposal envisions a network of **multiple-use modules,** or conservation units with a fully protected core area surrounded by zones of natural areas utilized in progressively more intense fashion for recreation, timber production, and other uses. These modules would represent "nodes" in a network tied together by habitat corridors adequate to allow movement of animals between modules. On the Georgia-Florida border, for example, 15 or so existing natural areas could be linked by habitat corridors, some

of them centered on major rivers, to create a network extending from Okefenokee National Wildlife Refuge in Georgia to the Florida Gulf Coast. Such a network should be adequate for the Florida panther, for example, and could allow reintroduction of the species to this region. On a continental scale, this approach is the core of an effort, The Wildlands Project, to create a network of interconnected wilderness areas that would include viable areas of all major North and Central American ecosystems (*See* Reading 27.1).

On a more limited scale, several heavily urbanized regions in Florida, Texas, and California have attempted to develop comprehensive conservation networks to protect most of the species of the region's distinctive ecosystems. These programs are being developed with the cooperation of local, state, and federal agencies. They are designed both to protect endangered or sensitive species and to reduce the conflicts between government agencies and landowners that have become frequent under the species-by-species approach that has been followed.

In California, the Natural Communities Conservation Planning Act laid the groundwork for a plan designed to protect the flora and fauna of the coastal scrub ecosystem of southern California (Reid & Murphy 1995). Plans have been completed for three major units of a preserve system covering about 97,000 hectares in Orange and San Diego Counties (Fig. 27.4). These preserves would consist of core units and corridors that would allow even the larger, wide-ranging species such as the mountain lion to maintain viable populations in the region. In return for dedicating land to this system, developers would receive assurance that they would be free from future endangered species restrictions on use of their property. The target date for final agreement on this plan is late 1996.

In Texas, the Balcones Canyonlands Conservation Plan proposes a similar preserve system of about 30,000 hectares to protect the unique species of the Texas hill country in Travis County. Acquisition of over 12,000 hectares in the Austin area was completed in 1996.

NEW TECHNOLOGIES FOR DESIGN OF PRESERVE SYSTEMS

Ideally, systems of preserves should include substantial populations of most of the species in a region, and thus should minimize the decline of species to endangered status (Scott et al. 1987a). This means that preserves should be located where major populations of endemic species are concentrated. Identifying areas of high diversity and endemicity is thus an important first step in preserve system design. The use of Rapid Biodiversity Assessment teams to evaluate sites for preserves in the Amazon Basin (*See* Chapter 9) is an example of this approach. Another example is provided by the island of New Guinea, divided between the independent nation of Papua, New Guinea and Irian Jaya, part of Indonesia. In this still largely undeveloped region, centers of endemism and biotic diversity have been identified and used as a major criterion for location of preserves (Diamond 1986).

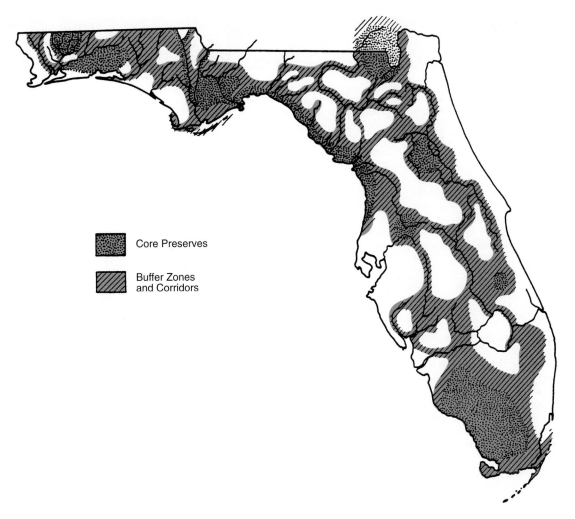

FIGURE 27.3

An initial proposed design for the Wildlands Project core zones, buffer zones, and corridors for Florida.

More advanced technology is also being brought to bear on biodiversity assessment, especially in more developed regions. Geographic information systems technology promises to assist in the identification of new preserve sites (Scott et al. 1987a). A **geographic information system (GIS)** consists of computer hardware and software that enables the input, storage, manipulation, and output of data referenced by a common system of spatial coordinates. A GIS permits data on many sets of variables over a specified geographic area to be combined or manipulated to answer questions about spatial relationships. Five major classes of questions can be addressed:

What condition or feature exists at a particular location?
Where does a particular feature or condition occur?
How has a particular condition or feature changed over time?
What spatial patterns exist over the landscape?
How will conditions or features respond to an impact or development?

The GIS enables the results of such analyses, as well as any of the data sets themselves, to be printed out in map form. Vogelmann (1995), for example, used GIS technology to trace the increase in forest fragmentation that has occurred in southern New England as human population density has increased.

Many types of data can be included in a GIS. For conservation purposes, the data might include physical characteristics such as elevation, rainfall, soil type, and slope; biotic characteristics such as vegetation type and the presence or abundance of individual species; and cultural variables such as land ownership or land use. Data on each variable are stored in separate computer files, or *layers,* using a common system of coordinates, so that values of any combination of variables can be determined for a given location.

Spatial information can be stored in two basic formats, which are combined in more sophisticated GISs. The *raster* format divides an area into cells or pixels, with conditions in each cell being specified by a value or index. Data on plant biomass might be recorded in this manner, for example, using a scale of

The Wildlands Project, also known as the North American Wilderness Recovery Project, is an effort to both protect biodiversity and restore wilderness conditions to an interconnected network of natural ecosystems spanning all of North America. This network is thus envisioned as extending from the Isthmus of Panama to the Arctic, and to include coastal marine environments as well as terrestrial ecosystems. The project was conceived in 1991 by a group of conservation biologists and environmentalists, one of whom, Reid Noss, characterizes it as "a positive vision for the future." In other words, it is an effort to be proactive in the fight for biodiversity, rather than mounting only a retreating battle against extinction losses.

In ecological terms, the overall objective of the system would be fourfold. First, all natural ecosystem types, including their various successional stages, would be incorporated. Second, the system should provide habitat for viable populations of all native species. Third, the system must maintain natural ecological and evolutionary processes. Finally, design and management must enable ecosystems and their member species to adjust to short-term changes, such as those caused by natural disturbances, and long-term environmental changes, such as that due to global warming.

The design of the Wildlands system includes three main elements: core reserves, corridors, and buffer zones. Large core areas of full wilderness habitat would be connected by broad corridors. The core reserves would include the full range of physical landscapes, habitat gradients, and ecosystems that occur in the major biotic regions of North America. The corridors would allow large, wide-ranging animals to migrate seasonally and to disperse throughout the system, preventing genetic deterioration of isolated subpopulations. They would also permit gradual migrations of other species in response to climatic change. Only very limited human activities—those compatible with wilderness conditions—would be allowed in core areas and central zones of corridors. The core and corridor areas would be surrounded by multiple-use buffer areas in which natural ecosystems would be protected or restored. In buffer areas sustainable exploitation of certain resources, such as selective-harvest forestry, could also be practiced. Intensive agriculture, industrial activi-

ties, and urban development would be restricted to areas outside the wildlands network.

A key feature of the system is the creation of core reserves that are large enough to be "minimum dynamic areas," that is, areas in which natural disturbance and recovery processes can occur entirely within the reserve. For this to be possible, some ecologists estimate that the reserve must be 50–100 times as large as the patches produced on average by disturbing forces, such as fires or hurricanes. These core areas, at least when linked together by major dispersal corridors, must also be capable of supporting minimum viable populations of large carnivores, such as grizzly bears, mountain lions, and gray wolves. The minimum size of core reserves should thus be hundreds to thousands of square kilometers.

The structure of corridors must vary with their purpose. Those that must permit long-distance dispersal of large carnivores between core reserves such as the Greater Yellowstone Ecosystem and neighboring core reserves several hundred kilometers distant, for example, may need to be 45 or so kilometers wide.

Preliminary schemes of core reserves, corridors, and buffer zones have been sketched out for several regions of North America, including Florida, coastal Oregon, the southern Appalachians, the northern U. S. Rocky Mountains, and other areas. For the Florida region (See Fig. 27.3), core reserves might be centered on the Everglades, Okefenokee Swamp, the upper Gulf coast, Apalachicola National Forest, Blackwater River Forest, and several other smaller areas. Corridors would follow many of the major river systems.

Reed Noss (1992), one of the project founders and a member of its board of directors, envisions the system as ultimately incorporating about half of North America in core preserves and the central zones of corridors. Most of the remaining half of the land would become part of the buffer zone, so that ultimately as much as 95% of North America would be managed as natural or near-natural landscape.

Supporters of this wilderness network realize that implementation would require several to many decades, and view its full development as perhaps requiring 100–200 years. They also emphasize that develop-

ment will be through voluntary actions that are designed not to displace people involuntarily or cause massive economic disruption.

Regionalization is a basic development strategy of the Wildlands Project. The planning needs in a region such as peninsular Florida are quite different than those in coastal southern California, and are best pursued by people familiar with and dedicated to conservation within those regions. The Wildlands Project organization thus has the goal of coordinating the activities of regional organizations dedicated to the establishment and protection of wilderness. Its emphasis is to promote the overall concept, to provide technical and logistic assistance to groups involved in regional planning, and to help integrate regional wildlands networks into the continent-wide system.

The general strategy is one of building on existing wilderness areas by expanding and linking them by incorporation of lands that have high biodiversity value. This would require use of biodiversity data bases being developed by state and provincial natural heritage programs, gap analysis programs, and other surveys to identify high-priority areas for addition to such core preserves. Corridor connections must be planned in consideration of the needs for local habitat linkages, seasonal migration routes, dispersal of mobile animals between core preserves, and long-term migration of species in response to climate shifts. The actions that would be necessary to acquire and restore the lands must also be considered. Areas designated for core and corridor classification may need to be restored by removal of roads, dams, power lines, and other features incompatible with wilderness.

Predictably, some of the response to this proposal has been negative. Critics state that it will displace people from productive land, restrict human access to vast areas, and expose humans to close contact with dangerous animals. They view the plan as totally impractical. Proponents respond that the plan is one that will be implemented very gradually, and by voluntary action. They contend that buffer zones will provide adequate security for people in urban areas. Impracticality, they suggest, is hard to project into the future: How many people in 1992 would have thought that the Soviet Union would disintegrate or that Nelson Mandela would become President of the Republic of South Africa by 1995?

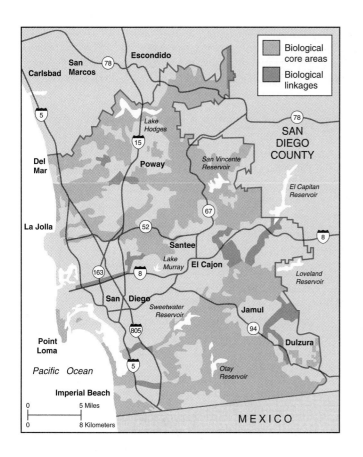

	Biological core areas
	Biological linkages

San Marcos
Escondido
San Carlsbad
78
Del Mar
5
Lake Hodges
15
Poway
San Vincente Reservoir
SAN DIEGO COUNTY
78
El Capitan Reservoir
La Jolla
67
52
Santee
8
163
Lake Murray
El Cajon
8
Loveland Reservoir
San Diego
805
Sweetwater Reservoir
Jamul
94
Point Loma
Dulzura
Pacific Ocean
5
Otay Reservoir
Imperial Beach

0   5 Miles
0   8 Kilometers

MEXICO

values from zero to some maximum. The *vector* format uses lines, their positions referenced by X and Y coordinates, to show the boundaries of areas of similar conditions. The distribution of soil types might, for example, be recorded in this manner.

From such a data base, many specific questions about where species that need protection are concentrated can be answered (Fig. 27.5), an effort now known as **gap analysis** (Scott et al. 1988). With modern GIS systems, it is possible to develop comprehensive data bases for vegetation types, together with their vertebrate animals, higher plants, and certain groups of invertebrates, such as butterflies, of the scale of geographical units 200 hectares in area (Scott et al. 1987a).

GIS systems are well adapted to using data from remote sensing sources. Detailed data on the actual vegetation of a geographical area are difficult to obtain from traditional vegetation maps, which often show the potential climax vegetation thought to characterize a region, rather than the vegetation actually present. With improved techniques of satellite imagery and analysis, detailed data on the vegetation that actually exists can soon be determined on a grid scale such as that indicated above.

FIGURE 27.4

Proposed design of the San Diego County Multiple Species Conservation Area.

(Source: Multiple Species Conservation Program draft.)

FIGURE 27.5

Outline of a geographic information systems (GIS) data base for evaluation of adequacy of existing preserves and the best sites for new preserves. *a.* Data on various physical and biotic features of a region are stored in computer data files. *b.* Information from the data base is used to create new maps that evaluate preserve adequacy by revealing gaps in the protection of species or ecosystems by existing preserves.

(Modified from Scott et al. 1987a)

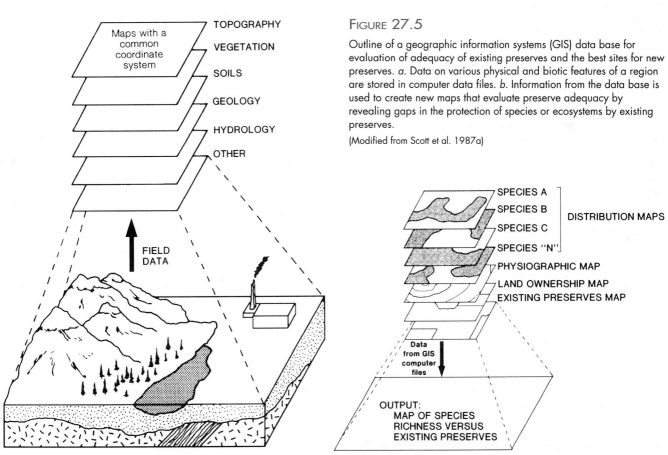

Maps with a common coordinate system

TOPOGRAPHY
VEGETATION
SOILS
GEOLOGY
HYDROLOGY
OTHER

FIELD DATA

a.

SPECIES A
SPECIES B
SPECIES C
SPECIES "N"
DISTRIBUTION MAPS
PHYSIOGRAPHIC MAP
LAND OWNERSHIP MAP
EXISTING PRESERVES MAP

Data from GIS computer files

OUTPUT: MAP OF SPECIES RICHNESS VERSUS EXISTING PRESERVES

b.

This GIS approach was utilized in an elementary fashion in evaluating conservation reserves for endemic forest birds in Hawaii (Scott et al. 1987b). Over a period of 8 years, surveys of the occurrence of forest birds were made at almost 10,000 sites throughout the main islands—a project known as the Hawaii Forest Bird Survey (Scott et al. 1986, 1989). These data enabled isolines of occurrence of different numbers of native species to be drawn. Locations where the greatest numbers of species occurred together were then identified and used as a guideline for the acquisition of new preserve lands. More than 17 square kilometers of new forest preserves, or easements to protect existing forests, resulted from this organized effort.

Gap analysis programs have now been undertaken in many states in order to assess the effectiveness of state biodiversity protection (Davis et al. 1990). In Idaho, for example, a statewide analysis of vegetation types compared to land ownership and management status revealed that 11 of 71 vegetation types had little or no protection (Caicco et al. 1995). For Utah, gap analysis revealed that over 13% of mammal species, but only 1.5% of reptile species, were well protected in the sense that at least 10% of their habitat occurred within protected areas such as national parks, wilderness areas, and Nature Conservancy preserves (Edwards et al. 1995). In Brazilian Amazonia, Fearnside and Ferraz (1995) conducted a gap analysis of natural vegetation types, and found that only a third of 111 zones representing 28 types had some portion of their area under protection.

Summary

1. Simple island biogeographic theory predicts that large preserves are preferable to small preserves, a single large preserve better than several small preserves with the same total area, preserves closely spaced or connected by corridors better than isolated ones, and a compact shape that minimizes edge influences better than an elongate shape.

2. The SLOSS controversy (Is a single large preserve better than several small preserves of the same area?) has been resolved to the extent that large preserves are clearly essential to large, wide-ranging species, but several small preserves usually contain and may often guarantee survival of more total species.

3. For at least some kinds of animals, corridors are capable of linking preserve units so that a population larger in number and less vulnerable to stochastic extinction is created.

4. By developing integrated biodiversity management plans and by linking existing preserves with habitat corridors, conservation networks that allow viable populations of large or scarce animals to be maintained can be created in many parts of North America.

5. Geographic information systems are now being used to store geographically explicit data on the distribution of species, habitat characteristics, and human land-use patterns. These data enable the conduct of gap analyses to determine where species that need protection are concentrated.

Questions for Discussion

1. Should corridors be used only to connect areas that are similar in their biota and had historical habitat connections, or should they be designed to enable species to migrate into areas to which they did not originally have access?

2. What do you think would be the important considerations in designing a system of preserves for lake ecosystems? River ecosystems? Marine ecosystems?

3. What changes in human populations, economic systems, resource use patterns, and attitudes do you think would be necessary to permit The Wildlands Project to achieve its goal over the next 200 or so years?

Suggested Reading

Beatley, T. 1994. *Habitat conservation planning: Endangered species and urban growth.* Univ. of Texas Press, Austin. An examination of habitat conservation plans designed to protect regional biodiversity, with particular attention to plans under development in Florida, Texas, Nevada, and California.

Hudson, W. E. (Ed.). 1991. *Landscape linkages and biodiversity.* Island Press, Washington, DC. Chapters by leading conservationists describe strategies for protecting areas of high biodiversity and linking them so that species can move among preserves.

Shafer, C. L. 1990. *Nature reserves: Island theory and conservation practice.* Smithsonian Institution Press, Washington, DC. The application of ecological theory to the design of biodiversity preserves, with examination of case studies from around the world.

eight

Conservation and Society

The earth's biodiversity will only be preserved if
society at large believes that it is a natural
heritage of both ethical standing and economic
importance to humanity. In this section we shall
examine the rapidly emerging fields of conservation
ethics and conservation economics. Finally, we shall
examine the prospects for developing a sustainable
relationship of humanity within the global ecosystem.

chapter

28

Conservation Ethics

O ne branch of philosophy concerns itself with the study of **ethics,** moral standards of human conduct based on beliefs about what is right or just. Over the centuries, most discussions of ethics have focused on standards of conduct among humans, their institutions, and their possessions. In recent decades, however, philosophers and environmentalists have become concerned with moral standards of conduct that relate humans to their nonhuman environment, particularly the other living organisms and the ecosystems of which both they and humans are part. **Environmental ethics** has emerged as a distinct field on the interface between philosophy and the environmental sciences (Des Jardins 1993; Jordan 1995). Environmental ethics considers the moral relations between humans and their natural environment (Des Jardins 1993).

What are the ehtical duties of humans to other organisms, species, and natural ecosystems?

Basic Concepts of Environmental Ethics

Ethics comprises what many persons might think of as "natural law"—a code of conduct that most people in a human society regard as basic to their conduct and interaction. Ethical codes in many primal human societies are unwritten; in some societies they are codified through formal religions, as, for example, in the case of the Mosaic Ten Commandments. Ethical principles underlie most formal law, but ethics may apply in many situations not covered by formal law. In some situations, too, ethical principles held by many people may stand in opposition to laws or governmental regulations.

Ethics can be examined from several perspectives (Des Jardins 1993). First, **descriptive ethics** involves simply summarizing the ethical beliefs that are held by society or various groups within society. What, for example, is the prevailing view about whether or not it is ethical to hunt mountain lions for sport? How do views about this question differ among different groups within society? A second perspective is that of **normative ethics,** the advocacy of particular ethical viewpoints. For example, many people argue, "It is unethical to pursue mountain lions with dogs and then shoot them in trees in which they have taken refuge." Quite different ethical positions are often advocated by different groups within society, and increasingly these involve issues with a biological and ecological basis. Above these levels, however, lies **philosophical ethics,** a domain in which conflicting ethical positions are identified and examined logically. What, for example, are the implications of ethical judgments about the mountain lion for ethical positions on other animals such as deer, and for the ecosystems that deer and mountain lions occupy? Our examination of environmental ethics will concentrate on the level of philosophical ethics. Nevertheless, we must remember that the issues dealt with at this level are constantly being debated in the public arenas of normative ethics.

Environmental ethics is concerned with whether or not humans have moral obligations toward nonhuman organisms or other entities, such as species, ecosystems, or the biosphere at large. For example, is it immoral to kill a whale? A rabbit? A poisonous snake? A snail? Is it morally just to cause environmental changes that drive a species to extinction? That destroy the last virgin tall-grass prairie in Kansas? That contribute to the depletion of the stratospheric ozone layer? If the answer to any of these questions is "sometimes," what circumstances define when humans should or should not carry out the action in question?

Environmental ethics is also concerned with whether or not living humans have responsibilities about the environment that we leave to future human generations. For example, is it morally permissible to drive to extinction thousands or millions of species, some of which might have proven to be useful to humans in the future? To release vast quantities of greenhouse gases into the atmosphere, leading to altered climates and sea level that will affect humans far into the future? If such responsibilities exist, how should they be balanced against the needs of living humanity?

Figure 28.1

A species such as the African lion may be the object of a primary ethic, if it is considered to have an intrinsic right to exist without persecution, and a secondary ethic, if it has instrumental value to humans, such as economic value as an object of ecotourism.
(Photo by Darla G. Cox)

The entities to which ethical principles apply are said to have **moral standing.** This means simply that an entity such as an organism, species, ecosystem, or the like, qualifies as something to which, or regarding which, humans have duties (Fig. 28.1). The distinction of "to which" or "regarding which" is important. A **primary ethic** involves a duty *to* something because it has **intrinsic moral standing.** If, for example, the gray whale is considered to have intrinsic moral standing, primary ethical principles might be: "One should not kill a gray whale," or "One should not harass parent gray whales and their newborn young in their breeding lagoons." A **secondary ethic** involves a duty *regarding* something because it has **utilitarian** or **instrumental value** to humans or to other entities to which primary ethical duties pertain. For example, the Pacific yew tree may not be considered to have intrinsic moral standing, but it has utilitarian value to humans because of the medicinal value of taxol from its bark. A secondary ethical principle in this case might be,

> "It is unethical to destroy Pacific yew trees in clear-cut logging, because they are an important source of taxol for the treatment of cancer in humans."

Or, for gray whale populations that are objects of tourist interest, a secondary principle might be,

> "It is unethical to convert gray whale breeding lagoons to salt evaporating ponds, because the livelihood of many people depends on income from ecotourism based on the whales."

Many secondary ethical duties that relate to humans thus have economic value, and should be considered in comprehensive efforts to establish economic values for ecological resources.

A HISTORICAL OVERVIEW OF ENVIRONMENTAL ETHICS

It seems likely that much of our view about the extent to which nonhuman entities deserve moral standing depends on how we have come to view the relation of humans to the rest of the biosphere. In the Western cultural tradition, this view is strongly anthropocentric. Its basis can be traced to Greek philosophy. Aristotle (384–322 B.C.), for example, considered humans (and especially *male* humans!) to be at the highest level in a natural order, with plants and animals existing for the sake of their utility to humans.

In Jewish tradition, as described in the creation account in the Biblical Book of Genesis, man is described as being created in God's image and given dominion over every living thing. The Roman Catholic theologian, Thomas Aquinas (A.D. 1225–1274), affirmed this dominion, and specifically noted that no injustice was involved in killing animals or forcing them to do work. He did consider, however, that compassion toward animals was a quality that benefitted humans by making them more compassionate toward each other.

Lynn White (1967), Professor of History at UCLA, integrated these ideas with later historical developments into a modern statement of the **Anthropocentric Ethic** in a lecture to the American Association for the Advancement of Science in 1966. White argued that four factors were responsible for the ecological crisis of the 1960s (a crisis that continues to the present!). First, the Judeo-Christian view of human dominion over nature predisposed humans to exploit nature with little concern. Second, the democratic revolutions of the 1700s and 1800s, which overturned feudal societies in which control over productive resources was held by an aristocratic minority, gave greater freedom to people at large to exploit nature for their material well-being. Third, the industrial revolution of the 1800s and 1900s gave them the power to exploit productive resources on a vastly greater scale. And fourth, the colonial expansion of European culture carried this form of unrestrained exploitation of nature worldwide.

The dichotomy of humans and nature is the basis of the Anthropocentric Ethic, which considers that humans have primary ethical duties only to other humans, and that other living organisms and natural entities do not have intrinsic moral standing. White's thesis, particularly his indictment of the Judeo-Christian idea of dominion over nature, has had a strong influence on the fields of science history and environmental ethics. Many consider the Judeo-Christian view a strong determinant of how we treat nonhuman entities, and, indeed, this rationale is offered by some people as a rationalization of their exploitation of the environment. Others consider that the biblical term "dominion" implies stewardship, and that this responsibility pervades Judeo-Christian philosophy as thoroughly as does that idea that humans are different

from, and superior to, the rest of nature. Indeed, in the Biblical Book of Genesis, God put Adam

"into the Garden of Eden to dress it and to keep it."

This idea, which can be recognized in Islam and some other religions, has become known as the **Environmental Stewardship Ethic.**

Other philosophers note that many of the Eastern religions, such as Buddhism, Hinduism, and Taoism hold very different views of the relation of humans and the rest of nature. These religions generally view humans as sharing some type of spiritual "oneness" with other life, and counsel a respectful or harmonious relationship with nonhuman life. This general view is also attributed to many primal peoples—tribal peoples who live technologically simple lives in close association with nature. American Indians, for example, viewed humans as an integral part of nature (Callicott 1982; Nash 1989) although the sophistication of their views has often been enhanced by Hollywood script writers.

Still other philosophers believe that these religious views, both Western and Eastern, are little more than abstract concepts that have relatively little influence on day-to-day activities of human life. Most humans, they contend, do what they can to maintain or better their material way of life, regardless of the formal religion to which they subscribe. Just as Europeans deforested western landscapes and drove species to extinction, ancient Chinese deforested much of eastern Asia, and American Indians quickly adopted horses and sheep to increase their abilities to exploit hunting and grazing resources.

Regardless of its philosophical origins, modern human society is dominated by a **Pragmatic-Utilitarian Ethic.** This ethic is primarily anthropocentric, is founded on the success of Western technology and resource exploitation, and embodies the assumption that technological competence will permit resource substitutability so as to allow unlimited economic growth (Ehrenfeld 1978). It recognizes strong intrinsic moral standing for humans, and extends intrinsic moral standing to some other animals, particularly domestic animals and "charismatic" species such as whales, dolphins, and other popular wildlife species. For the most part, however, other animals and plants, together with the ecosystems of which they are parts, have only secondary ethical standing based on their utilitarian value. The dominant question that arises about an endangered species of animal or plant that is proposed for protection is, "What is it good for?"

The Pragmatic-Utilitarian Ethic has been characterized, in its extreme, as the **Frontier Ethic** or the **Cowboy Ethic.** It assumes, in effect, that productive resources are unlimited, much as land, forests, and other resources on the North American western frontier seemed to be in the 1700s and 1800s. In this sense, this ethic dominates the politics of countries such as Brazil and Indonesia, where policy emphasizes colonization and development of the Amazon Basin and unoccupied forest regions of Borneo and Iryan Jaya.

Other Paradigms of Environmental Ethics in Western Society

Aldo Leopold (1933), in what might be considered the founding essay of modern environmental ethics, proposed the extension of ethics to human relations with land. He viewed successful civilization as involving a harmonious relation to the land,

> "a state of *mutual and interdependent cooperation* between human animals, other animals, plants, and soils . . ."

His concept of land included the total ecosystem, a term he would likely have used except that it had not yet been coined. He termed the concept the **Conservation Ethic** or, later (1949) the **Land Ethic.** Leopold believed that the roots of such an ethic existed, even though human mistreatment of the natural environment was rampant. He observed that human nature revolted against the idea of letting erosion go unchecked, even on land that had been abused to the point that its productivity was negligible. In his words, people simply would not look at such a piece of land threatened by erosion and say, "let her wash."

Kenneth Boulding (1966) proposed **Spaceship Ethics** as the reasonable alternative to Frontier or Cowboy Ethics. Spaceship ethics is a space-age update of Leopold's ethic, and assumes that humanity and the biosphere are inseparably linked, so that the welfare of each depends on the other. The support systems of spaceship earth maintain the health and well-being of its crew, but these systems depend in turn on their careful maintenance by the pilots and crew. The spaceship has a finite capacity, and, except for energy, is an almost completely closed system. The problem, as several writers have noted, is that the crew must create their own flight manual, word by word, to assure the continued success of the operation.

Proponents of Spaceship Ethics recognize the need to stabilize the human population and achieve a steady-state pattern of resource use by emphasis on frugality, efficiency, use of renewable resources and recycling of nonrenewable materials. Their belief is that this must be achieved worldwide, and that an equitable system of allocation of resources among people also must be achieved in order to prevent violent confrontation between "haves" and "have-nots." Further, they believe that although this state requires fundamental changes in human values, it must be achieved without harsh coercion. They affirm that change can be achieved through enlightened leadership that emphasizes education and incentives.

Garrett Hardin (1974) concluded that worldwide practice of Pragmatic-Utilitarian Ethics was leading toward catastrophe, and that the only real hope of saving something was through a radical new approach, which he termed **Lifeboat Ethics.** Hardin believed that many of the developed countries were in a position to achieve a stable relationship with the environment, but that many developing nations, with rapidly growing populations and destructive patterns of land use, were not. Thus, he reasoned that the developed regions were like uncrowded lifeboats and the developing regions like crowded lifeboats, from which passengers spilled and pleaded for rescue by the un-

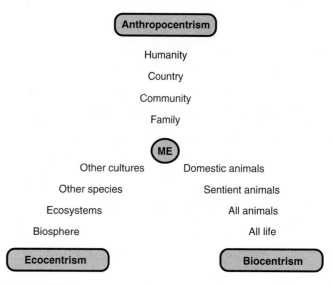

Figure 28.2

Ethical systems can be viewed as spreading various distances along axes of anthropocentrism, biocentrism, and ecocentrism.

crowded lifeboats. Were the uncrowded lifeboats to attempt to rescue everybody, the result would be that all lifeboats would ultimately become overcrowded and sink.

Lifeboat Ethics, in reality, places the ultimate health of the global ecosystem ahead of the well-being, or even survival, of people in developing countries. Placing this philosophy in specific terms, it argues against humanitarian food and medical aid to people in drought-ravaged Somalia or violence-torn Rwanda, since these actions will only delay an ultimate catastrophe on even greater scale, with even worse damage to the regional environment.

Arne Naess (1973) founded the ecophilosophy, or "ecosophy," of **Deep Ecology,** which holds that all life and the ecological systems that living organisms form have intrinsic value, and integrates this view with aspects of Leopold's Conservation Ethic and Spaceship Ethics (*See* Reading 28.1).

Progress Toward a Comprehensive Environmental Ethic

From the preceding, it should be clear that most people believe that both humans and at least some nonhuman entities have moral standing, and that quite a few primary and secondary ethical duties exist. Views about the objects and extent of these duties can be arrayed along three main axes (Fig. 28.2). Some of the basic ethical questions are:

1. What entities should be accorded moral standing? Do aspects of biodiversity have moral standing? That is, do individual organisms, species, and ecosystems have rights comparable in any way to those of humans?

2. When conflicts appear in duties to different entities, how should they be resolved? How should duties to nonhuman entities be balanced with human well-being

Deep Ecology is an ecophilosophy proposed by Arne Naess, a Norwegian philosopher, in 1973. As a philosophy, it is not derived by logic from other concepts of environmental ethics, nor is it derived inductively from the science of ecology. As Naess put it, Deep Ecology is

"suggested, inspired, and fortified"

by ecology, but it also owes much to the popular environmentalism that emerged in the 1960s. Deep Ecology contrasts, however, with **Shallow Ecology,** which Naess considers to be forms of environmentalism that are concerned with overexploitation of resources, environmental pollution, and the like, as they reduce the quality of life in developed countries. Shallow ecology stresses ways to clean up the environment, recycle materials, and use resources more efficiently to improve human health, fight poverty, and create a clean living environment.

The key concept of Deep Ecology, however, is acceptance of a total unity of humans with the biosphere, and adoption of a way of life that respects this unity. The platform of Deep Ecology consists of eight principles intended as commonalities about which people with deep concerns about humans and their relation to the biosphere can unite:

1. All human and nonhuman life has intrinsic value, in addition to any utilitarian values that may exist.
2. The richness and diversity of life also have intrinsic value, again beyond their utilitarian values.
3. Humans should not destroy other life except when necessary to meet human vital needs.
4. For human life and culture, as well as nonhuman life, to flourish, a decrease in the size of the human population is required.
5. At present, human disruption and destruction of other life is excessive and rapidly worsening.
6. Major changes must therefore occur in policies affecting economics, technology, and ideology.
7. Ideology must change to emphasize quality of the total life system, rather than just the material standard of human life.
8. People who accept these principles should work to bring them to reality.

Naess (1991) regards ecological sustainability at all levels from local to global as one of the main Deep Ecology goals for new policies directed at economics, technology, and ideology. He considers maximizing biodiversity as an essential requirement for sustainability. Species egalitarianism, a basic tenet of Deep Ecology, is the ethical key for preservation of biodiversity. Stabilizing and reducing global energy use, particularly energy from nonrenewable sources, is another requirement for sustainability. The additional two major goals of Deep Ecology are peace and social justice. Achieving goals of sustainability, peace, and social justice—what can be termed a "Green Society"—will require political action, and Naess considers the Green political movement as generally representing an appropriate effort to this end.

in present-day conflicts? For example, should biodiversity be preserved when human suffering might be alleviated, at least for a while, by exploiting it as fully as we can?

3. What duties do we have to future generations of humans in terms of the environment that we pass on to them? For example, does the present generation of humans have a moral obligation to save biodiversity for future generations? And if so, how should we balance this obligation to the future against the need to alleviate human problems today?

DO ASPECTS OF BIODIVERSITY HAVE MORAL STANDING?

Considering individual organisms first of all, most philosophers agree that at least some nonhuman animals have intrinsic moral standing, and that humans have ethical duties to them. This view constitutes a **Biocentric Ethic.** In general, animals that are most often accorded moral standing tend to be those that seem to show human emotions and high levels of humanlike intelligence—chimpanzees, whales, and the like. Most agree, as well, that such standing extends in at least some degree to all *sentient* animals, that is, those that can (or appear to be able to) experience pleasure, satisfaction, pain, or terror. Because of these capacities, such animals can be said to have *interests,* that is, behavioral drives to obtain comfort, satisfy hunger, avoid injury, and flee agents that might kill them. The active memberships of organizations concerned with humane treatment of animals, particularly domestic species, and with animal rights in general, attest to the strength of this belief in human society. In this case, ethical principles are essentially a direct extension of human ethics; animals should not be starved, unduly confined, tortured, or killed without reason, just as humans should not.

Other organisms, such as invertebrates and plants, lack sentience, or at least appear to lack it. Thus, many people feel that it is inappropriate to extend human ethical principles to them because mistreatment does not appear to cause suffering like it does to humans. Others disagree. Johnson (1991), for example, asserts that since all living things have biological functions that work to sustain life, all must also have interests and thus intrinsic value.

A more difficult question concerns species (including subspecies, races, and the like) and systems of which they are members, such as biotic communities and ecosystems. Do evolutionary lineages have intrinsic ethical values? Do such categories possess interests? Can groupings designated by systemetists and ecologists have moral standing? The view that such entities have intrinsic value is termed an **Ecocentric Ethic.**

Perhaps most difficult is the question of whether or not communities and ecosystems have moral standing. The biotic community is an even more arbitrary entity than the species. Indeed, many ecologists feel that communities are totally arbitrary, and that assemblages of different species vary along a continuum of composition and structure. Ecosystems, to make things worse, include nonliving aspects of the environment. How can such categories have interests, moral standing, or intrinsic ethical value?

Philosophical rejections of the concept of intrinsic ethical value for species, communities, and ecosystems typically hinge on the logical difficulty in extending traditional human ethics directly to these entities. Support for the recognition of intrinsic value for such entities, by the same token, tends to come from outside the mainstream of traditional ethics.

Feinberg (1974) and others argue that a species is simply a collection of individuals, and as such cannot have vital interests any more than a crowd of people can have vital interests distinct from those of individual humans. In other words, a "crowd," as such, does not have intrinsic moral standing. (Individuals, however, are usually considered to have an intrinsic right of peaceful assembly.) Thus, these philosophers believe that any values of species must be utilitarian, not intrinsic. On the other hand, Johnson (1991) argues that a species is not just a "crowd" of individuals, but an ongoing genetic lineage that constitutes a living system. As a living system, it is like an individual in requiring certain conditions in its environment and tending to maintain a state of homeostatic adaptation to the environment (Fig. 28.3). Thus he argues that species, like individuals, have interests and therefore intrinsic value. Rolston (1988) agrees, arguing that the species' lineage is a more fundamental living entity than the individual, which is nothing more than the species' way of perpetuating itself. If Johnson's and Rolston's argument is accepted, it follows that species do have intrinsic moral standing and those that become endangered deserve priority when ethical conflicts arise.

Leopold (1933) suggested extending ethics to ecosystems. Most people, environmentalists and nonenvironmentalists alike, probably agree that some conditions of ecosystems have utilitarian value to humans, both now and in the future. But do ecosystems have intrinsic value beyond that of the individual organisms and species that are their living members? Some believe that they do not. Cahen (1988) argues that the complex of organisms and nonliving habitat cannot in any sense be considered to have interests, and thus cannot hold intrinsic ethical value. The current ecological view that the ecosystem (including the living community of organisms) is an entity formed simply by the species that have been able to survive in a given habitat, and is not in any sense a "supra-organism," can be taken to support this view. Others disagree. Rolston (1988) points out that the species is inseparable from its ecological niche, by which he means its natural habitat and way of life. Thus, species are in a sense what they are because of their ecological context, that of the ecosystem. He argues further that ecosystems are not random assemblages subject to patterns of accidental variability in everything imaginable. Organization does exist, as evident by nonrandom patterns of succession and other processes, so that the complex is a sort of "vital field." Because of this organization, an ecosystem does have a degree of unity and continuity,

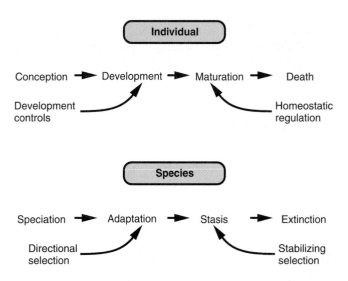

FIGURE 28.3

Like an individual organism to which moral standing is admitted, a species may be viewed as having a sort of "life history," during which regulatory processes shape its characteristics in a manner analogous to the way developmental and homeostatic controls regulate the individual.

with gradual change, in ecological and evolutionary time, just as do species. Then perhaps it is logical to view an ecosystem as having intrinsic value. If so, as for species, endangered ecosystem types might deserve special ethical priority.

A different basis for ethical relationships with other species, biodiversity, and ecosystems is offered by diZerega (1995), who supports the view that ethics is founded in the patterns of respect that have proven essential for functioning of communities. The particular patterns of respect that are important, however, differ with the type of community. The human family as a community, for example, functions not because of rigid rights and duties of its members, but because of caring relationships based on ties of concern, respect, and love among the individuals involved. Successful political communities, such as the United States, exhibit more rigid codes of ethics based on respect of individuals for basic "constitutional" rights, which enable people to interact and cooperate efficiently. Similarly, a successful ecological community, in which humans are a part, requires respect for appropriate characteristics, not so much the "constitutional" rights of individual humans transferred to entities of nature, but respect for the other species and the total ecological systems in which humans and nonhumans alike are enmeshed.

HOW SHOULD CONFLICTING ETHICAL DUTIES BE BALANCED?

Even within the human sphere, ethical obligations come into conflict. If intrinsic moral standing extends to nonhuman entities, many additional conflicts must arise.

Resolving ethical conflicts between humans and other animals that hold moral standing requires that one adopt a view about the degree of equality of the human and nonhuman individuals, and a scale of importance of the interests of each. Such a scale might extend from the most *basic interest,* continuing to live, to the most *peripheral interest,* perhaps some sort of immediate pleasure. VanDeVeer (1979) has identified degrees of equality that translate into different ways of resolving conflicts. Three of the patterns he identifies are:

Extreme Speciesism. Assumes a great inequality of humans and nonhumans, so that any human interest, however peripheral, takes precedence over even the most basic animal interest. This philosophy would, for example, consider trophy hunting of elephants or the killing of infant harp seals for their fur to be fully ethical.

Interest-Sensitive Speciesism. Assumes a basic priority of human interests over similar animal interests, so that a basic human interest takes precedence over a basic animal interest, but a peripheral human interest does not. This philosophy would not condone trophy hunting or killing of animals for fur, but would support subsistence hunting and fishing or the killing of a mountain lion that was acting aggressively toward humans.

Species Egalitarianism. Assumes absolute equality of humans and nonhumans, so that the more basic interest of either takes precedence. This view does not support the killing of animals for food or protection, but would support the raising of domestic animals that give milk or eggs, and the capture and relocation of an aggressive mountain lion to a remote location.

Naess (1973) and Taylor (1986) have championed the principle of species egalitarianism, but as French (1995) points out, both recognize some sort of hierarchy in treatment of species by humans. Naess, for example, noted that as a matter of practicality, some killing, exploitation, and suppression of other life must occur to protect vital human interests. Taylor outlined a series of principles to resolve conflicts between species' interests. These include the principles that self-defense by humans is permissible, that nonbasic human interests should not override basic interests of other species unless very important human values are involved, and even that human exploitation of other species for food is permissible, since humans have no obligation to give greater value to other species than to humans. Johnson (1991), who recognizes inherent value as existing for all organisms, regards that resolution of ethical conflicts between humans and other life must consider

"the degree of complexity of the life system in question and its degree of coherent, integrated, functional organic unity."

Shrader-Frechette and McCoy (1993) have suggested a somewhat different approach to resolving ethical conflicts. They have recommended the application of first-order and second-order principles to resolve such conflicts. First-order principles state simply the objective that should be pursued in a given situation. For example, if California mountain lions, as well as the human residents of California, are considered to have intrinsic moral standing, a first-order principle might be

"Mountain lions should not be killed or persecuted to the point of extinction."

For people, however, a first-order principle might be

"People should not be subject to being killed by mountain lions."

In some places, as recent events have shown, these two first-order principles come into conflict.

Thus, second-order principles are needed to provide guidelines about the resolution of conflicts involving first-order principles. Second-order principles specify what considerations take precedence in a given situation. Shrader-Frechette and McCoy (1993) suggest several possible principles:

1. Overall long-term value to society at large takes precedence over short-term economic benefits to parts of society.
2. The will of the public at large takes precedence over that of specific interest groups.
3. Repairing past injustice takes precedence over present human property rights.
4. Practical efforts take precedence over impractical efforts.

Even given that this list of second-order principles is tentative and incomplete, one can see that ethical conflicts will not often be easy to resolve. For the California condor, it might turn out (1) that long-term values override short-term economic benefits only in some wilderness areas, (2) that the public supports recovery in the wild, (3) that human impacts unjustly caused the decline of the species, and (4) that reintroduction to the wild in California is impractical but introduction somewhere else might be practical.

WHAT ETHICAL DUTIES DO WE HAVE TO FUTURE GENERATIONS?

Environmentalists fear that environmental damage and loss of biodiversity in the present will leave a degraded global environment for future generations. This raises the issue of **intergenerational equity,** i.e., the equality or inequality of opportunities available to humans of present and future generations. If living humans do have substantial duties to future generations, these must involve secondary ethical values—aspects of the environment that are likely to be of instrumental value to future humans. What ethical responsibilities do we have to our descendants? Two issues are involved. First, do future humans, perhaps even those dozens of generations removed from the present, deserve moral consideration equal to living humans? Second, will aspects of environmental quality and biodiversity be of utilitarian importance to future generations of humans?

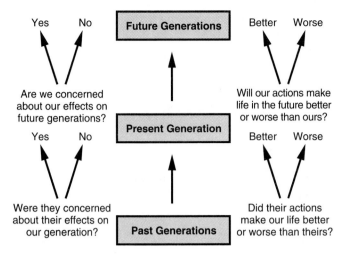

FIGURE 28.4

Questions of intergenerational equity.

Arguments have been raised both for and against members of future generations having rights equal to those of humans presently living (Shrader-Frechette 1981). Some contend that although living humans can have concern about those in the future, they can have no obligation to future humans. This view rests on two main arguments. First, it can be argued that a duty to another person is, in effect, a contract that involves mutual obligations, and that future humans cannot hold obligations to humans living today. On the other hand, some argue that the fact that living humans can and do benefit from actions of their forebears gives them an obligation to aid future generations. Also, one can argue that not all social contracts involve mutual obligations. Parents, for example, are clearly regarded as having obligations to their children, the nearest future generation, even though the children did not agree to any obligations to their parents and may or may not assume any.

Second, considering advances in technology that are occurring all the time, it is difficult for living humans to know what will be useful to future humans. Thus, saving what we consider resources now might hurt living humans more than it will benefit future humans. On the other hand, it may be argued that people do know much about how the earth's environment can be harmed long into the future, as by toxic chemical pollution, depletion of the ozone shield, and impoverishment of biodiversity. This knowledge, therefore, gives living humans an obligation not to leave a degraded earth to future generations.

To some extent, we can visualize the issues of intergenerational equity by considering how past generations viewed their actions as affecting the future, and how we view our actions as affecting the future (Fig. 28.4). Most people probably do believe that we owe future generations opportunities in life at least as good as ours. Most probably believe, as well, that we are managing the environment so as to make this possible. On the other hand, many environmentalists believe that biodiversity is one of the most important resources of the biosphere, and that it will become even more valuable in the future. To them, its loss will likely degrade the quality of life possible for future generations.

SUMMARY

1. A primary ethic involves duties to something that has intrinsic moral standing, whereas a secondary ethic involves duties to something because it has utilitarian or instrumental value to humans.

2. Modern human society is dominated by a Pragmatic-Utilitarian Ethic that admits intrinsic moral standing only to humans and certain other animals with humanlike traits, while considering other species and ecological systems to have only secondary moral standing based on their utilitarian value.

3. The view that species, communities, and ecosystems have intrinsic value, the Ecocentric Ethic, requires that equality of human and nonhuman interests must be judged on some scale, as from the most basic interest, continuing to exist, to the most peripheral interest, such as immediate satisfaction or pleasure.

4. Questions of ethical duties to future humans concern the balance of these duties relative to those to living humans, the degree to which present biodiversity is likely to be of utilitarian importance to people in the future.

5. Deep Ecology is an ecophilosophy that holds that all life and the ecological systems that living organisms form have intrinsic value, and that people should adopt a way of life that respects the unity of humans with the biosphere.

QUESTIONS FOR DISCUSSION

1. What are the primary and secondary ethical values that would be in conflict in the case of proposed logging of an old-growth redwood stand?

2. Should our uncertainty about the value of biodiversity to people in the future be considered a strong reason to preserve biodiversity, or justification for not worrying about its rate of decline?

3. What changes would Deep Ecology prescribe for people living the lifestyle current now in the United States?

SUGGESTED READING

Des Jardins, J. R. 1993. *Environmental ethics: An introduction to environmental philosophy*. Wadsworth Publishing Co., Belmont, CA. An introduction to the rapidly growing field of environmental philosophy, covering topics from Aristotelian natural law to Deep Ecology and Ecofeminism.

Pierce, C. and D. VanDeVeer (Eds.). 1995. *People, penguins, and plastic trees. Second edition*. Wadsworth Publishing Co., Belmont, CA. An anthology of published articles dealing with the major issues of environmental ethics, with section introductions by the editors.

Sessions, George (Ed.). 1995. *Deep ecology for the 21st century*. Shambala, Boston, MA. An anthology of readings, including several unpublished articles, tracing the history of the Deep Ecology movement from the 1960s to the present.

chapter

29

Conservation Economics and Public Policy

As human populations and their use of resources grow, conflicts over the use of the environment are growing exponentially. Questions arise about the value of aspects of nature that have previously had no market value, but that some people feel are of infinite value. How does the value of a population of endangered least Bell's vireos in riparian habitat in southern California stack up against the value of the same land as the site for a water-storage reservoir? Unless the vireos and their associated riparian ecosystem can be shown to have real and substantial value, the scales are tipped in favor of the ultimate construction of a reservoir. If such values do not exist, or cannot be shown, laws that now protect vireos are likely to be changed, or perhaps manipulated in devious ways, to permit alternative values to be realized. Thus, a full and accurate economic valuation of ecological resources—natural ecosystems and their member species—becomes more and more necessary.

Public policy in a democratic society is intended to reflect the interests of the public at large. Historically, public environmental policy has been dominated by societal interests concerned with consumptive or industrial use of resources such as minerals, timber, grazing resources, and water. Since the 1960s, however, concern has grown over the fact that such uses have degraded land and water environments at great public cost, as through air and water pollution, and reduced many associated values, as by the effects of hydroelectric dams and clear-cutting on stream ecology. Clearly, public policy must be based on a broader assessment of the costs and benefits of all possible uses of the environment.

Natural environments possess many recreational, aesthetic, and potential resources that are undervalued by marketplace economics.

Outline

FIGURE 29.1

Migratory snow geese, breeding in the Canadian arctic and wintering at Klamath Basin National Wildlife Refuge in the United States, and the scenic landscapes they often frequent are examples of common property resources.

(U.S. Fish and Wildlife Service photo by Paul Benvenuti)

FIGURE 29.2

Yosemite Valley and its surrounding mountains possess an aesthetic quality that is regarded as a national treasure, and thus a common property resource belonging to all Americans.

(Photo by G. W. Cox)

CONSERVATION ECONOMICS

Determining the full value of ecological resources is difficult, however, since they range so widely in character. Some resources, such as timber trees, are fixed in location and are easily measured; others, such as the great whales, range over vast areas of international oceans and are difficult even to count. Still other environmental features, such as pure air or a scenic natural landscape, are difficult to value in monetary terms. All of these resources, in addition, possess values that are real, but lie largely outside the market systems that define the values of goods and services in modern societies. But if the full economic value of such resources is not determined, these resources may easily fall victim to unregulated exploitation or destruction.

The objective of the field of conservation economics is to develop ways of determining the full economic value of ecological resources, and enabling the consideration of these values within the political and economic systems of the world's nations. In this chapter, we shall examine the different sorts of economic values that are associated with ecological resources, and consider some of the new approaches to their comprehensive valuation.

Many ecological resources, such as natural ecosystems, their member species, and their properties, are examples of **common property resources.** Resources of this type are owned by society at large, and typically are not fully valued in market transactions that affect them. That is, the value that is paid by someone who exploits them is often only a fraction of their total value, when the interests of society at large are considered.

Examples of common property resources include whales, migratory birds, wilderness areas, scenic landscapes, biotic diversity, and clean air and water. Whales and migratory birds, for example, occupy international environments, in the first case largely lying outside any national borders, and in the second case

often crossing international borders (Fig. 29.1). Who owns these animal populations? To the extent that the concept of ownership applies, the answer is "humanity at large." These kinds of animals were once exploited by whalers and market hunters, however, with the only considerations being the cost of killing, processing, and marketing them relative to the income received by their hunters. Wilderness areas and scenic landscapes are similar in their basic nature (Fig. 29.2). Who owns the wilderness areas of the Sierra Nevada Mountains or the scenic grandeur of Lake Tahoe? The answer, likewise, is "society at large." Specific groups, however, can propose uses that exploit or damage these areas, such as ski resort developments in pristine parts of the Sierras, or mining developments that could cause waste discharges into Lake Tahoe. These detrimental impacts, which are real costs of the activities, are spread over society at large. Pure air and water are also resources that rightly belong to humanity at large. Yet again, exploitation of these resources, or their contamination by waste discharges by certain groups of humans, affects their value for everybody.

Biotic diversity—the richness of life on earth—is likewise a common property resource, and belongs to the human race as a whole. Much of this diversity resides in\the tropical forests of a few countries, where its destruction through the exploitation of land and timber for short-term profits by those countries is imminent. Global concern over tropical deforestation, and other activities that drive plants and animals toward extinction, is an indication that important values are not being fully considered in our management of the biosphere.

Analysis of common property resources is very difficult. Resources of this sort often have both **consumptive** and **nonconsumptive values,** some of the latter involving **off-site** and **nonuse values** (Bishop 1987). Consumptive values are those that involve the actual destruction or degradation of the resources, such as the cutting of timber, the harvest of marine fish, or the pollution of pure water. Nonconsumptive values are those that can be obtained without depleting the resource, as through bird-watching, catch-and-release fishing, or recreational use of waters in ways that do not pollute them. Off-site values relate to the value of books, films, and the like to people who do not use resources where they exist, and nonuse values to satisfaction that people feel simply knowing that the resources exist.

In addition, the costs associated with common property resources are often borne in one location, and the benefits realized in another area. This is the case with migratory waterfowl, for example, many of which are reared in Canadian marshes, but killed by hunters in the United States (Adamowicz et al. 1986). Similarly, the benefits of using the atmosphere as a cheap location for the discharge of combustion wastes are reaped by urban-industrial regions, whereas many of the costs of such discharge appear as acid precipitation in regions far downwind. Unpaid costs like these, that are imposed on society, are termed **external costs** or **externalities.**

Further complications arise because natural ecosystems possess a variety of common property resources of very diverse nature. An undeveloped landscape may, for example, have pristine scenic value, populations of harvestable wildlife, watershed values, minerals that can be exploited, and several other resource values that are of interest to different segments of society. This is almost always true of areas of United States National Forests, for which the concept of multiple uses has explicitly been mandated.

Common property resources are subject to what Garrett Hardin (1968) termed **the tragedy of the commons.** The commons, or communal area of pasture land set aside for the use of the inhabitants of English and early American villages, was an area owned by the villagers at large. The purpose of the commons was to provide a grazing area for the cattle, horses, and other livestock that were the livelihood of small farming villages. The tragedy of this institution was that individuals were free to increase their exploitation of the commons by grazing more and more animals on it, without paying all of the added costs of keeping the commons in a productive condition. These external costs were spread over all the villagers, whereas the benefits all accrued to the individual who owned the added animals. The result was overgrazing and deterioration of the resource.

Common property resources present many problems of management. Often, as in the case of old-growth stands of trees in national forests, the resource is enormously valuable to a small portion of society. The ultimate loss of that resource, on the other hand, will reduce the quality of life a small degree for a much larger portion of society. When such a trade-off becomes perceived, moreover, protecting the resource may require an immediate, very large expenditure of funds, as in compensation of timber owners or retraining of those who depended on harvesting of old-growth timber.

ECONOMIC VALUES OF ECOLOGICAL RESOURCES

Ecological resources have many kinds of specific values (Fig. 29.3), some actual and represented in the marketplace, and some potential or conceptual (Myers 1983; Oldfield 1989). The traditional value of many ecological resources is their **harvestable value,** that is, the value of the usable products that can be taken from nature to those who harvest them. Harvests of wild plant foods, game animals, marine and freshwater fish and shellfish, timber and plant chemicals, skins of wild animals, and special materials such as ivory all fall into this category. Worldwide, trade in timber equals about $77 billion (Reid & Miller 1989). In the United States, the total annual harvest of wild-grown foods has an annual value of about $2.8 billion (Pimentel et al. 1980). The harvestable values of elephant ivory, rhinoceros horns, and the skins of large cats, of course, support the insidious poaching networks for these animals.

Subsistence hunting-and-gathering activities of groups such as the native peoples of Canada and Alaska are particularly important examples of the use of harvestable resources (Fig. 29.4). About 100,000 subsistence hunters in Alaska and 50,000 in northern Canada harvest caribou, moose, seals, salmon, and waterfowl in more or less traditional fashion (Cooch 1986). Regulation of these harvests has tended to be piecemeal, and enforcement of existing regulations often lax. Subsistence harvests, sometimes illegal, combined with sport or commercial harvests for some species, have placed increased pressure on some of the harvested species. In Alaska, for example, populations of several geese that were harvested by natives of the Yukon-Kuskokwim Delta declined from 1965 through 1985, creating a critical need for comprehensive management of their harvest (Pamplin 1986). Aboriginal whaling, practiced in several coastal Inuit villages in Alaska, is another such activity, but one that has attracted controversy in recent years (See Chapter 19).

Game cropping from wildlife ecosystems in Africa or elsewhere is still another example of harvestable resource value. Dasmann (1964) compared production of meat by cattle and game animals on a ranch in Zimbabwe, and concluded that the net profit from game cropping was six times that from cattle ranching. In Africa, the ecological diversification of native ungulates, their adaptation to arid environments, and their resistance to diseases that affect livestock have led many workers to suggest that game cropping is not only economically practical, but could be a strong force for preservation of the native wildlife. Crawford and Crawford (1974) estimated that game cropping could allow Africa to become one of the world's major producers and exporters of meat. Others, however, have been less optimistic about the potential for game cropping. Caughley (1976), for example, pointed out that most seemingly successful schemes have been short-lived, and that the difficulties of sustained harvest of wild species with varied social systems and population structure, over which humans have little control, are enormous (See Chapter 24).

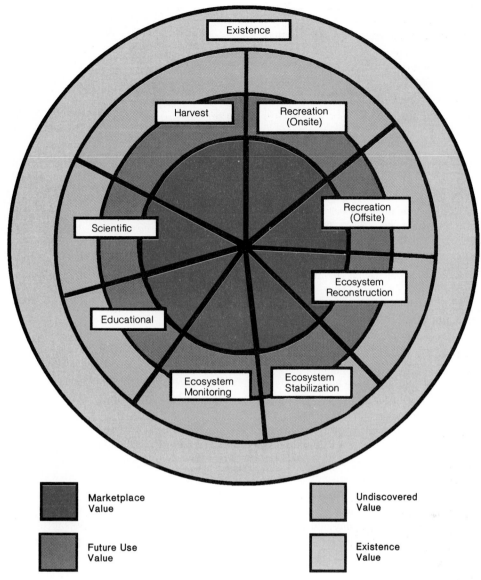

FIGURE 29.3

Ecological resources have many types of values, some of them realized in the marketplace, others relating to potential future use or undiscovered utility, and still others relating to the desire of many humans to assure the continued existence of natural ecosystems and their members. In this diagram, marketplace value of recreational resources includes the consumer's surplus.

Organisms that yield useful chemicals or genes are examples of harvestable resources, as well. Almost half of all drugs used in medicine were originally derived from plants, fungi, and microorganisms (National Science Board 1989). The Madagascar rosy periwinkle, *Catharanthus roseus*, for example, is the source of the drugs *vincristine* and *vinblastine* that are used in the treatment of leukemia and Hodgkin's disease. These drugs have a commercial value of about $100 million per year (Myers 1983). In addition, since 1950 about 3,000 antibiotic chemicals have been isolated from actinomycetes alone (Demain & Solomon 1981). A bacterial gene that produces a toxin to many insect larvae has now been incorporated in several crop plants, giving them protection against such pests.

Undiscovered value is a category that can be illustrated clearly for harvestable resources; it actually exists for all "slices" of

the value "pie" for ecological resources (*See* Fig. 29.3). As Bishop (1980) has observed, the components of natural ecosystems form a set of potential resources that become actual resources as social, economic, and technological conditions change. Many plants and animals exist that will at some future time be found to have valuable food, chemical, or other product values. Mendelsohn and Balick (1995), for example, estimated that tropical forests contain about 327 undiscovered pharmaceutical chemicals that have a potential value of about $147 billion to society. Advances in genetic engineering now make genes and co-adapted gene complexes of almost any organism transferable to species that are useful to humans. Wild relatives and local races of crop plants serve as potential sources of genes for resistance against diseases and pests (Myers 1983). The search for genetic resistance is a continuing effort, since disease and pest organisms themselves evolve

FIGURE 29.4

The harvest of wildlife, such as this bearded seal by Inuit hunters, is still an important part of the economy of Native American villages in many parts of the Arctic.

(U.S. National Park Service photo by Robert Belous)

FIGURE 29.5

Tourism, much of it related to wildlife, is Kenya's largest source of foreign exchange.

(Photo by Darla G. Cox)

to overcome such defenses. The importance of such sources of genetic variability is greater now than ever, considering the tendency for modern agricultural practices to emphasize the cultivation of a few varieties with very uniform genetic characteristics (National Academy of Sciences 1972). Genes from these same sources are valuable raw material for the improvement of domestic plants and animals, as well. Most environmental scientists agree that the yet-to-be-discovered values of this sort are enormous, and by themselves justify the preservation of biotic diversity.

The conservation of genetic variability of wild species is best achieved in nature (Frankel & Soulé 1981). Collections of certain types of plants can be maintained in botanical gardens, and small populations of certain animals in zoological gardens. Where the species in question have short generation times, the potential for genetic change due to the artificiality of the captive environment is substantial, however (*See* Chapter 25). Some plants, both wild and domesticated, can be placed in seed banks, where the seeds are kept in cold storage for long periods. Even here, however, the plants must be propagated at intervals to obtain new seed to replace that which has gradually deteriorated in storage. Thus, natural ecosystems have a real economic value related to their maintenance of gene pools of plants and animals.

Increasingly, **recreational value** is being recognized for many ecological resources. Indeed, recreational value is complex enough that a special field of recreational resource economics is now recognized. Three distinct components of recreational value can be identified.

The first component of recreational value is **on-site use value,** the value produced in the location where the resource exists. Part of this is the **on-site economic impact,** the monetary value of recreational activity itself. For recreational fishing, for example, this includes the money spent for travel, room, board, tackle, boat rental, and so forth in locations visited by fishermen. In developed countries, on-site activities of this sort are a major component of the national economy. In the United States, for example, recreational marine fishing was estimated to have an on-site recreational value of about $7.5 billion in 1980 (Anon. 1983). Wildlife viewing in the United States in 1991 was

estimated by the U.S. Fish and Wildlife Service to involve on-site economic impact of $18.1 billion, $5.2 billion alone for bird watching. At Point Pelee National Park, Ontario, visitors who came for bird watching spent $5.4 million annually, and $3.2 million in the immediate park area (Hvenegaard et al. 1989).

In Kenya in 1989, tourism centered on beaches and wildlife parks (Fig. 29.5) brought in $419 million, about 50% of which was related to wildlife (Norton-Griffiths & Southey 1995). When the costs associated with this income were taken into account, however, the net income from wildlife tourism was much less, and probably less than the income that could be gained by converting game parks to livestock ranching. Nevertheless, substantial potential appears to exist for **ecotourism,** or ecologically sensitive tourist activities, as a means to enable developing countries to benefit from ecological resources without destroying them.

On-site use value also includes a more nebulous component, the value of the experience to the recreationist over and above the costs paid. This value is usually termed the **consumer's surplus.** Measuring it is difficult, and is usually done by determining the willingness of the recreationist to pay more to carry out the activity, or to sell the right to pursue the activity. At Point Pelee National Park, for example, bird-watching visitors indicated that they would have been willing to pay more than twice as much for their visits to the area (Hvenegaard et al. 1989). A consumer's surplus of $450 million was estimated for foreign wildlife viewers in Kenya (Moran 1994), suggesting that ecotourism there has more potential for national income than is being realized. Norton (1988) has argued that contact with nature has major value in transforming and uplifting the human spirit, a special sort of consumer's surplus. Because the consumer's surplus does not involve an actual financial transaction, some people might tend to think that the value does not exist. When the use of a resource is abridged, however, the magnitude of the consumer's surplus often becomes forcibly expressed.

The second major category of recreational value is **off-site use value,** or the economic impact and any consumer's surplus that occurs away from the location of the resource itself. Examples of this category include books, films, paintings, radio and television programs, and the like that relate to the natural resource. The many such products that describe the Serengeti Plains of Tanzania demonstrate that off-site values can be large. Off-site use values also include the value of display of animals in zoos and aquaria, or of plants in botanical gardens.

A third major category of recreational value is **future use value.** This is simply the value attached to the opportunity to use the resource at a later time. Many people support the protection of the Florida Everglades, for example, because they wish to visit this area some day. Or, individuals may value the opportunity for their children to see something, such as the grizzlies of Glacier National Park, that they once saw as inhabitants of a wilderness park. Again, future use value actually applies to all categories of ecological resources (*See* Fig. 29.3).

Several additional categories of value can be recognized. **Ecosystem stabilization value** exists wherever organisms contribute to the maintenance of basic ecological functions (Westman 1977; Talbot 1987). Microorganisms and invertebrates carry out the processing of many kinds of wastes produced by human activity, and are the major agents of nutrient recycling. Larger animals perform many important ecological functions relating to soil formation and restoration, plant pollination and seed dispersal, plant and animal population regulation, and the control of habitat structure and community composition. A number of recent studies have shown that diversity of composition promotes inertia, resilience, productivity in model ecosystems (Tilman & Downing 1994; Naeem et al. 1994; Tilman 1996). Many species of animals exert natural biological control within crop, rangeland, and forest ecosystems (e.g., Marquis & Whelan 1994). Naturally occurring predatory and parasitic insects that prevent or moderate outbreaks of many agricultural pests, for example, exemplify this value. That a real economic value exists is shown by the fact that secondary outbreaks of pests that often follow crop treatments with broad-spectrum insecticides, which destroy the natural biological control agents (Van den Bosch 1978).

The living members of natural ecosystems also promote stability of abiotic features of the landscape, an influence that can also be very valuable. The damage to forests in the San Bernardino Mountains, east of Los Angeles, California, by photochemical smog, for example, results in tree death that in turn promotes increased soil erosion. Westman (1977) estimated that loss of 50% of the trees on an area of 4,000 hectares would lead to an annual increase of $27 million for sediment removal from streets, sewers, and debris basins in the urbanized areas affected by increased erosion.

Habitat reconstruction value represents the value of species for the restoration or reconstruction of natural ecosystems. Efforts are now being made, for example, to restore or recreate native grassland ecosystems in the Great Plains, California, and other areas. The success of these efforts depends in part on the availability of the native species to reconstitute grassland

communities. Similarly, restoration of coastal wetlands has become an important effort in areas such as California and the northeastern United States, where original wetland areas have been greatly disturbed, and in many cases, destroyed. Revegetation of land disturbed by mining and highway construction, creation of new habitat as mitigation for habitat lost to development, and reclamation of lands damaged by erosion and desertification are major challenges for ecotechnology. These activities will draw increasingly on native species with specific ecological capabilities.

Certain species possess **environmental monitoring value,** in that they can serve as indicators of environmental stress. The California Mussel Watch program, begun in 1977, has proven to be a practical way of monitoring chemical pollution of coastal marine waters. In this program, samples of mussels of the genus *Mytilus,* or of the freshwater clam, *Corbicula fluminea,* are collected at designated locations for tissue analysis for pollutant chemicals. These animals are filter feeders, and thus pass large volumes of water through a filtering apparatus within their bodies. Any chemical that is present in significant amounts, and that tends to enter living tissues, will be accumulated in mussel tissues. In 1986–1987, for example, 135 locations were sampled on the open coast, in bays and estuaries, and in fresh waters, and analyzed for 14 trace elements and 42 synthetic organic compounds (Stevens 1988). The California Mussel Watch Program has detected hotspots of pollution by pesticides, PCBs, heavy metals, tributyltin, and several other toxic chemicals. Over 15 years (1977–1992) clear trends, mostly showing declines in concentration, of chlorinated hydrocarbon pesticides and PCBs have also been noted (Stephenson et al. 1995). A national mussel watch program is operated by the United States National Oceanic and Atmospheric Administration (O'Connor et al. 1994), and an international program has been proposed (Kistner 1984).

Beyond these values, natural ecosystems and their members possess **educational value** and **scientific research value.** The intrinsic interest of people, particularly young people, in the lives of plants and animals provides a powerful tool for environmental education (Hair & Pomerantz 1987). The study of organisms in their natural ecosystems can be used to promote knowledge and an enlightened attitude toward the global environment—one that translates into socially responsible actions. First-hand contact with nature is an indispensable component of environmental education, and networks of natural areas are thus a key resource for education. Similarly, the availability of natural ecosystems for scientific study is essential to increasing our understanding of natural processes. Studies of the sea otters in coastal marine ecosystems, for example, helped reveal the concept of the keystone predator, and have promoted increased understanding of the role of predators in the organization of biotic communities. Networks of researched natural areas, such as the global system of Biosphere Reserves being developed by UNESCO (*See* Chapter 30), are thus absolutely essential to progress in ecological research.

Finally, an **existence value,** the value many people attach to what might be termed "right-of-survival," regardless of

CONTINGENT VALUATION AND THE EXXON VALDEZ

The oil spill from the *Exxon Valdez* tanker in Prince William Sound, Alaska (*See* Chapter 22), raised issues about the value that Americans place on maintaining a healthy environment, as well as on the penalty that should be imposed on *Exxon* for causing environmental damage. Damages to fishing and tourism in the region affected by the spill could be assessed in a relatively direct manner, by determining the income lost after the spill. Federal court rulings related to damage assessments under the Clean Water and Superfund Acts supported the concept that nonuse values (also termed passive-use values) should be considered, and that the contingent valuation technique was a justified approach to estimating such values.

The State of Alaska funded a $3 million contingent valuation study of the willingness of U.S. citizens at large to pay for preventing future accidents like that of the *Exxon Valdez* (Carson et al. 1995). This study involved face-to-face interviews with 1,042 people in all 50 states. The interviewers described the effects of the spill, as known in the fall of 1990, and then outlined an escort-ship program that would make it virtually certain that another tanker accident would not occur over the subsequent 10 years, during which all single-hulled tankers would be replaced by double-hulled tankers. The individuals interviewed were then told that this program would be paid for by a one-time tax on oil companies using Alaskan oil, and a one-time tax on U.S. households. They were then told that the one-time household tax would be a specific amount (either $10, $30, $60, or $120), and asked if they would vote for this spill-prevention program. Those who said they would vote for the program were given a follow-up choice involving a higher cost, and those who said they would vote against the program, a follow-up choice involving a lower cost.

The median willingness-to-pay was found to be $31, meaning that half the people interviewed were willing to pay that much or more. Detailed statistical analyses of the results suggested that the collective willingness-to-pay by U.S. households was between $2.8 and $8.6 billion. A penalty of $5 billion was assessed on *Exxon,* but is under appeal.

The use of contingent valuation in the context of damage assessment has stirred criticism from some economists, some supported by funds from *Exxon.* Contingent valuation has been characterized as "junk economics," because the process does not actually require people to pay the amounts they indicate that they are willing to pay, and because the results of some contingent valuation studies have seemed extreme or inconsistent (e.g., Mead 1993). But many other economists believe that the technique is being refined rapidly, and holds promise of being useful in assessing many nonmarket values (Hanemann 1994; Portney 1994).

---

whether the individual cares or expects to use the resource, must be recognized. Many people, for example, believe that the wildlife resources of the Antarctic should be left unexploited, as an example of undisturbed wilderness (Ehrenfeld 1976). Charles Elton (1958) put it this way:

> "There are some millions of people in the world who think that animals have a right to exist and be left alone, or at any rate that they should not be persecuted or made extinct as species. Some people will believe this even when it is quite dangerous to themselves."

Organisms and ecosystems clearly have existence value, in addition to other values that might be recognized (Fig. 29.3). Californians, for example, indicated in a survey that they would be willing to pay an average of over $13 per household just to assure the survival of the sea otter along the state coastline (Hageman 1985). Strong empirical evidence for existence value comes from the willingness of people to join organizations devoted to protection of nature (Madariaga & McConnell 1987).

In several of the above cases, translating the particular value into monetary terms is very difficult. These include the consumer's surplus, future use, and existence values. One technique for determining the consumer's surplus for on-site use value of ecological resources is the **travel cost method** (Rosenthal et al. 1984). This technique assumes that the difference between the willingness-to-pay for different numbers of visits and the average costs of visiting the site is a measure of the consumer's surplus for the entire population using the site. Data on willingness-to-pay are obtained by determining the cost and number of visits per capita by individuals living at different distances from a site, the assumption being that the maximum willingness-to-pay is that by individuals who come the farthest for a single visit.

A second technique used to estimate many nonmarket values, including future use and existence values, is the **contingent valuation method.** This technique uses a question-and-answer approach to determine the willingness to people to pay or pay extra to guarantee some environmental value. The contingent valuation approach is now being used in many situations (*See* Reading 29.1). In one of the earliest analyses, Hammack and Brown (1974) evaluated the consumer's surplus for duck hunting in the United States in 1968, using a questionnaire that determined what duck hunters actually spent, and what they would have been willing to spend before foregoing their usual hunting activity. The actual expense averaged $322, but the average hunter was willing to pay $247 more to pursue this sport. Thus, individual hunters received pleasure equal to $247 beyond the actual costs of the hunting trip. At Point Pelee National Park, Ontario, bird-watching visitors spent, on average, $224 per trip, but indicated a willingness to pay extra of $256 per trip before they would forego the visit (Hvenegaard et al. 1989). A willingness-to-sell analysis would involve a similar approach, as, for example, determining the price for which individuals would be willing to sell permits, drawn through a lottery, for a limited hunt of certain big game animals.

The contingent valuation method has been used to estimate future use value, as in relation to a proposal to locate a major, coal-burning power plant in the Glen Canyon National Recreation Area (Brookshire et al. 1976). Public reaction to this plan was evaluated by a questionnaire made available at a location from which the power plant site could be viewed. A display allowed visitors to see the site as it existed without the plant, and to view depictions of the site with A) a power plant without visible air pollution output, and B) a power plant with substantial air pollutant discharge (Fig. 29.6). The visitors were then asked to indicate how much extra they would be willing to pay as an entrance fee to prevent situations A and B. The results (Table 29.1) showed that response differed for different categories of visitors, with individuals who were interested in remote-area camping being willing to pay the most, and local residents the least. Over the 35-year lifetime of such a plant, however, the overall willingness to pay extra amounted to $12.2 million to prevent the construction of a plant building, and $21.2 million to prevent the construction of a plant giving off substantial visible pollution.

Boyle and Bishop (1987) also present an example dealing with the existence value of particular endangered species. A sample of Wisconsin taxpayers were given a survey that posed a hypothetical situation involving funding of programs for protection of the bald eagle (Fig. 29.7), a state endangered species, and the striped shiner, a species that had been reduced to a critically low level. The conditions of the hypothetical situation were that 1) all state funding for conservation of these species had been terminated, 2) the species would decline to extinction if active conservation programs did not continue, and 3) independent private foundations, supported by individual memberships, were created to manage these species. The persons surveyed were told that memberships would cost a certain amount, and asked if they would join. The results of the study, extrapolated to the state population, showed that annual membership income would be about $26.7 million for the bald eagle program and $12.0 million for the striped shiner program. Per capita, persons who expressed actual interest in seeing wild eagles showed a greater willingness to pay for preservation of the species, but the majority of the population showed willingness to support conservation of the bald eagle, and also of the rather inconspicuous striped shiner simply for the satisfaction of knowing that these species survived in Wisconsin.

PUBLIC ENVIRONMENTAL RESOURCE POLICY

In North America, resource-use policy was originally designed to promote the rapid exploitation of resources such as minerals, timber, water, and agricultural land for the building of a nation (Losos et al. 1995). Natural resource laws and policies were designed to encourage and subsidize such exploitation. In many developing nations, such as Brazil, as well as nations with large lightly populated regions, such as Russia, this is still the case. With the high level of development that has occurred

Power Plant with Heavy Pollution

Power Plant with Light Pollution

Pristine Landscape

FIGURE 29.6

The contingent valuation technique used to evaluate the impact of a power plant on Glen Canyon National Recreation Area involved presenting visitors with sketches similar to these of the proposed site, the site with a plant releasing only slight pollution, and the site with a heavily polluting plant.

(Modified from Brookshire et al. 1976)

FIGURE 29.7

A study using the contingent valuation technique, showed that Wisconsin taxpayers placed an annual existence value of about $26.7 million on survival of the bald eagle in the state.

(U.S. Fish and Wildlife Service photo by Sue Matthews)

TABLE 29.1

Willingness of Visitors to Pay Extra to Prevent Construction of a Nonpolluting Power Plant (A) or a Polluting Plant (B) in Glen Canyon National Recreation Area (1975)		
	EXTRA ENTRANCE FEE THAT WOULD BE PAID TO PREVENT A OR B	
VISITOR GROUP	A	B
Local Residents	$0.87	$1.75
Remote Area Campers	2.11	3.38
Developed Campground Users	1.08	2.60
Motel Visitors	1.94	3.11
Average Visitor	1.58	2.77
Total for 1975	**$414,000.00**	**$727,000.00**
Total over 35-year plant life*	**$12,174,000.00**	**$21,182,000.00**

*Present value with 5% discount per year over plant life
Source: Data from D. S. Brookshire, et. al., "The Valuation of Aesthetic Preferences" in *Journal of Environmental Economics and Management,* **3:**325–46, 1976, Academic Press, Orlando, FL.

throughout North America, however, has come the realization that values beyond those of extractable resources are basic to sustaining the quality of life. The challenge to policymakers now is to incorporate these values into a more comprehensive environmental policy.

The primary reason for efforts to identify and quantify values associated with ecological resources is to permit a fair comparison of the costs and benefits associated with environmental policies and actions. One approach to such evaluation is **benefit:cost analysis,** the weighing of the values gained or preserved by an action against the costs incurred or value of alternative benefits that are foregone (Randall 1986). This sort of comparison is useful, but may be inadequate. For example, benefit:cost analyses often consider only local costs weighed against short-term benefits. In addition, benefit:cost analysis often considers the public interest to be the simple summation of immediate individual interests, which are highly subjective and changing through time. Designing a benefit:cost analysis that takes a long-term view is difficult. Liberal versus conservative estimates of future or undiscovered use values can differ greatly, leading to benefit:cost analyses that favor conservation, on the one hand, or the alternative action, on the other. The logistic practicalities of determining the ecological functions of species in natural ecosystems may also tend to favor their undervaluation (Norton 1986).

Nevertheless, determining the benefits and costs associated with a particular action, together with those of alternative actions, is essential to sound environmental policy. One of the major mechanisms for such evaluation now is **environmental impact assessment.** Environmental impact assessment, as practiced in the United States, follows procedures specified by the **National Environmental Policy Act (NEPA)** of 1969 and state laws with similar intent. NEPA requires that projects involving federal support or approval be examined to determine their short-term and long-term effects, and that alternatives to the proposed project also be evaluated. Particular attention is devoted to threatened and endangered species (*See* Chapter 26).

Environmental policy also requires decisions at higher levels about governmental action over the long term. A major difficulty associated with formulating policy at this level is uncertainty about the importance of environmental conditions and ecological resources, and about how they are likely to be affected by alternative policies and actions. Uncertainty is great, for example, about the importance of species diversity in ecosystem functions such as productivity and recovery from disturbance. It is also great in regard to prediction of how global change will affect biodiversity several decades into the future.

Other difficulties in the formulation of environmental policy relate to the history of environmental laws, and how this history has shaped economic and political power structure. For example, efforts to promote development in western North America have created a legacy of laws and institutions that still subsidize mining, timber harvest, grazing, and water development on public lands (Losos et al. 1995). These subsidies total more than $1 billion annually in the United States, and lead to biodiversity degradation costs that exceed $100 million annually.

Strong political constituencies make revising law and policy relating to these activities very difficult, in spite of the burden of these subsidies and costs on the public at large.

What is the role of conservation biology in the development of public environmental policy? As scientists, conservation biologists are obliged to hold to strict standards of scientific investigation. Scientists also have an obligation to make their results available to policymakers and the public at large. Coupled with this is a responsibility to educate policymakers and the public about the nature of science. Many politicians and administrators view science more as a product than a process, for example (Brosnan 1995). Finally, scientists have a right to advocate that their expertise and findings become part of the policy development process.

Scientists can participate in development of environmental policy through their integration into interdisciplinary teams created to examine policy relating to specific topics. Brosnan (1995) describes how such a team examines issues relating to human use of marine resources in the intertidal and near-offshore waters of Oregon. Scientists contributed by conducting studies such as experimental evaluation of visitor impact on biodiversity in intertidal areas. Scientists, together with other individuals with interests and responsibilities relating to coastal ecology, worked with policymakers to develop the Oregon Territorial Sea Plan. This plan defines several shore management categories ranging from protected refuges to shores open to recreational use.

Working in such teams, scientists can also help define policy alternatives, such as those developed by the Forest Ecosystem Management Assessment Team (Franklin 1995). This team developed 10 alternative plans for harvesting timber in national forests in the Pacific Northwest, where issues of protection of the northern spotted owl had revealed the inadequacy of existing harvest practices. Analyses such as this are not science, but the use of the best knowledge available to project the most likely consequences of alternative policies. The choice among these options is not a choice made by scientists, but by the elected and appointed representatives of the public.

SUMMARY

1. Many ecological resources, such as biodiversity and unpolluted air and water, are examples of common property resources that are owned by society at large, and typically have not been valued fully in national economics.

2. Ecological resources have both consumptive and nonconsumptive values of many sorts, ranging from those that are handled in the marketplace to those of potential future use, those as yet undiscovered, and those that involve the simple existence of the resource.

3. Many nonmarket values, such as consumer's surplus, future use value, and existence value, can be assessed by the contingent valuation method, which determines what people would be willing to spend or spend extra to assure the continued access to or existence of a resource.

4. Environmental impact assessment, which involves assessing the benefits and costs associated with a particular action and its alternatives, has become a central tool in formulating environmental policy.

5. The role of science in the development of environmental policy may be improved by integrating scientists into interdisciplinary groups responsible for studying resource use issues and developing policy recommendations.

QUESTIONS FOR DISCUSSION

1. Do you think biodiversity is really an aspect of the global commons, owned by humanity at large, or is it the property of countries and states in whose lands and territorial waters it is located?

2. Do you think that attempting to quantify the dollar value of ecological resources, including such aspects as nonconsumptive use, future use, and existence values, will aid their preservation or hasten their destruction?

3. Do you think that conservation biologists, speaking as scientists, should advocate specific policies of managing ecological resources, or should they limit their professional activities strictly to making the results of their studies available to society?

SUGGESTED READING

Alverson, W. S., D. M. Waller, and W. Kuhlman. 1994. *Wild forests: Conservation biology and public policy*. Island Press, Washington, DC. An analysis of reforms needed in public forest land management to sustain forestry and preserve biodiversity.

Gowdy, J. and S. O'Hara. 1995. *Economic theory for environmentalists*. St. Lucie Press, Delray Beach, FL. A primer of how neoclassical economic theory addresses the environment, and a discussion of new directions that are being explored in ecological economics.

Swanson, T. M. (Ed.). 1995. *The economics and ecology of biodiversity decline: The forces driving global change*. Cambridge Univ. Press, New York. Thirteen articles by leading ecologists and economists, most nonmathematical, on the economic forces driving biodiversity decline, with concluding advice by each group to the other about the realities of the global system.

chapter

30

Sustainable Living in the Biosphere

Outline

A s human populations continue to grow, and as national economies continue to expand and develop complex patterns of international trade, the problems of conservation increasingly become international issues. Many countries, for example, now possess marine fishing fleets capable of exploiting fisheries anywhere in the world oceans. Intensified land use in dozens of countries from the tropics to the arctic affects the populations of migratory birds. Consumer demand for lumber, ivory, skins, and many other products in countries in one part of the world threatens species populations, and in some cases entire ecosystems, in faraway areas of the world. Pollution from urban-industrial regions creates acid deposition that spreads over broad geographic areas, without respect to international borders.

Recognition is growing, as well, that biotic diversity is a resource of the entire human race. The African savannas, with their rich diversity of wildlife, are the heritage of all humankind, not just of the countries in which they are located. Vast areas of healthy tropical forests are important to the future of the entire human race, not just the handful of countries in which the bulk of these forests is located. Sustained harvests of marine fisheries contribute to balancing the protein needs of human populations worldwide. These facts demand much more effective international coordination of conservation efforts.

In this chapter we shall describe some of the international organizations that are working to preserve biotic diversity and consider some of their major programs. Finally, we shall examine the issue of sustainability of human activity in the biosphere and the implications of achieving sustainability for environmental policy in the early years of the next millennium.

Biodiversity is a valuable global heritage that requires international cooperation to guarantee its sustainability.

TABLE 30.1

REGION	TOTAL AREA IN PRESERVES (PERCENT)	NUMBER OF BIOSPHERE RESERVES	NUMBER OF NATURAL WORLD HERITAGE SITES	NUMBER OF WETLANDS OF INTERNATIONAL IMPORTANCE
Distribution of Parks and Preserves of Major Importance to the Preservation of Natural Diversity, as of 1989				
North America	6.2	50	17	37
Central America	2.0	14	5	1
South America	4.6	23	8	4
Europe	6.6	81	11	277
USSR	0.9	19	0	12
Asia	2.3	34	8	34
Africa	3.4	40	20	31
Oceania	4.7	12	10	31
WORLD	**4.0**	**276**	**78**	**432**

Source: Data from World Conservation Monitoring Center, Cambridge, England, as appeared in *World Resources 1990–1991*, World Resources Institute.

BIODIVERSITY UNDER PROTECTION

About 4% of the world's land area is now in national parks and preserves in which wildlife species, and usually all natural ecosystems, are protected from exploitation (Table 30.1). In North America and Europe, more than 6% of land is so protected, while in Asia, Central America, and Russia much smaller fractions of land are preserved. In Central America, South America, and Africa, individual countries have made major commitments to preservation of natural ecosystems. Costa Rica, for example, has placed 12.0% of its land in parks, Chile 16.0%, and Tanzania 13.4%. The survival of parks in countries like these, however, is threatened by human population growth and the desperate efforts of impoverished peoples to meet their daily needs. In some countries, such as Uganda and Mozambique, prolonged civil wars have led to decimation of wildlife in many parks.

Still other areas are protected by private organizations or local governments, or are exploited under sustained-yield management. In the United States, for example, national forests are supposed to be managed for sustained yields of timber, wildlife, water, and recreation. The extent of natural ecosystems managed in sustainable fashion is unknown, but is probably greater in developed than in developing parts of the world.

Conservation ecologists agree that the areas of strictly protected ecosystems that now exist or are likely to be established in the near future are inadequate to preserve the biotic diversity on which a healthy global ecosystem depends. Thus, international conservation organizations are striving to find ways to exploit ecosystems, particularly those that retain much of their natural structure, in sustainable fashion.

These programs must also develop ways to match costs and benefits of conservation on an international scale. For example, the citizens and government of Costa Rica have supported the development of a comprehensive network of tropical parks and preserves—a system of international benefit and significance (Fig. 30.1). But should this nation be expected to bear the entire cost of maintaining this system, subsidizing the rest of the world in the process?

INTERNATIONAL CONSERVATION ORGANIZATIONS

Three agencies of the United Nations are active in global conservation. The **United Nations Educational, Scientific, and Cultural Organization (UNESCO)** and the **Food and Agricultural Organization (FAO)** were founded in 1945 as original branches of the United Nations, and the **United Nations Environment Program (UNEP)** was organized in 1972.

UNESCO, headquartered in Paris, has been active from its beginning in environmental affairs such as park development and scientific research. UNESCO is the coordinating organization for the **Man and the Biosphere Program (MAB),** initiated in 1971 (di Castri 1981). MAB has emphasized more applied studies of natural and exploited ecosystems, with the goal of improving their management for human welfare. MAB's activities have been supported from sources within individual countries. Its efforts have been centered on the Biosphere Reserve System, which is examined in detail in Reading 30.1. UNESCO also administers the World Heritage Program, which is described in detail in Reading 30.2.

The FAO, with head offices in Rome, Italy, is responsible for United Nations activities relating to food and fiber production, including not only crop agriculture but also the status and yields of world fisheries, forests, and rangelands. In addition to sponsoring technical studies of fisheries and their management, FAO monitors world fisheries harvests, publishing these statistics annually. The FAO also supports forestry research, and monitors the status of world forests and the rate of deforestation. In addition, the FAO is the front-line agency for United Nations activities relating to parks and protected areas. FAO activities in this last area have largely consisted of aid for planning preserves, training personnel, and developing national agencies to manage such areas.

UNEP, headquartered in Nairobi, Kenya, focused most of its early efforts on environmental factors bearing closely on human health, particularly nutrition, disease, and pollution by toxic chemicals. In recent years, as the recognition of ecosystem disruption and its relation to human welfare has heightened,

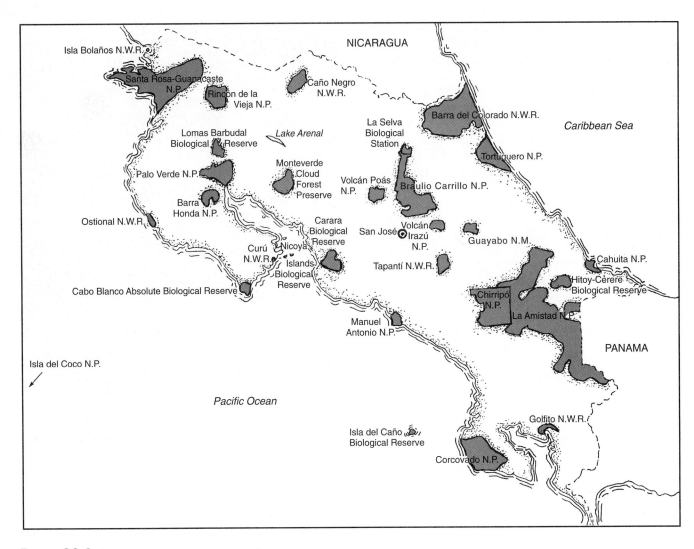

FIGURE **30.1**

Costa Rica has placed about 12% of its land in national parks and biodiversity preserves.

UNEP has become involved with more basic problems of environmental quality, such as desertification, acid deposition, marine pollution, and global climatic change. In this arena, one of UNEP's major programs is the **Global Environmental Monitoring System (GEMS).** GEMS is designed to feed information into a global resource data base, using geographical information systems technology (*See* Chapter 26). These data cover climate, oceanic conditions, pollution levels, human health, and natural resources, including biotic diversity. Using this data base, UNEP has issued reports on the state of the world environment, the most recent in 1987 (United Nations Environment Programme 1987). Desertification continues to be a major focus of activity, and UNEP supports a wide variety of projects to alleviate desertification in subSaharan Africa. UNEP also sponsors programs dealing with international aspects of biodiversity, including endangered species, migratory species, and marine mammals. A UNEP secretariat administers the **Convention on**

International Trade in Endangered Species (*See* Chapter 25). In 1992, UNEP initiated a project of **Global Biodiversity Assessment.** This project combines a survey of the state of knowledge of biodiversity with evaluation of its values and the adequacy of measures to conserve it (Heywood 1995).

Private organizations also play an important role in global conservation efforts. The **World Conservation Union,** formerly known as the **International Union for the Conservation of Nature and Natural Resources (IUCN),** is one of the most active. Founded in 1948 and headquartered in Switzerland, the IUCN coordinates conservation activities on behalf of its member organizations. The membership of the IUCN in 1989 included 61 countries, 121 governmental agencies of many countries, and 394 private organizations. Altogether, 120 countries participate in IUCN efforts through this membership. Participation in IUCN activities by developing countries in Africa, Asia, and Latin America is still relatively weak, however.

The annual IUCN budget of about $12.2 million supports a variety of activities. The IUCN plays a key role in the formation of global conservation policy, largely by coordinating the planning and operation of programs that are supported financially by United Nations agencies and by private groups such as the World Wide Fund for Nature. It furnishes technical aid to countries and other conservation organizations, emphasizing ways to integrate conservation with economic development. Regional offices are maintained in Latin America, Africa, and Asia to facilitate these efforts. The IUCN also maintains an **Environmental Law Center** that offers technical assistance in the development of laws and conventions and serves as a clearinghouse of information on national and international regulations relating to endangered species.

Under support from UNEP, the IUCN also operates the **World Conservation Monitoring Center (WCMC),** in Cambridge, England. The WCMC publishes the *U.N. List of National Parks and Other Protected Areas,* a compilation of natural preserves that meet basic standards of protection and financial support. The WCMC also oversees publication of the various Red Data Books on endangered species of animals and plants (*See* Chapter 26).

In 1975, UNESCO, FAO, UNEP, and the IUCN established a coordinating committee, the **Ecosystem Conservation Group,** to assist their joint conservation efforts. Recently, the **International Board for Plant Genetic Resources,** the **United Nations Development Program,** and the **World Bank** have become members of the committee. This coordinating committee thus has a very broad perspective on preservation of biotic diversity, both in nature and at off-site locations, and on the means to support this effort.

Several private organizations with individual membership are also active in international conservation. The **World Wildlife Fund (WWF),** better known internationally as the **World Wide Fund for Nature,** is the largest such organization. Historically, WWF has primarily been concerned with endangered plants and animals, and has supported research activities designed to promote their survival and recovery (Fig. 30.2). **The Nature Conservancy,** an outgrowth of the conservation arm of the Ecological Society of America, was organized in 1951. Its international program, established in 1980, has concentrated on the encouragement of **Conservation Data Centers (CDCs)** in various Latin American countries. These CDCs are designed to build inventories of natural diversity for individual countries, and serve as centers of action for the growth of a comprehensive network of preserves to protect this diversity. **Conservation International,** established in 1987, has similar objectives.

Conservation International pioneered the use of **debt-for-nature swaps** to support conservation efforts in developing countries. This technique involves the purchase of portions of the foreign debt owed by a nation to international banks, and the cancellation of this debt in exchange for the government's action in establishing preserves or supporting conserva-

FIGURE 30.2

The World Wide Fund for Nature (WWF) has supported studies of many endangered species. Here, a snow leopard is being fitted with a radio-collar to enable investigators to follow its movements.

(Photo by Rodney Jackson, courtesy of the WWF)

tion in other ways. In 1987, Conservation International arranged for the purchase of $650,000 of Bolivian foreign debt at a cost of $100,000. In exchange for the cancellation of this debt, the Bolivian government agreed to establish over 16,000 square kilometers of preserves, including the 1,336 square kilometer Beni Biosphere Reserve, an area of wet savannas on the Rio Beni, an Amazon tributary. As of 1990, Conservation International, TNC International, and WWF had arranged debt-for-nature swaps in five countries, retiring about $100 million in their foreign debt.

THE INTERNATIONAL GEOSPHERE-BIOSPHERE PROGRAM

The **International Geosphere-Biosphere Program (IGBP)** was initiated in 1986 by the International Council of Scientific Unions, the coordinating committee for national academies and international unions of science (World Resources Institute 1987). The headquarters for the program was established in Stockholm, Sweden. The IGBP is envisioned as a 10–20-year program to determine how conditions of the biosphere influence and are influenced by their interaction with the earth's physical and chemical environment. Foremost

THE BIOSPHERE RESERVE SYSTEM

T he Biosphere Reserve System is a global network of sites that combine preservation with research on sustainable management for human welfare. This system is being developed as one aspect of UNESCO's Man and the Biosphere Program (Batisse 1986). Most biosphere reserves are chosen to be ecologically representative of major types of regional ecosystems, and include large protected areas of these ecosystems. Many biosphere reserves, however, include ecosystems that have been modified or exploited by humans, such as rangelands, subsistence farmlands, or areas used for hunting or fishing. A few biosphere preserves include human settlements and environments that have been exploited extensively by traditional patterns of land use.

Schematically, the design of a typical biosphere reserve (Reading Fig. 30.1A) includes a core area where the natural ecosystem is strictly protected, and scientific activity involves only monitoring of conditions.

Outside this, in a buffer zone, nondestructive forms of research, education, and tourism are permitted. Finally, in an outer transition zone, regulated activities such as traditional land use by resident peoples, experimental research involving ecosystem manipulations, and major restoration efforts can be carried out (Batisse 1986). Sakaerat Biosphere Reserve in Thailand, for example, is designed to conserve and study the tropical monsoon forests of southeastern Asia. The goal of the research at

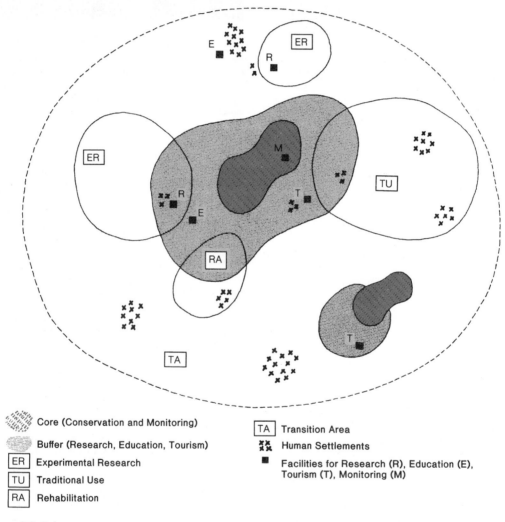

READING FIGURE 30.1A

Typical structure of a biosphere reserve, including core, buffer, and transition zones. The core, typically a protected area such as a national park, is designed to provide for conservation of biotic diversity, with research being limited to monitoring activities. The buffer zone is a region in which research, tourism, educational activities, and traditional subsistence activities of indigenous peoples are emphasized. The transition zone, its outer limit not rigidly defined, constitutes an area where major manipulative research, ecosystem restoration, and application of research to the management of exploited ecosystems are carried out.

(From Nature and Resources, Vol. XXII, 1986. Copyright © The UNESCO Press, Paris, France. Reprinted by permission.)

this location is sustainable management of these valuable forests. In Czechoslovakia, Trebon Basin Biosphere Reserve includes a 7,000 square kilometer area of forests, croplands, pastures, and ponds that have been managed continuously for food and fiber production since the 14th century. Australia's Hattah-Kulkyne Biosphere Reserve includes large areas of grassland and savanna that are being rehabilitated from overgrazing.

The biosphere reserve program has several objectives (Office of Technology Assessment 1988). These include the *conservation* of biological resources and the monitoring of ecological changes due to natural processes or human activities. These functions are fulfilled largely by the core areas of preserves, where natural systems are fully protected. The second major goal of the biosphere reserve network as a whole is *logistic*—to foster ecosystem research and monitoring efforts on a global scale. The objective of this research is to gain an understanding of how ecosystems function, and of how human activities may be changing patterns of function on a global scale. Research is thus a major activity in almost all reserves, and most reserves designate major parts of their buffer and transition zones for experiments, including major ecosystem manipulations. Research extends to the study of traditional forms of land use that have proven to be ecologically sustainable. The third objective is *development*, by using knowledge to develop sound, sustained ways to manage ecosystems for human benefit. The pursuit of research with direct application to human needs has, in fact, gained much greater emphasis in biosphere reserve activities in recent years.

The goal of the biosphere reserve program is to create one or more reserves in each of the world's important ecosystem types. For this, 14 major terrestrial biome types, such as temperate broadleaf forest, were first designated. The various world regions in which each of these biome types is represented were then identified. The result, for all 14 biomes, was a total of 193 ecosystem types.

Development of the biosphere reserve system has progressed continuously since 1976. More than 100 countries are now participants in the program (Von Droste 1989). By 1990, 285 reserves had been established in 72 countries (Table 30.1), 46 of these being in the United States. About two-thirds of the 193 world ecosystem types are now represented in this

READING FIGURE 30.1B

The San Dimas Experimental Forest in southern California is one of the U.S. biosphere reserves for Mediterranean-type ecosystems.

(U.S. Forest Service photo by David Larson)

READING FIGURE 30.1C

Fray Jorge National Park in Chile is an example of a biosphere reserve for Mediterranean-type ecosystems in a region outside North America.

(Photo by G. W. Cox)

system. For example, Biome Type No. 6, Evergreen Sclerophyll Forest, Woodland, and Shrubland, includes the various Mediterranean ecosystems of California, which is Region No. 1 of this biome. In California, the San Dimas Experimental Forest in the San Gabriel Mountains contains a range of communities ranging from chaparral at low elevations to oak woodlands and conifer forest at higher elevations (Reading Fig. 30.1B). A second reserve, the San Joaquin Experimental Range in Fresno County, contains grasslands and oak-pine woodlands typical of the San Joaquin Valley. A third reserve, Channel Islands National Park, contains varied ecosystems typical of the coastal region. Eight other regions of this biome are recognized in Chile, Australia, South Africa, and the region around the Mediterranean Sea. La Vallee du Fango on the French island of Corsica and Fray Jorge National Park in Chile (Reading Fig. 30.1C) are examples of biosphere reserves in two of the other regions having this biome type.

The concept of biosphere reserves is gradually being extended to include coastal and marine ecosystems, although no exclusively marine reserves have yet been established. The Channel Islands Biosphere Reserve, for example, includes the Channel Islands National Marine Sanctuary, an area much larger than the islands themselves (*See* Chapter 17). Marine biosphere reserves are likely to differ from terrestrial reserves, how-

ever. The spatial concepts of core, buffer, and transition areas are likely to be inappropriate for truly marine ecosystems. The need for strict protection of specific areas, other than reefs and islands, to guarantee the survival of marine life is probably secondary to the need for research into the dynamics of large units of the oceans defined by continental coastlines, submarine topography, and current systems (Kenchington & Agardy 1990). A tentative system of classification of oceanic realms and regions, comparable to that used for terrestrial biomes and regions, has been proposed (Ray & McCormick-Ray 1989). Several specific coastal and marine biosphere preserves have already been suggested. A Cape and Banks Biosphere Reserve, for example, has been proposed for the coastal and oceanic region extending from Cape Cod, Massachusetts northeast to the waters of the outer coast of Nova Scotia (Agardy & Broadus 1989). This preserve would include the Nantucket Shoals, George's Bank, and the Scotian Shelf, some of the most productive fisheries regions of the northwest Atlantic.

Growth of the network of terrestrial biosphere reserves has been rapid, and greater attention is now being devoted to integrating efforts among sites in the system. In 1990 a plan was approved to integrate 20 to 40 biosphere reserves into a system for long-term monitoring and research on global change. Standardized methods will be used at these sites, and the information integrated into one global data set.

among the objectives of the program is evaluating the role of human activities in modifying the global environment. The IGBP is the most ambitious and expensive international program yet undertaken, and will require funding from many national and international sources to achieve its goals. Major emphasis is being placed on the use of remote sensing and geographic information systems to gain global data on physical and biological conditions.

By 1994, six major core projects were under way, involving studies of atmospheric chemistry, the hydrologic cycle, land-ocean interactions, ocean biogeochemistry, past climatic and environmental change, and the effects of present global change on terrestrial ecosystems (IGBP 1994). Additional major studies dealing with land use/vegetational cover and the euphotic zone of the oceans are in the planning.

WORLD CONSERVATION STRATEGY

Developed by the IUCN, with support from the WWF, FAO, UNESCO, and UNEP, the **World Conservation Strategy (WCS)** was launched in 1980 (Allen 1980a). The WCS was a response to the gap that has existed between conservation and economic development (Halle 1985). The strategy had three basic objectives: 1) to maintain the ecological processes on which life depends, 2) to ensure the sustained utilization of ecosystems and their member species, and 3) to conserve genetic diversity (Talbot 1984). It recognized that economic development is necessary for successful conservation, and that failure to protect ecological resources limits the level of economic development that can be achieved. The WCS focused on the developing nations, where resources are being depleted rapidly, and where institutions to manage their exploitation are weak or absent.

The WCS, first of all, documented the need for protection of basic ecosystem processes, for the preservation of genetic diversity, and for the development of sustained use management of natural ecosystems. It also outlined ways to incorporate conservation planning as a positive component in national development plans, rather than, as often portrayed, a force opposing development. The rationale of the WCS has been outlined in detail for the general public (Allen 1980b). The IUCN has worked with various countries in using these materials to develop national conservation strategies. About 30 national strategies were completed or under development by the late 1980s.

The revised WCS, entitled "Caring for the Earth: A Strategy for Sustainable Living," focuses much more directly on ways to achieve sustainable resource development, and emphasizes the need to integrate economic and ecological efforts fully, both at the local and global levels. A major emphasis is the need to foster a global ethic of sustainable interaction with the biosphere.

THE HUMAN POPULATION, SUSTAINABILITY, AND BIODIVERSITY

The human population, now approaching 6 billion, may well have exceeded the capacity of the earth to support people at a healthy standard of living for the coming few centuries. The ultimate carrying capacity of the earth cannot be stated, however, as it depends on the way that human populations interact, by choice or default, with the earth's living and nonliving resources (Cohen 1995). The carrying capacity for a population with a high material standard of living, for example, would be less than for a population with a base subsistence lifestyle. Published estimates tend mostly to fall between 4 and 16 billion (Cohen 1995).

Conservation efforts must come to grips with both human population and economic growth, addressing the relation of their demands to the capacity of the biosphere to meet them. Sustainability and sustainable development have become the focal concepts in debate over this relationship. These terms are subject to various definitions, depending on points of view based in sociology, economics, and ecology (Holdren et al. 1995). In the ecological sense, **sustainability** implies a stable relationship between the human population and the capacity of the biosphere to provide resources and process wastes (Goodland 1995). **Sustainable development** is the process leading toward sustainability, in which economic development does not reduce the capacity of the biosphere to sustain humanity.

Are economic growth and the health of the biosphere compatible? In 1980, the World Conservation Strategy linked conservation and economic development. In 1987, the World Commission on Environment and Development asserted that economic growth was compatible with protection of the natural environment (Clark 1995). The **United Nations Conference on Environment and Development,** held in Brazil in 1992, outlined **Agenda 21,** an action plan that assumes that economic growth is environmentally sustainable. Some economists argue that as per-capita income increases, environmental degradation first increases, and then, at some threshold, begins to decrease as people become affluent enough to begin to protect the environment (Arrow et al. 1995).

Scientific organizations have also focused on sustainability, although not embracing the idea that it can occur in the face of unlimited economic growth. The Ecological Society of America (1991) developed the **Sustainable Biosphere Initiative (SBI),** which identifies priority areas of basic ecological research dealing with global environmental change, preservation of biotic diversity, and sustainable management of ecological systems. The SBI calls for the society to participate actively in education and in policy development at all national and international levels.

On a global scale, sustainability and sustainable development face many real and conceptual conflicts that are frequently ignored (Dovers & Handmer 1993). These include the following:

1. Technology. Although a great deal of human impact on the biosphere results from technology, the future application of technology to almost all ends seems virtually unquestioned.

2. Knowledge. Despite an enormously increased mass of information about the biosphere, our ability to predict the consequences of human actions seems to become increasingly uncertain.

3. Equity. Sustainability implies achieving intragenerational and intergenerational equity in resource use, yet the redistribution of wealth needed to achieve equity seems politically impractical.

4. Growth. Indefinitely continued economic growth is argued to be necessary to provide for basic human needs, but the finite global environment assures that some limit must exist.

5. Individuality. Freedom of choice by individuals is considered to be a basic right, yet environmental stress often results from the collective choices made by individuals.

6. Democracy. Self-determination by local human societies is considered to be fundamental, yet the decisions of such groups may be contrary to sustainability at the global level.

7. Institutional stability. The need for governing institutions that are strong and stable is assumed, yet such institutions are likely to resist the changes needed to achieve sustainability.

8. Optimization. Prevailing economic theory supports maximum use of resources, yet full use may exhaust the resilience of the biosphere and thus compromise sustainability.

Many of these conflicts reflect the tendency of economists and policymakers to treat ecological resources like economic resources (O'Neill et al. 1996). Prevailing economic theory treats economic resources as having unlimited substitutability; if petroleum runs out, coal can be processed to yield liquid fuels, for example. Unlike such resources, ecosystem functions are not substitutable, nor are they amenities that can be sacrificed for economic growth. Eventually, such sacrifice may reduce the resilience of ecosystems to the point that they are forced into an altered, less productive state from which recovery in any meaningful time is impossible (Holling 1995). Even benefit:cost analyses and contingent valuation studies that strive to include environmental health tend to be biased against maintaining essential ecosystem functions (Jaeger 1995).

Nevertheless, the biodiversity crisis cannot be solved without solving social and economic crises; only policies that attack this triad of crises have a chance of success (Holling 1995). Adaptive management may be the most appropriate approach: a diversity of institutions with flexible regulatory abilities, each acting, monitoring the results of its actions, and responding correctively to these results.

In any case, success in preserving the earth's natural diversity depends on the establishment of institutions that hold a long-term view of the values of natural diversity, and possess the strength to protect these values. These institutions must be flexible enough to adjust their basic policies, but stable enough to withstand the challenges of extreme political pressures at local, regional, and national levels. They must, for example, be able to adjust management practices in designated preserves, and also to protect such preserves in the face of efforts to delist or destructively exploit them. To do this, they must be able to bring strong social, economic, and political pressure to bear on the side of preservation.

Coordination among groups concerned with conservation of biodiversity is also essential. Global conservation efforts, pursued by various international agencies and private groups, still remain poorly coordinated. National park development still depends primarily on the initiative of individual countries, and is still financed by the countries involved. Under the Biosphere Reserve and World Heritage Programs, some parks have gained a degree of international recognition, but very little financial support comes from these arrangements. Some international organizations focus their efforts on species, others on ecosystems. Much can be accomplished by coordination of the activities of the varied organizations through sharing of data bases, identification of critical issues for joint action, and publicizing conservation issues in coordinated fashion.

To be successful, conservation efforts must make reality out of the contention that biotic diversity is a valuable resource (Western et al. 1989). Viable conservation strategies must be developed at national, regional, and local levels. These strategies must show ways of utilizing natural diversity to improve economic and social conditions, yet protect the natural diversity resource itself. Using and protecting this resource requires education, both as part of basic school curricula and through extension programs. The educational aspects of conservation have not yet received adequate attention.

The tools for global integration of this effort are rapidly becoming available: satellite systems for remote sensing of environmental change, computer networks for storing and sharing ecological data, communications systems for transmitting materials to people throughout the world, and a growing, international community of ecological scientists and environmental activists.

T he concept of the World Heritage Program evolved out of events surrounding the construction of the Aswan Dam in Egypt in the early 1960s. When the design of the dam and reservoir was finalized, it became apparent that the reservoir would flood the tombs of Abu Simbel, one of the most remarkable of the monuments to the royalty of ancient Egypt (Reading Fig. 30.2A). This threat stimulated an international effort to save the essential features of the site, particularly the giant statues, carved in cliff rock, that guarded the tomb entrances. These figures were disassembled and moved to a new location, above the reservoir level, where the statues were reassembled and the monument recreated.

This event led to the realization that many features, both cultural and natural, are of major significance to all of humanity, but that the countries in which they exist sometimes are unable or unwilling to protect them. Through the sponsorship of UNESCO, the **World Heritage Program** for the protection of such cultural and natural features was developed (Slatyer 1983). This plan, established by the **Convention Concerning the Protection of the World Cultural and Natural Heritage,** was adopted in 1972, but the program did not begin operating until 1976.

By 1990, the World Heritage Program had been ratified by 111 countries. Nations that are members of the program may nominate sites for the **World Heritage List,** which included 233 cultural sites, 73 natural sites, and 16 sites with major values that are both cultural and natural in 1990 (Table 30.1). To gain approval, sites must already be protected to a reasonable degree. Membership in the program includes a commitment to protect designated World Heritage Sites, both within that country and in the system at large. In the United States, natural sites include Yellowstone, Yosemite, Grand Canyon, Everglades, Redwood, and Olympic National Parks (Reading Fig. 30.2B), and cultural sites include the Mesa Verde Indian ruins in Colorado, Independence Hall in Philadelphia, and the Statue of Liberty in New York City. In France, on the other hand, almost all sites are of a cultural nature, such as the Palace of Versailles and Chartres Cathedral. Elsewhere in the world, natural sites include the Galápagos National

READING FIGURE 30.2 A

READING FIGURE 30.2 A

The relocation of the temples of Abu Simbel in Egypt led to the establishment of the World Heritage Program.

(Photo by Deborah Dexter)

Park (Ecuador), the Serengeti National Park and Ngorongoro Conservation Area (Tanzania), and the Great Barrier Reef (Australia). Kakadu National Park in Australia, with extensive areas of monsoon forest and wetlands (Reading Fig. 30.2C) and rocky escarpments containing rock formations with numerous Aboriginal pictographs (Reading Fig. 30.2D), exemplifies sites with both cultural and natural values.

In addition, the **World Heritage Fund** is maintained by contributions of member countries and private donors to assist the protection of sites, including emergency assistance for those that become threatened. In 1988, about $2.7 million were spent for these efforts. Threatened sites are placed on the **List of the World Heritage in Danger.** Several sites have been placed on this list, including Ngorongoro Conservation Area in Tanzania (Reading Fig. 30.2E), Garamba National Park in Zaire, and Djoudj National Park, Senegal. Ngorongoro and Garamba lie in the savanna zone of eastern and central Africa. Ngorongoro Conservation Area has suffered from a shortage of funds for management and research, as well as from poaching and increase

in the numbers of pastoral people and their herds. Garamba National Park holds the last wild population of the northern white rhinoceros, but lies in an area difficult to protect against poaching. Djoudj National Park encompasses seasonally flooded wetlands of the delta of the Senegal River, and supports an estimated three million migratory water birds. It is being threatened by construction of an upstream dam on the Senegal River. In 1995, Yellowstone National Park in the United States was added to the list, based on the threat posed by a large-scale mining project proposed just outside the park.

A continuing issue within the World Heritage Program has been the balance of cultural versus natural sites that are admitted (Hales 1984). The 3:1 ratio of cultural to natural sites reflects the dominance of nations with primarily cultural natural interests on the committee that evaluates proposed sites. Efforts have been made in recent years to improve the representation of natural areas by adopting new guidelines for structure and organization of the review committee, and for systematic identification of types of areas that are under-represented on the World Heritage List.

READING FIGURE 30.2 B

Grand Canyon National Park is one of the U.S. World Heritage sites.
(Photo by G. W. Cox)

READING FIGURE 30.2 D

Kakadu National Park was designated a World Heritage Site partly because of its rich wildlife and partly because of the cultural importance of the area to aborigines.
(Photo by G. W. Cox)

READING FIGURE 30.2 C

Kakadu National Park, in the Northern Territory of Australia, contains large areas of woodland, riparian forest, and seasonal wetlands representative of the ecosystems of northern Australia's region of monsoonal climate.
(Photo by G. W. Cox)

READING FIGURE 30.2 E

Ngorongoro Crater in Tanzania is a World Heritage Site that has been placed on the World Heritage in Danger List due to management problems relating to land use by pastoralist peoples and high levels of poaching of animals such as the black rhinoceros.
(Photo by Darla G. Cox)

SUMMARY

1. Only 4% of the world's land area is now in relatively secure preserves for biodiversity, and it is clear that strategies of sustained use of ecosystems outside such preserves will be necessary to maintain biodiversity for its values to humanity.

2. The Biosphere Reserve System, a part of UNESCO's Man and the Biosphere Program, is a global network of sites that combine preservation with research on sustainable management for human welfare.

3. Under the UNESCO World Heritage Program, sites of cultural and natural significance to all of humanity are placed on the World Heritage List and those that become threatened on the World Heritage in Danger List, which qualifies them for funds to assure their protection.

4. The World Conservation Strategy, developed by the World Conservation Union, is directed at protection of basic ecosystem processes, the preservation of genetic diversity, and the development of sustained use management of natural ecosystems.

5. Several agencies and scientific groups are promoting strategies at national, regional, and local levels to encourage the use of biodiversity to improve economic and social conditions, yet protect the health of the biosphere.

QUESTIONS FOR DISCUSSION

1. Considering the difficulties of protecting biodiversity on public lands, such as national forests in the United States, do you think that sustainable use of biodiversity can be achieved outside strict preserves in developing countries?

2. Do you think that organizations such as the United Nations, whose activities are constrained by national political pressures, or private organizations, which are dependent on funding from various agencies and private donors, will be the more effective agents for conservation of biodiversity over the next few decades?

3. Do you think that global sustainability in relations of humans to the rest of the biosphere can be obtained without the attainment of economic equality among the world's peoples?

SUGGESTED READING

Cohen, J. E. 1995. *How many people can the earth support?* W. W. Norton Co., New York, NY. A detailed examination of growth of the human population, efforts to predict its future growth, and the ecological, economic, and cultural relationships that will define the earth's carrying capacity.

Fautin, D. G., D. J. Futuyuma, and F. C. James (Eds.). 1995. Special section on sustainability issues. *Ann. Rev. Ecol. Syst.* **26**:1–248. A forum including articles on virtually all aspects of the issues of sustainability and sustainable development.

Munasinghe, M. and W. Shearer (Eds.). 1995. *Defining and measuring sustainability: The biogeophysical foundations.* The World Bank, Washington, DC. In 27 chapters by leading conservation scientists, the basic concept of sustainability is examined and applied to the management of terrestrial, freshwater, and marine ecosystems.

literature
cited

Abu-Jafar, M. Z. and C. Hays-Shadin. 1988. Reintroduction of the Arabian oryx into Jordan. Pp. 35–40 *in* A. Dixon and D. Jones (Eds.), *Conservation and biology of desert antelopes*. Christopher Helm, London.

Adamowicz, W. L., W. E. Phillips, and W. S. Pattison. 1986. The distribution of economic benefits from Alberta duck production. *Wildl. Soc. Bull.* **14:**396–98.

Adams, J. A., A. S. Endo, L. H. Stolzy, P. G. Rowlands, and H. B. Johnson. 1982. Controlled experiments on soil compaction produced by off-road vehicles in the Mojave Desert, California. *J. Appl. Ecol.* **19:**167–217.

Addessi, L. 1994. Human disturbance and long-term changes on a rocky intertidal community. *Ecol. Applications* **4:**786–97.

Addison, R. F. 1989. Organochlorines and marine mammal reproduction. *Can. J. Fish. Aq. Sci.* **46:**360–68.

Agardy, T. and J. M. Broadus. 1989. Coastal and marine biosphere reserve nominations in the Acadian Boreal Region: Results of a cooperative effort between the U.S. and Canada. Pp. 98–105 *in* W. P. Gregg, S. L. Krugman, and J. D. Wood (Eds.), *Proceedings of the symposium on biosphere reserves*. USDI National Park Service, Atlanta, GA.

Albertson, F. W., G. W. Tomanek, and A. Reigel. 1957. Ecology of drought cycles and grazing intensity of grasslands of central Great Plains. *Ecol. Monogr.* **27:**27–44.

Aldrich, J. W. and C. S. Robbins. 1970. Changing abundance of migratory birds in North America. *Smithsonian Contrib. Zool.* **26:**17–26.

Alexander, V. 1986. Arctic ocean pollution. *Oceanus* **29(1):**31–35.

Alexander, V. 1992. Arctic marine ecosystems. Pp. 221–32 *in* R. L. Peters and T. E. Lovejoy (Eds.), *Global warming and biological diversity*. Yale University Press, New Haven, CT.

Alho, C. J. R., T. E. Lacher, Jr., and H. C. Goncalves. 1988. Environmental degradation in the Pantanal ecosystem. *BioScience* **38:**164–71.

Allen, D. L. 1954. *Our wildlife legacy*. Funk and Wagnalls, New York.

Allen, D. L. 1979. *The wolves of Minong*. Houghton-Mifflin, Boston, MA.

Allen, J. D. and A. S. Flecker. 1993. Biodiversity conservation in running waters. *BioScience* **43:**32–43.

Allen, K. R. 1980. *Conservation and management of whales*. University of Washington Press, Seattle, WA. ix + 107 pp.

Allen, R. 1980a. The world conservation strategy: What it is and what it means for parks. *Parks* **5(2):**1–5.

Allen, R. 1980b. *How to save the world: Strategy for world conservation*. Kogan Page, London.

Allendorf, F. W. 1986. Genetic drift and the loss of alleles versus heterozygosity. *Zoo Biol.* **5:**181–90.

Allendorf, F. W. and R. F. Leary. 1986. Heterozygosity and fitness in natural populations of animals. Pp. 57–76 *in* M. E. Soulé (Ed.), *Conservation biology: The science of scarcity and diversity*. Sinauer Associates, Sunderland, MA.

Alvarez, L. W., W. Alvarez, F. Asaro, and H. V. Michel. 1980. Extraterrestrial cause for the Cretaceous-Tertiary extinction. *Science* **208:**1095–1108.

Alverson, D. L. 1978. Commercial fishing. Pp. 67–85 *in* H. P. Brokaw (Ed.), *Wildlife in America*. U.S. Government Printing Office, Washington, DC.

Alverson, D. L. 1990. Roe stripping in the Alaskan pollock fishery. *Fisheries* **15(3):**14–15.

Alverson, D. L., M. H. Freeberg, S. A. Murawski, and J. G. Pope. 1994. *A global assessment of fisheries bycatch and discards*. FAO Tech. Paper No. 339.

Alverson, W. S., D. M. Waller, and W. Kuhlman. 1994. *Wild forests: Conservation biology and public policy*. Island Press, Washington, DC.

Alverson, W. S., D. M. Waller, and S. L. Solheim. 1988. Forests to deer: Edge effects in northern Wisconsin. *Cons. Biol.* **2:**348–58.

Ambuel, B. and S. A. Temple. 1982. Area-dependent changes in the bird communities and vegetation of southern Wisconsin forests. *Ecology* **64:**1057–68.

Anderson, D. R. and K. P. Burnham. 1978. Effect of restrictive and liberal hunting regulations on annual survival rates of the mallard in North America. *Trans. N. Am. Wildl. Nat. Res. Conf.* **43:**181–86.

Anderson, D. W. and F. Gress. 1983. Status of a northern population of California brown pelicans. *Condor* **85:**79–88.

Anderson, D. W. and J. O. Keith. 1980. The human influence on seabird nesting success: Conservation implications. *Cons. Biol.* **18:**65–80.

Anderson, J. E. and M. L. Shumar. 1986. Impacts of black-tailed jackrabbits at peak population densities on sagebrush-steppe vegetation. *J. Range Manag.* **39:**152–56.

Anonymous. 1983. *Marine recreational fishing—big business*. SFI Bulletin No. 348.

Anthony, R. G., M. G. Garrett, and C. A. Schuler. 1993. Environmental contaminants in bald eagles in the Columbia River estuary. *J. Wildl. Manag.* **57:**10–19.

APIOS (Acidic Precipitation in Ontario Study). 1991. *Annual program report 1989/1990*. Ontario Ministry of the Environment, Toronto.

Archibald, J. D. and L. J. Bryant. 1990. Differential Cretaceous/Tertiary extinctions of nonmarine vertebrates; evidence from northeastern Montana. *Geol. Soc. Amer. Special Paper* **247:**549–62.

Arrigo, K. R. 1994. Impact of ozone depletion on phytoplankton growth in the Southern Ocean: Large-scale spatial and temporal variability. *Mar. Ecol. Prog. Ser.* **114:**1–12.

Arrow, K., B. Bolin, R. Costanza, P. Dasgupta, C. Folke, C. S. Holling, B-O. Jansson, S. Levin, K-G. Maler, C. Perrings, and D. Pimentel. 1995. Economic growth, carrying capacity, and the environment. *Science* **268**:520–21.

Art, H. W. 1976. *Ecological studies of the Sunken Forest, Fire Island National Seashore, New York.* U.S. National Park Service, Washington, DC.

Atkins, N. and B. Heneman. 1987. The dangers of gill netting to seabirds. *Amer. Birds* **41**:1395–1403.

Atkinson, I. A. E. 1988. Opportunities for ecological restoration. *New Zealand J. Ecol.* **11**:1–12.

Avise, J. C. and J. L. Hamrick (Eds.). 1995. *Conservation genetics: Case histories from nature.* Chapman and Hall, New York.

Babbitt, B. 1995. Science: Opening the next chapter of conservation history. *Science* **267**:1954–55.

Baden, J. A. and T. O'Brien. 1994. Economics and ecosystems: Coevolution in the Northwest. *Illahee* **10**:192–204.

Bailey, R. G. 1978. *Ecoregions of the United States.* U.S. Forest Service, Ogden, Utah.

Bailey, R. G. 1995. *Ecosystem geography.* Springer-Verlag, New York.

Bainbridge, D. A. and R. A. Virginia. 1990. Restoration in the Sonoran Desert of California. *Restoration & Management Notes* **8**:3–14.

Baker, H. G. 1974. The evolution of weeds. *Ann. Rev. Ecol. Syst.* **5**:1–24.

Baker, J. T. and L. H. Allen Jr. 1994. Assessment of the impact of rising carbon dioxide and other potential changes on vegetation. *Env. Pollut.* **83**:223–35.

Balcomb, R., C. A. Bowen, D. Wright, and M. Law. 1984. Effects on wildlife of at-planting corn applications of granular carbofuran. *J. Wildl. Manag.* **48**:1353–59.

Baldwin, M. F., M. Leslie, and E. H. Clark II. 1990. Government programs inducing wetlands alterations. Pp. 191–210 *in* G. Bingham, E. H. Clark II, L. V. Haygood, and M. Leslie (Eds.), *Issues in wetland protection: Background papers prepared for the National Wetlands Policy Forum.* The Conservation Foundation, Washington, DC.

Ballard, W. B., J. S. Whitman, and C. L. Gardner. 1987. Ecology of an exploited wolf population in south-central Alaska. *Wildl. Monogr.* **98**:1–54.

Ballou, J. D. 1991. Management of genetic variation in captive populations. Pp. 602–10 *in* E. C. Dudley (Ed.), *The unity of evolutionary biology. Vol. II.* Dioscorides Press, Portland, OR.

Bancroft, G. T. 1989. Status and conservation of wading birds in the Everglades. *Amer. Birds* **43**:1258–65.

Barbosa, A. C., A. A. Boischio, G. A. East, I. Ferrari, A. Gonçalves, P. R. M. Silva, and T. M. E. da Cruz. 1995. Mercury contamination in the Brazilian Amazon. Environmental and occupational aspects. *Water, Air, Soil Pollut.* **80**:109–21.

Barrett, S. W. 1980. Conservation in Amazonia. *Biol. Cons.* **18**:209–35.

Barthem, R. B., M. C. L. de Brito Ribeiro, and M. Petrere, Jr. 1991. Life strategies of some long-distance migratory catfish in relation to hydroelectric dams in the Amazon Basin. *Biol. Cons.* **55**:339–45.

Bartmann, R. M., G. C. White, and L. H. Carpenter. 1992. Compensatory mortality in a Colorado mule deer population. *Wildl. Monogr.* **121**:1–39.

Barton, K. 1987. Federal wetlands protection programs. Pp. 179–98 *in* R. L. Di Silvestro (Ed.), *Audubon wildlife report 1987.* Academic Press, Orlando, FL.

Bartonek, J. C. and D. N. Nettleship (Eds.). 1979. *Conservation of marine birds of North America.* U.S. Department of Interior, Fish and Wildlife Service Wildlife Research Report 11. ix + 319 pp.

Batisse, M. 1986. Developing and focusing the biosphere reserve concept. *Nature and Resources* **22**:1–10.

Bawa, K. S. and A. Markham. 1995. Climate change and tropical forests. *Trends Ecol. Evol.* **10**:348–49.

Bayley, P. B. 1995. Understanding large river-floodplain ecosystems. *BioScience* **45**:153–58.

Bean, M. J. 1987. The federal endangered species program. Pp. 147–60 *in* R. L. Di Silvestro, W. J. Chandler, K. Barton, and L. Labate (Eds.), *Audubon wildlife report 1987.* Academic Press, Orlando, FL.

Beatley, T. 1994. *Habitat conservation planning: Endangered species and urban growth.* University of Texas Press, Austin, TX.

Beddington, J. R. and R. M. May. 1980. Maximum sustainable yields in systems subject to harvesting at more than one trophic level. *Mathematical Biosciences* **51**:261–81.

Beddington, J. R. and R. M. May. 1982. The harvesting of interacting species in a natural ecosystem. *Sci. Amer.* **247(5)**:62–69.

Bednarz, J. C., D. Klem, Jr., L. J. Goodrich, and S. E. Senner. 1990. Migration counts of raptors at Hawk Mountain, Pennsylvania, as indicators of population trends, 1934–1986. *Auk* **107**:96–109.

Beier, P. 1993. Determining minimum habitat areas and habitat corridors for cougars. *Cons. Biol.* **7**:94–108.

Bell, P. R. F. and I. Elmetri. 1995. Ecological indicators of large-scale eutrophication in the Great Barrier Reef lagoon. *Ambio* **24**:208–15.

Belnap, J. 1993. Recovery rates of cryptobiotic crusts: Inoculant use and assessment methods. *Great Basin Nat.* **53**:89–95.

Belovsky, G. E. 1987. Extinction models and mammalian persistence. Pp. 34–57 *in* M. E. Soulé (Ed.), *Viable populations for conservation.* Cambridge University Press, Cambridge, England.

Berger, W. H., V. S. Smetacek, and G. Wefer. 1989. Ocean productivity and paleoproductivity—an overview. Pp. 1–34 *in* W. H. Berger, V. S. Smetacek, and G. Wefer (Eds.), *Productivity of the ocean: Present and past.* John Wiley and Sons, New York.

Bergerud, A. T. 1987. Increasing the numbers of grouse. Pp. 686–731 *in* A. T. Bergerud and M. W. Gratson (Eds.), *Adaptive strategies and population ecology of northern grouse.* University of Minnesota Press, Minneapolis, MN.

Bergerud, A. T. 1988. Caribou, wolves, and man. *Trends Ecol. Evol.* **3**:68–72.

Berner, E. K. and R. A. Berner. 1987. *The global water cycle.* Prentice-Hall, Englewood Cliffs, NJ.

Bertram, B. C. R. 1988. Reintroducing scimitar-horned oryx into Tunisia. Pp. 136–45 *in* A. Dixon and D. Jones (Eds.), *Conservation and biology of desert antelopes.* Christopher Helm, London.

Bertram, D. F. 1995. The roles of introduced rats and commercial fishing in the decline of ancient murrelets on Langara Island, British Columbia. *Cons. Biol.* **9**:865–72.

Bertram, R. C. 1984. *North Kings deer herd study.* California Dept. of Fish and Game, Sacramento, CA.

Bethke, R. W. and T. D. Nudds. 1995. Effects of climate change and land use on duck abundance in Canadian prairie-parklands. *Ecol. Applications* **5**:588–600.

Bildstein, K. L., G. T. Bancroft, P. L. Dugan, D. H. Gordon, R. M. Erwin, E. Nol, L. X. Payne, and S. E. Senner. 1991. Approaches to the conservation of coastal wetlands in the Western Hemisphere. *Wilson Bull.* **103**:218–54.

Billings, W. D. 1973. Arctic and alpine vegetations: Similarities, differences, and susceptibility to disturbance. *BioScience* **23**:697–704.

Billings, W. D. 1990. *Bromus tectorum,* a biotic cause of ecosystem impoverishment in the Great Basin. Pp. 301–22 *in* G. M. Woodwell (Ed.), *The earth in transition.* Cambridge University Press, Cambridge, England.

Billings, W. D. and K. M. Peterson 1992. Some possible effects of climatic warming on arctic tundra ecosystems of the Alaskan north slope. Pp. 233–43 *in* Peters and Lovejoy (1992).

Billington, H. L. 1991. Effects of population size on genetic variation in a dioecious conifer. *Cons. Biol.* **5**:115–19.

Birkeland, C. 1982. Terrestrial runoff as a cause of outbreaks of *Acanthaster planci* (Echinodermata: Asteroidea). *Mar. Biol.* **69**:59–64.

Bishop, R. C. 1980. Endangered species: An economic perspective. *Trans. N. A. Wildl. Nat. Res. Conf.* **45**:208–18.

Bishop, R. C. 1987. Economic values defined. Pp. 24–33 *in* D. J. Decker and G. R. Goff (Eds.), *Valuing wildlife: Economic and social perspectives.* Westview Press, Boulder, CO.

Bishop, S. C. and F. S. Chapin III. 1989*a.* Patterns of natural revegetation on abandoned gravel pads in arctic Alaska. *J. Ecol.* **26**:1073–81.

Bishop, S. C. and F. S. Chapin III. 1989*b.* Establishment of *Salix alaxensis* on a gravel pad in arctic Alaska. *J. Ecol.* **26**:575–83.

Biswell, H. H. 1989. *Prescribed burning in California wildlands vegetation management.* University of California Press, Berkeley, CA.

Bjorge, R. R. 1987. Bird kill at an oil industry flare stack in northwest Alberta. *Can. Field Nat.* **101**:346–50.

Bjorndal, K. A. (Ed.). 1995. *Biology and conservation of sea turtles.* Revised edition. Smithsonian Institution Press, Washington, DC.

Blake, J. G. 1983. Trophic structure of bird communities in forest patches in east-central Illinois. *Wilson Bull.* **95**:416–30.

Blake, J. G. 1986. Species-area relationship of migrants in isolated woodlots. *Wilson Bull.* **98**:291–96.

Blake, J. G. and J. R. Karr. 1984. Species composition of bird communities and the conservation benefit of large versus small forests. *Biol. Cons.* **30**:173–87.

Blaustein, A. R., P. D. Hoffman, D. G. Hokit, J. M. Kiesecker, S. C. Walls, and J. B. Hays. 1994. UV repair and resistance to solar UV-B in amphibian eggs: A link to population declines? *Proc. Nat. Acad. Sci. USA* **91**:1791–95.

Bloom, D. E. 1995. International public opinion on the environment. *Science* **269**:354–58.

Blus, L. J. 1982. Further interpretation of the relation of organochlorine residues in brown pelican eggs to reproductive success. *Env. Pollut. (Series A):* **28**:15–33.

Blus, L. J., C. S. Staley, C. J. Henny, G. W. Pendleton, T. H. Craig, E. H. Craig, and D. K. Halford. 1989. Effects of organophosphorus insecticides on sage grouse in southeastern Idaho. *J. Wildl. Manag.* **53**:1139–46.

Bock, J. H. and C. E. Bock. 1995. The challenges of grassland conservation. Pp. 199–222 *in* A. Joern and K. H. Keeler (Eds.), *The changing prairie.* Oxford University Press, New York.

Bock, C. E. and L. W. Lepthien. 1976. Population growth of the cattle egret. *Auk* **93**:164–66.

Boecklen, W. J. and N. J. Gotelli. 1984. Island biogeographic theory and conservation practice: Species-area or specious-area relationships? *Biol. Cons.* **29**:63–80.

Boecklen, W. J. and D. Simberloff. 1986. Area-based extinction models in conservation. Pp. 247–76 *in* D. K. Elliott (Ed.), *Dynamics of extinction.* John Wiley and Sons, New York.

Bohning-Gaese, K., M. L. Taper, and J. H. Brown. 1994. Avian community dynamics are discordant in space and time. *Oikos* **70**:121–26.

Bolze, D. A. and M. B. Lee. 1989. Offshore oil and gas development: Implications for wildlife in Alaska. *Marine Policy* **13**:231–48.

Bond, W. and P. Slingsby. 1984. Collapse of an ant-plant mutualism: The Argentine ant (*Iridomyrmex humilis*) and myrmecochorous Proteaceae. *Ecology* **65**:1031–37.

Bonnell, M. L. and R. K. Selander. 1974. Elephant seals: Genetic variation and near extinctions. *Science* **184**:908–9.

Borgese, E. M. 1983. The law of the sea. *Sci. Amer.* **246(3)**:42–49.

Bormann, F. H. and G. E. Likens. 1967. Nutrient cycling. *Science* **155**:424–29.

Botkin, D. B. 1990. *Discordant harmonies: A new ecology for the twenty-first century.* Oxford University Press, New York.

Botkin, D. B. and R. A. Nisbet. 1992. Projecting the effects of climatic change on biological diversity in forests. Pp. 277–93 *in* Peters and Lovejoy (1992).

Botton, M. L., R. E. Loveland, and T. R. Jacobsen. 1994. Site selection by migratory shorebirds in Delaware Bay, and its relationship to beach characteristics and abundance of horseshoe crab (*Limulus polyphemus*) eggs. *Auk* **111**:605–16.

Boughey, A. S. 1963. Interaction between animals, vegetation, and fire in Southern Rhodesia. *Ohio J. Sci.* **63**:193–209.

Boulding, K. 1966. The economics of coming spaceship earth. Pp. 3–14 *in* H. Jarrett (Ed.), *Environmental quality in a growing economy.* Johns Hopkins University Press, Baltimore, MD.

Bourliere, F. and M. Hadley. 1983. Present-day savannas: An overview. Pp. 1–17 *in* F. Bourliere (Ed.), *Tropical savannas.* Elsevier, Amsterdam.

Boyce, M. S. 1990. Wolf recovery for Yellowstone National Park: A simulation model. Pp. 3:3–58 *in* *Wolves for Yellowstone? A report to the United States Congress.* Volume II. U.S. Government Printing Office, Billings, MT.

Boyce, M. S. 1992. Population viability analysis. *Ann. Rev. Ecol. Syst.* **23**:481–506.

Boyle, K. J. and R. C. Bishop. 1987. Valuing wildlife in benefit-cost analyses: A case study involving endangered species. *Water Res. Res.* **23**:943–50.

Bradley, R. S., H. F. Diaz, J. K. Eischeid, P. D. Jones, P. M. Kelly, and C. M. Goodess. 1987. Precipitation fluctuations over northern hemisphere land areas since the mid-19th century. *Science* **237**:171–75.

Braham, H. W. 1989. Eskimos, Yankees, and bowheads. *Ambio* **32**:54–62.

Braithwaite, R. W., W. M. Lonsdale, and J. A. Estbergs. 1989. Alien vegetation and native biota in tropical Australia: The impact of *Mimosa pigra*. *Biol. Cons.* **48**:189–210.

Brattstrom, B. H. and M. C. Bondello. 1983. Effects of off-road vehicle noise on desert vertebrates. Pp. 167–206 in R. H. Webb and H. G. Wilshire (Eds.), *Environmental effects of off-road vehicles*. Springer-Verlag, New York.

Brett, J. J. 1991. *The mountain and the migration. Revised edition.* Cornell University Press, Ithaca, NY.

Bricklemyer, E. C., Jr., S. Iudicello, and H. J. Hartmann. 1989. Discarded catch in U.S. commercial marine fisheries. Pp. 258–95 in *Audubon wildlife report 1989–90.* Academic Press, San Diego, CA.

Bridgeham, S. D., C. A. Johnson, J. Pastor, and K. Updegraff. 1995. Potential feedbacks of northern wetlands on climate change. *BioScience* **45:**262–74.

Brockie, R. E., L. L. Loope, M. B. Usher, and O. Hamann. 1988. Biological invasions of island nature reserves. *Biol. Cons.* **44:**9–36.

Brooke, J. 1993. Brazilian rain forest yields most diversity for species of trees. *New York Times,* 30 March.

Brooks, M. 1994. Chairman's report: African rhino specialist group. *Pachyderm* **18:**16–18.

Brooks, P. M. 1989. Proposed conservation plan for the black rhinoceros *Diceros bicornis* in South Africa, the TBVC states, and Namibia. *Koedoe* **32(2):**1–30.

Brookshire, D. S., B. C. Ives, and W. D. Schulze. 1976. The valuation of aesthetic preferences. *J. Env. Econ. Manag.* **3:**325–46.

Brower, L. P. and S. B. Malcolm. 1991. Animal migrations: Endangered phenomena. *Amer. Zool.* **31:**265–76.

Brown, J. H. 1989. Patterns, modes, and extents of invasions by vertebrates. Pp. 85–109 in J. A. Drake, H. A. Mooney, F. di Castri, R. H. Groves, F. J. Kruger, M. Rejmanek, and M. Williamson (Eds.), *Biological invasions: A global perspective.* John Wiley and Sons, Chichester, England.

Brown, K. S. and G. G. Brown. 1992. Habitat alteration and species loss in Brazilian forests. Pp. 119–42 in T. C. Whitmore and J. A. Sayer (Eds.), *Tropical deforestation and species extinction.* Chapman and Hall, New York.

Brown, L. R. 1985. Maintaining world fisheries. Pp. 73–96 in L. Starke (Ed.), *State of the world: 1985.* W. W. Norton Co., New York.

Brown, R. W., R. S. Johnston, and D. A. Johnson. 1978. Rehabilitation of alpine tundra disturbances. *J. Soil Water Cons.* **33:**154–60.

Brown, S. and A. E. Lugo. 1990. Tropical secondary forests. *J. Trop. Ecol.* **6:**1–32.

Brownell, R. L., Jr., K. Ralls, and W. F. Perrin. 1989. The plight of the 'forgotten' whales. *Ambio* **32:**5–11.

Browning, R. J. 1980. *Fisheries of the North Pacific.* Alaska Northwest Pub. Co., Anchorage, AK.

Brussard, P. F. 1991. The role of ecology in biological conservation. *Ecol. Applications* **1:**6–12.

Brussard, P. F. 1995. The president's column: Thoughts from an outgoing president. *Soc. Cons. Biol. Newsletter* **2(2):**1.

Bucher, E. H. 1992. The causes of extinction of the passenger pigeon. *Current Ornithology* **9:**1–36.

Bunnell, F. L. and L. A. Dupuis. 1995. Conservation Biology's literature revisited: Wine or vinaigrette? *Wild. So. Bull.* **23:**56–62.

Burger, J. (Ed.). 1994. *Before and after an oil spill: The Arthur Kill.* Rutgers University Press, New Brunswick, NJ.

Burgis, M. J. and P. Morris. 1987. *The natural history of lakes.* Cambridge University Press, Cambridge, England.

Burkey, T. V. 1995. Faunal collapse in East African game reserves revisited. *Biol. Cons.* **71:**107–10.

Bury, R. B. and P. S. Corn. 1995. Have desert tortoises undergone a long-term decline in abundance? *Wild. Soc. Bull.* **23:**41–47.

Bury, R. B., T. C. Esque, L. A. DeFalco, and P. A. Medica. 1994. Distribution, habitat use, and protection of the desert tortoise in the eastern Mojave Desert. *USDI National Biol. Survey, Fish and Wild.Res. Rep.* **13:**57–72.

Bury, R. B., R. A. Luckenbach, and S. D. Busack. 1977. *Effects of off-road vehicles on vertebrates in the California desert.* U.S. Dept. Int., Fish and Wildl. Serv., Wildl. Res. Rep. No. 8.

Busch, D. E. 1995. Effects of fire on southwestern riparian plant community structure. *Southwestern Nat.* **40:**259–67.

Busch, D. E. and S. D. Smith. 1993. Effects of fire on water and salinity relations of riparian woody taxa. *Oecologia* **94:**186–94.

Busch, D. E. and S. D. Smith. 1995. Mechanisms associated with decline of woody species in riparian ecosystems of the southwestern U.S. *Ecol. Monogr.* **65:**347–70.

Butler, V. 1995. Is this the way to save Africa's wildlife? *Int. Wildlife* **25(2):**38–43.

Cahen, H. 1988. Against the moral considerability of ecosystems. *Env. Ethics* **10:**195–216.

Caicco, S. L., J. M. Scott, B. Butterfield, and B. Csuti. 1995. A gap analysis of the management status of the vegetation of Idaho (U.S.A.). *Cons. Biol.* **9:**498–511.

Cairns, J., Jr. 1991. The status of the theoretical and applied science of restoration ecology. *The Env. Professional* **13:**186–94.

Cairns, J., Jr. and J. R. Pratt. 1995. The relationship between ecosystem health and delivery of ecosystem services. Pp. 63–76 in D. J. Rapport, C. L. Gaudet, and P. Calow (Eds.), *Evaluating and monitoring the health of large-scale ecosystems.* Springer-Verlag, Berlin, Germany.

Caldwell, M. M., A. H. Teramura, M. Tevini, J. F. Bornman, L. O. Bjorn, and G. Kulandaivelu. 1995. Effects of increased solar ultraviolet radiation on terrestrial plants. *Ambio* **24:**166–73.

Callicott, J. B. 1982. Traditional American Indian and Western European attitudes toward nature: An overview. *Env. Ethics* **4:**293–318.

Campbell, F. T. 1988. The desert tortoise. Pp. 567–81 in W. J. Chandler (Ed.), *Audubon wildlife report 1988/1989.* Academic Press, San Diego, CA.

Carbyn, L. N., S. H. Fritts, and D. R. Seip (Eds.). 1996. *Ecology and conservation of wolves in a changing world.* Canadian Circumpolar Press Institute, University of Alberta, Edmonton, Alberta, Canada.

Carlquist, S. 1965. *Island life.* Natural History Press, New York.

Carlquist, S. 1980. *Hawaii: A natural history.* Pacific Tropical Botanic Garden, Honolulu, HI.

Carlton, J. T. and J. B. Geller. 1993. Ecological roulette: The global transport of nonindigenous marine organisms. *Science* **261:**78–82.

Carothers, S. W., M. E. Stitt, and R. R. Johnson. 1976. Feral asses on public lands: An analysis of biotic impact, legal considerations, and management alternatives. *Trans. N.A. Wildl. Conf.* **41:**396–40

Carothers, S. W. and R. Dolan. 1982. Dam changes on the Colorado River. *Nat. Hist.* **91(1)**:75–83.

Carothers, S. W. and R. R. Johnson. 1983. Status of the Colorado River Ecosystem in Grand Canyon National Park and Glen Canyon National Recreation Area. Pp. 139–60 in V. D. Adams and V. A. Lamarra (Eds.), *Aquatic resources management of the Colorado River ecosystem.* Ann Arbor Science Publishers, Ann Arbor, MI.

Carpenter, S. R., S. G. Fisher, N. B. Grimm, and J. F. Kitchell. 1992. Global change and freshwater ecosystems. *Ann. Rev. Ecol. Syst.* **23**:119–39.

Carr, A. 1967. *So excellent a fishe.* Natural History Press, Garden City, NY.

Carr, A. 1987. New perspectives on the pelagic stage of sea turtle development. *Cons. Biol.* **1**:103–21.

Carson, R. 1951. *The sea around us.* Oxford University Press, New York.

Carson, R. 1962. *Silent spring.* Houghton-Mifflin, New York.

Carter, H. R. 1986. Rise & fall of the Farallon common murre. *Point Reyes Bird Observatory Newsletter* **72**:1–3, 11.

Case, T. J. 1975. Species numbers, density compensation, and colonizing ability of lizards in the Gulf of California. *Ecology* **56**:3–18.

Case, T. J., D. T. Bolger, and A. D. Richman. 1992. Reptilian extinctions: The last 10,000 years. Pp. 91–125 in P. L. Fiedler and S. K. Jain (Eds.), *Conservation biology.* Chapman and Hall, New York.

Cauble, C. 1977. The great grizzly grapple. *Nat. Hist.* **86(7)**:74–81.

Caughley, G. 1976. Wildlife management and the dynamics of ungulate populations. Pp. 183–245 in T. H. Coker (Ed.), *Applied Biology,* Vol. I. Academic Press, New York.

Caughley, G. H. 1984. Harvesting of wildlife: Past, present, and future. Pp. 3–14 in S. L. Beasom and S. F. Roberson (Eds.), *Game harvest management.* Texas A & I University, Kingsville, TX.

Cederwall, H. and R. Elmgraen. 1990. Biological effects of eutrophication in the Baltic Sea, particularly the coastal zone. *Ambio* **19**:109–12.

Champ, M. A. and F. L. Lowenstein. 1987. TBT: The dilemma of high-technology antifouling paints. *Oceanus* **30(3)**:69–77.

Chapdelaine, G., P. Laporte, and D. N. Nettleship. 1987. Population, productivity, and DDT contamination trends of Northern Gannets (*Sula bassanus*) at Bonaventure Island, Quebec, 1967–84. *Can. J. Zool.* **65**:2922–26.

Chen, J., J. F. Franklin, and T. A Spies. 1992. Vegetation responses to edge environments in old-growth Douglas-fir forests. *Ecol. Appl.* **2**:387–96.

Chevone, B. I. and S. N. Linzon. 1988. Tree decline in North America. *Env. Pollut.* **50**:87–99.

Cisneros-Mata, M. A., G. Montemayor-López, and M. J. Román-Rodríguez. 1995. Life history and conservation of *Totoaba macdonaldi. Cons. Biol.* **9**:806–14.

Clark, J. G. 1995. Economic development vs. sustainable societies: Reflections on the players in a crucial contest. *Ann. Rev. Ecol. Syst.* **26**:225–48.

Clark, K. E., L. J. Niles, and J. Burger. 1993. Abundance and distribution of migrant shorebirds in Delaware Bay. *Condor* **95**:694–705.

Coblentz, B. E. 1978. The effects of feral goats (*Capra hircus*) on island ecosystems. *Biol. Cons.* **13**:279–86.

Cohen, J. E. 1995. *How many people can the earth support?* W. W. Norton Co., New York.

Cohn, J. P. 1988. Halting the rhino's demise. *BioScience* **38**:740–44.

Cohn, J. P. 1990. Elephants: Remarkable and endangered. *BioScience* **40**:10–14.

Cole, M. M. 1986. *The savannas.* Academic Press, San Diego, CA.

Collier, C. R., R. J. Pickering, and J. J. Musser (Eds.) 1970. Influences of strip mining on the hydrologic environment of parts of Beaver Creek Basin, Kentucky, 1955–66. *Geol. Surv. Prof. Paper* 427-C, viii + 80 pp.

Collins, M. 1990. *The last rain forests.* Oxford University Press, New York.

Collins, S. L. and L. L. Wallace (Eds.). 1990. *Fire in North American tallgrass prairies.* University of Oklahoma Press, Norman, OK.

Commission on Life Sciences. 1992. *Dolphins and the tuna industry.* National Academy Press, Washington, DC.

Conant, S. 1988. Saving endangered species by translocation. *BioScience* **38**:254–57.

Connell, J. H. and R. O. Slatyer. 1977. Mechanisms of succession in natural communities and their role in community stability and organization. *Amer. Nat.* **111**:1119–44.

Cooch, F. G. 1986. The current status of goose populations in Canada. *Trans. N. A. Wildl. Nat. Res. Conf.* **51**:480–86.

Cook, R. S. (Ed.). 1993. *Ecological issues on reintroducing wolves into Yellowstone National Park.* USDI National Park Service Scientific Monograph NPS/NRYELL/NRSM-93/22.

Cooper, C. F. 1990. Recreation and wildlife. Pp. 329–39 in P. E. Waggoner (Ed.), *Climate change and U.S. water resources.* John Wiley and Sons, New York.

Cooper, C. F. and B. S. Stewart. 1983. Demography of northern elephant seals, 1911–82. *Science* **219**:969–71.

Cooper, S. R. 1995. Chesapeake Bay watershed historical land use: Impact on water quality and diatom communities. *Ecol. Applications* **5**:703–23.

Cooperrider, A. 1985 The desert bighorn. Pp. 473–85 in R. L. Di Silvestro (Ed.), *Audubon wildlife report 1985.* National Audubon Society, New York.

Copeland, B. J. 1966. Effects of decreased river flow on estuarine ecology. *J. Water Pollut. Control Fed.* **38**:1831–39.

Corke, D. 1992. The status and conservation needs of the terrestrial herpetofauna of the Windward Islands (West Indies). *Biol. Cons.* **62**:47–58.

Corn, P. S. 1994. Recent trends of desert tortoise populations in the Mojave Desert. USDI National Biol. Survey, *Fish and Wildlife Research Report* **13**:85–93.

Costanza, R. (Ed.). 1991. *Ecological economics: The science and management of sustainability.* Columbia University Press, New York.

Cottam, G. 1987. Community dynamics on an artificial prairie. Pp. 257–70 in W. R. Jordan III, M. E. Gilpin, and J. D. Aber (Eds.), *Restoration ecology: A synthetic approach to ecological research.* Cambridge University Press, New York.

Cowardin, L. M., V. Carter, F. C. Golet, and E. T. LaRoe. 1979. *Classification of wetlands and deepwater habitats of the United States.* U.S. Fish and Wildlife Service Office of Biological Services, Washington, DC.

Cowles, H. C. 1899. The ecological relationships of the vegetation on sand dunes of Lake Michigan. *Bot. Gaz.* **27**:95–117, 167–202, 281–308, 361–91.

Cox, G. W. 1982. *Systylis hoffi* (Protozoa: Ciliata) in California vernal pools. *Trans. Amer. Microscop. Soc.* **101**:294–98.

Cox, G. W. 1985. The evolution of avian migration systems between temperate and tropical regions of the New World. *Amer. Nat.* **126**:451–74.

Cox, G. W. 1990. Centers of speciation in the Galápagos land bird fauna. *Evol. Ecol.* **4**:130–42.

Cox, G. W. 1995. Galápagos Islands. Pp. 167–79 *in Encyclopedia of environmental biology,* Vol. 2. Academic Press, Orlando, FL.

Cox, G. W. 1996. *Laboratory manual of general ecology.* 7th ed. Wm. C. Brown Co., Dubuque, IA.

Cox, G. W. and M. D. Atkins. 1979. *Agricultural ecology.* W. H. Freeman Co., San Francisco.

Cox, G. W. and R. E. Ricklefs. 1977. Species diversity, ecological release, and community structuring in Caribbean land bird faunas. *Oikos* **28**:113–22.

Craighead, J. J., J. S. Sumner, and J. A. Mitchell. 1995. *The grizzly bears of Yellowstone: Their ecology in the Yellowstone ecosystem, 1959–1992.* Island Press, Washington, DC.

Crawford, R. L. 1981. Bird casualties at a Leon County, Florida TV tower: A 25-year migration study. *Bull. Tall Timbers Res. Stn.* **22**, 30 pp.

Crawford, S. M. and M. A. Crawford. 1974. An examination of systems of management of wild and domestic animals based on the African ecosystems. Pp. 218–34 *in* H. H. Cole and M. Ronning (Eds.), *Animal agriculture.* W. H. Freeman Co., San Francisco.

Cree, A., C. H. Daugherty, and J. M. Hay. 1995. Reproduction of a rare New Zealand reptile, the tuatara *Sphenodon punctatus,* on rat-free and rat-inhabited islands. *Cons. Biol.* **9**:373–83.

Cringan, A. T. 1957. History, food habits and range requirements of the woodland caribou of continental North America. *Trans. N. A. Wildl. Conf.* **22**:485–501.

Crouse, D. T., L. B. Crowder, and H. Caswell. 1987. A stage-based population model for loggerhead sea turtles and implications for conservation. *Ecology* **68**:1412–23.

Crowe, M. 1995. Troubled waters: Ocean contaminants—including PCBs—are linked to harbor seal die-off. *Audubon* **97(4)**:14,16.

Croxall, J. P., P. G. H. Evans, and R. W. Schreiber. 1984. *Status and conservation of the world's seabirds.* Int. Council for Bird Preservation, Cambridge, England.

Cummins, K., D. Bodkin, H. Regier, M. Sobel, and L. Talbot. 1995. *Status and future of salmon of western Oregon and northern California: Management of the riparian zone for the conservation and production of salmon.* Center for the Study of the Environment, Santa Barbara, CA.

Curtis, J. T. 1956. The modification of mid-latitude grasslands and forests by man. Pp. 721–26 *in* W. L. Thomas (Ed.), *Man's role in changing the face of the earth.* University of Chicago Press, Chicago, IL.

Dahl, F. H. and R. B. McDonald. 1980. Effects of control of the sea lamprey (*Petromyzon marinus*) on migratory and resident fish populations. *Can. J. Fish. Aq. Sci.* **37**:1886–94.

Dahlsten, D. L. 1986. Control of invaders. Pp. 276–302 *in* H. A. Mooney and J. A. Drake (Eds.), *Ecology of biological invasions of North America and Hawaii.* Ecological Studies, Vol. 58. Springer-Verlag, New York.

Darwin, C. R. 1859. *The origin of species by means of natural selection or the preservation of favored races in the struggle for life.* Murray, London.

Dasmann, R. F. 1964. *African game ranching.* Macmillan Co., New York.

Davies, B. R. and K. F. Walker (Eds.). 1986. *The ecology of river systems.* W. Junk, Dordrecht, The Netherlands.

Davis, F. W., D. M. Stoms, J. E. Estes, J. Scepan, and J. M. Scott. 1990. An information systems approach to the preservation of biological diversity. *Int. J. Geogr. Inform. Syst.* **4**:55–78.

Davis, M. B. 1992. Changes in geographical range resulting from greenhouse warming: Effects on biodiversity in forests. Pp. 297–308 *in* Peters and Lovejoy (1992).

Davis, S. M. and J. C. Ogden (Eds). 1994. *Everglades: The ecosystem and its restoration.* St. Lucie Press, Delray Beach, FL.

D'Elia, C. F. 1987. Nutrient enrichment of the Chesapeake Bay. *Environment* **29**:6–11, 30–33.

Demain, A. L. and N. A. Solomon. 1981. Industrial microbiology. *Sci. Amer.* **245(3)**:66–75.

DeMaster, D., D. Miller, J. R. Henderson, and J. M. Coe. 1985. Conflicts between marine mammals and fisheries off the coast of California. Pp. 111–18 *in* J. R. Beddington, R. J. H. Beverton, and D. M. Lavigne (Eds.), *Marine mammals and fisheries.* George Allen and Unwin, London.

Des Jardins, J. R. 1992. *Environmental ethics.* Wadsworth Publishing Co., Belmont, CA.

Diamond, J. M. 1975. The island dilemma: Lessons of modern biogeography studies for the design of nature preserves. *Biol. Cons.* **7**:129–46.

Diamond, J. M. 1976. Island biogeography and conservation: Strategy and limitations. *Science* **193**:1027–29.

Diamond, J. M. 1986. The design of a nature reserve system for Indonesian New Guinea. Pp. 485–503 *in* M. E. Soulé (Ed.), *Conservation Biology: The science of scarcity and diversity.* Sinauer Associates, Sunderland, MA.

Diamond, J. M., J. Terborgh, R. F. Whitcomb, J. F. Lynch, P. A. Opler, and C. S. Robbins. 1976. Island biogeography and conservation: Strategy and limitations. *Science* **193**:1027–33.

Diamond, J. M. and C. R. Veitch. 1981. Extinctions and introductions in the New Zealand avifauna: Cause and effect? *Science* **211**:499–501.

di Castri, F., M. Hadley, and J. Damlamian. 1981. MAB: The Man and the Biosphere Program as an evolving system. *Ambio* **10**:52–57.

Dirksen, S., T. J. Boudewijn, L. K. Slager, R. G. Mes, M. J. M. van Shaick, and P. de Voogt. 1995. Reduced breeding success of cormorants (*Phalacrocorax carbo sinensis*) in relation to persistent organochlorine pollution of aquatic habitats in the Netherlands. *Env. Pollut.* **88**:119–32.

diZerega, G. 1995. Individuality, human and natural communities, and the foundations of ethics. *Env. Ethics* **17**:23–37.

Doak, D. 1989. Spotted owls and old growth logging in the Pacific Northwest. *Cons. Biol.* **3**:389–96.

Dodd, C. K., Jr. and R. A. Siegel. 1991. Relocation, repatriation, and translocation of amphibians and reptiles: Are they conservation strategies that work? *Herpetologica* **47**:336–50.

Dodd, J. L. 1994. Desertification and degradation in sub-Saharan Africa. *BioScience* **44**:28–34.

Dolan, R., P. J. Godfrey, and W. E. Odum. 1973. Man's impact on the barrier islands of North Carolina. *Amer. Sci.* **61:**152–62.

Dolan, R., B. Hayden, and H. Lins. 1980. Barrier islands. *Amer. Sci.* **68:**16–25.

Dovers, S. R. and J. W. Handmer. 1993. Contradictions in sustainability. *Env. Cons.* **20:**217–22.

Dregne, H. E. 1983. *Desertification of arid lands.* Harwood Academic Publishers, New York.

Duedall, I. W. 1990. A brief history of ocean disposal. *Oceanus* **33(2):**29–38.

Duggins, D. O. 1980. Kelp beds and sea otters: An experimental approach. *Ecology* **61:**447–53.

Dunlap, T. R. 1981. *DDT: Scientists, citizens, and public policy.* Princeton University Press, Princeton, NJ.

Dunlap, T. R. 1985. The coyote itself: Ecologists and the value of predators, 1900–1972. Pp. 594–618 *in* K. E. Bailes (Ed.), *Environmental history: Critical issues in comparative perspective.* University Press of America, Lanham, MD.

Dunning, J. B., Jr., D. J. Stewart, B. L. Danielson, B. R. Noon, T. L. Root, R. H. Lamberson, and E. E. Stevens. 1995. Spatially explicit population models: Current forms and future uses. *Ecol. Applications* **5:**3–11.

East, R. 1981. Species-area curves and populations of large mammals in African savanna reserves. *Biol. Cons.* **21:**111–26.

East, R. 1983. Application of species-area curves to African savannah reserves. *Afr. J. Ecol.* **21:**123–28.

Eckhardt, R. C. 1972. Introduced plants and animals in the Galápagos Islands. *BioScience* **22:**585–90.

Ecological Society of America. 1991. The sustainable biosphere initiative: An ecological research agenda. *Ecology* **72:**371–412.

Ecological Society of America Committee on Endangered Species. 1996. Strengthening the use of science in achieving the goals of the Endangered Species Act: An assessment by the Ecological Society of America. *Ecol. Applications* **6:**1–11.

Edmondson, W. T. and J. T. Lehman. 1981. The effect of changes in the nutrient income on the condition of Lake Washington. *Limnol. Oceanogr.* **26:**1–29.

Edwards, P. J., R. M. May, and N. R. Wegg (Eds.). *Large-scale ecology and conservation biology.* Blackwell Scientific Publications, London, England.

Edwards, T. C., Jr., C. G. Homer, S. D. Bassett, A. Falconer, R. D. Ramsey, and D. W. Wight. 1995. *Utah gap analysis: An environmental information system.* Utah State University, Logan, UT.

Egerton, F. N. 1987. Pollution and aquatic life in Lake Erie: Early scientific studies. *Env. Review* **11:**189–205.

Ehrenfeld, D. W. 1976. The conservation of non-resources. *Amer. Sci.* **64:**648–56.

Ehrenfeld, D. W. 1978. *The arrogance of humanism.* Oxford University Press, New York.

Ehrenfeld, D. W. 1993. *Beginning again. People and nature in the new millennium.* Oxford University Press, Oxford, England.

Ehrlich, P. R. 1968. *The population bomb.* Ballantine, New York.

Eisma, D. (Ed.). 1995. *Climate change: Impact on coastal habitation.* Lewis Publishers, Boca Raton, FL.

Elfring, C. 1990. Conflict in the Grand Canyon: How should Glen Canyon Dam be operated? *BioScience* **40:**709–11.

Elliott, K. J. and W. T. Swank. 1994. Changes in tree species diversity after successive clearcuts in the southern Appalachians. *Vegetatio* **115:**11–18.

Ellis, J. E. and D. M. Swift. 1988. Stability of African pastoral ecosystems: Alternate paradigms and implications for development. *J. Range Manag.* **41:**450–59.

El-Sayed, S. Z. 1988. The BIOMASS program. *Oceans* **31:**75–79.

Elton, C. S. 1958. *The ecology of invasions by animals and plants.* Methuen, London.

Emanuel, W. R., H. H. Shugart, and M. P. Stevenson. Climatic change and the broad-scale distribution of terrestrial ecosystem complexes. *Clim. Change* **7:**29–43.

Emerson, J. W. 1971. Channelization: A case study. *Science* **173:**325–26.

Emmons, L. H. 1995. Mammals of rain forest canopies. Pp. 199–223 *in* Lowman and Nadkarni (1995).

Endean, R. 1982. Crown-of-thorns starfish on the Great Barrier Reef. *Endeavour* **6:**10–14.

Errington, P. L. 1946. Predation and vertebrate populations. *Quart. Rev. Biol.* **21:**144–77, 221–45.

Erwin, T. L. 1982. Tropical forests: their richness in Coleoptera and other arthropod species. *Coleopterists Bull.* **36:**74–75.

Erwin, T. L. 1983. Tropical forest canopies: The last biotic frontier. *Bull. Ent. Soc. Amer.* **29:**14–19.

ESA Committee on Endangered Species. 1996. Strengthening the use of science in achieving the goals of the Endangered Species Act: An assessment by the Ecological Society of America. *Ecol. Applications* **6:**1–11.

Esseen, P-A. 1994. Tree mortality patterns after experimental fragmentation of an old-growth conifer forest. *Biol. Cons.* **68:**19–28.

Estes, J. A. and D. O. Duggins. 1995. Sea otters and kelp forests in Alaska: Generality and variation in a community ecological paradigm. *Ecol. Monogr.* **65:**75–100.

Estes, J. A. and J. F. Palmisano. 1974. Sea otters: Their role in structuring nearshore communities. *Science* **185:**1058–60.

Evans, F. C. 1956. Ecosystem as the basic unit in ecology. *Science* **123:**1127–28.

Evans, R. A. 1989. Response of limnetic insect populations of two acidic, fishless lakes to liming and brook trout (*Salvelinus fontinalis*). *Can. J. Fish. Aq. Sci.* **46:**342–51.

Everson, I., J. L. Watkins, D. G. Bone, and K. G. Foote. 1990. Implications of a new acoustic target strength for abundance estimates of Antarctic krill. *Nature* **345:**338–40.

Ewel, J. J. 1986. Invasibility: Lessons from South Florida. Pp. 214–30 *in* H. A. Mooney and J. A. Drake (Eds.), *Ecology of biological invasions of North America and Hawaii.* Springer-Verlag, New York.

Falk, D. A. 1990. Integrated strategies for conserving plant genetic diversity. *Ann. Missouri Bot. Gard.* **77:**38–47.

Falkengren-Grerup, U. 1986. Soil acidification and vegetation changes in deciduous forests in southern Sweden. *Oecologia* **70:**339–47.

FAO. 1993. *Forest resources assessment 1990. Tropical countries.* FAO Forestry Paper 112.

Fautin, D. G., D. J. Futuyma, and F. C. James (Eds.). 1995. Special section on sustainability issues. *Ann. Rev. Ecol. Syst.* **26:**1–248.

Fearnside, P. M. 1987. Deforestation and international economic development projects in Brazilian Amazonia. *Cons. Biol.* **1:**214–21.

Fearnside, P. M. 1989. Extractive reserves in Brazilian Amazonia. *BioScience* **39**:387–93.

Fearnside, P. M. 1995. Hydroelectric dams in the Brazilian Amazon as sources of "greenhouse" gases. *Env. Cons.* **22**:7–19.

Fearnside, P. M. and J. Ferraz. 1995. A conservation gap analysis of Brazil's Amazonian vegetation. *Cons. Biol.* **9**:1134–47.

Feazel, C. T. 1990. *White bear.* Henry Holt and Co., New York.

Feeney, A. 1989. The Pacific Northwest's ancient forests: Ecosystems under siege. Pp. 93–153 in W. J. Chandler (Ed.), *Audubon wildlife report 1989/1990.* Academic Press, San Diego, CA.

Feinberg, J. 1974. Rights of animals and unborn generations. Pp. 55–56 in W. T. Blackstone (Ed.), *Philosophy and environmental crisis.* University of Georgia Press, Athens, GA.

Felix, N. A. and M. K. Reynolds. 1989. The effects of winter seismic trails on tundra vegetation in northeastern Alaska, U.S.A. *Arctic Alpine Res.* **21**:188–202.

Felix, N. A., M. K. Reynolds, J. C. Jorgenson, and K. E. DuBois. 1992. Resistance and resilience of tundra plant communities to disturbance by winter seismic vehicles. *Arctic Alpine Res.* **22**:69–77.

Field, C. B., F. S. Chapin III, P. A. Matson, and H. A. Mooney. 1992. Responses of terrestrial ecosystems to the changing atmosphere: A resource-based approach. *Ann. Rev. Ecol. Syst.* **23**:201–35.

Findholt, S. L. 1984. Organochlorine residues, eggshell thickness, and reproductive success of snowy egrets nesting in Idaho. *Condor* **86**:163–69.

Fisher, J., N. Simon, and J. Vincent. 1969. *Wildlife in danger.* Viking Press, New York.

Flader, S. 1974. *Thinking like a mountain.* University of Missouri Press, Columbia, MO.

Flader, S. 1987. Aldo Leopold and the evolution of a land ethic. Pp. 3–24 in T. Tanner (Ed.), *Aldo Leopold: The man and his legacy.* Soil Conservation Society of America, Ankeny, IA.

Flather, C. H. and J. R. Sauer. 1996. Using landscape ecology to test hypotheses about large-scale abundance patterns in migratory birds. *Ecology* **77**:28–35.

Fleck, R. F. 1985. *Henry Thoreau and John Muir among the Indians.* Archon Books, Hamden, CT.

Forman, L., D. G. Kleiman, R. M. Bush, J. M. Dietz, J. D. Ballou, L. G. Phillips, A. F. Coimbra Filho, and S. J. O'Brien. 1986. Genetic variation within and among lion tamarins (*Leontopithecus rosalia rosalia*). *Amer. J. Phys. Anthropol.* **71**:1–12.

Forman, R. T. T., A. E. Galli, and C. F. Leck. 1976. Forest size and avian diversity in New Jersey woodlots with some land-use implications. *Oecologia* **26**:1–8.

Forman, R. T. T. and M. Godron. 1986. *Landscape ecology.* John Wiley and Sons, New York.

Forsburg, C. 1994. The large-scale flux of nutrients from land to water and the eutrophication of lakes and marine waters. *Mar. Pollut. Bull.* **29**:409–13.

Fortes, M. D. 1988. Mangrove and seagrass beds of East Asia: Habitats under stress. *Ambio* **17**:207–13.

Fosberg, F. R. 1948. Derivation of the flora of the Hawaiian Islands. Pp. 107–19 in E. C. Zimmerman (Ed.), *Insects of Hawaii.* University of Hawaii Press, Honolulu, HI.

Foster, M. S. and D. R. Schiel. 1985. The ecology of giant kelp forests in California: A community profile. *U.S. Fish Wildl. Serv. Biol. Rep.* **85**(7.2).

Foster, M. S. and D. R. Schiel. 1988. Kelp communities and sea otters: Keystone species or just another brick in the wall? Pp. 92–115 in G. R. VanBlaricom and J. A. Estes (Eds.), *The community ecology of sea otters.* Springer-Verlag, Berlin, Germany.

Foster, N. M. and J. H. Archer. 1988. The National Marine Sanctuary Program—policy, education, and research. *Oceanus* **31**(1):4–12.

Foster, N. W. 1989. Acidic deposition: What is fact, what is speculation, what is needed? *Water, Air, Soil Pollut.* **48**:299–306.

Fox, G. A., P. Mineau, B. Collins, and P. C. James. 1989. *The impact of the insecticide carbofuran (furadan 480F) on the burrowing owl in Canada.* Technical Report Series No. 72, Canadian Wildlife Service, Ottawa, Canada.

Fox, M. D. and B. J. Fox. 1986. The susceptibility of natural communities to invasion. Pp. 57–66 in R. H. Groves and J. J. Burdon (Eds.), *Ecology of biological invasions.* Cambridge University Press, Cambridge, England.

Frankel, O. H. and M. E. Soulé. 1981. *Conservation and evolution.* Cambridge University Press, Cambridge, England.

Frankel, O. H., A. D. H. Brown, and J. J. Burdon. 1995. *The conservation of plant biodiversity.* Cambridge University Press, Cambridge, England.

Franklin, I. R. 1980. Evolutionary change in small populations. Pp. 135–50 in M. E. Soulé and B. A. Wilcox (Eds.), *Conservation biology: An evolutionary-ecological perspective.* Sinauer Associates, Sunderland, MA.

Franklin, J. F., F. L. Swanson, M. E. Harmon, D. A. Perry, T. A. Spies, V. H. Dale, A. McKee, W. K. Ferrell, J. E. Means, S. V. Gregory, J. A. Lattin, T. D. Schowalter, and D. Larsen. 1992. Effects of global climate change on forests in northwestern North America. Pp. 258–76 in Peters and Lovejoy (1992).

Freda, J. 1986. The influence of acidic pond water on amphibians. *Water, Air, Soil Pollut.* **30**:439–50.

Freemark, K. and C. Boutin. 1995. Impacts of agricultural herbicide use on terrestrial wildlife in temperate landscapes: A review with special reference to North America. *Agriculture, Ecosystems & Environment* **52**:67–91.

French, H. F. 1990. Clearing the air. Pp. 98–118 in L. Starke (Ed.), *State of the world 1990.* W. W. Norton Co., New York.

French, W. C. 1995. Against biospherical egalitarianism. *Enviro. Ethics* **17**:39–57.

Fryxell, J. M., J. Greever, and A. R. E. Sinclair. 1988. Why are migratory ungulates so abundant? *Am. Nat.* **131**:781–98.

Fuller, T. K. 1989. Population dynamics of wolves in north-central Minnesota. *Wildl. Monogr.* **105**:1–41.

Furness, R. W. and D. G. Ainley. 1984. Threats to seabird populations presented by commercial fisheries. Pp. 701–8 in *Status and conservation of the world's seabirds.* International Council for Bird Preservation, Cambridge, England.

Gagne, W. C. and C. C. Christensen. 1985. Conservation status of native terrestrial invertebrates in Hawaii. Pp. 105–41 in C. P. Stone and J. M. Scott (Eds.), *Hawaii's terrestrial ecosystems: Preservation and management.* University of Hawaii, Honolulu, HI.

Galazii, G. 1991. Lake Baikal reprieved. *Endeavour* **15**:13–17.

Galli, A. E., C. F. Leck, and R. T. T. Forman. 1976. Avian distribution patterns in forest islands of different sizes in central New Jersey. *Auk* **93**:356–64.

Gambell, R. 1976. Population biology and the management of whales. *Applied Biol.* **1**:247–343.

Gambell, R. 1990. The International Whaling Commission—*quo vadis? Mammal Rev.* **20**:31–43.

Garner, J. H. B., T. Pagano, and E. B. Cowling. 1989. *An evaluation of the role of ozone, acid deposition, and other airborne pollutants in the forests of eastern North America.* USDA Forest Service Gen. Tech Pap. SE-59.

Gasaway, W. C., R. D. Boertje, D. V. Grangaard, D. G. Kelleyhouse, R. O. Stephenson, and Douglas G. Larsen. 1992. The role of predation in limiting moose at low densities in Alaska and Yukon and implications for conservation. *Wildl. Monogr.* **120**:1–59.

Gasaway, W. C., R. O. Stevenson, J. L. Davis, P. E. K. Shepherd, and O. E. Burris. 1983. Interrelationships of wolves, prey, and man in interior Alaska. *Wildl. Monogr.* **84**:1–50.

Gaskin, D. E. 1982. *The ecology of whales and dolphins.* Heinemann, Exeter, NH. 472 pp.

Gaston, K. J. 1991. The magnitude of global insect species richness. *Cons. Biol.* **5**:283–96.

Gaston, K. J. and E. Hudson. 1994. Regional patterns of diversity and estimates of global insect species richness. *Biodiversity and Conservation* **3**:493–500.

Gauthreaux, S. A. 1992. The use of weather radar to monitor long-term patterns of trans-Gulf migration in spring. Pp. 96–100 *in* J. M. Hagan and D. W. Johnston (Eds.), *Ecology and conservation of neotropical migrant landbirds.* Smithsonian Institution Press, Washington, DC.

Gee, C. K., R. S. Magelby, W. R. Bailey, R. L. Gum, and L. M. Arthur. 1977. *Sheep and lamb losses to predators and other causes in the western United States.* USDA Agr. Econ. Rep. No. 369. 41 pp.

Gentry, A. H. 1986. Endemism in tropical versus temperate plant communities. Pp. 153–81 *in* M. E. Soulé (Ed.), *Conservation biology: The science of scarcity and diversity.* Sinauer Associates, Sunderland, MA.

Gentry, A. H. 1988. Tree species of upper Amazonian forests. *Proc. Natl. Acad. Sci. U.S.A.* **85**:156.

Gentry, A. H. and J. Lopez-Parodi. 1980. Deforestation and increased flooding of the upper Amazon. *Science* **210**:1354–56.

Geraci, J. R. and T. D. Williams. 1990. Physiologic and toxic effects on sea otters. Pp. 211–21 *in* J. R. Geraci and D. J. St. Aubin (Eds.), *Sea mammals and oil: Confronting the risks.* Academic Press, San Diego, CA.

Getz, W. M. and R. G. Haight. 1989. *Population harvesting: Demographic models of fish, forest, and animal resources.* Princeton University Press, Princeton, NJ.

Gibbons, E. F., B. S. Durrant, and J. Demarest (Eds.). 1995. *Conservation of endangered species in captivity.* State University of New York Press, Albany, NY.

Gilbert, F. S. 1980. Food web organization and conservation of neotropical diversity. Pp. 11–34 *in* M. Soulé and B. Wilcox (Eds.), *Conservation biology: An evolutionary-ecological perspective.* Sinauer Associates, Sunderland, MA.

Gillespie, G. D. 1985. Hybridization, introgression, and morphometric differentiation between mallard (*Anas platyrhynchos*) and grey duck (*Anas superciliosa*) in Otago, New Zealand. *Auk* **102**:459–69.

Gilpin, M. E. and M. E. Soulé. 1986. Minimum viable populations: Processes of species extinction. Pp. 19–34 *in* M. E. Soulé (Ed.), *Conservation biology: The science of scarcity and diversity.* Sinauer Associates, Sunderland, MA.

Ginsberg, J. R., K. A. Alexander, S. Creel, P. W. Kat, J. W. McNutt, and M. G. L. Mills. 1995. Handling and survivorship of African wild dog (*Lycaon pictus*) in five ecosystems. *Cons. Biol.* **9**:665–74.

Glantz, M. H. (Ed.). 1994. *Drought follows the plow.* Cambridge University Press, Cambridge, England.

Godfrey, P. J., S. P. Leatherman, and P. A. Buckley. 1980. ORVs and barrier beach degradation. *Parks* **5**:5–11.

Goldberg, E. D. 1986. TBT: An environmental dilemma. *Environment* **28**(8):17–20, 42–44.

Goldman, C. R. 1989. Lake Tahoe: Preserving a fragile ecosystem. *Environment* **31**:6–11, 27–31.

Goodland, R. 1995. The concept of environmental sustainability. *Ann. Rev. Ecol. Syst.* **26**:1–24.

Goodman, D. 1980. The maximum yield problem: Distortion of the yield curve due to age structure. *Theor. Pop. Biol.* **18**:160–74.

Goodman, D. 1987. The demography of chance extinction. Pp. 11–34 *in* M. E. Soulé (Ed.), *Viable populations for conservation.* Cambridge University Press, Cambridge, England.

Goulding, M. 1980. *The fishes and the forest.* University of California Press, Berkeley, CA.

Goulding, M. 1985. Forest fishes of the Amazon. Pp. 267–76 *in* G. T. Prance and T. E. Lovejoy (Eds.), *Amazonia.* Pergamon Press, Oxford, England.

Gowdy, J. and S. O'Hara. 1995. *Economic theory for environmentalists.* St. Lucie Press, Delray Beach, FL.

Graber, J. W. and R. R. Graber. 1983. Feeding rates of warblers in spring. *Condor* **85**:139–50.

Grabherr, G., M. Gottfried, and H. Pauli. 1994. Climate effects on mountain plants. *Nature* **369**:448.

Gradwohl, J. and R. Greenberg. 1988. *Saving the tropical forests.* Island Press, Washington, DC.

Gradwohl, J. and R. Greenberg. 1989. Conserving nongame migratory birds: A strategy for monitoring and research. Pp. 297–328 *in* W. J. Chandler (Ed.), *Audubon wildlife report 1989/1990.* Academic Press, San Diego, CA.

Grant, P. R. 1986. *Ecology and evolution of Darwin's finches.* Princeton University Press, Princeton, NJ.

Grassle, J. F. 1991. Deep-sea benthic biodiversity. *BioScience* **41**:464–69.

Green, G. M. and R. W. Sussman. 1990. Deforestation history of the eastern rain forests of Madagascar from satellite images. *Science* **248**:212–15.

Gregg, S. S. 1988. Of soup and survival: The plight of the sea turtles. *Sea Frontiers* **34**:297–302.

Greig, J. C. 1979. Principles of genetic conservation in relation to wildlife management in southern Africa. *S. Afr. J. Wildl. Res.* **9**:57–78.

Grier, J. W. 1982. Ban of DDT and subsequent recovery of reproduction in bald eagles. *Science* **218**:1232–35.

Griffith, B., J. M. Scott, J. W. Carpenter, and C. Reed. 1989. Translocation as a species conservation tool: Status and strategy. *Science* **245**:477–80.

storage of conversion of old-growth to young forests. *Science* **247**:699–702.

Grimm, N. B. 1993. Implications of climatic change for stream communities. Pp. 293–314 *in* P. R. Kareiva, J. G. Kingsolver, and R. B. Huey (Eds.), *Biotic interactions and global change.* Sinauer Associates, Sunderland, MA.

Grimmett, R. 1987. A review of the problems affecting Palearctic migratory birds in Africa. *Int. Council Bird Pres., Study Rep.* 22.

Grosholz, E. D. and G. M. Ruiz. 1995. Spread and potential impact of the recently introduced European green crab, *Carcinus meanas,* in central California. *Marine Biol.* **122**:239–47.

Grue, C. E., W. J. Fleming, D. G. Busby, and E. F. Hill. 1983. Assessing hazards of organophosphate pesticides to wildlife. *Trans. N. A. Wildl. Nat. Res. Conf.* **48**:200–20.

Gulland, J. A. 1980. Open ocean resources. Pp. 347–78 *in* R. T. Lackey and L. A. Nielson (Eds.), *Fisheries management.* John Wiley and Sons, New York.

Gullion, G. W. 1977. Forest manipulation for ruffed grouse. *Trans. N. A. Wildl. Nat. Res. Conf.* **42**:449–58.

Gum, R. L., L. M. Arthur, and R. S. Magleby. 1978. *Coyote control: A simulation evaluation of alternative strategies.* USDA Agr. Econ. Rep. No. 408. 49 pp.

Gunn, J. M. and W. Keller. 1990. Biological recovery of an acid lake after reductions in industrial emissions of sulphur. *Nature* **345**:431–33.

Haas, C. A. 1995. Dispersal and use of corridors by birds in wooded patches on an agricultural landscape. *Cons. Biol.* **9**:845–54.

Hader, D-P., R. C. Worrest, H. D. Kumar, and R. C. Smith. 1995. Effects of increased solar ultraviolet radiation on aquatic ecosystems. *Ambio* **24**:174–80.

Hagan, J. M. 1995. Environmentalism and the science of conservation biology. *Cons. Biol.* **9**:1995.

Hageman, R. K. 1985. *Valuing marine mammal populations: Benefit valuations in a multi-species ecosystem.* Administrative Report LJ-85-22, National Marine Fisheries Service, La Jolla, CA.

Haight, R. G. 1987. Evaluating the efficiency of even-aged and uneven-aged stand management. *Forest Sci.* **33**:116–34.

Haines, T. A. 1981. Acidic precipitation and its consequences for aquatic ecosystems: A review. *Trans. Amer. Fish. Soc.* **110**:669–707.

Haines, T. A. 1986. Fish population trends in response to surface water acidification. Pp. 300–34 *in* National Academy of Sciences. *Acid deposition: Long-term trends.* National Academy Press, Washington, DC.

Hair, J. D. and G. A. Pomerantz. 1987. The educational value of wildlife. Pp. 197–207 *in* D. J. Decker and G. R. Goff (Eds.), *Valuing wildlife: Economic and social perspectives.* Westview Press, Boulder, CO.

Hales, D. F. 1984. The World Heritage Convention: Status and directions. Pp. 744–50 *in* J. A. McNeely and K. R. Miller (Eds.), *National parks, conservation, and development: The role of protected areas in sustaining society.* Smithsonian University Press, Washington, DC.

Hall, G. A. 1984. Population decline of neotropical migrants in an Appalachian forest. *Amer. Birds* **38**:14–18.

Hall, R. J. and F. P. Ide. 1987. Evidence of acidification effects on stream insect communities in central Ontario between 1937 and 1985. *Can. J. Fish. Aq. Sci.* **44**:1652–57.

Halle, M. 1985. The World Conservation Strategy—An historical perspective. Pp. 241–57 *in* J. P. Hearn and J. K. Hodges (Eds.), *Advances in animal conservation.* Clarendon Press, Oxford, England.

Halliday, T. R. 1980. The extinction of the passenger pigeon *Ectopistes migratorius* and its relevance to contemporary conservation. *Biol. Cons.* **17**:157–62.

Hammack, J. and G. M. Brown, Jr. 1974. *Waterfowl and wetlands: Toward bioeconomic analysis.* Johns Hopkins University Press, Baltimore, MD.

Hanemann, W. M. 1994. Valuing the environment through contingent valuation. *J. Econ. Perspectives* **8**:19–43.

Hanes, T. L. 1977. Chaparral. Pp. 417–69 *in* M. G. Barbour and J. Major (Eds.), *Terrestrial vegetation of California.* John Wiley and Sons, New York.

Hanski, I., J. Poyry, T. Pakkala, and M. Kuussaari. 1995. Multiple equilibria in metapopulation dynamics. *Nature* **377**:618–21.

Hardin, G. 1968. The tragedy of the commons. *Science* **162**:1243–48.

Hardin, G. 1974. Living on a lifeboat. *BioScience* **24**:561–68.

Harmon, M. E., W. K. Ferrell, and J. F. Franklin. 1990. Effects on carbon

Harris, L. D. 1988. Landscape linkages: The dispersal corridor approach to wildlife conservation. *Trans. N. A. Wildl. Nat. Res. Conf.* **53**:595–607.

Harris, L. D. and G. Silva-Lopez. 1992. Forest fragmentation and the conservation of biological diversity. Pp. 197–237 *in* P. L. Fiedler and S. Jain (Eds.), *Conservation biology: The theory and practice of nature conservation, preservation, and management.* Chapman and Hall, New York.

Harte, J. and R. Shaw. 1995. Shifting dominance within a montane vegetation community: Results of a climate-warming experiment. *Science* **267**:876–80.

Hartshorn, G. S. 1990. Natural forest management by the Yanesha Forestry Cooperative in Peruvian Amazonia. Pp. 128–38 *in* A. B. Anderson (Ed.), *Alternatives to deforestation: Steps toward sustainable use of the Amazon rain forest.* Columbia University Press, New York.

Hartshorn, G. S. 1992. Possible effects of global warming on the biological diversity in tropical forests. Pp. 137–46 *in* Peters and Lovejoy (1992).

Haskell, D. G. 1995. A reevaluation of the effects of forest fragmentation on rates of bird-nest predation. *Cons. Biol.* **9**:1316–18.

Hawkins, A. S., R. C. Hanson, H. K. Nelson, and H. M. Reeves (Eds.). 1984. *Flyways: Pioneering waterfowl management in North America.* U.S. Fish and Wildlife Service, Washington, DC.

Hayden, B. P., G. C. Ray, and R. Dolan. 1984. Classification of coastal and marine environments. *Env. Cons.* **11**:199–207.

Hecht, S. B. 1989. Sacred cow in the Green Hell: Livestock and forest conversion in the Brazilian Amazon. *The Ecologist* **19**:229–34.

Hecht, S. and A. Cockburn. 1989. *The fate of the forest.* Verso, New York.

Hedgepeth, J. W. 1989. Commentary: The life and works of Aldo Leopold. *Quart. Rev. Biol.* **64**:169–73.

Hedgepeth, J. W. 1993. Foreign invaders. *Science* **261**:34–35.

Hedrick, P. W. 1992. Genetic conservation in captive populations and endangered species. Pp. 45–68 *in* S. K. Jain and L. W. Botsford (Eds.), *Applied population biology*. Klewer Academic Publishers, The Netherlands.

Hedrick, P. W. 1995. Gene flow and genetic restoration: The Florida panther as a case study. *Cons. Biol.* **9:**996–1007.

Heinselman, M. L. 1981. Fire and succession in the conifer forests of northern North America. Pp. 374–405 *in* D. C. West, H. H. Shugart, and D. B. Bodkin (Eds.), *Forest succession: Concepts and application.* Springer-Verlag, New York.

Hemley, G. 1994. *International wildlife trade: A CITES sourcebook.* Island Press, Washington, DC.

Hendrey, G. R., K. Baalsrud, T. S. Traaen, M. Laake, and G. R. Raddum. 1976. Acid precipitation: Some hydrobiological changes. *Ambio* **5:**224–27.

Hendrickson, J. R. 1980. The ecological strategies of sea turtles. *Amer. Zool.* **20:**597–608.

Henny, C. J. and G. B. Herron. 1989. DDE, selenium, mercury, and white-faced ibis reproduction at Carson Lake, Nevada. *J. Wildl. Manag.* **53:**1032–45.

Henricksen, A., L. Lien, B. O. Rosseland, T. S. Traaen, and I. S. Sevaldrud. 1989. Lake acidification in Norway: Present and predicted fish status. *Ambio* **18:**314–21.

Herbold, B. and P. B. Moyle. 1989. The ecology of the Sacramento-San Joaquin Delta: A community profile. *U.S. Fish Wildl. Serv. Biol. Rep.* **85**(7.22).

Herkert, J. R. 1994. The effects of habitat fragmentation on midwestern grassland bird communities. *Ecol. Applications* **4:**461–71.

Herkert, J. R. 1995. An analysis of midwestern breeding bird population trends: 1966–1993. *Am. Midl. Nat.* **134:**41–50.

Herlihy, A. T., P. R. Kaufman, M. E. Mitch, and D. D. Brown. 1990. Regional estimates of acid mine drainage impact on streams in the mid-Atlantic and southeastern United States. *Water, Air, Soil Pollut.* **50:**91–107.

Herrera, R. 1985. Nutrient cycling in Amazonian forests. Pp. 95–105 *in*

G. T. Prance and T. E. Lovejoy (Eds.), *Amazonia.* Pergamon Press, Oxford, England.

Herrero, S. 1970. Human injury inflicted by grizzly bears. *Science* **170:**593–98.

Herrero, S. 1985. *Bear attacks: Their causes and avoidance.* Winchester Press, Piscataway, NJ.

Hertz, O. and F. O. Kapel. 1986. Commercial and subsistence hunting of marine mammals. *Ambio* **15:**144–51.

Hess, G. R. 1994. Conservation corridors and contagious disease: A cautionary note. *Cons. Biol.* **8:**256–62.

Hey, D. L. and N. S. Philippi. 1995. Flood reduction through wetland restoration: The upper Mississippi River basin as a case history. *Restoration Ecol.* **3:**4–17.

Heywood, V. H. 1989. Patterns, extents, and modes of invasions by terrestrial plants. Pp. 31–60 *in* J. A. Drake, H. A. Mooney, F. di Castri, R. H. Groves, F. J. Kruger, M. Rejmanek, and M. Williamson (Eds.), *Biological invasions: A global perspective.* John Wiley and Sons, Chichester, England.

Heywood, V. H. (Ed.). 1995. *Global biodiversity assessment.* Cambridge University Press, Cambridge, England.

Hickman, T. J. 1983. Effects of habitat alteration by energy resource developments in the upper Colorado River basin on endangered species. Pp. 537–50 *in* V. D. Adams and V. A. Lamarra (Eds.), *Aquatic resources management of the Colorado River ecosystem.* Ann Arbor Science Publishers, Ann Arbor, MI.

Hill, N. P. and J. M. Hagan III. 1991. Population trends of some northeastern North American landbirds: A half-century of data. *Wilson Bull.* **103:**165–82.

Hinckley, B. S., R. M. Iverson, and B. Hallet. 1983. Accelerated water erosion in ORV-use areas. Pp. 97–109 *in* R. H. Webb and H. G. Wilshire (Eds.), *Environmental effects of off-road vehicles.* Springer-Verlag, New York.

Hinsley, S. A., P. E. Bellamy, and I. Newton. 1995. Bird species turnover and stochastic extinction in woodland fragments. *Ecography* **18:**41–50.

Hively, W. 1989. How bleak is the outlook for ozone? *Amer. Sci.* **77:**219–24.

Hobbs, R. J. 1992. The role of corridors in conservation: Solution or bandwagon? *Trends Ecol. Evol.* **7:**389–92.

Hobdy, R. 1993. Lana'i—a case study: The loss of biodiversity on a small Hawaiian island. *Pacific Sci.* **47:**201–10.

Hofman, R. J. 1989. The marine mammal protection act: A first of its kind anywhere. *Oceanus* **32(1):**21–25.

Hofman, R. J. 1990. Cetacean entanglement in fishing gear. *Mammal Rev.* **20:**53–64.

Holden, C. 1985. Hawaiian rain forest being felled. *Science* **228:**1073–74.

Holdren, J. P., G. C. Daily, and P. R. Ehrlich. 1995. The meaning of sustainability: Biogeophysical aspects. Pp. 3–17 *in* Munasinghe and Shearer (Eds.), *Defining and measuring sustainability: The biogeophysical foundations.* The World Bank, Washington, DC.

Holgate, M. W. and N. M. Wace. 1961. The influence of man on the floras and faunas of southern islands. *Polar Record* **10:**473–93.

Holland, R. and S. Jain. 1984. Vernal Pools. Pp. 515–33 *in* M. G. Barbour and J. Major (Eds.), *Terrestrial vegetation of California.* John Wiley and Sons, New York.

Holling, C. S. 1959. The components of predation as revealed by small mammal predation of the European pine sawfly. *Can. Entomol.* **91:**290–320.

Holling, C. S. 1995. Sustainability: The cross-scale dimension. Pp. 65–75 *in* Munasinghe, M. and W. Shearer (Eds.). 1995. *Defining and measuring sustainability: The biogeophysical foundations.* The World Bank, Washington, DC.

Holloway, M. 1990. Hot geese: Contaminated wildlife roam nuclear reservations. *Sci. Amer.* **263:**22.

Holmes, R. T. and T. W. Sherry. 1988. Assessing population trends of New Hampshire forest birds: Local vs. regional patterns. *Auk* **105:**756–68.

Hom, J. H. 1995. Climate and ecological relationships in northern latitude ecosystems. Pp. 75–88 *in* D. L. Peterson and D. R. Johnson (Eds.), *Human ecology and climate change: People and resources in the far north.* Taylor and Francis, Washington, DC.

Honegger, R. H. 1981. List of amphibians and reptiles either known or thought to have become extinct since 1600. *Biol. Cons.* **19:**141–58.

Hornocker, M. G. 1970. An analysis of mountain lion predation upon mule deer and elk in the Idaho Primitive Area. *Wildl. Monogr.* **21**:1–39.

Hosier, P. E. and T. E. Eaton. 1980. The impact of vehicles on dune and grassland vegetation on a southeastern North Carolina barrier beach. *J. Appl. Ecol.* **17**:173–82.

Houghton, J. T., G. J. Jenkins, and J. J. Ephraums (Eds.). 1990. *Climate change: The IPPC assessment.* Cambridge University Press, Cambridge, England.

Houston, D. B., E. G. Schreiner, and B. B. Moorhead. 1994. *Mountain goats in Olympic National Park: Biology and management of an introduced species.* Scientific Monograph NPS/NROLYM/NRSM-94/25, U.S. Dept. of Interior, National Park Service.

Howe, H. F. 1977. Bird activity and seed dispersal of a tropical wet forest tree. *Ecology* **58**:539–50.

Howell, E. A. and W. R. Jordan III. 1991. Tallgrass prairie restoration in the North American midwest. Pp. 395–414 *in* I. F. Spelerbery, F. B. Goldsmith, and M. G. Morris (Eds.), *The scientific amnagement of temperate communities for conservation.* Blackwell Scientific Publications, London, England.

Huckabee, J. W., J. S. Mattice, L. F. Pitelka, D. B. Porcella, and R. A. Goldstein. 1989. An assessment of the ecological effects of acidic deposition. *Arch. Environ. Contam. Toxicol.* **18**:3–27.

Hudson, W. E. (Ed.). 1991. *Landscape linkages and biodiversity.* Island Press, Washington, DC.

Humphrey, R. R. and L. A. Mehrhoff. 1958. Vegetation changes on a southern Arizona grassland range. *Ecology* **39**:720–26.

Hunt, C. E. 1988. *Down by the river.* Island Press, Washington, DC.

Hunter, C. L. and C. W. Evans. 1995. Coral reefs in Haneohe Bay, Hawaii: Two centuries of western influence and two decades of data. *Bull. Mar. Sci.* **57**:501–15.

Hutto, R. L. 1988. Is tropical deforestation responsible for the reported declines in neotropical migrant populations? *Amer. Birds* **42**:375–79.

Hutto, R. L. 1989. The effect of habitat alteration on migratory birds in a west Mexican tropical deciduous forest: A conservation perspective. *Cons. Biol.* **3**:138–48.

Hutto, R. L., S. Reel, and P. B. Landres. 1987. A critical evaluation of the species approach to biological conservation. *End. Spe. UPDATE* **4(12)**:1–4.

Hvenegaard, G. T., J. R. Butler, and D. K. Krystofiak. 1989. Economic values of bird watching at Point Pelee National Park, Canada. *Wildl. Soc. Bull.* **17**:526–31.

IGBP. 1994. *IGBP in action: Work plan for 1994–1998.* Global Change Report No. 28, Int. Council of Sci. Unions, Stockholm, Sweden.

Inman, D. L. and R. Dolan. 1989. The Outer Banks of North Carolina: Budget of sediment and inlet dynamics along a migrating barrier system. *J. Coastal Res.* **5**:193–237.

International Pacific Halibut Commission. 1987. *The Pacific halibut: biology, fishery, and management.* Int. Pac. Halibut Comm. Tech. Rep. No. 22.

IUCN/UNEP/WWF. 1991. *Caring for the earth. A Strategy for sustainability.* Gland, Switzerland.

Iverson, B. M., B. S. Hinckley, R. M. Webb, and B. Hallet. 1981. Physical effects of vehicular disturbance on arid landscapes. *Science* **212**:915–17.

Iverson, R. M. 1980. Processes of accelerated pluvial erosion on desert hillslopes modified by vehicular traffic. *Earth Surface Processes* **5**:369–88.

Jackson, M. H. 1985. *Galápagos: A natural history guide.* University of Calgary Press, Calgary, Alberta, Canada.

Jaeger, W. K. 1995. Is sustainability optimal? Examining the differences between economists and environmentalists. *Ecol. Economics* **15**:43–57.

James, F. C., C. E. McCullouch, and D. A. Wiedenfeld. 1996. New approaches to the analysis of population trends in land birds. *Ecology* **77**:13–27.

Janzen, D. H. and P. S. Martin. 1982. Neotropical anachronisms: The fruits the gomphotheres ate. *Science* **215**:19–27.

Jarvinen, O. 1982. Conservation of endangered plant populations: Single large or several small reserves? *Oikos* **38**:301–7.

Jefferson, T. A. and B. E. Curry. 1994. A global review of porpoise (*Cetacea: Phocoenidae*) mortality in gillnets. *Biol. Cons.* **67**:167–83.

Jehl, J. R., Jr. 1988. Biology of the eared grebe and Wilson's phalarope in the nonbreeding season: A study of adaptations to saline lakes. *Studies in Avian Biology* No. 12.

Joern, A. and K. H. Keeler (Eds.). 1995. *The changing prairie: North American grasslands.* Oxford University Press, New York.

Johns, A. D. 1988. Economic development and wildlife conservation in Brazilian Amazonia. *Ambio* **17**:302–6.

Johnson, L. E. 1991. *A morally deep world.* Cambridge University Press, Cambridge, England.

Jordan, C. F. 1982. Amazon rain forests. *Amer. Sci.* **70**:394–401.

Jordan, C. F. 1985. *Nutrient cycling in tropical forest ecosystems.* John Wiley and Sons, New York.

Jordan, C. F. 1987. Permanent plots for agriculture and forestry. Pp. 76–89 *in* C. F. Jordan (Ed.), *Amazonian rain forests: Ecosystem disturbance and recovery.* Springer-Verlag, New York.

Jordan, C. F. 1995. *Conservation: Replacing quantity with quality as a goal for global management.* John Wiley and Sons, New York.

Josens, G. 1983. The soil fauna of tropical savannas. Pp. 505–24 *in* F. Bourliere (Ed.), *Tropical savannas.* Elsevier, Amsterdam.

Joseph, J. 1994. The tuna-dolphin controversy in the eastern Pacific Ocean: Biological, economic, and political impacts. *Ocean Dev. and Int. Law* **25**:1030.

Junk, W. J. 1984. Ecology, fisheries, and fish culture in Amazonia. Pp. 443–76 *in* H. Sioli (Ed.), *The Amazon: Limnology and landscape ecology of a mighty tropical river and its basin.* W. Junk, Dordrecht, The Netherlands.

Kahn, J. R. and J. A. McDonald. 1995. Third-world debt and tropical deforestation. *Ecol. Econ.* **12**:107–23.

Kareiva, P. M., J. G. Kingsolver, and R. B. Huey (Eds.). 1993. *Biotic interactions and global change.* Sinauer Associates, Sunderland, MA.

Karr, J. R. 1982. Avian extinction on Barro Colorado Island, Panama: A reassessment. *Amer. Nat.* **119**:220–39.

Kassas, M. 1995. Desertification: A general review. *J. Arid Env.* **30**:115–28.

Kaufman, L. 1991. A fish faunal conservation program: The Lake Victoria cichlids. *End. Sp. UPDATE* **8**:72–75.

Kaufman, L. and K. Mallory (Eds.). 1993. *The last extinction.* 2nd ed. The MIT Press, Cambridge, MA.

Keeley, J. E. and C. C. Swift. 1995. *Biodiversity and ecosystem functioning in Mediterranean-climate California.* Pp. 121–83 *in* G. W. Davis and D. M. Richardson (Eds.), *Mediterranean-type ecosystems: The function of biodiversity.* Springer-Verlag, New York.

Keiper, R. R. 1985. Are sitka deer responsible for the decline of white-tailed deer on Assateague Island, Maryland? *Wildl. Soc. Bull.* **13:**144–46.

Keith, L. B. 1963. *Wildlife's ten-year cycle.* University of Wisconsin Press, Madison, WI.

Keith, L. B. 1981. The role of food in hare population cycles. *Oikos* **40:**385–95.

Keith, L. B. 1983. Population dynamics of wolves. Pp. 66–77 *in* L. N. Carbyn (Ed.), *Wolves in Canada and Alaska.* Can. Wildl. Rep. Ser. 45.

Keith, L. B., J. R. Cary, O. J. Rongstad, and M. C. Brittingham. 1984. Demography and ecology of a declining snowshoe hare population. *Wildl. Monogr.* **90:**1–43.

Keller, W., D. P. Dodge, and G. M. Booth. 1990. Experimental lake neutralization program: Overview of neutralization studies in Ontario. *Can. J. Fish. Aq. Sci.* **47:**410–11.

Kellert, S. R. 1985. Social and perceptual factors in endangered species management. *J. Wildl. Manag.* **49:**528–36.

Kenchington, R. A. and M. T. Agardy. 1990. Achieving marine conservation through biosphere reserve planning and management. *Env. Cons.* **17:**39–44.

Kendeigh, S. C. 1982. Bird populations in east-central Illinois: Fluctuations, variations, and development over a half-century. *Ill. Biol. Monogr.* **52:**1–136.

Khan, M. K. and T. J. Foose. 1994. Chairman's report: Asian rhino specialist group. *Pachyderm* **No. 18,** pp. 3–8.

Kilgore, B. M. 1973. The ecological role of fire in Sierran conifer forests: Its application to national park management. *J. Quaternary Res.* **3:**496–513.

Kilgore, B. M. and R. W. Sando. 1975. Crown-fire potential in a sequoia forest after prescribed burning. *Forest Sci.* **21:**83–87.

King, F. W. 1981. Historical review of the decline of the green turtle and the hawksbill. Pp. 183–88 *in* K. A. Bjorndal (Ed.), *Biology and conservation of sea turtles.* Smithsonian Institution Press, Washington, DC.

King, W. B. 1980. Ecological basis of extinction in birds. *Proc. XVII Int. Orn. Congr.* **1978:**905–11.

Kirch, P. 1982. The impact of the prehistoric Polynesians on the Hawaiian ecosystem. *Pacific Sci.* **36:**1–14.

Kistner, W. 1984. International mussel watch. *Oceans* **17(6):**64–67.

Kitchener, D. J., A. Chapman, and B. J. Muir. 1980*a*. Lizard assemblage and reserve size and structure in the western Australian wheatbelt—some implications for conservation. *Biol. Cons.* **17:**25–62.

Kitchener, D. J., A. Chapman, and B. J. Muir. 1980*b*. The conservation value for mammals of reserves in the western Australian wheatbelt. *Biol. Cons.* **18:**179–207.

Kitchener, D. J., J. Dell, and B. G. Muir. 1982. Birds in western Australian wheatbelt reserves—implications for conservation. *Biol. Cons.* **22:**127–63.

Klein, D. R. 1968. The introduction, increase, and crash of reindeer on St. Matthew Island. *J. Wildl. Manag.* **32:**350–67.

Klein, D. R. 1982. Fire, lichens, and caribou. *J. Range Manag.* **35:**390–95.

Klein, D. R. 1988. The establishment of muskox populations by translocation. Pp. 298–318 *in* L. Nielsen and R. D. Brown (Eds.), *Translocation of wild animals.* Wisconsin Humane Society, Madison, WI, and Caesar Kleberg Wildlife Research Institute, Kingsville, TX.

Klein, R. M. and T. D. Perkins. 1988. Primary and secondary causes and consequences of contemporary forest decline. *Bot. Rev.* **54:**1–43.

Kline, V. M. and E. A. Howell. 1987. Prairies. Pp. 75–83 *in* W. R. Jordan III, M. E. Gilpin, and J. D. Aber (Eds.), *Restoration ecology: A synthetic approach to ecological research.* Cambridge University Press, New York.

Klopatek, J. M., R. J. Olson, C. J. Emerson, and J. L. Jones. 1979. Land-use conflicts with natural vegetation in the United States. *Env. Cons.* **6:**191–99.

Knick, S. T. and J. T. Rotenberry. 1995. Landscape characteristics of fragmented shrubsteppe habitats and breeding passerine birds. *Cons. Biol.* **9:**1059–71.

Knox, G. A. 1984. The key role of krill in the ecosystem of the southern ocean with special reference to the Convention on the Conservation of Antarctic Marine Living Resources. *Ocean Manag.* **9:**113–56.

Kobayashi, S. 1985. Species diversity preserved in different numbers of nature reserves of the same total area. *Res. Pop. Ecol.* **27:**137–43.

Kohlhorst, D. W. 1980. Recent trends in the white sturgeon population in California's Sacramento-San Joaquin estuary. *Calif. Fish and Game* **66:**210–19.

Koopowitz, H. and H. Kaye. 1983. *Plant extinction: A global crisis.* Winchester Press, Piscataway, NJ.

Kubiak, T. J., H. J. Harris, L. M. Smith, T. R. Schwartz, D. L. Stalling, J. A. Trick, L. Sileo, D. E. Docherty, and T. C. Erdman. 1989. Microcontaminants and reproductive impairment of the Forster's tern on Green Bay, Lake Michigan—1983. *Arch. Env. Contam. Toxicol.* **18:**706–27.

Lacroix, G. L. 1987. Fish community structure in relation to acidity in three Nova Scotia rivers. *Can. J. Zool.* **65:**2908–15.

Lacy, R. C. 1987. Loss of genetic diversity from managed populations: Interacting effects of drift, mutation, immigration, selection, and population subdivision. *Cons. Biol.* **1:**143–58.

Lacy, R. C. 1988. A report on population genetics in conservation. *Cons. Biol.* **2:**245–48.

Lacy, R. C. 1989. Analysis of founder representation in pedigrees: Founder equivalents and founder genome equivalents. *Zoo Biol.* **8:**111–23.

Lamprey, H. F. 1983. Pastoralism yesterday and today: The overgrazing problem. Pp. 643–66 *in* F. Bourliere (Ed.), *Tropical savannas.* Elsevier, Amsterdam.

Lande, R. and G. F. Barrowclough. 1987. Effective population size, genetic variation, and their use in population management. Pp. 87–123 *in* M. E. Soulé (Ed.), *Viable populations for conservation.* Cambridge University Press, Cambridge, England.

Lathrop, E. W. 1983. Recovery of perennial vegetation in military maneuver areas. Pp. 265–77 *in* R. H. Webb and H. G. Wilshire (Eds.), *Environmental effects of off-road vehicles.* Springer-Verlag, New York.

Laws, R. M. 1985. The ecology of the southern ocean. *Amer. Sci.* **73**:26–40.

Lawson, D. E. 1986. Response of permafrost terrain to disturbance: A synthesis of observations from northern Alaska, U.S.A. *Arctic Alpine Res.* **18**:1–17.

Lawson, D. E., J. Brown, K. R. Everett, A. W. Johnson, V. Komarkova, B. M. Murray, D. F. Murray, and P. J. Webber. 1978. Tundra disturbance and recovery following the 1949 exploratory drilling, Fish Creek, northern Alaska. *U.S. Army Cold Regions Res. Eng. Lab. Rep.* 78–28.

Lawton, J. H. and R. M. May. 1995. *Extinction rates.* Oxford University Press, Oxford, England.

Laycock, W. A. 1991. Stable states and thresholds of range conditions on North American rangelands: A viewpoint. *J. of Range Manag.* **44**:427–33.

LeBaron, H. M. and J. Gressel (Eds.). 1982. *Herbicide resistance in plants.* John Wiley and Sons, New York.

Leck, C. F., B. G. Murray, Jr., and J. Swinebroad. 1988. Long-term changes in the breeding bird populations of a New Jersey forest. *Biol. Conser.* **46**:145–57.

Ledig, F. T. 1986. Heterozygosity, heterosis, and fitness in outbreeding plants. Pp. 77–104 *in* M. E. Soulé (Ed.), *Conservation biology: The science of scarcity and diversity.* Sinauer Associates, Sunderland, MA.

Lehman, J. T. and C. E. Caceres. 1993. Food-web responses to species invasion by a predatory invertebrate: *Bythotrepthes* in Lake Michigan. *Limnol. Oceanogr.* **38**:879–91.

Le Houerou, H. N. and H. Gillet. 1986. Conservation versus desertization in African arid lands. Pp. 444–61 *in* M. E. Soulé (Ed.), *Conservation biology: The science of scarcity and diversity.* Sinauer Associates, Sunderland, MA.

Leopold, A. 1933*a*. The conservation ethic. *J. Forestry* **31**:634–43.

Leopold, A. 1933*b*. *Game management.* Scribner and Sons, New York.

Leopold, A. 1949. *A sand county almanac.* Oxford University Press, New York.

Leopold, A. S. 1966. Adaptability of animals to habitat change. Pp. 66–75 *in* F. F. Darling and J. P. Milton (Eds.), *Future environments of North America.* Doubleday and Co., New York.

Lester, R. T. and J. P. Myers. 1989. Global warming, climate disruption, and biological diversity. Pp. 177–221 *in* W. J. Chandler (Ed.), *Audubon wildlife report 1989/1990.* Academic Press, San Diego, CA.

Lever, C. 1990. Lake Nakuru rhinoceros sanctuary. *Oryx* **24**:90–94.

Levin, S. S. (Ed.). 1993. Forum: Science and sustainability. *Ecol. Applications* **3**:545–89.

Liebhold, A. M., W. L. MacDonald, D. Bergdahl, and V. C, Mastro. 1995. Invasion by exotic forest pests: A threat to forest ecosystems. Society of American Foresters *Forest Science Monograph* **30**:1–49.

Limburg, K. E. 1986. PCBs in the Hudson. Pp. 83–130 *in* K. E. Limburg, M. A. Moran, and W. H. McDowell (Eds.), *The Hudson River ecosystem.* Springer-Verlag, New York.

Lincer, J. L. 1975. DDE-induced eggshell-thinning in the American Kestrel: A comparison of the field situation and laboratory results. *J. Appl. Ecol.* **12**:781–93.

Lincer, J. L., D. Zalkind, L. H. Brown, and J. Hopcraft. 1981. Organochlorine residues in Kenya's Rift Valley lakes. *J. Appl. Ecol.* **18**:157–71.

Lindzen, R. S. 1994. On the scientific basis for global warming. *Env. Pollut.* **83**:125–34.

Line, L. 1993. Silence of the songbirds. *Nat. Geogr.* **186(6)**:68–91.

Lodge, T. E. 1994. *The Everglades handbook.* St. Lucie Press, Delray Beach, FL.

Loftis, W. and O. Bass, Jr. 1992. Mercury threatens wildlife resources. *Park Sci.* **12(94)**:18–20.

Lomolino, M. V. 1994. An evaluation of alternative strategies for building networks of nature reserves. *Biol. Cons.* **69**:243–49.

Lomolino, M. V. and R. Channell. 1995. Splendid isolation: Patterns of geographical range collapse in endangered mammals. *Journal of Mammalogy* **76**:106–24.

Loope, L. L. and P. G. Scowcroft. 1985. Vegetation response within exclosures in Hawaii: A review. Pp. 377–402 *in* C. P. Stone and J. M. Scott (Eds.), *Hawaii's terrestrial ecosystems: Preservation and management.* University of Hawaii, Honolulu, HI.

Losos, E., J. Hayes, A. Philips, D. Wilcove, and C. Alkire. 1995. Taxpayer-subsidized resource extraction harms species. *BioScience* **45**:446–55.

Loughlin, T. R., J. L. Bengston, and R. L. Merrick. 1987. Characteristics of feeding trips of female northern fur seals. *Can. J. of Zool.* **65**:2079–84.

Louw, G. and M. Seely. 1982. *Ecology of desert organisms.* Longman, London.

Lovejoy, T. E. 1980. A projection of species extinctions. Pp. 328–31 *in The global 2000 report to the president.* Vol. 2. Council on Environmental Quality, Washington, DC.

Lovejoy, T. E. 1985. Amazonia, people and today. Pp. 328–38 *in* G. T. Prance and T. E. Lovejoy (Eds.), *Amazonia.* Pergamon Press, Oxford, England.

Lovejoy, T. E., R. O. Bierregaard, Jr., A. B. Rylands, J. R. Malcolm, C. E. Quintela, L. H. Harper, K. S. Brown, Jr., A. H. Powell, G. V. N. Powell, H. O. R. Shubart, and M. B. Hays. 1986. Edge and other effects of isolation on Amazon forest fragments. Pp. 257–85 *in* M. E. Soulé (Ed.), *Conservation biology: The science of scarcity and diversity.* Sinauer Associates, Sunderland, MA.

Lowman, M. D. and N. M. NadKarni (Eds.). 1995. *Forest canopies.* Academic Press, San Diego, CA.

Lowry, L. F. and K. J. Frost. 1985. Biological interactions between marine mammals and commercial fisheries in the Bering Sea. Pp. 41–61 *in* J. R. Beddington, R. J. H. Beverton, and D. M. Lavigne (Eds.), *Marine mammals and fisheries.* George Allen and Unwin, London.

Loya, Y. and B. Rinkevich. 1980. Effects of oil pollution on coral reef communities. *Mar. Ecol. Prog. Ser.* **3**:167–80.

Luckenbach, R. A. 1978. An analysis of off-road vehicle use on desert avifaunas. *N. A. Wildl. Nat. Res. Conf. Trans.* **43**:157–62.

Ludwig, D., R. Hilborn, and C. Walters. 1993. Uncertainty, resource exploitation, and conservation: Lessons from history. *Science* **260**:17, 36.

Lynch, J. F. 1989. Distribution of overwintering nearctic migrants in the Yucatan Peninsula, I. General patterns of occurrence. *Condor* **91**:515–44.

Lynch, J. F. and D. F. Whigham. 1984. Effects of forest fragmentation on breeding bird communities in Maryland, USA. *Biol. Cons.* **28**:287–324.

Mabbutt, J. A. 1984. A new global assessment of the status and trends of desertification. *Env. Cons.* **11**:103–13.

MacArthur, J. W. and E. O. Wilson. 1967. *The theory of island biogeography.* Princeton University Press, Princeton, NJ.

MacCall, A. D. 1986. Changes in the biomass of the California Current ecosystem. Pp. 33–54 *in* K. Sherman, and L. M. Alexander (Eds.), *Variability and management of large marine ecosystems.* Amer. Assoc. Adv. Sci., Washington, DC.

MacClintock, L., R. F. Whitcomb, and B. L. Whitcomb. 1977. Island biogeography and "habitat islands" of eastern forest. II. Evidence for the value of corridors and minimization of isolation in preservation of biotic diversity. *Amer. Birds* **31**:6–12.

MacDonald, G. J. 1985. *Climate change and acid rain.* MITRE Corporation, McLean, VA.

Macdonald, I. A. W. 1985. The Australian contribution to southern Africa's alien flora: An ecological analysis. *Proc. Ecol. Soc. Aust.* **14**:225–36.

Macdonald, I. A. W., L. Ortiz, J. E. Lawesson, and J. B. Nowal. 1988. The invasion of highlands in Galápagos by the red quinine tree *Cinchona succirubra. Env. Cons.* **15**:215–20.

Macdonald, I. A. W., L. L. Loope, M. B. Usher, and O. Hamman. 1989. Wildlife conservation and the invasion of nature reserves by introduced species: A global perspective. Pp. 215–55 *in* J. A. Drake, H. A. Mooney, F. di Castri, R. H. Groves, F. J. Kruger, M. Rejmanek, and M. Williamson (Eds.), *Biological invasions: A global perspective.* John Wiley and Sons, Chichester, England.

MacGregor, J. S. 1974. Changes in the amount and proportions of DDT and its metabolites, DDE and DDD, in the marine environment off southern California, 1949–1972. *Fishery Bull.* **72**:275–93.

Mack, R. N. 1981. Invasion of *Bromus tectorum* L. into western North America: An ecological chronicle. *Agro-Ecosystems* **7**:145–65.

Mack, R. N. 1989. Temperate grasslands vulnerable to plant invasions: Characteristics and consequences. Pp. 155–79 *in* J. A. Drake, H. A. Mooney, F. di Castri, R. H. Groves,

F. J. Kruger, M. Rejmanek, and M. Williamson (Eds.), *Biological invasions: A global perspective.* John Wiley and Sons, Chichester, England.

Mack, R. N. and J. N. Thompson. 1982. Evolution in steppe with few large, hoofed mammals. *Amer. Nat.* **119**:757–73.

MacKay, B. K. 1987. The arctic's abused bearer of white gold. *Mainstream* **18(3)**:32–34.

MacMahon, J. A. 1981. Successional processes: Comparisons among biomes with special reference to probable roles and influences on animals. Pp. 277–304 *in* D. C. West, H. H. Shugart, and D. B. Bodkin (Eds.), *Forest succession: Concepts and application.* Springer-Verlag, New York.

Madariaga, B. and K. E. McConnell. 1987. Exploring existence value. *Water Res. Res.* **23**:936–42.

Madenjian, C. P. 1995. Removal of algae by the zebra mussel (*Dreissena polymorpha*) population in western Lake Erie: A bioenergetics approach. *Can. J. Fish. Aq. Sci.* **52**:381–90.

Madronich, S., R. L. McKenzie, M. M. Caldwell, and L. O. Bjorn. 1995. Changes in ultraviolet radiation reaching the earth's surface. *Ambio* **24**:143–52.

Maehr, D. S. and J. A. Cox. 1995. Landscape features and panthers in Florida. *Cons. Biol.* **9**:1008–19.

Makarewicz, J. C. and P. Bertram. 1991. Evidence for the restoration of the Lake Erie ecosystem. *BioScience* **41**:216–23.

Makhijani, A. and K. R. Gurney. 1995. *Mending the ozone hole: Science, technology, policy.* MIT Press, Cambridge, MA.

Manire, C. A. and S. H. Gruber. 1990. Many sharks may be headed toward extinction. *Cons. Biol.* **4**:10–11.

Mann, C. C. and M. L. Plummer. 1993. The high cost of biodiversity. *Science* **260**:1868–71.

Mann, C. L. and M. L. Plummer. 1995. *Noah's choice. The future of endangered species.* Alfred A. Knopf, New York.

Manning, L. L. 1989. Marine mammals and fisheries conflicts: A philosophical dispute. *Ocean & Shoreline Manag.* **12**:217–32.

Mano, S. and M. O. Andrae. 1994. Emission of methyl bromide from biomass burning. *Science* **263**:1255–57

Margules, C. R., A. O. Nicholls, and R. L. Pressey. 1988. Selecting networks of reserves to maximize biological diversity. *Biol. Cons.* **43**:63–76.

Marquis, R. J. and C. J. Whelan. 1994. Insectivorous birds increase growth of white oak through consumption of leaf-chewing insects. *Ecology* **75**:2007–14.

Marsh, G. P. 1864. *Man and nature.* Charles Scribner, New York.

Marshall, D. B. 1989. The marbled murrelet. Pp. 434–455 *in* W. J. Chandler (Ed.), *Audubon wildlife report 1989/1990.* Academic Press, San Diego, CA.

Martin, E. B. and C. B. Martin. 1989. The Taiwanese connection—a new peril for rhinos. *Oryx* **23**:76–81.

Martin, P. S. 1973. The discovery of America. *Science* **179**:969–74.

Martin, P. S. 1984. Prehistoric overkill: The global model. Pp. 354–403 *in* P. S. Martin and R. G. Klein (Eds.), *Quaternary extinctions.* University of Arizona Press, Tucson, AZ.

Martinka, C. J. 1988. *An experiment in grizzly bear conservation: Glacier National Park.* Ecological Society of America Annual Meeting (Oral presentation), 15 August, Davis, CA.

Maser, C. 1994. *Sustainable forestry: Philosophy. science, and economics.* St. Lucie Press, Delray Beach, FL.

Maser, C. and J. R. Sedell. 1994. *From the forest to the sea: The ecology of wood in streams, rivers, estuaries, and the ocean.* St. Lucie Press, Delray Beach, FL.

Maxwell, J. 1996. *The salmon circle.* HarperCollins, New York.

May, R. M. 1988. How many species are there on earth? *Science* **241**:1441–49.

May, R. M. 1990. How many species? *Phil. Trans. R. Soc. London* **B330**:293–304.

May, R. M., J. H. Lawton, and N. E. Stork. 1995. Assessing extinction rates. Pp. 1–24 *in* J. H. Lawton and R. M. May (Eds.), *Extinction rates.* Oxford University Press, Oxford, England.

Mayfield, H. 1977. Brown-headed cowbird: Agent of extermination? *Amer. Birds* **31**:107–13.

McClenaghan, L. R., Jr. and A. C. Beauchamp. 1986. Low genic differentiation among isolated populations of the California fan palm (*Washingtonia filifera*). *Evolution* **40**:315–22.

McClenaghan, L. R., Jr., J. Berger, and H. D. Truesdale. 1990. Founding lineages and genic variability in plains bison (*Bison bison*) from Badlands National Park, South Dakota. *Cons. Biol.* **4**:285–89.

McCullough, D. R. 1979. *The George Reserve deer herd: Population ecology of a K-selected species.* University of Michigan Press, Ann Arbor, MI.

McCullough, D. R. 1984. Lessons from the George Reserve, Michigan. Pp. 211–42 *in* L. K. Halls (Ed.), *White-tailed deer: Ecology and management.* Stackpole Books, Harrisburg, PA.

McCullough, D. R. 1990. Detecting density dependence: Filtering the baby from the bathwater. *Trans. N. A. Wildl. Nat. Res. Conf.* **55**:534–43.

McCullough, D. R., D. S. Pine, D. L. Whitmore, T. M. Mansfield, and R. H. Decker. 1990. Linked sex harvest strategy for big game management with a test case on black-tailed deer. *Wild. Monogr.* No. 112.

McDade, A. and R. Emmott. 1995. Stage set for black bear restoration at Big South Fork. *Park Sci.* **15**(3):24–26.

McDonnell, M. J. 1981. Trampling effects on coastal dune vegetation in the Parker River National Wildlife Refuge, Massachusetts, USA. *Biol. Cons.* **21**:289–301.

McIntosh, R. P. 1995. H. A. Gleason's "Individualistic Concept" and theory of animal communities: A continuing controversy. *Biol. Rev.* **70**:317–57.

McKnight, B. N. (Ed.). 1993. *Biological pollution: The control and impact of invasive exotic species.* Indiana Academy of Science, Indianapolis, IN.

McMillen, J. L. 1988. Conservation of North American cranes. *Amer. Birds* **42**:1212–21.

McNaughton, S. J. 1985. Ecology of a grazing ecosystem: The Serengeti. *Ecol. Monogr.* **55**:259–94.

McNaughton, S. J. 1989. Ecosystems and conservation in the twenty-first century. Pp. 109–20 *in* D. Western and M. Pearl (Eds.), *Conservation for the twenty-first century.* Oxford University Press, New York.

McNaughton, S. J. 1990. Mineral nutrition and seasonal movements of African migratory ungulates. *Nature* **345**:613–15.

McNaughton, S. J. and G. A. Sabuni. 1988. Large African mammals as regulators of vegetation structure. Pp. 339–54 *in* M. J. A. Werger, P. J. M. van der Aart, and J. T. A. Verhoeven (Eds.), *Plant form and vegetation structure.* SPB Academic Publishing, The Hague, The Netherlands.

Mead, W. J. 1993. Review and analysis of state-of-the-art contingent valuation studies. Pp. 305–32 *in* J. A. Hausman (Ed.), *Contingent valuation: A critical assessment.* Elsevier Science Publishers, New York.

Mech, L. D. 1979. *The wolf: Ecology and behaviour of an endangered species.* Nat. Hist. Press, New York.

Meffe, G. K. and C. R. Carroll. 1994. *Principles of conservation biology.* Sinauer Associates, Sunderland, MA.

Meith, N. 1984. Saving the small cetaceans. *Ambio* **13**:2–13.

Mendelsohn, R. and M. J. Balick. 1995. The value of undiscovered pharmaceuticals in tropical forests. *Econ. Bot.* **49**:223–28.

Merton, D. V. 1975. The saddleback: Its status and conservation. Pp. 61–64 *in* R. D. Martin (Ed.), *Breeding endangered wildlife in captivity.* Academic Press, London.

Metcalfe, S. 1994. The Zimbabwe Communal Areas Management Programme for Indigenous Resources (CAMPFIRE). Pp. 161–92 *in* D. Western and R. M. Wright (Eds.), *Natural connections: Perspectives in community-based conservation.* Island Press, Washington, DC.

Meyer, C. H. 1989. Western water and wildlife: The new frontier. Pp. 59–91 *in* W. J. Chandler (Ed.), *Audubon wildlife report 1989/90.* Academic Press, San Diego, CA.

Michelmore, F., K. Beardsley, R. F. W. Barnes, and I. Douglas-Hamilton. 1994. A model illustrating the changes in forest elephant numbers caused by poaching. *Afr. J. Ecol.* **32**:89–99.

Miller B., G. Ceballos, and R. Reading. 1994. The prairie dog and biotic diversity. *Cons. Biol.* **8**:677–81.

Miller, D. G. M. and I. Hampton. 1989. Biology and ecology of the Antarctic krill (*Euphausa superba* Dana): A review. *Biomass* **1**:1–166.

Miller, R. I. 1978. Applying island biogeographic theory to an East African reserve. *Env. Cons.* **5**:191–95.

Miller, R. I. and L. D. Harris. 1977. Isolation and extirpations in wildlife reserves. *Biol. Cons.* **12**:311–15.

Miller, R. R., J. D. Williams, and J. E. Williams. 1989. Extinctions of North American fishes during the past century. *Fisheries* **14**:22–38.

Miller, T. O., G. D. Weatherford, and J. E. Thorson. 1986. *The salty Colorado.* The Conservation Foundation, Washington, DC.

Mills, E. L., J. H. Leach, J. T. Carlton, and C. L. Secor. 1994. Exotic species and the integrity of the Great Lakes. *BioScience* **44**:666–76.

Mills, J. A., W. G. Lee, and R. B. Lavers. 1989. Experimental investigations of the effects of takahe and deer grazing on *Chionochloa pallens* grassland, Fiordland, New Zealand. *J. Appl. Ecol.* **26**:397–417.

Mills, L. S. 1995. Edge effects and isolation: Red-backed voles on forest remnants. *Cons. Biol.* **9**:395–403.

Mills, L. S., M. E. Soulé, and D. F. Doak. 1993. The keystone-species concept in ecology and conservation. *BioScience* **43**:219–24.

Minckley, W. L. 1995. Translocation as a tool for conserving imperiled fishes: Experiences in the western United States. *Biol. Cons.* **72**:297–309.

Minckley, W. L. and J. E. Deacon. 1968. Southwestern fishes and the enigma of "endangered species." *Science* **159**:1424–32.

Mineau, P. 1988. Avian mortality in agro-ecosystems. 1. The case against granular insecticides in Canada. Pp. 3–12 *in Environmental effects of pesticides.* BCPC Monogr. No. 40.

Mintzer, I. 1988. Global climate change and its effects on wild lands. Pp. 56–67 *in* V. Martin (Ed.), *For the conservation of earth.* Fulcrum, Inc., Golden, CO.

Mitchell, B. A. 1989. Acid rain and birds: How much proof is needed? *Amer. Birds* **43**:234–41.

Mitsch, W. J. and J. G. Gosselink. 1986. *Wetlands.* Van Nostrand Reinhold, New York.

Mittermeier, R. A., I. de G. Camara, M. T. J. Padua, and J. Blanck. 1990. Conservation in the pantanal of Brazil. *Oryx* **24**:103–12.

Moffett, M. W. 1993. *The high frontier. Exploring the tropical rain forest canopy.* Harvard University Press, Cambridge, MA.

Moment, G. B. 1968. Bears: The need for a new sanity in wildlife conservation. *BioScience* **18**:1105–08.

Moment, G. B. 1969. Bears and conservation: realities and recommendations. *BioScience* **19**:1019–20.

Mono Basin Ecosystem Study Committee. 1987. *The Mono Basin ecosystem.* National Academy Press, Washington, DC.

Moore, F. and P. Kerlinger. 1987. Stopover and fat deposition by North American wood-warblers (*Parulinae*) following spring migration over the Gulf of Mexico. *Oecologia* **74**:47–54.

Moors, P. J. and I. A. E. Atkinson. 1985. *Conservation of island birds: Case studies for the management of threatened island species.* Int. Council Bird Preservation, Cambridge, England.

Moran, D. 1994. Contingent valuation and biodiversity: Measuring the user surplus of Kenyan protected areas. *Biodiversity and Conservation* **3**:663–84.

Moran, P. 1987. The *Acanthaster* phenomenon. *Oceanogr. Mar. Biol. Ann. Rev.* **24**:379–480.

Moreau, R. E. 1972. *The Palearctic-African bird migration systems.* Academic Press, London.

Morgan, S. O. (Ed.). 1988. *Caribou.* Alaska Dept. Fish and Game, Juneau, AK.

Morris, G. C. 1983. The Great Barrier Reef Marine Park: A unique management concept. *Parks* **8(3)**:1–4.

Morrison, R. I. G. and B. A. Harrington. 1979. Critical shorebird resources in James Bay and eastern North America. *Trans. N. A. Wildl. Nat. Res. Conf.* **44**:498–507.

Mosimann, J. E. and P. S. Martin. 1975. Simulating overkill by paleoindians. *Amer. Sci.* **63**:304–13.

Moss, B. 1988. *Ecology of fresh waters: Man and medium.* 2nd ed. Blackwell Scientific Publications, Oxford, England.

Moulton, M. P. and S. L. Pimm. 1986. Species introductions to Hawaii. Pp. 231–49 *in* H. A. Mooney and J. A. Drake (Eds.), *Ecology of biological invasions of North America and Hawaii.* Springer-Verlag, New York.

Mountainspring, S. and J. M. Scott. 1985. Interspecific competition among Hawaiian forest birds. *Ecol. Monogr.* **55**:219–39.

Mueller-Dombois, D. 1973. A nonadapted vegetation interferes with water removal in a tropical rain forest area in Hawaii. *Trop. Ecol.* **14**:1–18.

Mueller-Dombois, D., K. W. Bridges, and H. L. Carson (Eds.). 1981. *Island ecosystems.* Academic Press, New York.

Mueller-Dombois, D. and L. L. Loope. 1990. Some unique ecological aspects of oceanic island ecosystems. *Monogr. Syst. Bot. Missouri Bot. Gard.* **32**:21–27.

Muir, J. 1894. *The mountains of California.* Century Co., New York.

Muir, J. 1901. *Our national parks.* Houghton Mifflin Co., Boston.

Munasinghe, M. and W. Shearer (Eds.). 1995. *Defining and measuring sustainability: The biogeophysical foundations.* The World Bank, Washington, DC.

Munn, C. A. and B. A. Loiselle. 1995. Canopy access techniques and their importance for the study of tropical forest canopy birds. Pp. 165–77 *in* Lowman and Nadkarni (1995).

Myers, J. P. 1983. Conservation of migrating shorebirds: Staging areas, geographic bottlenecks, and regional movements. *Amer. Birds* **37**:23–25.

Myers, J. P., R. I. G. Morrison, P. Z. Antas, B. A. Harrington, T. E. Lovejoy, M. Sallaberry, S. E. Senner, and A. Tarak. 1987. Conservation strategy for migratory species. *Amer. Sci.* **75**:19–26.

Myers, K. 1986. Introduced vertebrates in Australia, with emphasis on the mammals. Pp. 120–36 *in* R. H. Groves and J. J. Burdon (Eds.), *Ecology of biological invasions.* Cambridge University Press, Cambridge, England.

Myers, N. 1980. *Conversion of moist tropical forests.* National Academy of Sciences, Washington, DC.

Myers, N. 1981. The hamburger connection: How Central America's forests become North America's hamburgers. *Ambio* **10**:3–8.

Myers, N. 1983. *A wealth of wild species.* Westview Press, Boulder, CO.

Myers, N. 1988. Tropical deforestation and climatic change. *Env. Cons.* **15**:293–98.

Myers, N. 1988. Threatened biotas: "Hotspots" in tropical forests. *The Environmentalist* **8**:187–208.

Myers, N. 1989. *Deforestation rates in tropical forests and their climatic implications.* Friends of the Earth, London.

Myers, N. 1990. The biodiversity challenge: Expanded "hotspots" analysis. *The Environmentalist* **10**:243–56.

Nadkarni, N. M. 1981. Canopy roots: Convergent evolution in rain forest nutrient cycles. *Science* **214**:1023–24.

Nadkarni, N. M. 1988. Tropical rain forest ecology from a canopy perspective. Pp. 189–208 *in* F. Almeda and C. M. Pringle (Eds.), *Tropical rain forests: Diversity and conservation.* California Academy of Sciences, San Francisco, CA.

Nadkarni, N. M. 1991. Fine litter dynamics within the tree canopy of a tropical cloud forest. *Ecology* **72**:2071–82.

Naeem, S. L., J. Thompson, S. P. Lawlor, J. H. Lawton, and R. M. Woodfin. 1994. Declining biodiversity can alter the performance of ecosystems. *Nature* **368**:734–37.

Naess, A. 1973. The shallow and the deep, long-range ecology movement: A summary. *Inquiry* **16**:95–100.

Naess, A. 1995. Politics and the ecological crisis: An introductory note. Pp. 445–53 *in* G. Sessions (Ed.). *Deep ecology for the 21st century.* Shambala, Boston, MA.

Nash, R. 1989a. *American environmentalism: Readings in conservation history.* 3rd ed. McGraw-Hill, New York.

Nash, R. F. 1989. *The rights of nature: A history of environmental ethics.* University of Wisconsin Press, Madison, WI.

National Academy of Sciences. 1972. *Genetic vulnerability of major crops.* National Academy Press, Washington, DC.

National Academy of Sciences. 1985. *Oil in the sea: Inputs, fates, and effects.* National Academy Press, Washington, DC. 601 pp.

National Research Council. 1982. *Ecological aspects of development in the humid tropics.* National Academy Press, Washington, DC.

National Research Council. 1986. *Pesticide resistance: Strategies and tactics for management.* National Academy Press, Washington, DC.

National Research Council. 1986. *Acid deposition: Long-term trends.* National Academy Press, Washington, DC.

National Research Council. 1990. *Decline of sea turtles: Causes and prevention.* National Academy Press, Washington, DC.

National Research Council. 1995. *Clean ships, clean ports, clean oceans: Controlling garbage and plastic wastes at sea.* National Academy Press, Washington, DC.

National Research Council. 1995. *Science and the Endangered Species Act.* National Academy Press, Washington, DC.

National Science Board. 1989. *Loss of biological diversity: A global crisis requiring international solutions.* National Science Foundation, Washington, DC.

Naveh, Z. and A. S. Lieberman. 1984. *Landscape ecology. Theory and applications.* Springer-Verlag, New York.

Neal, D. L., G. N. Steger, and R. C. Bertram. 1987. *Mountain lions: Preliminary finding on home-range use and density in the central Sierra Nevada.* USDA Forest Service Res. Note PSW-392. 6 pp.

Neff, J. M. 1990. Composition and fate of petroleum and spill-related agents in the marine environment, Pp. 1–33 *in* J. R. Geraci and D. J. St. Aubin (Eds.), *Sea mammals and oil: Confronting the risks.* Academic Press, San Diego, CA.

Nehlsen, W., J. E. Williams, and J. A. Lichatowich. 1991. Pacific salmon at the crossroads: Stocks at risk from California, Oregon, Idaho, and Washington. *Fisheries* 16(2):4–21.

Newby, J. E. 1988. Aridland wildlife in decline: The case of the scimitar-horned oryx. Pp. 146–66 *in* A. Dixon and D. Jones (Eds.), *Conservation and biology of desert antelopes.* Christopher Helm, London.

Newby, J. E. 1990. The slaughter of Sahelian wildlife by Arab royalty. *Oryx* 24(1):6–8.

Newmark, W. D. 1986. *Mammalian richness, colonization, and extinction in western North American national parks.* Ph. D. Dissertation, University of Michigan, Ann Arbor, MI.

Newmark, W. D. 1986. Species-area relationships and its determinants for mammals in western North American national parks. *Biol. J. Linn. Soc.* 28:83–98.

Newmark, W. D. 1993. The role and design of wildlife corridors with examples from Tanzania. *Ambio* 22:500–504.

Newmark, W. D. 1995. Extinction of mammal populations in western North American national parks. *Cons. Biol.* 9:512–26.

Nicholls, A. O. and C. R. Margules. 1991. The design of studies to demonstrate the biological importance of corridors. Pp. 49–61 *in* D. A. Saunders and R. J. Hobbs (Eds.),

Nature conservation 2: The role of corridors. Surrey Beatty and Sons, Chipping Norton, Australia.

Nichols, F. H., J. E. Cloern, S. N. Luoma, and D. H. Peterson. 1986. The modification of an estuary. *Science* 231:567–73.

Nichols, J. D., M. J. Conroy, D. R. Anderson, and K. P. Burnham. 1984. Compensatory mortality in waterfowl populations: A review of the evidence and implications for research and management. *Trans. N. A. Wildl. Nat. Res. Conf.* 49:535–54.

Nichols, J. D., F. A. Johnson, and B. K. Williams. 1995. Managing North American waterfowl in the face of uncertainty. *Ann. Rev. Ecol. Syst.* 26:177–99.

Nicol, S. 1990. The age-old problem of krill longevity. *BioScience* 40:833–36.

Nicol, S. and W. de la Mare. 1993. Ecosystem management and the Antarctic krill. *Amer. Sci.* 81:36–47.

Nielsen, B. 1984. The global plan of action for the conservation, management, and utilization of marine mammals. *Ambio* 15:134–36.

Nijhoff, P. 1979. Lake Baikal endangered by pollution. *Env. Cons.* 6:111–15.

Nilsson, S. and W. Schopfhauser. 1995. The carbon-sequestration potential of a global afforestation program. *Climate Change* 30:267–93.

Nisbet, I. C. T. 1970. Autumn migration of the blackpoll warbler: Evidence for long flight provided by regional survey. *Bird-Banding* 41:207–40.

Norment, C. and C. L. Douglas. 1977. *Ecological studies of feral burros in Death Valley.* Coop. Nat. Park. Resource Studies Unit, University of Nevada, Las Vegas, NV, Contrib. No. 17. 132 pp.

Norse, E. A. (Ed.). 1993. *Global marine biological diversity.* Island Press, Washington, DC.

Norton, B. G. 1986. On the inherent danger of undervaluing species. Pp. 110–37 *in* B. G. Norton (Ed.), *The preservation of species.* Princeton University Press, Princeton, NJ.

Norton, B. G. 1988. *Why preserve natural variety?* University of Chicago Press, Chicago, IL.

Norton, D. A. 1991. *Trilepidea adamsii*: An obituary for a species. *Cons. Biol.* 5:52–57.

Norton, M. R. and V. G. Thomas. 1994. Economic analyses of 'crippling losses' of North American waterfowl

and their policy implications for management. *Env. Cons.* 21:347–53.

Norton-Griffiths, M. and C. Southey. 1995. The opportunity costs of biodiversity conservation in Kenya. *Ecol. Econ.* 12:125–39

Noss, R. F. 1992. The Wildlands Project: Land conservation strategy. *Wild Earth* (Special Issue):10–25.

Noss, R. F. and L. D. Harris. 1986. Nodes, networks, and MUMs: Preserving diversity at all scales. *Environ. Manag.* 19:299–309.

Noss, R. F., E. T. LaRoe III, and J. M. Scott. 1995. *Endangered ecosystems of the United States: A preliminary assessment of loss and degradation.* USDI National Biological Service Biol. Report 28.

Noss, R. F. and A. Y. Cooperrider (Eds.) 1994. *Saving nature's legacy: Protecting and restoring biodiversity.* Island Press, Washington, DC.

O'Brien, S. J., M. E. Roelke, L. Marker, A. Newman, C. A. Winkler, D. Meltzer, L. Colly, J. F. Evermann, M. Bush, and D. E. Wildt. 1985. Genetic basis for species vulnerability in the cheetah. *Science* 227:1428–34.

O'Brien, S. J., D. E. Wildt, M. Bush, T. Caro, C. Fitzgibbon, I. Aggundey, and R. E. Leakey. 1987. East African cheetahs: Evidence for two population bottlenecks? *Proc. Nat. Acad. Sci. USA.* 84:508–11.

O'Connor, T. P., A. Y. Cantillo, and G. G. Lauenstein. 1994. Monitoring of temporal trends in chemical contamination by the NOAA National Status and Trends Mussel Watch Project. Pp. 29–50 *in* K. J. M. Kramer (Ed.), *Biomonitoring of coastal waters and estuaries.* CRC Press, Boca Raton, FL.

Odum, E. P. 1980. The status of three ecosystem-level hypotheses regarding salt marsh estuaries: Tidal subsidy, outwelling, and detritus-based food chains. Pp. 485–96 *in* V. S. Kennedy (Ed.), *Estuarine perspectives.* Academic Press, New York.

Odum, H. T. and E. P. Odum. 1955. Trophic structure and productivity of a windward coral reef community on Eniwetok Atoll. *Ecol. Monogr.* 25:291–320.

Odum, W. T., C. C. McIvor, and T. J. Smith III. 1982. *The ecology of the mangroves of south Florida: A community profile.* U.S. Fish Wildl. Serv., Office of Biol. Services, Washington, DC. FWS/OBS-81/24.

Oechel, W. C., S. J. Hastings, G. Vourlitis, M. Jenkins, G. Reichers, and N. Grulke. 1993. Recent change of Arctic tundra ecosystems from a net carbon dioxide sink to a source. *Nature* **361**:520–23.

Office of Migratory Bird Management. 1990. *Conservation of avian diversity in North America.* U.S. Fish and Wildlife Service, Washington, DC.

Office of Pesticides and Toxic Substances. 1985. *Suspended, canceled, and restricted pesticides.* U.S. Environmental Protection Agency, Washington, DC.

Office of Technology Assessment. 1988. *Technologies to maintain biological diversity.* J. B. Lippincott Co., Philadelphia, PA.

Officer, C. B. and C. L. Drake. 1983. The Cretaceous-Tertiary transition. *Science* **219**:1384–90.

O'Hara, K. J. 1988. Plastic debris and its effects on marine wildlife. Pp. 395–434 *in* W. J. Chandler (Ed.), *Audubon wildlife report 1988/89.* Academic Press, San Diego, CA.

Ohlendorf, H. M. 1989. Bioaccumulation and effects of selenium in wildlife. Pp. 133–77 *in Selenium in agriculture and the environment.* Soil Science Society of America, Madison, WI.

Ohlendorf, H. M. and W. J. Fleming. 1988. Birds and environmental contaminants in San Francisco and Chesapeake Bays. *Marine Pollut. Bull.* **19**:487–95.

Ohlendorf, H. M., R. L. Hothem, and D. Welsh. 1989. Nest success, cause-specific nest failure, and hatchability of aquatic birds at selenium-contaminated Kesterson Reservoir and a reference site. *Condor* **91**:787–96.

Ohlendorf, H. M., R. W. Risebrough, and K. Vermeer. 1978. Exposure of marine birds to environmental pollutants. *U.S. Fish and Wildl. Service Wildl. Res. Report No. 9,* 40 pp.

Ojasti, J. 1983. Ungulates and large rodents of South America. Pp. 427–39 *in* F. Bourliere (Ed.), *Tropical savannas.* Elsevier, Amsterdam.

Oksanen, L. 1990. Exploitation ecosystems in seasonal environments. *Oikos* **57**:14–24.

Oldemeyer, J. L. 1994. Livestock grazing and the desert tortoise in the Mojave Desert. USDI National Biol. Survey, *Fish and Wildlife Research Report* **13**:95–103.

Oldfield, M. L. 1989. *The value of conserving genetic resources.* Sinauer Associates, Sunderland, MA.

Olson, R. R. 1987. In situ culturing as a test of the larval starvation hypothesis for the crown-of-thorns starfish, *Acanthaster planci. Limnol. Oceanogr.* **32**:895–904.

Olson, S. L. and H. F. James. 1984. The role of Polynesians in the extinction of the avifauna of the Hawaiian Islands. Pp. 768–80 *in* P. S. Martin and R. G. Klein (Eds.), *Quaternary extinctions.* University of Arizona Press, Tucson, AZ.

O'Neill, R. V. 1996. Economic growth and sustainability: A new challenge. *Ecol. Applications* **6**:23–24.

Opler, P. A., H. G. Baker, and G. W. Frankie. 1977. Recovery of tropical lowland forest ecosystems. Pp. 379–421 *in* J. Cairns, Jr., K. L. Dickson, and E. E. Herricks (Eds.), *Recovery and restoration of damaged ecosystems.* University of Virginia Press, Charlottesburg, VA.

Oren, D. C. 1987. Grande Carajas, international financing agencies, and biological diversity in southeastern Brazilian Amazonia. *Cons. Biol.* **1**:222–27.

Orth, R. J. and K. A. Moore. 1983. Chesapeake Bay: An unprecedented decline in submerged aquatic vegetation. *Science* **222**:51–53

O'Shea, T. J., W. J. Fleming, and E. Cromartie. 1980. DDT contamination at Wheeler National Wildlife Refuge. *Science* **209**:509–10.

Owen-Smith, R. N. 1988. *Megaherbivores: The influence of very large size on ecology.* Cambridge University Press, Cambridge, England.

Owen-Smith, R. N. 1989. Megafaunal extinctions: The conservation message from 11,000 years B.P. *Cons. Biol.* **3**:405–12.

Packer, C., A. E. Pusey, H. Rowley, D. A. Gilbert, J. Martenson, and S. J. O'Brien. 1991. Case study of a population bottleneck: Lions of the Ngorongoro Crater. *Cons. Biol.* **5**:219–30.

Paddock, R. C. 1993. Iron Mountain mines defy efforts to stop toxic flow. *Los Angeles Times,* 10 April 1993.

Pamplin, W. L., Jr. 1986. Cooperative efforts to halt population declines of geese nesting on Alaska's Yukon-Kuskokwim Delta. *Trans N. A. Wildl. Nat. Res. Conf.* **51**:487–506.

Park, C. C. 1987. *Acid rain: Rhetoric and reality.* Methuen, London.

Pastor, J., R. J. Naiman, B. Dewey, and P. McInnes. 1988. Moose, microbes, and the boreal forest. *BioScience* **38**:770–77.

Paton, P. W. C. 1994. The effect of edge on avian nest success: How strong is the evidence? *Cons. Biol.* **8**:17–26.

Patrick, R., V. P. Binetti, and S. G. Halterman. 1981. Acid lakes from natural and anthropogenic causes. *Science* **211**:446–48.

Patterson, J. H. 1994. The North American Waterfowl Management Plan and Wetlands for the Americas programmes: A summary. *Ibis* **137**:S215–S218.

Pauly, D. and V. Christensen. 1995. Primary production required to sustain global fisheries. *Nature* **374**:255–57.

Payne, I. 1987. A lake perched on piscine peril. *New Scientist* **115(1575)**:50–54.

Payne, R. 1995. *Among whales.* Charles Scribner's Sons, New York.

Peakall, D. B. 1990. Prospects for the peregrine falcon, *Falco peregrinus,* in the nineties. *Can. Field Nat.* **104**:168–73.

Peixun, C. and W. Ding. 1988. The Chinese river dolphin, *Lipotes vexillifer. Endeavour* **12**:176–78.

Pellew, R. A. P. 1983. Impacts of elephant, giraffe, and fire upon the *Acacia tortilis* woodlands of the Serengeti. *Afr. J. Ecol.* **21**:41–74.

Perrin, W. F. 1989. *Dolphins, porpoises, and whales. An action plan for the conservation of biological diversity: 1988–1992.* IUCN, Gland, Switzerland.

Peters, R. L. 1991. Consequences of global warming for biological diversity. Pp. 99–118 *in* R. L. Wyman (Ed.), *Global climate change and life on earth.* Routledge, Chapman and Hall, New York.

Peters, R. L. and T. E. Lovejoy (Eds.). 1992. *Global warming and biological diversity.* Yale University Press, New Haven, CT.

Peterson, R. O. and R. E. Page. 1988. The rise and fall of Isle Royale wolves, 1975–1986. *J. Mammal.* **69**:89–99.

Peterson, R. O., R. E. Page, and K. M. Dodge. 1984. Wolves, moose, and the allometry of population cycles. *Science* **224**:1350–52.

Petraitis, P. S., R. E. Latham, and R. A. Niesenbaum. 1989. The maintenance of species diversity by disturbance. *Quart. Rev. of Biol.* **64**:393–418.

Petts, G. E., H. Moller, and A. L. Roux (Eds.). 1989. *Historical change of large alluvial rivers: Western Europe.* John Wiley and Sons, New York.

Philander, G. 1989. El Niño and La Niña. *Amer. Sci.* **77**:451–59.

Philippart, J. C. 1995. Is captive breeding an effective solution for the preservation of endemic species? *Biol. Cons.* **72**:281–95.

Phillips, M. K. 1990. *Restoration of endangered red wolves in northeastern North Carolina*. National Meeting of the Society for Ecological Restoration, Chicago, IL.

Phillips, O. L. and A. H. Gentry. 1994. Increasing turnover through time in tropical forests. *Science* **263**:954–58.

Pickett, S. T. A., S. L. Collins, and J. J. Armesto. 1987. Models, mechanisms and pathways of succession. *Bot. Rev.* **53**:335–71.

Pierce, C. and D. VanDeVeer (Eds.). 1995. *People, penguins, and plastic trees. Second Edition*. Wadsworth Publishing Co., Belmont, CA.

Pilkey, O. H. 1990. Barrier islands: Formed by fury, they roam and fade. *Sea Frontiers* **36**:(6):30–36.

Pimentel, D. and H. Lehman (Eds.). 1993. *The pesticide question: Environment, economics, and ethics*. Chapman and Hall, New York.

Pimentel, D., E. Garnick, A. Berkowitz, S. Jacobson, S. Napolitano, P. Black, S. Valdes-Cogliano, B. Vinzant, E. Hudes, and S. Littman. 1980. Environmental quality and natural biota. *BioScience* **30**:750–55.

Pimlott, D. H. 1967. Wolf predation and ungulate populations. *Amer. Zool.* **7**:267–78.

Pimlott, D. H. and P. W. Joslin. 1968. The status and distribution of the red wolf. *Trans. N. A. Wildl. Nat. Res. Conf.* **33**:373–89.

Pimm, S. L., H. L. Jones, and J. Diamond. 1988. On the risk of extinction. *Amer. Nat.* **132**:757–85.

Pimm, S. L., G. J. Russell, J. L. Gittleman, and T. M. Brooks. 1995. The future of biodiversity. *Science* **269**:347–50.

Pitcher, K. W. 1990. Major decline in number of harbor seals, *Phoca vitulina richardsi*, on Tugidak Island, Gulf of Alaska. *Marine Mammal Sci.* **6**:121–34.

Pitcher, T. J. and P. J. B. Hart. 1982. *Fisheries ecology*. Croom Helm, London, England.

Pitelka, L. F. and D. J. Raynal. 1989. Forest decline and acidic deposition. *Ecology* **70**:2–10.

Pollock, N. C. 1969. Some observations on game ranching in southern Africa. *Biol. Cons.* **2**:18–23.

Ponce, V. M. 1995. *Hydrologic and environmental impact of the Panana-Paraguay waterway on the Pantanal of Mato Grosso, Brazil: A reference study*. Report for the Charles Stewart Mott Foundation, Flint, MI.

Porcella, D. B. 1989. Lake acidification mitigation project (LAMP): An overview of an ecosystem perturbation experiment. *Can. J. Fish. Aq. Sci.* **46**:246–48.

Portney, P. R. 1994. The contingent valuation debate: Why economists should care. *J. Econ. Perspectives* **8**:3–17.

Powell, A. N. and F. J. Cuthbert. 1993. Augmenting small populations of plovers: An assessment of cross-fostering and captive rearing. *Cons. Biol.* **7**:160–68.

Powell, G. V. N. and J. H. Rappole. 1986. The hooded warbler. Pp. 827–53 *in* R. L. DiSilvestro (Ed.), *Audubon wildlife report 1986*. Nat. Audubon Soc., NY.

Powell, J. R. and J. P. Gibbs. 1995. A report from Galápagos. *Trends Ecol. Evol.* **10**:351–54.

Prabhu, M. A. 1988. International pesticide regulatory programs. *Environment* **30**(9):43–45.

Primack, R. B. 1993. *Essentials of conservation biology*. Sinauer Associates, Sunderland, MA.

Prose, D. V. 1985. Persisting effects of armored military maneuvers on some soils of the Mojave Desert. *Environ. Geol. Water Sci.* **7**:163–70.

Punsly, R. G., P. K. Tomlinson, and A. J. Mullen. 1994. Potential tuna catches in the eastern Pacific Ocean from schools not associated with dolphins. *Fishery Bull.* **92**:132–43.

Quinn, J. F. and S. P. Harrison. 1988. Effects of habitat fragmentation and isolation on species richness: Evidence from biogeographic patterns. *Oecologia* **75**:132–40.

Quinn, J. F. and A. Hastings. 1987. Extinction in subdivided habitats. *Cons. Biol.* **1**:198–208.

Quintela, C. E. 1990. An SOS for Brazil's beleaguered Atlantic forest. *Nature Conservancy Magazine* **40**(2):14–19.

Ralls, K. and J. Ballou. 1986. Preface to the proceedings of the workshop on genetic management of captive populations. *Zoo. Biol.* **5**(2):81–86.

Ralls, K., J. D. Ballou, and A. Templeton. 1988. Estimates of lethal equivalents and the cost of inbreeding in mammals. *Cons. Biol.* **2**:185–92.

Randall, A. 1986. Human preferences, economics, and the undervaluing of species. Pp. 79–109 *in* B. G. Norton (Ed.), *The preservation of species*. Princeton University Press, Princeton, NJ.

Randi, E. 1993. Effects of fragmentation and isolation on genetic variability of the Italian populations of wolf *Canis lupus* and brown bear *Ursus arctos*. *Acta Theriologica* **38, Suppl. 2**:113–20.

Rappole, J. H. 1995. *The ecology of migrant birds: A neotropical perspective*. Smithsonian Institution Press, Washington, DC.

Rappole, J. H. and D. W. Warner. 1978. Migratory bird population ecology: Conservation implications. *Trans. N. A. Wildl. Nat. Res. Conf.* **43**:235–40.

Rappole, J. H. and M. V. McDonald. 1994. Cause and effect in population declines of migratory birds. *Auk* **111**:652–60.

Ratcliffe, D. A. 1967. Decrease in eggshell weight in certain birds of prey. *Nature* **215**:208–10.

Raveling, D. G. and M. E. Heitmeyer. 1989. Relationships of population size and recruitment of pintails to habitat conditions and harvest. *J. Wildl. Manag.* **53**:1088–1103.

Raven, P. H. 1988. Our diminishing tropical forests. Pp. 119–22 *in* E. O. Wilson and F. M. Peter (Eds.), *Biodiversity*. National Academy Press, Washington, DC.

Ray, G. C. 1991. Coastal-zone biodiversity patterns. *BioScience* **41**:490–96.

Ray, G. C., B. P. Hayden, A. J. Bulger, Jr., and M. G. McCormick-Ray. 1992. Effects of global warming on the biodiversity of coastal-marine zones. Pp. 91–104 *in* R. L. Peters and T. E. Lovejoy (Eds.), *Global warming and biological diversity*. Yale University Press, New Haven, CT.

Ray, G. C. and M. G. McCormick-Ray. 1989. Coastal and marine biosphere reserves. Pp. 68–78 *in* W. P. Gregg, S. L. Krugman, and J. D. Wood (Eds.), *Proceedings of the symposium on biosphere reserves*. USDI National Park Service, Atlanta, GA.

Read, A. J. and D. E. Gaskin. 1988. Incidental catch of harbor porpoises by gill nets. *J. Wildl. Manag.* **52**:517–23.

Redford, K. H. 1992. The empty forest. *BioScience* **42**:412–22.

Reed, J. M., P. D. Doerr, and J. R. Walters. 1988. Minimum viable population size of the red-cockaded woodpecker. *J. Wildl. Manag.* **52**:385–91.

Rees, M. D. 1989. Red wolf recovery effort intensifies. *End. Sp. Tech. Bull.* **14(1–2)**:3.

Reeves, R. R., B. S. Stewart, and S. Leatherwood. 1991. *The Sierra Club handbook of seals and sirenians.* Sierra Club Books, San Francisco.

Regan, T. 1983. *The case for animal rights.* University of California Press, Berkeley, CA.

Regier, H. A. 1979. Changes in species composition of Great Lakes fish communities caused by man. *Trans. N. A. Wildl. Nat. Res. Conf.* **44**:558–66.

Regier, H. A. and W. L. Hartman. 1973. Lake Erie's fish community: 150 years of cultural stress. *Science* **180**:1248–55.

Regier, H. A. and K. H. Loftus. 1972. Effects of fisheries exploitation on salmonid communities of oligotrophic lakes. *J. Fish. Res. Bd. Canada* **29**:959–68.

Reichman, O. J. 1987. *Konza Prairie: A tallgrass natural history.* University Press of Kansas, Lawrence, KS.

Reid, T. S. and D. D. Murphy. 1995. Providing a regional context for local conservation action. *BioScience* **45(Supplement)**:84–90.

Reid, W. V. and K. R. Miller. 1989. *Keeping options alive: The scientific basis for conserving biodiversity.* World Resources Institute, Washington, DC.

Riedman, M. 1990. *The pinnipeds: Seals, sea lions, and walruses.* University of California Press, Berkeley, CA.

Reidman, M. L. and J. A. Estes. 1988. A review of the history, distribution and foraging ecology of sea otters. Pp. 4–21 *in* G. R. VanBlaricom and J. A. Estes (Eds.), *The community ecology of sea otters.* Springer-Verlag, Berlin.

Reijnders, P. J. H. 1986. Reproductive failure in common seals feeding on fish from polluted coastal waters. *Nature (Lond.)* **324**:456–57.

Repetto, R. 1985. *Paying the price: Pesticide subsidies in developing countries.* World Resources Institute Research Report 2.

Reuther, R. 1994. Mercury accumulation in sediment and fish from rivers affected by alluvial gold mining in the Brazilian Madeira River basin, Amazon. *Env. Monitor. Assess.* **32**:239–58.

Rex, M. A., C. T. Stuart, R. R. Hessler, J. A. Allen, H. L. Sanders, and G. D. F. Wilson. 1993. Global-scale latitudinal patterns of species diversity in the deep-sea benthos. *Nature* **365**:636–39.

Reynolds, R. E. and J. R. Sauer. 1991. Changes in mallard breeding populations in relation to production and harvest rates. *J. Wildl. Manag.* **55**:483–87.

Ricklefs, R. E. and G. W. Cox. 1972. The taxon cycle in the West Indian avifauna. *Amer. Nat.* **106**:195–219.

Robards, M. D., J. F. Piatt, and K. D. Wohl. 1995. Increasing frequency of plastic particles ingested by seabirds in the subarctic North Pacific. *Mar. Pollut. Bull.* **30**:151–57.

Robbins, C. S., D. K. Dawson, and B. A. Dowell. 1989. Habitat area requirements of breeding forest birds of the Middle Atlantic States. *Wildl. Monogr.* **103**:1–34.

Robbins, C. S., J. R. Sauer, R. S. Greenberg, and S. Droege. 1989. Population declines in North American birds that migrate to the neotropics. *Proc. Natl. Acad. Sci. USA* **86**:7658–62.

Robel, R. J., A. D. Dayton, F. R. Henderson, R. L. Meduna, and C. W. Spaeth. 1981. Relationships between husbandry methods and sheep losses to canine predators. *J. Wildl. Manag.* **45**:894–911.

Roberts, C. M. 1995. Effects of fishing on the ecosystem structure of coral reefs. *Cons. Biol.* **9**:988–95.

Robins, C. R. 1991. Regional diversity among Caribbean fish species. *BioScience* **41**:458–59.

Robinson, S. K. 1992. Population dynamics of breeding neotropical migrants in a fragmented Illinois landscape. Pp. 408–418 *in* J. M. Hagan and D. W. Johnston (Eds.), *Ecology and conservation of neotropical migrant landbirds.* Smithsonian Institution Press, Washington, DC.

Rodhe, H. 1990. A comparison of the contribution of various gases to the greenhouse effect. *Science* **248**:1217–19.

Rodhe, H. and R. Herrera (Eds.). 1988. *Acidification in tropical countries.* John Wiley and Sons, New York.

Roemmich, D. and J. McGowan. 1995. Climatic warming and the decline of zooplankton in the California Current. *Science* **267**:1324–26.

Rolston, H., III. 1988. *Environmental ethics: Duties to and values in the natural world.* Temple University Press, Philadelphia, PA.

Romme, W. H. and D. G. Despain. 1989. The Yellowstone fires. *Sci. Amer.* **261(5)**:37–46.

Ronald, K. and J. L. Dougan. 1982. The ice lover: Biology of the harp seal (*Phoca groenlandica*). *Science* **215**:928–33.

Rosen, Y. 1991. Alaska fisheries depleted by foreign fleets. *Christian Science Monitor,* 5 November 1991.

Rosenberg, A. A., M. J. Fogarty, M. P. Sissenwine, J. R. Beddington, and J. G. Shepherd. 1993. Achieving sustainable use of renewable resources. *Science* **262**:828–29.

Rosencrantz, A. and A. Scott. 1992. Siberia's threatened forests. *Nature* **355**:293–94.

Rosenthal, D. H., J. B. Loomis, and G. L. Peterson. 1984. *The travel cost model: Concepts and applications.* USDA Forest Service Ben. Tech. Rep. RM-109.

Ross, J. P. 1981. Historical decline of Loggerhead, Ridley, and Leatherback sea turtles. Pp. 189–95 *in* K. A. Bjorndal (Ed.), *Biology and conservation of sea turtles.* Smithsonian Institution Press, Washington, DC.

Royte, E. and C. Jones. 1995. Hawaii's vanishing species. *Nat. Geogr.* **188(3)**:3–37.

Rudd, R. L. 1964. *Pesticides and the living landscape.* University of Wisconsin Press, Madison, WI.

Rudnicky, T. C. and M. L. Hunter, Jr. 1993. Reversing the fragmentation perspective: Effects of clearcut size on bird species richness in Maine. *Ecol. Applications* **3**:357–66.

Ruesink, J. L., I. M. Parker, J. J. Groom, and P. M. Karieva. 1995. Reducing the risk of nonindigenous species introductions. *BioScience* **45**:465–77.

Ruess, R. W. 1987. Herbivory. The role of large herbivores in nutrient cycling of tropical savannas. Pp. 67–91 *in* B. H. Walker (Ed.), *Determinants of tropical savannas.* Int. Union. Biol. Sci., Paris, France.

Russell, C. E. 1987. Plantation forestry. Pp. 76–89 *in* C. F. Jordan (Ed.), *Amazonian rain forests: Ecosystem disturbance and recovery.* Springer-Verlag, New York.

Ryther, J. H. 1969. Photosynthesis and fish production in the sea. *Science* **166**:72–76.

Ryther, J. H. and W. M. Dunstan. 1971. Nitrogen, phosphorus, and eutrophication in the coastal marine environment. *Science* **171**:1008–13.

Sagoff, M. 1988. *The economy of the earth.* Cambridge University Press, Cambridge, England.

Sahrhage, D. 1989. Antarctic krill fisheries: Potential resources and ecological concerns. Pp. 13–33 in J. F. Caddy (Ed.), *Marine invertebrate fisheries: Their assessment and management.* John Wiley and Sons, New York.

Salwasser, H. 1974. Coyote scats as an indicator of the time of fawn mortality in the North Kings deer herd. *Calif. Fish and Game* **60**:84–87.

Salwasser, H., C. Schonewald-Cox, and R. Baker. 1987. The role of interagency cooperation in managing for viable populations. Pp. 159–73 in M. E. Soulé (Ed.), *Viable populations for conservation.* Cambridge University Press, Cambridge, England.

Sanders, H. L., J. F. Grassle, G. R. Hampson, L. S. Morse, S. Garner-Price, and C. C. Jones. 1980. Anatomy of an oil spill: Long-term effects of the grounding of the barge *Florida* off West Falmouth, Massachusetts. *J. Mar. Res.* **38**:265–380.

Santiapillai, C. and W. S. Ramono. 1987. Tiger numbers and habitat evaluation in Indonesia. Pp. 85–91 in R. L. Tilson and U. S. Seal (Eds.), *Tigers of the world: The biology, biopolitics, management, and conservation of an endangered species.* Noyes Publications, Park Ridge, NJ.

Sarmiento, G. 1984. *The ecology of neotropical savannas.* Harvard University Press, Cambridge, MA.

Saunders, D. A., R. J. Hobbs, and C. R. Margules. 1991. Biological consequences of ecosystem fragmentation. A review. *Cons. Biol.* **5**:18–32.

Savidge, J. A. 1987. Extinction of an island forest avifauna by an introduced snake. *Ecology* **68**:660–68.

Schaller, G. B. 1972. *The Serengeti lion.* University of Chicago Press, Chicago, IL.

Scheffer, V. B. 1989. Environmentalism: Its articles of faith. *Northwest Env. J.* **5**:99–109.

Scheffer, V. B. 1991. *The shaping of environmentalism in America.* University of Washington Press, Seattle, WA.

Scheffer, V. B. and K. W. Kenyon. 1989. The rise and fall of a seal herd. *Animal Kingdom* **92(2)**:14–19.

Schindler, D. W. 1974. Eutrophication and recovery in experimental lakes: Implications for lake management. *Science* **184**:897–99.

Schindler, D. W. 1988. Effects of acid rain on freshwater ecosystems. *Science* **239**:149–56.

Schindler, D. W., K. H. Mills, D. F. Malley, D. L. Findlay, J. A. Shearer, I. J. Davies, M. A. Turner, G. A. Lindsay, and D. R. Cruikshank. 1985. Long-term ecosystem stress: The effects of years of experimental acidification on a small lake. *Science* **228**:1395–1401.

Schindler, D. W., K. G. Beaty, E. J. Fee, D. R. Cruikshank, E. R. DeBruyn, D. L. Findley, G. A. Linsey, J. A. Shearer, M. P. Stainton, and M. A. Turner. 1990. Effects of climatic warming on lakes of the central boreal forest. *Science* **250**:967–70.

Schlesinger, W. H. 1991. *Biogeochemistry: An analysis of global change.* Academic Press, San Diego, CA.

Schlesinger, W. H., J. F. Reynolds, G. L. Cunningham, L. F. Huenneke, W. M. Jarrell, R. A. Virginia, and W. G. Whitford. 1990. Biological feedbacks in global desertification. *Science* **247**:1043–48.

Schlosser, D. W. and T. F. Nalepa. 1994. Dramatic decline of unionid bivalves in offshore waters of western Lake Erie after infestation by the zebra mussel. *Dreissena polymorpha. Can. J. Fish. Aq. Sci.* **51**:2234–42.

Schneider, S. H. 1992. The climatic response to greenhouse gases. *Adv. Ecol. Res.* **22**:1–32.

Schofield, E. K. 1989. Effects of introduced plants and animals on island vegetation: Examples from the Galápagos archipelago. *Cons. Biol.* **3**:227–38.

Schomer, N. S. and R. D. Drew. 1982. *An ecological characterization of the lower Everglades, Florida Bay, and the Florida Keys.* U.S. Fish and Wildl. Serv., Office of Biol. Services, Washington, DC. FWS/OBS-82/58.1.

Schrader-Frechette, K. S. (Ed.). 1981. *Environmental ethics.* Boxwood Press, Pacific Grove, CA.

Schrader-Frechette, K. S. and E. D. McCoy. 1993. *Method in ecology: Strategies for conservation.* Cambridge University Press, Cambridge, England.

Schrivener, J. H., W. E. Howard, A. H. Murphy, and J. R. Hays. 1985. Sheep losses to predators on a California range, 1973–1983. *J. Wildl. Manag.* **38**:418–21.

Scott, J. M. and J. L. Sincock. 1985. Hawaiian birds. Pp. 549–62 in R. L.

Di Silvestro (Ed.), *Audubon wildlife report 1985.* National Audubon Society, New York.

Scott, J. M., S. Mountainspring, F. L. Ramsey, and C. B. Kepler. 1986. *Forest bird communities of the Hawaiian Islands: Their dynamics, ecology, and conservation.* Studies in Avian Biology, No. 9.

Scott, J. M., B. Csuti, J. D. Jacobi, and J. E. Estes. 1987. Species richness: A geographical information systems approach to the protection of biodiversity. *BioScience* **39**:782–88.

Scott, J. M., C. B. Kepler, P. Stine, H. Little, and K. Taketa. 1987b. Protecting endangered forest birds in Hawaii: The development of a conservation strategy. *Trans. N. A. Wildl. Nat. Res. Conf.* **52**:348–63.

Scott, J. M., B. Csuti, K. Smith, J. E. Estes, and S. Caicco. 1988. Beyond endangered species: An integrated conservation strategy for the preservation of biological diversity. *End. Sp. UPDATE* **5**:43–48.

Scott, J. M., C. B. Kepler, C. van Riper III, and S. I. Fefer. 1988. Conservation of Hawaii's vanishing avifauna. *BioScience* **38**:238–53.

Scott, J. M., F. Davis, B. Csuti, R. Noss, B. Butterfield, C. Groves, H. Anderson, S. Caicco, F. D'Erchia, T. C. Edwards, Jr., J. Ulliman, and R. G. Wright. 1993. Gap analysis: A geographic approach to protection of biological diversity. *Wild. Monogr.* **123**:1–41.

Scully, R. T., W. Y. Brown, and B. S. Manheim. 1986. The Convention for the Conservation of Antarctic Marine Living Resources: A model for large marine ecosystem management. Pp. 281–86 in K. Sherman and L. M. Alexander (Eds.), *Variability and management of large marine ecosystems.* Amer. Assoc. Adv. Sci., Washington, DC.

Seagle, S. W., R. A. Lancia, D. A. Adams, M. R. Lennartz, and H. A. Devine. 1987. Integrating timber and red-cockaded woodpecker habitat management. *Trans. N. A. Wildl. Nat. Res. Conf.* **52**:41–52.

Sears, P. B. 1935. *Deserts on the march.* University of Oklahoma Press, Norman, OK.

Seegmiller, R. F. and R. D. Ohmart. 1981. Ecological relationships of feral burros and desert bighorn sheep. *Wildl. Monogr.* **78**:1–58.

Selander, R. K. 1983. Evolutionary consequences of inbreeding. Pp. 201–15 in C. M. Schoenewald-Cox, S. M. Chambers, B. MacBryde, and L. Thomas (Eds.), *Genetics and conservation*. Benjamin/Cummings, Menlo Park, CA.

Seliger, H. H., J. A. Boggs, and W. H. Biggley. 1985. Catastrophic anoxia in the Chesapeake Bay in 1984. *Science* **228**:70–73.

Senner, S. E. 1979. An evaluation of the Copper River delta as critical habitat for migrating shorebirds. *Studies in Avian Biology* **2**:131–45.

Sessions, George (Ed.). 1995. *Deep ecology for the 21st century*. Shambala, Boston, MA.

Shafer, C. L. 1990. *Nature reserves: Island theory and conservation practice*. Smithsonian Institution Press, Washington, DC.

Shafer, C. L. 1995. Values and shortcomings of small reserves. *BioScience* **45**:80–88.

Shaffer, M. 1987. Minimum viable populations: Coping with uncertainty. Pp. 69–86 in M. E. Soulé (Ed.), *Viable populations for conservation*. Cambridge University Press, Cambridge, England.

Shaffer, M. L. 1981. Minimum population sizes for species conservation. *BioScience* **31**:131–34.

Shaughnessy, G. 1986. A case study of some woody plant introductions to the Cape Town area. Pp. 37–43 in I. A. W. Macdonald, F. J. Kruger, and A. A. Ferrar (Eds.). 1986. *The ecology and management of biological invasions in southern Africa*. Oxford University Press, Cape Town, South Africa.

Shaver, D. 1992. Kemp's Ridley research continues at Padre Island National Seashore. *Park Sci.* **12(4)**:26–27.

Shaver, G. R., W. D. Billings, F. S. Chapin III, A. E. Giblin, K. N. Nadelhoffer, W. C. Oechel, and E. B. Rastetter. 1992. Global change and the carbon balance of arctic ecosystems. *BioScience* **42**:433–41.

Sherman, K. 1986. Large marine ecosystems as tractable units for measurement and management. Pp. 3–7 in K. Sherman and L. M. Alexander (Eds.), *Variability and management of large marine ecosystems*. Amer. Assoc. Adv. Sci., Washington, DC.

Sherry, T. W. and R. T. Holmes. 1996. Winter habitat quality, population limitation, and conservation of neotropical-nearctic migrant birds. *Ecology* **77**:36–48.

Shukla, J., C. Nobre, and P. Sellers. 1990. Amazon deforestation and climate change. *Science* **247**:1322–25.

Siegel, V. and V. Loeb. 1995. Recruitment of Antarctic krill *Euphausia superba* and possible causes for its variability. *Mar. Ecol. Prog. Ser.* **123**:45–56.

Siegfried, W. R. 1985. Krill: The last unexploited food resource. *Optima* **33**:67–79.

Simberloff, D. 1986. Are we on the verge of a mass extinction in tropical rain forests? Pp. 165–80 in D. K. Elliott (Ed.), *Dynamics of extinction*. John Wiley and Sons, New York.

Simberloff, D. 1987. The spotted owl fracas: Mixing academic, applied, and political ecology. *Ecology* **68**:766–72.

Simberloff, D. and L. G. Abele. 1976. Island biogeography theory and conservation practice. *Science* **191**:285–86.

Simberloff, D. and L. G. Abele. 1982. Refuge design and island biogeographic theory: Effects of fragmentation. *Amer. Nat.* **120**:41–50.

Simberloff, D., J. A. Farr, J. Cox, and D. W. Mehlman. 1992. Movement corridors: Conservation bargains or poor investments? *Cons. Biol.* **6**:493–504.

Simonich, S. L. and R. A. Hites. 1995. Global distribution of persistent organochlorine compounds. *Science* **269**:1851–54.

Simons, T., S. K. Sherrod, M. W. Collopy, and M. A. Jenkins. 1988. Restoring the bald eagle. *Amer. Sci.* **76**:253–60.

Simpson, P. W., J. R. Newman, M. A. Keirn, R. M. Matter, and P. A. Guthrie. 1982. *Manual of stream channelization impacts on fish and wildlife*. Biological Services Program, U.S. Fish and Wildlife Service (FWS/OBS-82/24), Washington, DC.

Sinclair, A. R. E. 1979. Dynamics of the Serengeti ecosystem. Pp. 1–30 in A. R. E. Sinclair and M. Norton-Griffiths (Eds.), *Serengeti: Dynamics of an ecosystem*. University of Chicago Press, Chicago, IL.

Sinclair, A. R. E. 1995. Serengeti past and present. Pp. 3–30 in Sinclair and Arcese (1995).

Sinclair, A. R. E. and P. Arcese (Eds.). 1995. *Serengeti II: Dynamics, management, and conservation of an ecosystem*. University of Chicago Press, Chicago, IL.

Singer, F. J. 1990. Some predicted effects concerning a wolf recovery into Yellowstone National Park. Pp. 4:3–34 in *Wolves for Yellowstone? A report to the United States Congress*. Volume II. U.S. Government Printing Office, Billings, MT.

Siniff, D. B. 1991. An overview of the ecology of antarctic seals. *Amer. Zool.* **31**:143–49.

Skinner, J. D. 1985. Wildlife management in practice: Conservation of ungulates through protection or utilization. Pp. 25–46 in J. P. Hearn and J. K. Hodges (Eds.), *Advances in animal conservation*. Clarendon Press, Oxford, England.

Skole, D. and C. Tucker. 1993. Tropical deforestation and habitat fragmentation in the Amazon: Satellite data from 1978 to 1988. *Science* **260**:1905–10.

Slatyer, R. O. 1983. The origin and evolution of the World Heritage Convention. *Ambio* **12**:138–45.

Smit, C. J., R. H. D. Lambeck, and W. J. Wolff. 1987. *Threats to coastal wintering and staging areas of waders*. Wader Study Group **49 (Supplement):** 105–13.

Smith, B. R. and J. J. Tibbles. 1980. Sea lamprey (*Petromyzon marinus*) in Lakes Huron, Michigan, and Superior: History of invasion and control, 1936–1978. *Can. J. Fish. Aq. Sci.* **37**:1780–1801.

Smith, C. W. 1985. Impact of alien plants on Hawaii's native biota. Pp. 180–250 in C. P. Stone and J. M. Scott (Eds.), *Hawaii's terrestrial ecosystems: Preservation and management*. University of Hawaii, Honolulu, HI.

Smith, C. W. 1989. Non-native plants. Pp. 60–69 in C. P. Stone and D. B. Stone (Eds.), *Conservation biology in Hawaii*. University of Hawaii Press, Honolulu, HI.

Smith, F. D. M., R. M. May, R. Pellew, T. H. Johnson, and K. S. Walter. 1993. Estimating extinction rates. *Nature* **364**:494–96.

Smith, G. J. 1987. Pesticide use and toxicology in relation to wildlife: Organophosphorus and carbamate compounds. *U.S. Dept. Int., Fish Wildl. Ser., Res. Publ.* 170.

Smith, G. W. and R. E. Reynolds. 1992. Hunting and mallard survival, 1979–88. *J. Wildl. Manag.* **56**:306–16.

Smith, J. N. M., C. J. Krebs, A. R. E. Sinclair, and R. Boonstra. 1988. Population biology of snowshoe hares. II. Interactions with winter food plants. *J. Anim. Ecol.* **57:**269–86.

Smith, J. R. 1990. Coyote diets associated with seasonal mule deer activities in California. *Calif. Fish and Game* **76:**78–82.

Smith, K. A. 1976. *The natural resources of the Nipomo Dunes and wetlands.* Coastal Wetlands Series 15, Calif. Dept. Fish and Game, Sacramento, CA.

Smith, K. G., J. D. Clark, and P. S. Gipson. 1991. History of black bears in Arkansas: Over-exploitation, elimination, and successful reintroduction. *Eastern Workshop on Black Bear Research and Management* **10:**5–14.

Smith, R. A., R. B. Alexander, and M. G. Wolman. 1987. Water-quality trends in the nation's rivers. *Science* **235:**1607–15.

Smith, S. V. and R. W. Buddemeier. 1992. Global change and coral reef ecosystems. *Ann. Rev. Ecol. Syst.* **23:**89–118.

Snyder, N. F. R. and H. A. Snyder. 1989. Biology and conservation of the California condor. *Current Ornithology* **6:**175–267.

Solbrig, O. T. and M. D. Young. 1992. Toward a sustainable and equitable future for savannas. *Environment* **34(3):**6–15, 32–35.

Soulé, M. E. 1985. What is conservation biology? *BioScience* **35:**727–34.

Soulé, M. E. 1987. History of the Society for Conservation Biology: How and why we got here. *Cons. Biol.* **1:**4–5.

Soulé, M. E. 1987. Where do we go from here? Pp. 175–83 *in* M. E. Soulé (Ed.), *Viable populations for conservation.* Cambridge University Press, Cambridge, England.

Soulé, M. E., D. T. Bolger, A. C. Alberts, J. Wright, M. Sorice, and S. Hill. 1988. Reconstructed dynamics of rapid extinctions of chaparral-requiring birds in urban habitat islands. *Cons. Biol.* **2:**75–92.

Soulé, M. E., M. Gilpin, W. Conway, and T. Foose. 1986. The millennium ark: How long a voyage, how many staterooms, how many passengers? *Zoo Biol.* **5:**101–13.

Soulé, M. E. and M. E. Gilpin. 1991. The theory of wildlife corridor capability. Pp. 3–8 *in* D. A. Saunders and R. J. Hobbs (Eds.), *Nature conservation 2: The role of corridors.* Surrey Beatty and Sons, Chipping Norton, Australia.

Soulé, M. E., B. A. Wilcox, and C. Holtby. 1979. Benign neglect: A model of faunal collapse in the game reserves of East Africa. *Biol. Cons.* **15:**259–72.

Sparks, R. E. 1995. Need for ecosystem management of large rivers and their floodplains. *BioScience* **45:**168–82.

Spear, L. B., D. G. Ainley, and C. A. Ribie. 1995. Incidence of plastic in seabirds from the tropical Pacific, 1984–91: Relation with distribution of species, sex, age, season, year, and body weight. *Mar. Env. Res.* **40:**123–46.

Spencer, C. N., B. R. McClelland, and J. A. Stanford. 1991. Shrimp stocking, salmon collapse, and eagle displacement. *BioScience* **41:**14–21.

Spitzer, P. R. and A. F. Poole. 1980. Coastal ospreys between New York City and Boston: A decade of reproductive recovery, 1969–1979. *Amer. Birds* **34:**234–41.

Sprugel, D. G. 1991. Distance, equilibrium, and environmental variability: What is "natural" vegetation in a changing environment? *Biol. Cons.* **58:**1–18.

Stanley Price, M. R. 1989. *Animal reintroductions: The Arabian oryx in Oman.* Cambridge University Press, Cambridge, England.

Stark N. M. and C. F. Jordan. 1978. Nutrient retention by the root mat of an Amazonian rain forest. *Ecology* **59:**434–37.

St. Aubin, D. J. and V. Lounsbury. 1990. Oil effects on manatees: Evaluating the risks. Pp. 241–51 *in* J. R. Geraci and D. J. St. Aubin (Eds.), *Sea mammals and oil: Confronting the risks.* Academic Press, San Diego, CA.

Steadman, D. W. 1986. *Holocene vertebrate fossils from Isla Floreana, Galápagos.* Smithsonian Contrib. Zool. No. 413.

Steadman, D. W. 1995. Prehistoric extinctions of Pacific island birds: Biodiversity meets zooarcheology. *Science* **267:**1123–31.

Stenson, J. A. E. and M. O. G. Eriksson. 1989. Ecological mechanisms important for the biotic changes in acidified lakes in Scandinavia. *Arch. Environ. Contam. Toxicol.* **18:**201–6.

Stephenson, M. D., M. Martin, and R. S. Tjeerdema. 1995. Long-term trends in DDT, polychlorinated biphenyls, and chlordane in California mussels. *Arch. Env. Contam. Toxicol.* **28:**443–50.

Sternin, V. and W. H. L. Alsopp. 1982. Strategies to avoid by-catch in shrimp trawling. Pp. 61–64 *in Fish by-catch— bonus from the sea.* International Development Research Centre, Ottawa, Canada.

Stevens, L. E., J. C. Schmidt, T. J. Ayers, and B. T. Brown. 1995. Flow regulation, geomorphology, and Colorado River marsh development in the Grand Canyon, Arizona. *Ecol. Applications* **5:**1025–39.

Stevens, T. P. 1988. *California state mussel watch marine water quality monitoring program 1986–87.* Calif. State Water Res. Control Board, Sacramento, CA.

Stevens, W. K. 1995. Experts confirm human role in global warming. *New York Times,* 10 September 1995.

Stewart, C. and J. A. J. Thompson. 1994. Extensive butyltin contamination in southwestern coastal British Columbia, Canada. *Mar. Pollut. Bull.* **28:**601–6.

Stewart, J. M. 1990. The great lake is in great peril. *New Scientist* **126(1723):**58–62.

Stirrup, M. 1990. A sea lion mystery: Why are the numbers dwindling in the North Pacific? *Sea Frontiers* **36(2):**46–53.

Stoker, S. W. and I. I. Krupnik. 1993. Subsistence whaling. Pp. 579–629 *in* J. J. Burns, J. J. Montague, and C. J. Cowles (Eds.), *The bowhead whale.* Society for Marine Mammalogy, Lawrence, KS.

Stone, C. P. 1985. Alien animals in Hawaii's native ecosystems: Toward controlling the adverse effects of introduced vertebrates. Pp. 251–97 *in* C. P. Stone and J. M. Scott (Eds.), *Hawaii's terrestrial ecosystems: Preservation and management.* University of Hawaii, Honolulu, HI.

Stone, C. P. and L. L. Loope. 1987. Reducing negative effects of introduced animals on native biotas in Hawaii: What is being done, what needs doing, and the role of national parks. *Env. Cons.* **14:**245–58.

Stone, E. C. 1965. Preserving vegetation in parks and wilderness. *Science* **150:**1261–67.

Storch, I. 1989. Condition in chamois populations under different harvest levels in Bavaria. *J. Wildl. Manag.* **53:**925–28.

Stork, N. E. 1991. The composition of the arthropod fauna of Bornean lowland rain forest trees. *J. Trop. Ecol.* **7:**161–80.

Stork, N. E. 1993. How many species are there? *Biodiversity and Conservation* **2:**215–32.

Stouffer, P. C. and R. O. Bierregaard, Jr. 1995. Effects of forest fragmentation on understory hummingbirds in Amazonian Brazil. *Cons. Biol.* **9:**1085–94.

Strong, D. H. 1988. *Dreamers & defenders: American conservationists.* University of Nebraska Press, Lincoln, NE.

Sukumar, R. 1989. *The Asian elephant: Ecology and management.* Cambridge University Press, Cambridge, England.

Swanson, T. M. (Ed.). 1995. *The economics and ecology of biodiversity decline: The forces driving global change.* Cambridge University Press, New York.

Taber, R. D. and R. F. Dasmann. 1957. The black-tailed deer of the chaparral. *Calif. Dept. Fish Game Wildl. Bull. No. 8,* 163 pp.

Talbot, L. M. 1984. The role of protected areas in the implementation of the World Conservation Strategy. Pp. 15–19 *in* J. A. McNeely and K. R. Miller (Eds.), *National parks, conservation, and development: The role of protected areas in sustaining society.* Smithsonian University Press, Washington, DC.

Talbot, L. M. 1987. The ecological value of wildlife to the well-being of human society. Pp. 179–86 *in* D. J. Decker and G. R. Goff (Eds.), *Valuing wildlife: Economic and social perspectives.* Westview Press, Boulder, CO.

Tanabe, S. 1988. PCB problems in the future: Foresight from current knowledge. *Env. Pollut.* **50:**5–28.

Tanabe, S., H. Tanaka, and R. Tatsukawa. 1984. Polychlorophenyls, DDT, and hexachlorocyclohexane isomers in the western North Pacific ecosystem. *Arch. Environ. Contam. Toxicol.* **13:**731–38.

Tannock, J., W. W. Howells, and R. J. Phelps. 1983. Chlorinated hydrocarbon pesticide residues in eggs of some birds in Zimbabwe. *Env. Pollut., Ser. B.* **5:**147–55.

Tansley, A. G. 1935. The use and abuse of vegetation concepts and terms. *Ecology* **16:**284–307.

Taylor, J. N., W. R. Courtney, Jr., and J. A. McCann. 1984. Known impacts of exotic fishes in the continental United States. Pp. 322–73 *in* W. R. Courtney, Jr. and J. R. Stauffer, Jr. (Eds.), *Distribution, biology, and management of exotic fishes.* Johns Hopkins University Press, Baltimore, MD.

Taylor, P. W. 1986. *Respect for nature: A theory of environmental ethics.* Princeton University Press, Princeton, NJ.

Teal, J. M. 1993. A local oil spill revisited. *Oceanus* **36(1):**65–70

Teal, J. M. and R. W. Howarth. 1984. Oil spill studies: A review of ecological effects. *Env. Manag.* **8:**27–44.

Tear, T. H., J. M. Scott, P. H. Hayward, and B. Griffith. 1995. Recovery plans and the Endangered Species Act: Are criticisms supported by data? *Cons. Biol.* **9:**182–95.

Teer, J. G. 1988. Conservation biology. The science of scarcity and diversity. (Book review). *J. Wild. Manag.* **52:**570–72.

Tegner, M. J. 1989. The California abalone fishery: Production, ecological interactions, and prospects for the future. Pp. 401–20 *in* J. F. Caddy (Ed.), *Marine invertebrate fisheries: Their assessment and management.* John Wiley and Sons, New York.

Temple, S. A. 1986. The problem of avian extinctions. *Current Ornithology* **3:**453–85.

Temple, S. A. 1987. Do predators always capture substandard individuals disproportionately from prey populations? *Ecology* **68:**669–74.

Templeton, A. R. 1986. Coadaptation and outbreeding depression. Pp. 105–16 *in* M. E. Soulé (Ed.), *Conservation biology: The science of scarcity and diversity.* Sinauer Associates, Sunderland, MA.

Terborgh, J. W. 1974. Preservation of natural diversity: The problem of extinction prone species. *BioScience* **24:**715–22.

Terborgh, J. W. 1975. Faunal equilibria and the design of wildlife preserves. Pp. 369–80 *in* F. B. Golley and E. Medina (Eds.), *Tropical ecological systems: Trends in terrestrial and aquatic research.* Springer-Verlag, New York.

Terborgh, J. W. 1980. The conservation status of neotropic migrants: Present and future. Pp. 21–30 *in* A. Keast and E. S. Morton (Eds.), *Migrant birds in the Neotropics: Ecology, behavior, distribution, and conservation.* Smithsonian Institution Press, Washington, DC.

Terborgh, J. W. 1986. Keystone plant resources in the tropical forest. Pp. 330–44 *in* M. E. Soulé (Ed.), *Conservation biology: The science of scarcity and diversity.* Sinauer Associates, Sunderland, MA.

Terborgh, J. W. 1989. *Where have all the birds gone?* Princeton University Press, Princeton, NJ.

Thayer, G. W., D. A. Wolfe, and R. B. Williams. 1975. The impact of man on seagrass systems. *Amer. Sci.* **63:**288–96.

Thayer, G. W., W. J. Kenworthy, and M. S. Fonseca. 1984. *The ecology of eelgrass meadows of the Atlantic coast: A community profile.* U.S. Fish Wildl. Serv. FWS/OBS-84/02.

Thomann, R. V. 1972. The Delaware River—a study in water quality management. Pp. 99–129 *in* R. T. Oglesby, C. A. Carlson, and J. A. McCann (Eds.), *River ecology and man.* Academic Press, New York.

Thomas, J. W., L. F. Ruggiero, R. W. Mannan, J. W. Schoen, and R. A. Lancia. 1988. Management and conservation of old-growth forests in the United States. *Wildl. Soc. Bull.* **16:**252–62.

Thomas, L. 1989. Pesticides and endangered species: New approaches to evaluation impacts. *End. Sp. Tech. Bull.* **14(1–2):**1, 7.

Thomas, L. 1996. Monitoring long-term population change: Why are there so many analysis methods? *Ecology* **77:**49–58.

Thoreau, H. D. 1849. *A week on the Concord and Merrimack Rivers.* James Munroe, Boston, MA.

Thoreau, H. D. 1854. *Walden.* Ticknor and Fields, Boston.

Tilman, D. and J. A. Downing. 1994. Biodiversity and stability in grasslands. *Nature* **367:**363–65.

Tilman, D., R. M. May, C. L. Lehman, and M. A. Nowak. 1994. Habitat destruction and the extinction debt. *Nature* **371:**65–66.

Tiner, R. W., Jr. 1984. *Wetlands of the United States: Current status and recent trends.* U.S. Dept. Interior, Fish and Wildl. Serv., Washington, DC.

Tobin, R. J. 1990. *The expendable future: U.S. politics and the protection of biological diversity.* Duke University Press, Durham, NC.

Torrey, B. and F. H. Allen. 1906. *The journal of Henry D. Thoreau.* Vol. 1–14. Houghton Mifflin Co., Boston, MA.

Tracy, C. R. and T. L. George. 1992. On the determinants of extinction. *Ameri. Nat.* **139:**102–22.

Trostel, K., A. R. E. Sinclair, C. J. Walters, and C. J. Krebs. 1987. Can predation cause the 10-year hare cycle? *Oecologia* **74**:185–92.

Turnbull, R. E., F. A. Johnson, M. A. Hanandez, W. B. Wheeler, and J. P. Toth. 1989. Pesticide residues in fulvous whistling-ducks from south Florida. *J. Wildl. Manag.* **53**:1052–57.

Turner, B. J. 1983. Genic variation and differentiation of remnant natural populations of the desert pupfish, *C. macularius. Evolution* **37**:690–700.

Tuttle, M. D. 1979. Status, causes of decline, and management of the endangered gray bat. *J. Wildl. Manag.* **43**:1–17.

Udvardy, M. D. F. 1975. *A classification of the biogeographical provinces of the world.* IUCN Occasional Paper No. 18. IUCN, Morges, Switzerland.

United Nations Environment Programme. 1987. *The state of the world environment.* UNEP/GC.14/6, Nairobi, Kenya.

UN FAO. 1987*a. World fisheries situation and outlook.* United Nations Food and Agriculture Organization, Rome.

UN FAO. 1987*b. Review of the state of the world fishery resource.* United Nations Food and Agriculture Organization, Rome.

U.S. Fish and Wildlife Service. 1986. *North American waterfowl management plan.* U.S. Department of Interior, Washington, DC. and Environment Canada, Ottawa, Canada.

U.S. Fish and Wildlife Service. 1987. *Florida panther (Felis concolor coryi) recovery plan.* Florida Panther Interagency Commission, Atlanta, GA.

U.S. Fish and Wildlife Service. 1995. *Reintroduction of the Mexican wolf within its historic range in the southwestern United States.* Draft Environmental Impact Statement. USDA Fish and Wildlife Service, Washington, DC.

Van Blarcum, S. C., J. R. Miller, and G. L. Russell. High latitude river runoff in a doubled CO_2 climate. *Climatic Change* **30**:7–26.

Vance, D. R. and R. L. Westemeier. 1979. Interactions of pheasants and prairie chickens in Illinois. *Wildl. Soc. Bull.* **7**:221–25.

Van Cleave, K. 1977. Recovery of disturbed tundra and taiga surfaces in Alaska. Pp. 422–55 *in* J. Cairns, Jr., K. L. Dickson, and E. E. Herricks (Eds.), *Recovery and restoration of damaged ecosystems.* University of Virginia Press, Charlottesville, VA.

Van den Bosch, R. 1978. *The pesticide conspiracy.* University of California Press, Berkeley, CA.

van Riper III, C., S. G. van Riper, M. L. Goff, and M. Laird. 1986. The epizootiology and ecological significance of malaria in Hawaiian land birds. *Ecol. Monogr.* **56**:327–44.

Veblen, T. T., M. Mermoz, C. Martin, and E. Ramilo. 1989. Effects of exotic deer on forest regeneration and composition in northern Patagonia. *J. Appl. Ecol.* **26**:711–24.

Verstraete, M. M. and S. A. Schwartz. 1991. Desertification and global change. *Vegetatio* **91**:3–13.

Villard, M-A. and B. A. Maurer. 1996. Geostatistics as a tool for examining hypothesized declines in migratory songbirds. *Ecology* **77**:59–68.

Vitousek, P. M. 1990. Biological invasions and ecosystem processes: Toward an integration of population biology and ecosystem studies. *Oikos* **57**:7–13.

Vitousek, P. M. 1992. Global environmental change: An introduction. *Ann. Rev. Ecol. Syst.* **23**:1–14.

Vitousek, P. M. and L. R. Walker. 1989. Biological invasion by *Myrica faya* in Hawaii: Plant demography, nitrogen fixation, ecosystem effects. *Ecol. Monogr.* **59**:247–65.

Vogel, J. 1994. *Genes for sale: Privatization as a conservation policy.* Oxford University Press, New York.

Vogelmann, H. W., G. J. Badger, M. Bliss, and R. M. Klein. 1985. Forest decline on Camels Hump, Vermont. *Bull. Torrey Bot. Club* **112**:274–87.

von Broembsen, S. L. 1989. Invasions of natural ecosystems by plant pathogens. Pp. 77–83 *in* J. A. Drake, H. A. Mooney, F. di Castri, R. H. Groves, F. J. Kruger, M. Rejmanek, and M. Williamson (Eds.), *Biological invasions: A global perspective.* John Wiley and Sons, Chichester, England.

Von Droste, B. 1989. The role of biosphere reserves at a time of increasing globalization. Pp. 1–6 *in* W. P. Gregg, S. L. Krugman, and J. D. Wood (Eds.), *Proceedings of the symposium on biosphere reserves.* USDI National Park Service, Atlanta, GA.

Vrijenhoek, R. C. 1989. Population genetics and conservation. Pp. 89–98 *in* D. Western and M. Pearl (Eds.), *Conservation for the twenty-first century.* Oxford University Press, New York.

Vrijenhoek, R. C., M. E. Douglas, and G. K. Meffe. 1985. Conservation genetics of endangered fish

populations in Arizona. *Science* **229**:400–402.

Wade, P. R. 1995. Revised estimates of incidental kill of dolphins (Delphinidae) by the purse-seine tuna fishery in the eastern tropical Pacific, 1959–1972. *Fishery Bull.* **93**:345–54.

Wagner, F. H. 1977. Species vs. ecosystem management: Concepts and practices. *Trans. N. A. Wildl. Nat. Res. Conf.* **42**:14–24.

Wagner, W. L., D. R. Herbst, and R. S. N. Yee. 1985. Status of the native flowering plants of the Hawaiian Islands. Pp. 23–74 *in* C. P. Stone and J. M. Scott (Eds.), *Hawaii's terrestrial ecosystems: Preservation and management.* University of Hawaii, Honolulu, HI.

Waid, J. (Ed.) 1986. *PCBs and the environment.* Vols. I–III. CRC Press, Boca Raton, FL.

Wakimoto, R. H. 1990. The Yellowstone fires of 1988: Natural process and natural policy. *Northwest Sci.* **64**:239–42.

Walker, B. 1989. Diversity and stability in ecosystem conservation. Pp. 121–30 *in* D. Western and M. Pearl (Eds.), *Conservation for the twenty-first century.* Oxford University Press, New York.

Wallace, M. P. and S. A. Temple. 1987. Releasing captive-reared Andean condors to the wild. *J. Wildl. Manag.* **51**:541–50.

Warner, R. E. 1968. The role of introduced diseases in the extinction of the endemic Hawaiian avifauna. *Condor* **70**:101–20.

Watt, W. B. 1983. Adaptation at specific loci. II. Demographic and biochemical elements in the maintenance of the *Colias* PGI polymorphism. *Genetics* **103**:691–724.

Wayne, R. K., N. Lehman, D. Girman, P. J. P. Gogan, D. A. Gilbert, K. Hansen, R. O. Peterson, U. S. Seal, A. Eisenhawer, L. D. Mech, and R. J. Krumenaker. 1991. Conservation genetics of the endangered Isle Royale gray wolf. *Cons. Biol.* **5**:41–51.

Webb, R. H. 1983. Compaction of desert soils by off-road vehicles. Pp. 51–79 *in* R. H. Webb and H. G. Wilshire (Eds.), *Environmental effects of off-road vehicles.* Springer-Verlag, New York.

Webb, R. H., H. G. Wilshire, and M. A. Henry. 1983. Natural recovery of soils and vegetation following human disturbance. Pp. 279–302 *in* R. H. Webb and H. G. Wilshire (Eds.), *Environmental effects of off-road vehicles.* Springer-Verlag, New York.

Weber, M. 1986. Federal marine fisheries management. Pp. 267–344 in R. L. Di Silvestro (Ed.), *Audubon wildlife report 1986.* National Audubon Society, New York.

Weber, M. and J. Gradwohl. 1995. *The wealth of oceans.* W. W. Norton Co., New York.

Weiss, J. S. 1988. Action on antifouling paints. *BioScience* **38**:90.

Weller, M. W. 1987. *Freshwater marshes.* 2nd ed. University of Minnesota Press, Minneapolis, MN.

Wemmer, C., J. L. D. Smith, and H. R. Mishra. 1988. Tigers in the wild: The biopolitical challenges. Pp. 396–404 in R. L. Tilson and U. S. Seal (Eds.), *Tigers of the world: The biology, biopolitics, management, and conservation of an endangered species.* Noyes Publications, Park Ridge, NJ.

Werner, P. A. (Ed.). 1990. Savanna ecology and management. *J. Biogeogr.* **17**:341–557.

West, N. E., J. M. Stark, D. W. Johnson, M. M. Abrams, J. R. Wright, D. Heggem, and S. Peck. 1994. Effects of climatic change on the edaphic features of arid and semiarid lands of western North America. *Arid Soil Res. and Rehab.* **8**:307–51.

Western, D. 1989. The ecological role of elephants in Africa. *Pachyderm* **12**:42–45.

Western, D., M. C. Pearl, S. L. Pimm, B. Walker, I. Atkinson, and D. S. Woodruff. 1989. An agenda for conservation action. Pp. 304–23 in D. Western, and M. Pearl. *Conservation for the twenty-first century.* Oxford University Press, New York.

Western, D. and J. Ssemakula. 1981. The future of the savannah ecosystems: Ecological islands or faunal enclaves? *Afr. J. Ecol.* **19**:7–19.

Western, D. and R. M. White. 1994. *Natural connections. Perspectives in community-based conservation.* Island Press, Washington, DC.

Westman, W. E. 1977. How much are nature's services worth? *Science* **197**:960–64.

Westman, W. W. 1986. Resilience: Concepts and measures. Pp. 5–19 in B. Dell, A. J. M. Hopkins, and B. B. Lamont (Eds.), *Resilience in Mediterranean-type ecosystems.* W. Junk, Dordrecht, The Netherlands.

Westman, W. E. and G. P. Malanson. 1992. Effects of climate change on Mediterranean-type ecosystems in California and Baja California. Pp. 258–76 in Peters and Lovejoy (1992).

Wheeler, Q. D. 1995. Systematics and biodiversity. *BioScience* **45(Supplement)**:21–28.

Whitcomb, R. F. 1977. Island biogeography and "habitat islands" of eastern forest. *Amer. Birds* **31**:3–5.

Whitcomb, R. F., C. S. Robbins, J. F. Lynch, B. L. Whitcomb, M. K. Klimkiewicz, and D. Bystrak. 1981. Effects of forest fragmentation on avifauna of the eastern deciduous forest. Pp. 125–205 in R. L. Burgess and D. M. Sharpe (Eds.), *Forest island dynamics in man-dominated landscapes.* Springer-Verlag, New York.

White, D. H. and J. T. Seginak. 1994. Dioxins and furans linked to reproductive impairment in wood ducks. *J. Wildl. Manag.* **58**:100–6.

White, G. F. 1988. The environmental effects of the high dam at Aswan. *Environment* **30(7)**:5–11, 34–40.

White, L., Jr. 1967. The historical roots of our ecologic crisis. *Science* **155**:1203–7.

White, P. S. and S. B. Bratton. 1980. After preservation: Philosophical and practical problems of change. *Biol. Cons.* **18**:241–55.

White, W. A. and T. A. Tremblay. 1995. Submergence of wetlands as a result of human-induced subsidence and faulting along the upper Texas coast. *J. Coastal Res.* **11**:788–807.

Whitham, J. W., Jr. and M. L. Hunter, Jr. 1992. Population trends of neotropical migrant landbirds in northern coastal New England. Pp. 85–95 in J. M. Hagan and D. W. Johnston (Eds.), *Ecology and conservation of neotropical migrant landbirds.* Smithsonian Institution Press, Washington, DC.

Whitney, G. G. 1994. *From coastal wilderness to fruited plain: A history of environmental change in temperate North America 1500 to present.* Cambridge University Press, Cambridge, England.

Wiedemann, A. M. 1984. *The ecology of the Pacific Northwest coastal sand dunes: A community profile.* U.S. Fish Wildl. Serv. FWS/OBS-84/04.

Wilcove, D. S. 1985. Nest predation in forest tracts and the decline of migratory songbirds. *Ecology* **66**:1211–14.

Wilcove, D. S. 1994. Turning tangible goals into tangible results: The case of the spotted owl and old-growth forests. Pp. 313–29 in P. J. Edwards, R. M. May, and N. R. Webb (Eds.), *Large-scale ecology and conservation biology.* Blackwell Scientific Publications, London, England.

Wilcox, B. A. 1978. Supersaturated island faunas: A species-age relationship for lizards on post-Pleistocene land-bridge islands. *Science* **199**:996–98.

Wiley, K. 1990. Untying the western water knot. *Nature Conservancy Magazine* **40(2)**:5–13.

Willard, B. E. and J. W. Marr. 1970. Effects of human activities on alpine tundra ecosystems in Rocky Mountain National Park, Colorado. *Biol. Cons.* **2**:257–65.

Williams, B. K. and F. A. Johnson. 1995. Adaptive management and the regulation of waterfowl harvests. *Wildl. Soc. Bull.* **23**:430–36.

Williams, J. D., M. L. Warren, Jr., K. S. Cummings, J. L. Harris, and R. J. Neves. 1993. Conservation status of freshwater mussels of the United States and Canada. *Fisheries* **18(9)**:6–22.

Williams, J. E., J. E. Johnson, D. A. Hendrickson, S. Contreras-Balderas, J. D. Williams, M. Navarro-Mendoza, D. E. McAllister, and J. E. Deacon. 1989. Fishes of North America endangered, threatened, or of special concern: 1989. *Fisheries* **14(6)**:2–20.

Williams, M. L., R. L. Hotham, and H. M. Ohlendorf. 1989. Recruitment failure in American avocets and black-necked stilts at Kesterson Reservoir, California, 1984–1985. *Condor* **91**:797–802.

Williams, P. H. and K. J. Gaston. 1994. Measuring more of biodiversity: Can higher-taxon richness predict wholesale species richness? *Biol. Cons.* **67**:211–17.

Williamson, D., J. Williamson, and K. T. Ngwamotsoko. 1988. Wildebeest migration in the Kalahari. *Afr. J. Ecol.* **26**:269–80.

Willis, E. O. 1979. The composition of avian communities in remanescent woodlots in southern Brazil. *Papéis Avulsos de Zoologia* **33**:1–25.

Wilshire, H. G. 1983. The impact of vehicles on desert soil stabilizers. Pp. 31–50 in R. H. Webb and H. G. Wilshire (Eds.), *Environmental effects of off-road vehicles.* Springer-Verlag, New York.

Wilson, E. O. 1992. *The diversity of life.* Harvard University Press, Cambridge, MA.

Wilson, S. D. and J. W. Belcher. 1989. Plant and bird communities of native prairie and introduced Eurasian vegetation in Manitoba, Canada. *Cons. Biol.* **3**:39–44.

Wingate, D. B. 1985. The restoration of Nonesuch Island as a living museum of Bermuda's precolonial terrestrial biome. Pp. 225–38 *in* P. J. Moors (Ed.), *Conservation of island birds.* International Council for Bird Preservation, Tech. Pub. No. 3.

Woodruff, D. S. and O. A. Ryder. 1986. Genetic characterization and conservation of endangered species: The Arabian oryx and Pere David's deer. *Isozyme Bull.* **19**:35.

Woodwell, G. M., C. F. Wurster, Jr., and P. A. Isaacson. 1967. DDT residues in an East Coast estuary: A case of biological concentration of a persistent pesticide. *Science* **156**:821–23.

World Conservation Monitoring Centre. 1992. *Global biodiversity.* Status of the Earth's living resources. Chapman and Hall, London.

World Resources Institute. 1987. *World resources 1987.* Basic Books, Inc., New York.

World Resources Institute. 1988. *World resources 1988–89.* Basic Books, Inc., New York.

Worthington, E. B., and R. Lowe-McConnell. 1994. African lakes reviewed: Creation and destruction of biodiversity. *Env. Cons.* **21**:199–213.

Wunderlich, R. C., B. D. Winter, and J. H. Meyer. 1994. Restoration of the Elwha River ecosystem. *Fisheries* **19(8)**:11–19.

Wursig, B. 1989. Cetaceans. *Science* **244**:1550–57.

Wurster, C. F. 1969. Chlorinated hydrocarbon insecticides and the world ecosystem. *Biol. Cons.* **1**:123–29.

Wyman, R. L. 1991. Multiple threats to wildlife: Climate change, acid precipitation, and habitat fragmentation. Pp. 134–55 *in* R. L. Wyman (Ed.), *Global climate change and life on earth.* Routledge, Chapman and Hall, New York.

Yellowstone National Park. 1982. *Final environmental impact statement—grizzly bear management program.* U.S. National Park Service, Denver, CO.

Yodzis, P. 1994. Predator-prey theory and the management of multispecies fisheries. *Ecol. Applications* **4**:51–58.

Zangerl, R., L. P. Hendrickson, and J. R. Hendrickson. 1988. *A redescription of the Australian flatback sea turtle.* Bishop Museum Press, Honolulu, HI.

Zaret, T. M. and R. T. Paine. 1973. Species introduction in a tropical lake. *Science* **182**:449–55.

Zedler, J. B. 1982. *The ecology of southern California coastal salt marshes: A community profile.* U.S. Fish Wildl. Serv. Biol. Serv. Program, Washington, DC. FWS/OBS-81/54.

Zedler, P. H. 1987. The ecology of southern California vernal pools: A community profile. *USDI Fish and Wildlife Service Biological Report* **85**(7.11).

Zimmerman, J. L. 1990. *Cheyenne bottoms: Wetland in jeopardy.* University of Kansas Press, Lawrence, KS.

Zwinger, A. H. 1989. *The mysterious lands.* E. P. Dutton, New York.

glossary

A

Acid deposition Increased input of acids to ecosystems through rain, fog, snow, dry particulate fallout, and capture of gaseous compounds.

Acid mine drainage Pollution of streams with acids and heavy metals due to leaching of spoil material from coal or ore mines.

Acid precipitation Rain or snow with a pH less than 5.0.

Acre-foot Amount of water equal to one acre, or about the area of a football field, covered to a depth of one foot.

Adaptive management Approach in which major alternative management options are defined, one is applied, and the effect on system dynamics is monitored and used to refine options for subsequent management.

Adaptive radiation Evolutionary diversification of species for different ways of life.

Age-structured populations Populations in which reproductive potential differs for individuals of different age.

Agroforestry Management of multi-species, tree-dominated communities of plants that yield timber, fuelwood, basic foods, and industrial products such as rubber and oils.

Albedo Reflectivity of the earth's surface to incoming solar radiation.

Algal turfs Low mats of diverse species of algae that flourish in sheltered shallow waters of coral reefs.

Alleles Slightly different forms of a gene that consequently tend to produce different versions of the coded protein.

Alpha diversity See *diversity.*

Ambergris Substance formed in the intestine of the sperm whale; used as a chemical stabilizer in perfumes.

Anadromous Entering rivers from the ocean to spawn.

Anagenesis Evolutionary change in a species until it becomes so different that taxonomists no longer regard it as being in the same taxonomic unit.

Anthropocentric ethic Philosophical view that only humans and entities useful to humans have moral standing.

Aquaculture Husbandry of aquatic plants or animals in a fashion comparable to that of domesticated agricultural species.

Area-sensitive species Species that are likely to disappear when an insular habitat reaches some critical minimum size.

Area strip mining See *strip mining.*

Arribada Mass arrival of female sea turtles at nesting beaches, where they concentrate their egg-laying into a period of a few nights.

B

Baleen apparatus Set of fibrous plates suspended from the upper jaws of toothless whales that acts as a filter to capture small animal foods.

Barrier beaches See *barrier islands.*

Barrier islands Long, narrow islands formed of beach sediments and lying more or less parallel to the seacoast.

Barrier beaches are similar to barrier islands, differing only in that they are connected to the mainland at one end.

Benefit-cost analysis Evaluation of the economic values gained by an action in comparison to the costs of carrying out the action.

Beta diversity See *diversity.*

Bioaccumulation Concentration of a chemical from an organism's physical environment.

Biocentric ethic View that at least some nonhuman animals have intrinsic moral standing, and that humans have ethical duties to them.

Biodiversity Richness of the biosphere in genetically distinct organisms and the systems they compose.

Biodiversity hotspots Locations where unusual concentrations of species, and usually of endemic taxa, occur.

Biogeographic province Subdivision of a continent or its near-shore waters characterized by particular biomes or a distinctive physical landscape.

Biogeographic realms Major continental units that have been centers of independent evolution for long periods of geological time.

Biological control Establishment of a strong, negative biological interaction against a designated pest species.

Biological oxygen demand Quantity of oxygen needed to decompose the organic matter in one liter of water.

Biological species See *species.*

Biomagnification Increase of a chemical concentration in tissues of organisms from one link in a food chain to the next.

Biome Regional ecosystem type characterized by the distinctive appearance of the dominant plants.

Biosphere Zone of the earth's surface occupied or influenced by living organisms.

Biotic community Different species of organisms that occur together and interact with each other in an area of habitat.

Biotic interactions Actions between organisms of the same or different species, including predation, parasitism, competition, and mutual benefit.

Biotic relaxation Decline in the number of species on a newly formed island due to an excess of extinctions over colonizations by new species.

Biotic succession Change in the composition and structure of an ecosystem through time. In **primary succession,** change begins on a site previously unoccupied; in **secondary succession,** on a site from which a community was removed by some disturbance.

Blowout Wind erosion basin formed in a sandy substrate following disturbance of the stabilizing vegetation.

Bromeliads Members of a family of largely tropical flowering plants, many of which grow as epiphytes.

Brucellosis Bacterial disease of ungulates that can be very destructive to herds of livestock and can also be transmitted to humans.

Bunch grasses Species of grasses that grow in dense clumps or large patches rather than as a continuous turf.

C

Caliche Cemented horizon of calcium carbonate and other minerals deposited by leaching at a certain depth in soils of arid regions.

Calipee Cartilaginous material lining the inner side of the ventral shell of the green sea turtle; an essential ingredient of green turtle soup.

Canids Members of the dog family.

Captive propagation Breeding of species in confinement for their preservation, display in zoos, and reintroduction to the wild.

Carbamates Chemical group of pesticides that have very high toxicity, but a short environmental life.

Carnivores Organisms that feed exclusively on herbivorous animals; **top carnivores** feed exclusively on other carnivores.

Carrying capacity The level around which populations fluctuate under the influence of density-dependent regulation.

Catadromous Living in fresh water but returning to the sea to spawn.

Ceilometer Fixed-beam or rotating light used at airports to show the height of the cloud ceiling.

Channelization Straightening a meandering stream channel to speed flood water outflow.

Chemosynthesis Process by which some producer organisms use energy from chemical reactions for manufacturing organic molecules.

Chlorinated hydrocarbons A group of petroleum-derivative chemicals with a structure of carbon rings to which chlorine atoms are attached.

Chocolate mousse Heavier components of crude oil that have been churned into a viscous emulsion in sea water.

Cladocera Subclass of crustaceans, commonly known as water fleas, that are common in freshwater zooplankton.

Clearcutting Cutting of all trees and removal of all harvestable timber at one time.

Climax community Biotic community consisting of species that are able to reproduce and replace themselves under existing conditions.

Coevolution Adaptation of species to each other through patterns of mutual evolutionary change.

Colonization and extinction curves Graphs of the rates of establishment of new species and disappearance of established species on islands relative to the number of species present.

Commensal A species that lives in close association with another species without benefitting or hurting the latter.

Common property resource A resource owned by society at large that typically is not fully valued in market transactions affecting it.

Community The group of species that occur together in a particular area of habitat.

Community type An assemblage of species that has a characteristic composition and structure and can be recognized in different locations.

Compound 1080 Sodium fluoroacetate, a broad-spectrum poison sometimes used to kill mammals judged to be pests, such as coyotes.

Conservation networks Natural and semi-natural areas managed by different organizations or public agencies, but linked by habitat corridors to maximize effective total habitat area.

Continental islands Islands that once possessed a continental connection.

Contingent valuation method Technique of estimating nonmarket economic values by determining people's willingness to pay to preserve such values.

Contour strip mining See *strip mining*.

Continental shelf Areas of continental crust lying under shallow seas adjacent to exposed land portions of continental crust.

Copepods Subclass of small freshwater and marine crustaceans that commonly occur in zooplankton.

Coral bleaching Loss of symbiotic zooxanthellae by corals due to some type of stress; may lead to mortality.

Corridor Linear habitat feature that differs from its surroundings and connects units of similar habitat type.

Corridor capability Effectiveness of a corridor in delivering species from source to target areas.

Counteradaptation Evolutionary adjustments by the member species of a community to each other.

Critical habitat Areas within or outside the range of an endangered species that are formally designated as essential to its survival.

Cross-fostering Rearing of young of one species by adults of another species.

Cryogenic storage Long-term preservation of semen, ova, or embryos at subfreezing temperatures.

Cryptobiotic crust Algae, fungi, and lichens that form a fine stabilizing network of organic tissues at the surface of desert soils.

D

Debris flows Mass slippages of destabilized, saturated soil on hillsides.

Debt-for-nature swap Purchase and cancellation of portions of the foreign debt owed by a nation to international banks in exchange for a government's establishment of preserves or support of conservation.

Decomposers Organisms that carry out the breakdown of dead organic matter.

Decreasers Native rangeland plants that decline in abundance under heavy grazing by livestock.

Deep ecology Philosophy that holds that all life and the ecological systems that living organisms form have intrinsic value.

Demographic failure Extinction of a population by the death of all its members.

Demographic stochasticity See *stochastic variability*.

Density-dependent factors Agents that affect a population with an intensity determined by its density relative to carrying capacity.

Density-independent factors Agents that affect a population with an intensity unrelated to its density relative to carrying capacity.

Desertification Decline in the productivity of arid or semiarid lands due to human mismanagement.

Desert pavement Surface layer of stones in deserts that protects the underlying soil from wind erosion.

Direct nutrient cycling Passage of nutrients directly from dead matter into mycorrhizal fungal filaments and then into plant rootlets without entering the mineral soil.

Diversity Biotic richness of a community or region. The richness of individual communities in species is termed **alpha diversity. Beta diversity** is the degree of change in species from one community type to another. The richness of a region as a whole, **gamma diversity,** reflects the number of community types present, the alpha diversity of each, and the pattern of beta diversity from type to type.

Dominants Members of a biotic community that receive the impact of external conditions and determine the internal environment of an ecosystem.

Drift gill netting Open-sea fishing with gill nets that suspend from floats, hang downward in the surface waters, and extend for many kilometers.

Dynamic pool models Sets of equations describing numbers or biomass of different age (or size) classes as a function of variables that influence their growth, survival, and reproduction.

Dystrophic Having low nutrient availability.

E

Ecoclines See *ecogeographic variation*.

Ecogeographic variation Differentiation of local populations in genetic adaptations or responses to habitat conditions. Sometimes these are evident as relatively distinct **ecotypes** that occur in different habitats, but more often they appear as **ecoclines,** or gradients of genetic characteristics along gradients of habitat conditions.

Ecological niche Resources that a species exploits to survive and reproduce.

Ecological release Expansion of the ecological niche to include more habitats and more kinds of resources when competitor species are fewer.

Ecology The science concerned with relationships of organisms with each other and with their nonliving environment.

Ecoregions Major terrestrial ecosystem types, defined on the basis of climate, dominant plant life form, region of the country, and vegetational composition.

Ecosystem Unit of the environment, consisting of living and nonliving components that interact with exchanges of energy and nutrients.

Ecosystem analysis Branch of ecology that deals with the dynamics of energy flow and nutrient cycling in ecosystems.

Ecosystem dynamics Functional processes that occur in ecosystems, such as energy flow and nutrient cycling.

Ecosystem structure Physical, chemical, and biotic conditions that exist in an ecosystem at a given instant.

Ecotone Border zone between distinct ecosystems.

Ecotourism Ecologically sensitive tourist activities that enable a region to benefit from ecological resources without destroying them.

Ecotypes See *ecogeographic variation*.

El Niño Influx of warm, nutrient-poor tropical water into areas of normally cold, upwelling water in tropical or temperate oceans.

Emigration Movement of individuals out of a population to other locations.

Endangered species Species, subspecies, or populations determined to be in danger of extinction in all or a significant portion of their range.

Endemic species Species that have evolved in, and are restricted to, a particular area.

Energy flow Capture of solar energy by producers, its transfer along food chains, and its loss as heat energy from living organisms.

Environmental ethics Branch of philosophy concerned with moral values that relate to the natural environment.

Environmental impact assessment Process of examining a proposed project and its alternatives to determine ecological and other effects and ways that they can be avoided or mitigated.

Environmentalism Philosophy of striving to live in harmony within the global ecosystem.

Environmental stochasticity See *stochastic variability*.

Epilimnion Surface zone of warm, low-density water in a lake.

Epiphylls Thin layer of algae, lichens, and mosses covering leaves of tropical forest plants.

Epiphytes Plants that grow on the trunks and branches of larger trees and shrubs, especially in tropical forests.

Estuary Bodies of water occurring where rivers discharge into the ocean and having salinity regimes intermediate between fresh waters and the ocean.

Ethics Branch of philosophy dealing with moral standards of human conduct based on beliefs about what is right or just. **Descriptive ethics** involves simply summarizing the ethical beliefs that are held by society or various groups within society. **Normative ethics** is the advocacy of particular ethical viewpoints. **Philosophical ethics** is the domain in which conflicting ethical positions are identified and examined logically. A **primary ethic** involves a duty *to* something because it has intrinsic moral standing. A **secondary ethic** involves a duty *regarding* something because it has utilitarian value to humans or to other entities to which primary ethical duties pertain.

Eutrophic Having high nutrient availability and high primary productivity.

Eutrophication Increase in fertility and productivity of an ecosystem due to an increased rate of nutrient input.

Evolutionarily significant unit Geographically-isolated portion of the species population that also has a high level of genetic difference from other subpopulations of the species.

Evolutionary unit A population genetically isolated from other such units and having a shared evolutionary history and unique biological characteristics.

Ex situ care Maintenance of populations outside their native habitat, as in botanical gardens, zoological parks, and aquariums.

External costs See *externalities*.

Externalities Unpaid costs of some activity that are imposed on society; also termed **external costs.**

Extinction Termination of an evolutionary lineage due to death or genetic modification of all of its members.

Extractive reserves Areas of tropical forest that are protected from logging, but are open to harvesting of tree products such as rubber, oils, fruits, nuts, tubers, and other substances.

F

Facilitation Mechanism of biotic succession in which habitat changes caused by one species favor the establishment of other species.

Faunal relaxation Decline in number of species on a newly created island because their initial number exceeds the equilibrium that can be maintained by colonization-extinction relationships.

Feral populations Populations of domestic animals that live and reproduce in the wild.

Fitness Contribution of an individual to the genetic composition of subsequent generations in a population.

Flyways Major pathways of migration of birds, particularly the **Atlantic, Mississippi, Central,** and **Pacific** waterfowl flyways.

Forest decline Increased mortality and reduced growth of trees due to factors such as tree age and environmental stress.

Foundation species Lower trophic level species whose abundance determines much of the upper trophic level structure of the ecosystem.

Founder effect Influence of the small group of individuals forming a new wild or captive population on the genetic structure of the resulting population.

Functional response Change in the effort exerted by a predator in hunting a particular prey species in response to changed prey abundance.

Fynbos Evergreen shrubland ecosystem type characteristic of the Cape region of South Africa.

G

Game cropping Commercial harvesting of wildlife from unconfined populations.

Game farming Ranching operation that mixes livestock with selected species of wildlife.

Gamma diversity See *diversity*.

Gap analysis Determination of which species are inadequately protected by existing or planned sets of preserves.

Gene Unit of the genetic molecule, DNA (deoxyribose nucleic acid), that contains the coded instructions for assembly of particular proteins.

Gene flow Spread of alleles through populations by dispersal and interbreeding of individuals.

Gene pool Genetic makeup of a population in terms of number of alleles per gene and the frequency of heterozygous allele combinations.

General circulation models Mathematical models that simulate the global system of atmospheric and oceanic circulation.

Genetically effective population size Size of an ideal breeding population (1:1 sex ratio, random mating, randomly varying number of offspring per pair) to which an actual population is equivalent.

Genetic bottleneck Greatly reduced genetic variability in a population due to its past reduction to a very small size at which many alleles were lost by genetic drift.

Genetic drift Random fluctuation in the frequency of an allele due to accidents that affect the survival and reproduction of individuals.

Genetic swamping Extinction or loss of identity of a species through extensive interbreeding with one or more related forms.

Geographic information systems Computerized recording of diverse data sets for a region using geographic coordinates as the primary indexing system.

Grazing lawns Dense carpets of actively growing grasses that are kept in an actively growing, juvenile state by herbivore grazing.

Greenhouse effect Warming of the lower atmosphere due to increased concentration of gases that are transparent to incoming short-wave solar radiation, but intercept outgoing long-wave heat radiation.

Greenhouse gases Gases that act in a manner similar to the glass in a greenhouse roof, letting sunlight pass through but absorbing and reradiating some of the outgoing heat radiation back toward the earth's surface.

Gyres Large, circular systems of water circulation that occupy the major ocean basins.

H

Habitat Conditions of the site occupied by a population or community of organisms.

Hacking Reintroduction of captive-reared animals to the wild at acclimation sites where food is provided regularly.

Headstarting Raising young turtles to the age of about a year until their release in areas believed to be optimal habitat.

Herbivores Organisms that feed exclusively on green plants.

Heterozygosity Condition in which the two examples of a particular gene in an individual are of different alleles.

Highwall A cliff left as the result of strip mining on a hillside.

Homozygosity Condition in which both examples of a particular gene in an individual are of the same allele.

Hypolimnion Deep, cold, high-density bottom water zone in a lake.

I

Immigration Movement of individuals into a population from another location.

Inbreeding Mating of closely related individuals.

Inbreeding depression Poor survival or reproduction resulting from pairing of deleterious alleles in matings of close relatives.

Increasers Native rangeland plants that become more abundant under grazing pressure by livestock.

Inertia Ability of an ecosystem to resist change in the face of a disrupting external influence.

Integrated pest management Strategy of pest control that combines a variety of cultural, biological, and chemical approaches.

Intergenerational equity Ethical issue regarding the equality or inequality of opportunities available to humans of present and future generations.

Interglacial Period(s) between major episodes of continental glaciation.

Intertropical convergence Belt of most intense solar heating of the earth's surface, into which northern and southern trade winds flow.

Intrinsic moral standing See *moral standing.*

Invaders Non-native plant species that tend to colonize rangelands that are grazed by livestock.

Island biogeography Study of processes of colonization, evolution, and extinction in insular biotas.

K

Keystone exotics Foreign species capable of restructuring almost completely the ecosystems they invade.

Keystone mutualists Tropical forest plant species that are essential to the survival of many species of frugivorous birds and mammals.

Keystone predator Animal that can cause an almost total restructuring of the ecosystem by predation influences.

Kipukas Islands of native vegetation that exist in Hawaii where extensive lava flows have isolated them.

Krill Large marine zooplankton, primarily the small shrimp, *Euphausia superba,* that abound in cold upwelling waters.

L

Lampricide Pesticide used to kill lamprey larvae during their early life in streams.

Landscape connectivity Degree to which absolute isolation is prevented by features that allow organisms to move among preserve units.

Landscape ecology Branch of ecology that concentrates on how the interactions of the different types of ecosystems in a certain area influence their own structure and dynamics and that of the total area.

La Niña Stage in the long-term oscillation of water in tropical ocean basins when warm surface water becomes concentrated in the western part of the basin and upwelling dominates in the eastern.

Large marine ecosystems Ocean areas with more or less natural boundaries formed by surface currents and submarine topography.

Large woody debris Fallen trees and logs that obstruct the stream channels or create debris dams that form pools and trap gravel, thus creating microhabitats important to fish.

Lianas Massive vines that expose their foliage in the tropical forest canopy.

Load-On-Top (LOT) Procedure used by oil tanker ships to concentrate and retain oil residues when ballast water is discharged, so that the new oil cargo incorporates these residues.

Logistic equation Mathematical relationship describing the S-shaped growth curve of a population toward the limit set by resources.

M

Macroclimate General regional climate, or the climatic conditions prevailing outside an ecosystem.

Maximum sustained yield Greatest harvest that can be taken from a population indefinitely.

Mean heterozygosity Average percentage of the genes that are represented by different alleles in individuals of a population.

Megafauna Large-bodied animals, either living or extinct.

Metapopulation Overall regional population comprising many local populations in isolated patches of habitat.

M-44 device Metal tube that uses a shotgun shell to shoot cyanide into the mouth of an animal, such as a coyote, that pulls on the baited end of the tube.

Microclimate Climatic conditions prevailing within an ecosystem.

Minimum viable population Smallest number of breeding individuals that has a specified probability of surviving for a certain time without losing its evolutionary adaptability.

Mobile links Animals of tropical forests that carry pollen from individual to individual, or seeds from mature tree to potential establishment sites.

Monomorphism Condition in which a gene has only a single allele.

Moral standing Any entity to which ethical principles apply. An entity to which ethical principles apply regardless of whether it has utilitarian value has **intrinsic** moral standing. Moral standing also exists for entities with **utilitarian** or **instrumental value** to humans.

Multiple stable states Alternative, persistent conditions of structure and function among which strong environmental influences may cause the ecosystem to shift.

Multiple-use modules Conservation units with a fully protected core area surrounded by natural or semi-natural zones used in progressively more intense fashion.

Mutation An accident in replication of the chemical structure of a gene during the process of cell division.

Mutualism The interaction of two or more species in ways beneficial to both or all.

Mycorrhizae Symbiotic associations between filamentous fungi and the fine roots of higher plants.

N

Niche See *ecological niche*.

Nonessential experimental population A reintroduced population of an endangered species that is subject to artificial control to reduce detrimental impacts on other species or on humans.

Numerical response Change in the density of predators due to reproduction and mortality, resulting from change in prey abundance.

Nutrient capital Quantity of an essential element or compound that is in active circulation within an ecosystem at a given time.

Nutrient cycling Movement of an essential element or compound among living and nonliving components of an ecosystem.

O

Oceanic islands Islands that have never been connected to a continent.

Off-road vehicles All motorized vehicles designed for operation on roadless terrain.

Oligotrophic Ecosystem with low concentrations of nutrients and a low rate of primary productivity.

Optimum sustained yield Yield that provides the best compromise of all societal values.

Organophosphates Chemical group of pesticides that have very high toxicity, but a short environmental life.

Outbreeding depression Reduced fitness of offspring produced by hybridization of distantly related individuals.

Ozone hole Area of depleted stratospheric ozone that has developed over the Antarctic region during the Southern Hemisphere spring.

P

Pass Inlet between barrier islands that connects the open ocean and the sound on the inland side of the islands.

Pastoralism Subsistence food production system involving the herding of animals such as cattle, sheep, goats, and camels.

Penetrometer Device for measuring the resistance of soil to entry by a solid object.

Permafrost Permanently frozen subsoil zone that occurs extensively in the Arctic tundra.

Pesticide resistance Genetically-based ability of a pest to avoid, tolerate, or inactivate the pesticide chemical.

Phenoxy herbicides Plant-killing pesticides that are similar in structure to plant growth hormones.

Photosynthesis Process by which some producer organisms use light energy for manufacturing organic molecules.

pH scale Logarithmic index, ranging from 0 to 14, of concentration of acid hydrogen ions; acidity increases as the scale value decreases.

Physical and chemical weathering Changes in nonliving conditions of a habitat due to chemical reactions, fluctuations in temperature, and other abiotic processes.

Phytoplankton Tiny single-celled or colonial green plants that live free-floating in the waters of aquatic ecosystems.

Pioneer community Group of species that colonizes a new or disturbed habitat and initiate biotic succession.

Plank buttresses Flanged extensions of the trunk base that connect with lateral roots of tropical rain forest trees.

Playa Bed of a shallow, temporary desert lake.

Polychlorinated biphenyls Class of petroleum-derivative industrial chemicals related to the chlorinated hydrocarbon pesticides.

Polyethylene pellets Small spherules of polyethylene that are the raw material for manufacture of molded plastic objects.

Polymorphism Percentage of genes in a population that have more than one allele.

Polynuclear aromatic hydrocarbons Organic molecules consisting of two or more fused benzene rings.

Polystyrene pellets Small spherules that are the raw materials for manufacture of polystyrene objects.

Population Individuals of a species that occur together and interact with each other in a defined area of habitat.

Population regulation Action of direct density-dependent factors tends to maintain numbers of individuals within a narrow range of densities.

Population viability analysis Determination of the probability that a certain number of breeding individuals can survive for a specified time without losing their evolutionary adaptability.

Potholes Small prairie ponds and marshes of glacial origin.

Predator selectivity The extent to which a predator concentrates on substandard prey.

Primary forest Mature tropical stands that are free of signs of disturbance and exhibit high biomass and diversity.

Primary succession See *biotic succession*.

Prior appropriation doctrine Water rights based on the principle of first use by diversion of surface waters for a beneficial purpose.

Producers Organisms that store energy in new organic matter through the processes of photosynthesis or chemosynthesis.

Propagule Any form of reproductive body that serves as a dispersal unit for a plant or animal.

Public trust doctrine Concept that the public has an overriding right to natural surface waters that might be destroyed by excessive diversion.

R

Rain shadow Region of low rainfall on the lee side of a mountain range; caused by warming and decrease in relative humidity of air after it passes over the mountains.

Rapid resurgence Quick recovery of the population of a target pest after a pesticide treatment.

Recovery plans Procedures formally adopted to increase the abundance and distribution of endangered species so that they can be reclassified as nonendangered.

Recycling Removal of eggs or young from a pair of animals to stimulate a new breeding cycle, leading to a replacement clutch or litter.

Red Data Books Worldwide listings of endangered species, their ecology, and their conservation status prepared by the World Conservation Union.

Regional biota Plant and animal species that occur within a geographical area and are available to colonize its ecosystems.

Reserved water rights Concept that preservation of natural surface waters is implicit when areas are set aside as parks, protected forests, and wildlife areas.

Resilience Ability of an ecosystem to restore its normal structure after some disturbance.

Rinderpest Highly virulent virus disease of cattle and related wild ungulates.

Riparian Associated with streams or their floodplains.

Riparian doctrine Water rights based on the principle of preservation of undiminished and undefiled surface waters.

Roe stripping Fishing practice in which egg masses from female fish are removed and the rest of the body is discarded.

S

Salinization Accumulation of salts in a soil or ecosystem at large.

Secchi disk Black-and-white metal disk that is lowered into a lake to measure water transparency by the depth at which it can be seen.

Secondary forest Forest stands recovering from disturbances such as fire, hurricane impacts, or cutting by humans.

Secondary outbreak Population explosion of a species that was not a pest due to killing of natural biological control species by a pesticide treatment.

Secondary succession See *biotic succession*.

Seed-tree cutting Logging of all but a scattering of mature trees that are left to provide the seed for forest regeneration.

Selective harvesting Harvesting of individual mature trees at frequent intervals.

Serpentine soils Soils derived from igneous rocks rich in magnesium.

Shark finning Fishing practice in which the fins are removed and the remaining fish is discarded.

Shelterwood cutting Removal of mature trees by logging over a period of years so that a partial canopy favoring growth of new trees is maintained.

Shifting cultivation Farming systems in which a plot of tropical forest is cleared and farmed for two to several seasons and then abandoned to succession.

Sink populations Local populations that are maintained almost entirely by dispersal of individuals from other locations.

SLOSS argument Controversy over whether a single(S) large(L) preserve or(O) several(S) small(S) preserves of equal total area is most effective in long-term preservation of biotic diversity.

Soft release See *hacking*.

Soil crusts Surface layers of varied nature that act to reduce wind or water erosion in desert soils.

Sound Estuarine water area lying between a barrier island and the mainland.

Source populations Local populations that realize a positive population growth rate and a net emigration rate of individuals to other locations.

Spatially explicit models Mathematical models that describe the dynamics of individual subunits of a population or ecosystem, taking into account the location of these subunits relative to each other.

Speciation Separation of one evolutionary lineage into two or more daughter lineages.

Species Basic kinds of organisms, as determined by prevailing taxonomic concepts and criteria. The **biological species** is a group of actually or potentially interbreeding individuals that is reproductively isolated from other such groups. The **phylogenetic species** is a group of individuals that is distinct in its characteristics and has a common ancestry.

Species-area curve Graph of the numbers of species in relation to the sizes of the areas censused constructed to determine the rate at which species number changes with change in area of habitat.

Spermaceti Waxy material from the oil deposit in the head of the sperm whale; formerly used in the making of fine candles.

Stability Constancy of ecosystem conditions as influenced by internal interactions.

Staging areas Locations at which birds gather prior to the first major leg of migration to store body fat for migratory flights.

Stochastic variability Chance events that lead to variation in some quantitative feature of an ecosystem or its components. **Demographic stochasticity** refers to random variation in mortality and survivorship among individuals in a population; **environmental stochasticity** refers to random variation in habitat conditions.

Stopover areas Locations where birds stop to replenish their fat stores partway through migration.

Strip mining Gaining access to a coal or other mineral deposit by removal of overlying soil and rock. **Area strip mining** is removal of the soil and rock in a long trench across level terrain and dumping it into the adjacent trench that has just been mined. **Contour strip mining** is removal above an outcrop of coal or ore on a hillside, leaving a cliff or highwall.

Subspecies Geographically-defined subpopulation of a species that has been distinguished from other such populations by consistent taxonomic differences.

Survival habitat Habitat necessary to support an endangered population.

Sustainability State in which a stable relationship exists between the human population and the capacity of the biosphere to provide resources and process wastes.

Sustainable development Process leading toward sustainability, in which economic development does not reduce the capacity of the biosphere to support humanity.

Sustained yield Any level of harvest that can be taken from a population indefinitely.

Symbiotic The living of two or more species in intimate association with each other.

T

Territoriality Defense of an area by one or more individuals against others of the same, related, or ecologically similar species.

Thermal pollution Release of heated water from the cooling systems of power plants and certain industries into aquatic ecosystems.

Thermocline Relatively thin zone in a lake where temperature and density change rapidly with depth.

Threatened species Species, subspecies, or populations likely to become endangered within the foreseeable future.

Threshold of protection Average capacity, based on availability of food and cover, of a habitat to carry animals through an unfavorable season.

Top carnivores See *carnivores*.

Transcendentalism Philosophy of rejection of material goals and a seeking of harmony and beauty through the contemplation of nature.

Translocation Movement of individuals from one location to another to create or enlarge a wild population.

Tributyltin Organic tin-containing compound that is an ingredient of anti-fouling paints for marine vessels.

Trophic levels Groups of organisms that are the same number of feeding steps away from the energy input to an ecosystem.

Turnover of species Amount of change in the composition of an island community due to colonizations and extinctions during a certain time period.

Turtle excluder device Trap-door arrangement that allows turtles and other large animals to escape trawl nets of fishing boats.

U

Umbrella species Species whose protection also benefits many other species.

Upwellings Ocean areas, usually near the edges of continents, where deep, cold, nutrient-rich water rises to the surface.

W

Weathering Modification of terrestrial substrates and aquatic sediments by physical or chemical processes.

Z

Zooplankton Tiny, weakly-swimming animals that live in the open waters of aquatic ecosystems.

Zooxanthellae Single-celled symbiotic algae that live within the cells of coral animals.

credits

ILLUSTRATOR CREDITS
CARTO-GRAPHICS Figure 4.2, Reading 4.1, Figure 14.4, Reading 1.1A, Figure 27.4.

MARYLAND CARTOGRAPHICS Figures 6.1, 6.3, 7.1, 8.1, 9.2, 10.1, 16.6, 17.1, 21.6, and 23.2.

PRECISION GRAPHICS Figures 1.1, 1.2, 1.5, 3.1, 4.1, 4.3, 5.8, Reading 10.1B, Figures 14.6, 17.3, 17.5, 23.1, 24.3B, 28.2, 28.3, 28.4.

FIGURE CREDITS
CHAPTER 8
Figure 8.6 From H. E. Dregne, *Desertification of Arid Lands.* Copyright © 1983 Harwood Academic Publishers, Montreaux, Switzerland. Reprinted by permission.

CHAPTER 10
Figure 10.2 From S. Niewolt, *Tropical Climatology.* Copyright © 1977 John Wiley & Sons, London. Reprinted by permission of John Wiley & Sons, Ltd.

CHAPTER 11
Figure 11.10B Reprinted with permission from *Decline of the Sea Turtles: Causes and Prevention.* Copyright 1990 by the National Academy of Sciences. Courtesy of the National Academy Press, Washington, DC.

CHAPTER 14
Figure 14.1 Modified from T. K. Fuller, "Population Dynamic of Wolves in North-Central Minnesota" in *Wildlife Monographs,* 105:1–41, 1989 with permission of The Wildlife Society, Inc., Bethesda, MD.

CHAPTER 16
Figure 16.4 From P. Beaumont, 1978, "Man's Impact on River Systems: A World-wide View" in *Area,* 10:38–41. Copyright © 1978 Institute of British Geographers, London, England. Reprinted by permission.

CHAPTER 17
Figure 17.4 Based on a map in W. Roy Siegfried, 1985, "Krill: The Last Unexploited Food Resource" in *Optima,* 33(2):67–79. Reproduced courtesy of the author.

CHAPTER 18
Figure 18.4 From D'Elia, 1987, in *Environment,* 29:6–11, 30–33. Reprinted with permission of the Helen Dwight Reid Educational Foundation. Published by Heldref Publications, 1319 Eighteenth St., N.W., Washington, DC. 20036-1802. Copyright © 1987.

Figure 18.6 From "Tropic Structure and Productivity of a Windward Coral Reef Community on Eniwetok Atoll" by H. T. Odum and E. P. Odum, in *Ecological Monographs,* 1957, 25:291–320. Copyright © 1957 Ecological Society of America, Tempe, AZ. Reprinted by permission.

CHAPTER 20
Figure 20.4 From M. D. Fox and B. J. Fox, "The Susceptibility of Natural Communities to Invasion," R. H. Groves and J. J. Burdon, (editors), *Ecology of Biological Invasions.* Copyright © 1986 Cambridge University Press, New York, NY. Reprinted by permission.

CHAPTER 21
Figure 21.1 From J. P. Myers, et al., 1987, "Conservation Strategy for Migratory Species" in *American Scientist,* 75:19–26. Copyright © 1987 American Scientist, Research Triangle Park, NC. Reprinted by permission of *American Scientist,* Journal of Sigma Xi, The Scientific Research Society.

Figure 21.3 From F. Moore and P. Kerlinger, 1987, "Stopover and Fat Deposition by North American Wood-Warblers (*Parulinae*) Following Spring Migration Over the Gulf of Mexico" in *Oecologia,* 74:47–54. Copyright © 1987 Springer-Verlag, Heidelberg, Germany. Reprinted by permission.

CHAPTER 23
Figure 23.1 Copyright © 1988 by the World Resources Institute for Environment and Development in collaboration with the United Nations Environment Programme. Reprinted by permission of BasicBooks, a division of the HarperCollins Publishers, Inc.

CHAPTER 24
Figure 24.9 From D. R. McCullough, *The George Reserve Deer Herd: Population Ecology of a K-Selected Species.* Copyright © 1979 University of Michigan Press, Ann Arbor, MI. Reprinted by permission.

CHAPTER 25
Figure 25.9 From C. Santiapillai and W. S. Ramono, "Tiger Numbers and Habitat Evaluation in Indonesia" in R. L. Tilson and U. S. Seal, (editors), *Tigers of the World: The Biology, Biopolitics, Management, and Conservation of an Endangered Species.* Copyright © 1987 Noyes Publications, Park Ridge, NJ. Reprinted by permission.

CHAPTER 26

Figure 26.3 From M. Shaffer, 1987, "Minimum Viable Populations: Coping with Uncertainty" in M. E. Soule, (editor), *Viable Populations for Conservation*. Copyright © 1987 Cambridge University Press, New York, NY. Reprinted by permission.

Figure 26.4 From G. E. Belovsky, 1987, "Extinction Models and Mammalian Persistence" in M. E. Soule, (editor), *Viable Populations for Conservation*. Copyright © 1987 Cambridge University Press, New York, NY. Reprinted by permission.

CHAPTER 27

Figure 27.1 From J. M. Diamond, 1975, "The Island Dilemma: Lessons of Modern Biogeography Studies for the Design of Nature Preserves" in *Biological Conservation,* 7:129–146. Copyright © 1975 Elsevier Applied Science Publishers Ltd., Barking, Essex, England. Reprinted by permission.

Figure 27.2 From H. Salawasser, et. al., 1987, "The Role of Interagency Cooperation in Managing for Viable Populations" in M. E. Soule, (editor), *Viable Populations for Conservation*. Copyright © 1987 Cambridge University Press, New York, NY. Reprinted by permission.

Figure 27.3 From R. F. Noss, *The Woodlands Project. Wild Earth. Special Issue.* Copyright © Cenozoic Society @ Wild Earth, Richmond VT. Reprinted by permission of the publisher and the author.

Figure 27.5 From J. M. Scott, 1987, "Species Richness" in *BioScience,* 37:782–788. Copyright © 1987 American Institute of Biological Sciences, Washington, DC. Reprinted by permission.

CHAPTER 29

Figure 29.8 From D. S. Brookshire, B. Ives, and W. Schulze, 1976, "The Valuation of Aesthetic Preferences" in *Journal of Environmental Economics and Management,* Vol. 3, No. 4, pp. 325–346. Copyright © 1976 Academic Press, Orlando, FL. Reprinted by permission of the publisher and the authors.

CHAPTER 30

Figure 30.1 From D. R. Wallace, *The Quetzal and the McCaw: The Story of Costa Rica's National Parks.* Copyright © Sierra Club Books, San Francisco, CA. Reprinted by permission.

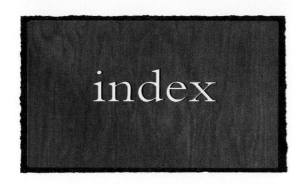

index

Note: Page numbers in italics indicate illustrations; those followed by "t" indicate tables.

carrying capacity of, 21
definition of, 21
demographic variability in, 32
ecological variability in, 32
evolutionarily significant unit in, 32
feral, 205
genetic variability in, 32, 258-66. *See also*
    Genetic variability
nonessential experimental, 142
regulation of, 21
sink, 128
subspecies in, 32
Population cycles
    food supply—predation hypothesis for, 65
    herbivory hypothesis for, 65
    predation hypothesis for, 65-66
    tundra, 65-66
Population dynamics, spatially explicit models
    of, 18
Population growth
    conservation activism and, 15-16
    sustainability and, 312-13
        density dependent factors in, 248
        density independent factors in, 248
        dynamic pool models for,
            252-53, *253*
        factors affecting, 21
        logistic models for, *248-52,* 249-50
        tropical deforestation and, 88
Population size
    genetically effective, 264
    genetic variability and, 259-63
Population viability analysis (PVA), 271,
    271, 272
Potholes, 147, 148
Pragmatic-Utilitarian Ethic, 290
Prairie
    mid-grass, 63
    palouse, 64
    short-grass, 63
    tall-grass, 63, *63*
Prairie dogs, 67, *67*
    black-footed ferrets and, 273
Predation hypothesis, 65-66
Predators, 132-42
    behavior of, 133
    coyotes as, 134-35
    ecology of, 133
    environmental influences on, 133-38
    functional response by, 133
    gray wolves as, 135-38
    human attacks by, 139, 140
    keystone, 184
    livestock depradation from, 138-39, 139t
    nest, 127, 129
    numerical response by, 133
    reintroduction of, 6, 140-41, 142
    selectivity of, 133, 134t
    territoriality of, 133, *133*
Prescribed natural fires, 59
Primary ethics, 289
Primary succession, 22
Prior appropriation doctrine, 164
Producer, 23
Public lands, management of, 5
Public policy, formulation of, 303-5
Public trust doctrine, 164
Purse seines, dolphin by-catch in, 193-95, *195*

R

Radioisotopes, pollution from, 229
Rain forests. *See* Moist tropical forests
Rain shadow, 72
Ramsar Convention, 155
Rapid Assessment Program, 90
Recreational value, 300-301
Recycling, in captive propagation, 273
Red Data Books, 270
Reefs, coral, 169, 184-86, *185*
    global warming and, 187
Religion, conservation and, 11
Reserved water rights, 164
Resilance, ecosystem, 26
Resources. *See* Ecological resources
Rhinoceroses, population decline in, 90,
    100-101, 101t
Rinderpest, 97
Riparian doctrine, 164
Riparian ecosystems, 156-65. *See also* Rivers
    and streams; Watersheds
Rivers and streams, 156-65
    channelization of, 160
    damming of, 159-60, *160*
        fish migration and, *161,* 161-63, *162*
    estuary inflow from, 179
    fish of, 156. *See also* Fish
    floodplains of, 157
    global change and, 164
    human impact on, 157-63
    management of, 164-65
    mining and, 159t, 159
    overfishing of, 158
    pollution of, 158-59
    siltation of, 157-58
    species decline in, 157-58
    tropical and subtropical, human impact on,
        163-64
    types of, 156
    water diversion from, 147-48, 157, 161
    watersheds of. *See* Watersheds
Roe stripping, 175
Royal Botanic Gardens, 266
Ruffed grouse, as early successional species, 56
Rule of 50, 264, 272

S

Sagebrush steppe, 64
Salmon migration, damming and, 159-60, *161,*
    161-63, *162*
Salt marshes, 179-80
San Francisco Bay estuary, 182-83
Sand dunes, 73, *73, 104,* 104-5
Sandy coastal ecosystems, 104-11
    animals of, *106,* 106-8, *107*
    barrier beaches and islands of, *105,* 105-6
    dunelands of, *104,* 104-5
    global change and, 111
    plants of, 106
    protection of, 109-11
Savannas. *See* Tropical savannas and woodlands
Scientific research value, 301
Scientists, in public policy formulation, 305
Sea lamprey, 150, 151
Sea lions, 191t, 196, 197

Sea otters, 183-84
    oil pollution and, 228
Sea turtles, *107,* 107-9, 108t, *109*
Seabirds, *106,* 106-7, *111,* 111, 190, 198.
        See also Birds
    oil spills and, 225-28, *227, 228*
    plastic pollution and, 230, *230*
    protection of, 198, 219
Seagrass beds, 180-81
Seals, 191t, 195-96, *196,* 197
Sears, Paul B., 15
Seasonal migration. *See* Migration
Secchi Disk, 34
Secondary ethics, 289
Secondary succession, 22
Seeds, cryogenic storage of, 266
Seed-tree cutting, 255-56
Selectivity, predator, 133, 134t
Selenium, pollution from, 229
Serengeti-Mara ecosystem, 95-97, *95-97*
Serpentine soils, 32
Sewage, water pollution from, 148, 158,
        164-65, 181
Shallow Ecology, 292
Shallow marine environments, 181-84
Shark finning, 175
Shellfish, harvesting of, 168t, 168-76, *170,*
        170t. *See also* Fishery harvesting
    in estuaries, 179
Shell-thinning
    acid deposition and, 240
    pesticide-related, 223-24
Shelterwood cutting, 255-56
Shifting cultivation, 87, 88
Ships, pollution from, 225-28, 230, 231
Short-distance migration, 212
Short-grass prairie, 63
Shrublands, temperate, 52-60. *See also*
        Temperate forests, woodlands, and
        shrublands
    biotic succession in, 53-56, 55t, 57
    human impact on, 56-57
Sierra Club, 13
Siltation, 157-58
Sink populations, 128
Sirenians, 190
SLOSS controversy, 280-81
Society for Conservation Biology, 6, 18
Soft release, 274
Soil Conservation Service, 15
Soil crusts, 73
Soil fertility
    in moist tropical forests, 85-87, 86t
    in temperate forests, 86t
Sonoran Desert, 72-73, *73*
Sounds, 105
Source populations, 128
Spaceship Ethics, 291
Spatially explicit models, 18, 128
Speciation, 41
Species
    area-sensitive, 127
    biological, 33
    early successional, 53-56, 55t
    endemic, 115, 116t
    establishment of
        adequacy of introduction unit
            and, 203-4